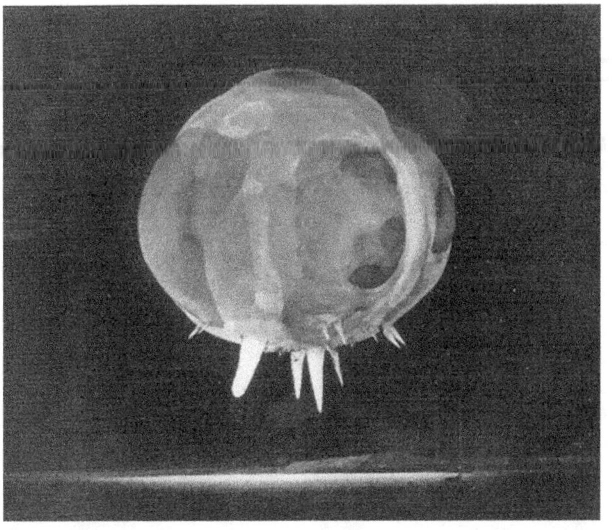

NUCLEAR WEAPONS TESTING At The NEVADA TEST SITE

The First Decade

John C. Hopkins
And
Barbara Germain Killian

2nd Edition - April, 2013

Cover photo description:
Cover photo was taken within microseconds of the detonation of a nuclear device by an Edgerton, Germeshausen, and Grier rapatronic camera. The spikes at the base of the explosion are due to radiation absorption by guy wires for the tower which held the device.

www.NuclearTestingNevada.com

ISBN: 978-0-9887679-0-4

TABLE OF CONTENTS

PAGE

ACKNOWLEDMENTS 15

STYLE 17

AUTHORS' NOTE 21

PART I. PRELUDE TO A CONTINENTAL TEST SITE 23
INTRODUCTION

CHAPTER 1. TRINITY 27
EARLY CONCEPTS FOR TESTING BEFORE DEPLOYMENT
REORGANIZATION TO EXPEDITE IMPLOSION WORK
PREPARATION FOR TRINITY
THE TRINITY TEST
RESULTS

CHAPTER 2. OPERATION CROSSROADS AND CREATION OF THE ATOMIC ENERGY COMMISSION (AEC) 35
OPERATION CROSSROADS
CREATION OF THE ATOMIC ENERGY COMMISION (AEC)

CHAPTER 3. LOS ALAMOS LABORATORY, ORGANIZATION AND LEADERSHIP: 1945-1950 41
DEVELOPMENT OF PEACETIME LOS ALAMOS
FIELD TEST DIVISION

CHAPTER 4. SANDIA LABORATORY 1945-1950 45
1945
FORMATION AND EARLY YEARS OF THE SANDIA CORPORATION

CHAPTER 5. ARMED FORCES SPECIAL WEAPONS PROJECT, ORGANIZATION AND LEADERSHIP: 1946-1950 51
1946 - FORMATION OF THE 2761st ENGINEERING BATTALION (SPECIAL)
FORMATION OF AFSWP
NATIONAL MILITARY ESTABLISHMENT AND THE DoD
AFSWP FIELD OPERATIONS
AFSWP 1950

CHAPTER 6. OPERATION SANDSTONE AND ITS INFLUENCE ON THE FUTURE OF TESTING 57

CHAPTER 7. PUSH FOR A CONTINENTAL TEST SITE 61
INTRODUCTION
POST SANDSTONE: SUMMER 1948 - JANUARY 1949
1949
1950
KOREA
IMPACT OF KOREA ON TESTING AND GREENHOUSE
BRADBURY WRITES TO DMA
COSTS AND SCHEDULES INFLUENCE CONTINUING DEBATE
BRADBURY WRITES AGAIN
SAFETY
Radiation Concerns
Maximum Yields
THE HOLMES AND NARVER REPORT
AEC AND PENTAGON BATTLE OVER OPERATION GREENHOUSE
PRESIDENT TRUMAN STEPS IN
DECEMBER 1950

CHAPTER 8. JANUARY 1951 - PREPARATION FOR THE FIRST NEVADA OPERATION - RANGER 81

PART I FIGURES 85
PART I PHOTOS 91

PART II: THE FIRST DECADE 99
INTRODUCTION

WEAPONS DEVELOPMENT
WEAPONS EFFECTS
DESERT ROCK - MILITARY EXERCISES
CIVIL DEFENSE
PLOWSHARE - PEACEFUL USES OF NUCLEAR EXPLOSIVES
PUBLIC IMPACT OF NUCLEAR TESTING
TEST BAN DISCUSSIONS
UNDERGROUND TESTING
MORATORIUM

CHAPTER 1. RANGER JANUARY 27 - FEBRUARY 6, 1951 105
INTRODUCTION
NEVADA TEST SITE ORGANIZATION (NTSO) FOR OPERATION RANGER
LOCATIONS OF RANGER ACTIVITIES
Las Vegas
Nellis Air Force Base
Indian Springs
Control Point
Ground Zero
TRANSPORTATION TO LAS VEGAS AND ITS GROWTH
SUPPORT ACTIVITIES AND THE REYNOLDS ELECTRIC & ENGINEERING, COMPANY (REECo)
TECHNICAL ACTIVITIES AND TESTS
TEST DIAGNOSTICS
Yield
Alpha
Gamma-Rays
Neutron Measurements
Transit Time
Detonation Location
Airblast
Other Experiments and Measurements
DoD NUCLEAR WEAPONS EFFECTS (NWE) MEASUREMENTS, OPERATION HOT ROD
FALLOUT
AIRDROP TECHNOLOGY
CONCLUSIONS AND LESSONS LEARNED FROM OPERATION RANGER
OBSERVATIONS ON THE SIGNIFICANCE OF OPERATION RANGER
RANGER FIGURES
RANGER PHOTOS

CHAPTER 2. OPERATION BUSTER-JANGLE:
OCTOBER 22 - NOVEMBER 29, 1951 129
INTRODUCTION
LOS ALAMOS' FIELD TEST (J-DIVISION) BECOMES PERMANENT
RANGER TO BUSTER-JANGLE
PERMANETIZATION
REAL ESTATE
SANTA FE OPERATIONS OFFICE (SFOO) MOVES TO ALBUQUERQUE
SITE NAMED NEVADA TEST SITE, JULY 8, 1951
PREPARATION FOR JANGLE
PREPARATION FOR BUSTER
NEVADA TEST SITE ORGANIZATION (NTSO) FOR OPERATION BUSTER-JANGLE
BUSTER-JANGLE TESTS
WEAPONS DIAGNOSTIC PROGRAM
DoD PARTICIPATION ON OPERATION BUSTER-JANGLE
DESERT ROCK OPERATIONS: I (DOG), II (SUGAR), and III (UNCLE)
DoD NWE PROGRAMS AND PROJECTS
BUSTER NWE PROGRAMS AND PROJECTS
Thermal and Nuclear Radiation
Blast Effects
Biomedical
Service Equipment and Operations
Long-Range Detection
Supporting Measurements
Federal Civil Defense Agency (FCDA) Collaboration With DoD On BUSTER
JANGLE NWE PROGRAMS AND PROJECTS
Blast and Shock
Radiological Phenomena
Blast Effects On Structures
Special Phenomena
Service Equipment and Operations
Long-Range Detection and Supporting Measurements
LIVING AND WORKING CONDITIONS AT THE SITE DURING BUSTER-JANGLE
BUSTER-JANGLE EPILOGUE
THE SITE: 1951-1952
Real Estate
Site Named Nevada Proving Ground, February 25, 1952
BUSTER-JANGLE FIGURES
BUSTER-JANGLE PHOTOS

CHAPTER 3. OPERATION TUMBLER-SNAPPER:
APRIL 1 - JUNE 5, 1952 177
PLANNING
NEVADA TEST ORGANIZATION FOR OPERATION TUMBLER-SNAPPER
REECo
CONSTRUCTION STATUS
OTHER CONTRACTORS AT THE SITE
LIVING CONDITIONS DURING TUMBLER-SNAPPER
SOME CHARACTERISTICS OF THE WORKING TEAMS
Los Alamos' J-Division Was A Team
The POGO Office and Staff Team
TUMBLER-SNAPPER TESTS
DIAGNOSTICS WEAPONS DEVELOPMENT EXPERIMENTS
Alpha
Transit Time
Yield
Film-Badge Measurements, Total Gamma Dose
Neutron Measurements
ADDITIONAL MEASUREMENTS
Investigation of Fluoresence Under High Gamma Fluxes
Hot-Spot Experiment
Light Absorption Around Bomb
Airblast Measurements
Sound Velocity
Dust Loading
Mass Motion Photography
DoD PARTICIPATION ON OPERATION TUMBLER-SNAPPER
EXERCISE DESERT ROCK IV
DoD NWE PROGRAMS AND PROJECTS
Airblast and Ground Motion Measurements
Structures
Long-Range Detection
DoD PARTICIPATION ON LOS ALAMOS WEAPONS
DEVELOPMENT TESTS
FEDERAL CIVIL DEFENSE ADMININSTRATION (FCDA)
MILITARY SIGNIFICANCE
TUMBLER-SNAPPER FIGURES
TUMBLER-SNAPPER PHOTOS

CHAPTER 4. A SECOND NUCLEAR WEAPONS LABORATORY
IS ESTABLISHED AT LIVERMORE, CALIFORNIA 209
A PERSONAL PERSPECTIVE ON THE FORMATION OF THE NUCLEAR
WEAPONS LABORATORY AT LIVERMORE, CALIFORNIA
By Herbert F. York
LIVERMORE PHOTOS

CHAPTER 5. ARMED FORCES SPECIAL WEAPONS PROJECT (AFSWP)/DEFENSE ATOMIC SUPPORT AGENCY (DASA) ORGANIZATION AND LEADERSHIP 219
1951 REORGANIZATIONS AT HEADQUARTERS
1951 AFSWP FIELD COMMAND ESTABLISHED
DIRECTORATE OF WEAPONS EFFECTS TESTS, FIELD COMMAND (DWET, FC)
1953
1954 - 1955
1957
1958
FORMATION OF DASA
DASA DURING THE MORATORIUM

CHAPTER 6. SANDIA LABORATORY DURING THE FIRST DECADE 227
PRE-1952
FIELD SUPPORT AND LOGISTICS DIVISION
1952 AND REORGANIZTION
1953
SANDIANS AT THE NEVADA TEST SITE
SUMMARY OF SANDIA'S ARMING AND FIRING (A&F) ACTIVITIES DURING THE 1950s
SANDIA SUPPORT FOR THE DoD ON UPSHOT-KNOTHOLE
1955
1956
1957
1958
SANDIA PHOTOS

CHAPTER 7. NUCLEAR WEAPONS 1952 - 1953 241
FALL 1952
British Nuclear Tests In Australia
Mike - The World's First Full-Scale Test Of A Thermonuclear Device
Operation DOMINO Tentatively Scheduled For Fall 1953
COMMITTEE ON OPERATIONAL FUTURE OF THE NPG
NUCLEAR WEAPONS TEST OBJECTIVES
Military Requirements
Nuclear Weapons Development and Effects Tests
One-Point Safety
Initiation Options
Composite Pits

Primaries: Thermonuclear Weapons Triggers
PHASES OF DEVELOPMENT FOR NUCLEAR WEAPONS

CHAPTER 8. OPERATION UPSHOT- KNOTHOLE: MARCH 17 - JUNE 4, 1953 251
PLANNING
NEVADA TEST ORGANIZATION FOR OPERATION UPSHOT-KNOTHOLE
PERSONNEL CONDITIONS AT THE NPG DURING UPSHOT-KNOTHOLE
UPSHOT-KNOTHOLE TESTS
WEAPONS DEVELOPMENT DIAGNOSTIC EXPERIMENTS
OVERVIEW OF TESTS AND DIAGNOSTICS
Annie, March 17
Nancy, March 24
Ruth, March 31
Ray, April 11
Dixie, April 6
Badger, April 18
Simon, April 25
Encore, May 8
Harry, May 19
METEOROLOGY FOR HARRY
FALLOUT FROM HARRY
Grable, May 25
Climax, June 4
DoD PARTICIPATION ON UPSHOT-KNOTHOLE
EXERCISE DESERT ROCK V
Camp Desert Rock Troop Support, AKA Permanent Party
Damage Effects Evaluation
Observers, Troop Orientation, and Indoctrination
Volunteer Officer Observers Program
Tactical Troop Maneuvers
Operational Helicopter Tests
DoD NWE PROGRAMS AND PROJECTS
Background
Blast and Shock Measurements
Structures, Material and Equipment
Railroad Stock
Lots More Conifers
Field Medical Facilities
The "Motels"
Aircraft Structures
SAC And TAC Operations
Long-Range Detection
Thermal Measurements and Effects
Technical Photography

DoD PARTICIPATION ON WEAPONS DEVELOPMENT
DIAGNOSTICS TESTS
AFSWC Support
Radiological Safety
DOORSTEP - FEDERAL CIVIL DEFENSE ADMINISTRATION (FCDA)
PARTICIPATION ON UPSHOT-KNOTHOLE
FCDA Observer Program - Open Shot
Field Training Exercises
FCDA NWE PROGRAMS AND PROJECTS
Two Typical American Homes
Home-Type Shelters
Civilian Vehicles
Other Projects
UPSHOT-KNOTHOLE SUMMARY
UPSHOT-KNOTHOLE FIGURES
UPSHOT-KNOTHOLE PHOTOS

CHAPTER 9. 1953-1954 303
SECOND COMMITTEE ON OPERATIONAL FUTURE OF THE
NPG - SUMMER 1953
Criteria For Future Continental Tests - Graves and Felt
Second Committee On Operational Future Of The NPG Meets Again
SOVIET AND U.S. TESTS OF THERMONUCLEAR DEVICES
1954 - CHANGES AT THE SANTA FE AND ALBUQUERQUE
AEC OFFICES
OPERATION CASTLE IN THE PACIFIC
THE NEVADA PROVING GROUND BECOMES THE NEVADA
TEST SITE AGAIN

CHAPTER 10. OPERATION TEAPOT:
FEBRUARY 18 - MAY 15, 1955 311
PLANNING
PUBLIC RELATIONS AND INFORMATION
NEVADA TEST SITE ORGANIZATION (NTSO) FOR
OPERATION TEAPOT
Technical Staff
AEC - Los Alamos Responsibilities
TEST OBJECTIVES FOR OPERATION TEAPOT
TEAPOT TESTS
ESS
Hadr (High altitude dry run)
Apple 1
PREPARATION AND DETONATION OF A DEVICE: 1955-1958
Transportation of Nuclear Devices
Device Assembly

Signal Dry Runs
Final Shot Preparation
Emplacement
Neutron Sources
Fuzing
LASL'S DIAGNOSTIC EXPERIMENTS
UCRL'S DIAGNOSTIC EXPERIMENTS
ON-SITE RAD-SAFE
DoD PARTICIPATION ON OPERATION TEAPOT
DoD NWE PROGRAMS AND PROJECTS
Drone Aircraft For Investigation Of Lethal Effects
Investigations Within The Fireball
MET - A *Tour de Force* On Non-Ideal Airblast
MEASUREMENTS ON MET'S 3 SURFACES
Other Measurements of Free-Air Pressure
The ESS Crater
FCDA PARTICIPATION ON TEAPOT
Open Shot
Field Training Exercises
FCDA NWE PROGRAMS AND PROJECTS
Houses
Shelters
Warehouses and Chemical Plant Control Room
Static vs Dynamic Loads
Non-Ideal Shock on Structure
Mobile Homes and Emergency Vehicles
Foods
Materials Found Near and In Homes
Heavy-Duty Machine Tools
Missiles
Electrical Equipment
Civilian Communications Equipment
Liquefied Petroleum Gas (LPG)
Natural Gas
Records, Materials, and Storage
Biomedical Animal Exposures
PROJECT SUNSHINE AND FALLOUT STUDIES
FCDA Fallout Projects
Off-Site Fallout
PERSONNEL CONDITIONS AT THE NTS DURING TEAPOT
Construction
Security
Safety
Telecommunications
Air Transportation
Field Command Support Unit
Mercury

TEAPOT FIGURES
TEAPOT PHOTOS

CHAPTER 11. 1955-1957 351
GROOM LAKE - AREA 51
1955
Planning For Future Operations
Hertford Becomes Manager Of SFOO
BALLOON SUSPENSION FOR TEST DEVICES
RADIATION AND CLEANUP
FALLOUT
WIGWAM
YEAR-ROUND OPERATION OF THE TEST SITE
COMMITTEE ON THE USE OF NEVADA TEST SITE
Pluto And Rover
Real Estate
GRAVES SUFFERS HEART ATTACK
OPERATION PROJECT 56
PROPOSAL FOR TESTING UNDERGROUND
FALLOUT CONCERNS
SECOND MEETING OF THE COMMITTEE ON THE USE OF NTS
OPERATION REDWING IN THE PACIFIC (MAY TO JUNE 1956)
PUBLIC RATIONALE FOR THE 1957 OPERATION PLUMBBOB
COMMON TEST ORGANIZATION
THE NTS PLANNING BOARD
REVIEW OF NTS OPERATING CRITERIA
1957

CHAPTER 12. OPERATION PLUMBBOB: MAY 28 - OCTOBER 7, 1957 373
OPERATION PILGRIM EVOLVES INTO OPERATION PLUMBOB
PLANNING
NEVADA TEST SITE ORGANIZATION (NTSO) FOR OPERATION PLUMBBOB
PLUMBBOB TESTS
Some Highlights of the Tests
The Grueling Schedule
PROGRAMS CONDUCTED ON PLUMBBOB
DoD NWE PROGRAMS AND PROJECTS - PROGRAMS 1 - 9
Blast and Shock
Airblast On Different Terrains
Ground Motion Spectra
Nuclear Radiation Effects
Effects On Structures
Biomedical
Aircraft Structures

An Example Of Advances In Nuclear Weapons And Delivery Systems
John
Minefields
Navigation Aids
Investigations Within The Fireball
Supporting Measurements
WEAPONS DEVELOPMENT - PROGRAMS 10 - 26
EG&G Support
ARMING AND FIRING DRY RUNS
FCDA PARTICIPATION ON PLUMBBOB
Open Shots
Field Training Exercises
FCDA NWE PROGRAMS AND PROJECTS - PROGRAMS 30 - 39
Domes
Parking Garage
Family Shelters
French and German Shelters
The Mosler Safe
Personnel Experience Shots in a Shelter
Flying Dummies
Burro and Monkey Exposures
SANDIA TEST GROUP - WEAPONS DEVELOPMENT PROGRAM, PROJECT 41
DESERT ROCK, EXERCISES VII AND VIII - PROGRAMS 50 - 53
A Military Critique of Desert Rock VII and VIII
SANDIA PROGRAMS 62, 63, 64
PROJECT 57 TEST GROUP-WEAPONS DEVELOPMENT PROGRAMS 71-74
MUSTER BADGES AND AREA SWEEPING
ENGINEERING AND CONSTRUCTION SUPPORT
AIRCRAFT PARTICIPATION
LIVING CONDITIONS DURING OPERATION PLUMBBOB
Test Director's Services
RADIOLOGICAL SAFETY
Eye Burn Safety Concerns
PLUMBBOB FIGURES
PLUMBBOB PHOTOS

CHAPTER 13. 1957 - 1958 407
PROJECTS 58 AND 58A
EARLY 1958
FRENCH VISIT THE NEVADA TEST SITE
ISSUES REGARDING OPERATIONS AT NTS
Continuous Operation of the NTS Revisited
Centralization of Operations
OPERATION HARDTACK PHASE I

OPERATION ARGUS
TEST BAN INITIATIVES
E. O. LAWRENCE DIES
THE RAMP-UP TO HARDTACK PHASE II
EISENHOWER ANNOUNCES U.S. MORATORIUM
THE SOVIET UNION AGREES TO START NEGOTIATIONS
READINESS ISSUES

CHAPTER 14. OPERATION HARDTACK PHASE II: SEPTEMBER 12 - OCTOBER 30, 1958 417
PLANNING AND RUNUP
NEVADA TEST SITE ORGANIZATION (NTSO) FOR OPERATION HARDTACK PHASE II
ADVISORY PANEL AND THE PREDICTION GROUP
PLANNING BOARD DURING HARDTACK PHASE II
HARDTACK PHASE II TESTS
Tunnel Tests
DoD PARTICIPATION ON OPERATION HARDTACK PHASE II
DoD NWE PROGRAMS AND PROJECTS
CIVIL DEFENSE PARTICIPATION
LIVING CONDITIONS AT THE SITE DURING HARDTACK PHASE II
HARDTACK II FIGURES

CHAPTER 15. NUCLEAR TEST MORATORIUM: 1958 - 1961 435
READINESS
WASHINGTON POLITICS
1960
1961
Test Resumption

EPILOGUE 443
Summary
Assessment

APPENDICIES
A. Basic Concepts of Nuclear Weapon Design 445
B. Los Alamos' Organization During WWII and Trinity 467
C. The Soviet Union's Semipalatinsk Test Site (STS) 477
D. Acquisition of Real Estate for the Nevada Test Site by Carole Schoengold 481
E. Environmental Conditions at NTS by Lawrence S.Germain

and Barbara Germain Killian 485
F. AEC and DoD Senior Leadership, Advisory Panels, and Planning Boards For NTS Orginization During Atmospheric Testing 491
G. Diagnostics Experiments Conducted by AEC Laboratories 1950-1958 Adapted from work by William E. Ogle (1985: 52-83) 497
H. Synopses of NWE Field Activities and Structures on JANGLE 503
I. A Brief History of U.S. Civil Defense Organizations 515
J. Location of Troops During Atmospheric Tests 519
K. 1954 Shot Criteria for NTS 523
L. Public Relations and Information 527
M. Desert Rock VI 533
N. Desert Rock VII and VIII 537

ATTACHMENTS

I. Early Working Relationships Between Laboratory, Military, and AEC Personnel by Robert Duff 551
II. Byron At Nevada Test Site by Byron Murphey 553
III. The Turk Device in TEAPOT by Raymond A. (Ray) Gilbert 557
IV. Excerpts from Interview with Major General Edward B. Giller 565
V. Arms Control: 1945 - 1961 by Milo Nordyke 571
VI. The Plowshare Program by Milo Nordyke 579
VII. Underground Testing and Early Containment Concepts by Cliff Olsen 593
VIII. The Vela Program by Milo Nordyke 595

ACRONYMS 601

REFERENCES 607

ABOUT THE AUTHORS 633

ACKNOWLEDGMENTS

The names of many people who conducted the numerous different types of activities at the site (such as miners, technicians, security personnel, secretaries, engineers, physicists, cooks, drivers, chemists, geologists, maintenance personnel and the support staff) have faded into fond memories. It was their positive attitude and professionalism that made the NTS so special. All of those who made a career at the NTS deserve special acknowledgment and it is the memories of these individuals that motivated us to write this book.

Numerous friends and colleagues assisted during our years of research for this book. This assistance was technical and enabling, encouragement, support, etc.

Former Los Alamos Director John Browne and Associate Directors Stephen Younger and Susan Seestrom provided JCH with encouragement and support throughout the years of research.

Los Alamos Group Leader, Henry Johnson, hosted JCH at the laboratory archives. This enabled us to have numerous productive discussions with historians and colleagues Alan Carr, Roger Meade, Betty Perkins and Linda Sandoval. They were always ready to listen, comment, offer suggestions, and research sources. Carr generously provided most of the non-DoD photos. Former weapons theoretical design Division Leader, Mike Henderson used his extensive experience to review and comment on Appendix A. James Cruz and Leonard Martinez produced the figures and Roger Liljestrand provided classification guidance and a review of the entire manuscript.

Former Los Alamos Test Directors Robert Campbell and Tom T. Scolman were invaluable sources of firsthand experience of the site operations in the 1950s. Former nuclear Test Alternate Division Leader Robert Brownlee provided insights into the early days of underground testing. Physician Payne Harris provided us with fascinating accounts of his experiences related to health physics and sampling aircraft at the NTS during the 1950s.

One of us (JCH) would also like to thank Edward Cornett for interesting discussions and anecdotes about his life at the test site as a young army officer in the early 1950s.

At the Defense Threat Reduction Information Analysis Center (DTRIAC) on Kirtland AFB, days and months evaporated in research that was made possible by the staff and their professionalism. They are a priceless national resource: Byron Ristvet, Dea Hunt, Fritz Roessler, Connie Salas, Joyce Housh, Janet Ortiz, Shri Theobald, Virginia Sedillo, Leona Abeita, Lisa Coca, Priscilla Diolazo, Hanna Benhalim, and Katrin Reliford. Byron's encouragement, enthusiasm, and ability to share his vast test experiences were greatly appreciated, though they usually led to even more useful research.

Herb Hoppe, provided the DoD classification review of this manuscript. Above and beyond just a classification review, he diligently uncovered innumerable editorial faux pas and graciously provided valuable suggestions, both of which we greatly appreciated.

As always, Sheldon Schuster and John Lewis were invaluable sources of information and tidbits regarding nuclear weapons effects tests.

From Livermore National Laboratory, John Rambo provided archival information and additional sources for information on Livermore test activities. Pete Moulthrop was a supporting source of information and encouragement. Maxine Trost of the archives staff cheerfully addressed our numerous requests, and Carol Gerich provided helpful insights into the history of the laboratory.

We would like to thank long-time REECo manager Harold Cunningham for sharing his memoirs and reminiscences with us.

Terrence R. Fehner and F. G. Gosling of the DoE in Washington, D.C. were very helpful in providing us with material that they had prepared for their publications on the Nevada Test site.

Martha De Marre and Jeff Gordon of the National Security Technologies (NST), LLC, in Las Vegas generously made available DOE records and photos. Maggie Townsend of the GIS Group at the NTS provided maps of the basin and range geology and was a supportive source of information.

Sandia's Al Chabai, Dick Lynch, Luke Vortman and Charlie Winter were major sources of information on the laboratory of the 1950s. Myra O'Canna graciously provided photographs from the Sandia archives.

The capabilities and ever ready help of Sheldon Liebman enabled us to retain some sanity while coping with the everyday vagaries of our PCs as well as providing invaluable guidance throughout all aspects of the publishing process.

BGK's husband, Larry Germain, was a prince during the years of writing by providing encouragement and accepting her long days at the Albuquerque DTRIAC Base or in their study. His enablers Emily Stewart, Pam Dailey, and Hank Ness made it possible for her to have the time at home and elsewhere to conduct this work.

We also want to acknowledge the following persons whom we interviewed during the course of our research for this manuscript: Harold M. Agnew, Nick Aquilina, F, L. Ashworth, John C. Clark, B. C. Diven, William Flangas, Jim Frank, Dale Frazer, William Harvey, Albert Lieber, Clive Marsh, Michael May, Pete Moulthrope, Jay Norman, Jake Perea, Byron L. Ristvet, Duane Sewell, Leon Smith, Frank Solaegui, Bruce Tarter, Troy Wade, Rich Wagner, Herbert F. York, and Peter Zavattaro.

STYLE

- Test operations that consist of a series of nuclear tests are specified in all capital letters, for instance: Operation BUSTER-JANGLE.

- Nuclear tests are specified as names, the first letter being capitalized, for instance: Fox

- Operation HARDTACK consisted of two phases: Phase I (in the Pacific) and Phase II (at NTS). The terms HARDTACK Phase I and simply HARDTACK I are used interchangeably, similarly HARDTACK Phase II and simply HARDTACK II are used interchangeably.

- The spelling of Enewetak that is used here was adopted in 1973 and is consistent with the written Marshallese. It means, "Island which points to the east." Ref: Philip Helfrich and Roger Ray, Chapter 1 of The Natural History of Enewetak Atoll, DOE/EV/00703-T1-Vol.1, (DE87006110) p.1

- Figures may be: maps, sketches, or engineering drawings. They are numbered as follows:
  1-1.1 means Part I-Chapter 1.Figure1
  1-1.2 means Part I-Chapter 1.Figure 2
  1-2.1 means Part I-Chapter 2.Figure 1

  2-1.1 means Part II-Chapter 1.Figure 1
  2-1.2 means Part II-Chapter 1.Figure 2
  2-2.1 means Part II-Chapter 2.Figure 1
  etc.

- Photos are not numbered. They are given at the end of most chapters with titles in roughly the order in which the person or objects were mentioned in the chapter. The photo of an individual is given in the chapter where he or she is first mentioned. Where possible, photos taken in the 1950s are used.

- When a person is first introduced in the text, his or her complete name, as it is usually written, appears. For instance: John (Jack) C. Clark. In subsequent references to an individual, just the last name might be given. Or, he might be referred to in the text as Jack Clark because he was called Jack most frequently. In the Name Index, he is referenced as John (Jack) C. Clark. The first introduction of a person in the text often contains a description of the individual's background. For instance: professor of physics at Stanford University.

- A few words about the level of detail are in order. Specifically we want to address our choice to use extensive quotations, particularly from the Los Alamos Director, Norris Bradbury. Bradbury, more than anyone else, established the post-War nature and character of the nuclear weapons R&D establishment. His influence largely carried over to Sandia, and later to Livermore when those laboratories were formed in 1949 and 1952 respectively. Bradbury's views on the role of the labs as research institutions encouraged the Department of Defense and the civil defense communities to shoulder the responsibilities for understanding the effects of nuclear weapons. This had, of course, a profound effect on the way that nuclear testing developed, both in Nevada and in the Pacific. We feel that the extensive quotations from Bradbury are necessary to accurately convey how and why the programs developed as they did.

- Our goal was to convey in the body of the text enough detail to provide an appreciation for the magnitude and diversity of the test operations. We chose to include additional details in attachments and appendices. We also included, as attachments, several firsthand accounts of various facets of the programs.

- Those programs or projects that the authors feel were the most significant in the 1950s and those that subsequently had the most importance on future work, got the most detail. This is, of course, rather subjective. In addition, we provide more discussion of those programs and projects that caught our personal interest; again, a subjective selection.

- In the discussions of programs and projects, emphasis is placed on what was done at the test site. Generally, results are not presented, but they can be found in the material referenced for that activity. Also, the extensive activities, ranging from basic research to shipping that were conducted "back at the lab", are not detailed. Our style might be related to reporting at the Olympics - the long hard years of preparation might be briefly mentioned, but the emphasis is on the event.

- We have written this book in a style that resembles a report format more than story telling. Our reasons for doing this are:
    - It represents the style in which testing information was documented; thus, it best captures the atmosphere in which the testing community worked.
    - A considerable amount of testing information is more effectively conveyed by tables or figures than by narrative.
    - It enables the reader to more readily locate a particular area of interest.

- Excellent archives related to the nuclear weapons effects programs reside at Kirtland Air Force Base in Albuquerque, NM. The accommodating staff of this facility can be contacted through the Public Affairs Office of the Defense Threat Reduction Agency (DTRA), Alexandria, VA.

- An explanation of archival references from Los Alamos is appropriate. The best, and most complete, archives for the weapons development programs reside at the Los Alamos National Laboratory. Many, if not most, of the relevant Livermore and Sandia documents are also in the Los Alamos archives. When we refer in the text to a memorandum from Los Alamos Director, Norris Bradbury, to the AEC on a specific date we do not usually provide additional reference information. It is implied that the memo is in the Bradbury files at the Los Alamos archives. The documents are easily found by the Los Alamos archival staff with simply the information that it was a Bradbury document and a date, or even an approximate date. The reference information cited in the text is often terse, but it is all that is required to obtain the document that is being referenced from Los Alamos archivists.

AUTHORS' NOTE

We embarked on this project for several reasons, all centered on our fascination with the history of nuclear weapons testing and our interest in making more of the details of these complex operations available to the general public.

We focused largely on the people. We have enormous respect and admiration for those who devoted major parts of their lives to the development and understanding of nuclear weapons. The esprit de corps, the camaraderie and the appreciation of the accomplishments contributed to making the participants feel good about what they were doing and proud of their work. From an intellectual standpoint, nuclear weapons test-related activities were both challenging and exciting.

It is worthwhile to stand back and ask what this was all about. What was the significance of what was done at the Nevada Test Site? The answer has two parts: The first is that the Nation was assured of a safe, reliable nuclear deterrent; and the second part is that the government acquired an understanding of the environment that would be faced in a nuclear war. How we got there required the efforts of tens of thousands of people and billions of dollars. We chronicle the details of that trek and occasionally touch on the geopolitical context of the period. It is clear to us that much more could be done to make the historical connections that would put the work of the test site into a broader perspective. This will have to come later.

While we both are veterans of many years at the Test Site, we did not participate in Nevada during the "golden age" of atmospheric testing of the 1950s. Our tenure came later; during the 1960s, '70s and '80s. This only increased our fascination with the first decade and further heightened our curiosity about the test operations with names such as TEAPOT and PLUMBBOB.

Lastly, and perhaps most importantly, the test site is a very special place. We recognized this as soon as we passed the front gate and approached the base camp, Mercury, for the first time in the 1960s. Obviously we entered a different world. It is a world that is isolated from the rest of our universe; miles from supermarkets, department stores, and home. It is a place where the people were truly inspired; a place where they fielded experiments that had major significance and represented years of planning and personal commitment. There was a lot at stake; perhaps the success or failure of a major weapons system. One all-pervasive feature was the atmosphere of team work; people working collaboratively together to get the job done. Our colleagues were not consumed with who did what or who got the most credit. People helped and trusted each other. This element of trust in the working relationships at the site was perhaps like those experienced in other work situations where one's safety and perhaps life depends on others.

Food was always plentiful, cheap, and usually good. Meals were eaten for recreation as well as for nourishment, and might be taken in a cafeteria, the Steak House, or in the forward areas. Sleeping accommodations improved over time, but even after private rooms became available, people still gathered after meals to continue work or to talk. There were endless discussions of the programs: What could be done better? What went wrong? Why was one approach used rather than another? ... As the planned shot date approached, very long work days were usually the norm, with rest available as cat naps in the forward areas. However, even during these hectic periods, time was occasionally made to appreciate a sunrise or a sunset; probably the most beautiful and wondrous moments in the desert.

John C. Hopkins
Barbara Germain Killian

PART I. PRELUDE TO A CONTINENTAL TEST SITE

INTRODUCTION

World War II ended on August 15, 1945, after two nuclear strikes on Japan. Two different bomb designs were used: an enriched uranium gun-type weapon (called "Little Boy") at Hiroshima on August 6, and a plutonium implosion weapon (nicknamed "Fat Man") at Nagasaki on August 9, 1945 (See Appendix A). Only the implosion device had been tested previously, in the Trinity event of July 16, 1945, in New Mexico.

Herbert York briefly characterized the environment as follows: "After World War II ended, the huge American effort that produced those bombs, the Manhattan Project, was largely, but not entirely, demobilized. The Soviet nuclear weapons program, which had begun at the same time as the American program but on a very much smaller scale, was revitalized and greatly expanded." (York 1995: 278) Simultaneously, the Soviets were tightening their grip on what had become their satellite states. Winston Churchill, in a March 1946 speech in Fulton, Missouri, coined a new phrase when he stated that the Soviets had drawn an "iron curtain" across Europe.

Immediately following the Japanese surrender, the U.S. Navy, proposed a series of tests. Those tests, called Operation CROSSROADS, were designed to examine the effects that nuclear weapons would have on ships and other military assets. At that time, the Navy had an enormous fleet of battle-scarred U.S., Japanese, and German ships with which to experiment.

Los Alamos, fresh from a post-war exodus of late 1945 and early 1946, was struggling to reconstitute its nuclear weapon design and development capability. Ultimately the laboratory did successfully support the military, by providing the nuclear devices that were to be detonated, the timing and firing, and the supporting diagnostics, during the summer of 1946 on Operation CROSSROADS at the Bikini Atoll in the Pacific. It took a toll, however. By the end of WWII, Los Alamos had several new fission weapon concepts. These had to be put on hold while the laboratory supported CROSSROADS.

The lessons from this experience crystallized the thoughts of the new Laboratory director, Norris Bradbury, on the future role of Los Alamos. Initially he was not convinced that the laboratory should be in the testing business at all (Roger Meade, private communication). He soon changed his mind and concluded that nuclear weapon testing was a fundamental part of weapon development. However, he did feel that weapons tests that examined the effects of nuclear weapons on materials, structures and systems should be the purview of the military.

The first opportunity to test the new designs came in Operation SANDSTONE; the three shot test series at Enewetak in 1948. In addition to weapon information, the lessons learned at SANDSTONE included how to field complex technical weapon diagnostic measurements far from home and how best to staff and organize a test division that had to work closely with a diverse group of military, governmental, and contracting organizations.

Meanwhile, the government management of nuclear weapons development was also undergoing a transformation. During the war, the U.S. Army, under the leadership of General Leslie Groves, managed the Manhattan Project. Following the war, there rapidly evolved a growing public sentiment throughout the country that nuclear weapons development should be under civilian control. On January 1, 1947, the Atomic Energy Act of 1946 came into effect, and the civilian-headed Atomic Energy Commission (AEC) assumed responsibility for nuclear weapons development and testing. Within the AEC, the responsibility for testing resided in the Division of Military Applications (DMA), which was headed by a military officer and staffed by both military and civilian personnel. While over the years the AEC evolved into the Energy Research and Development Administration (ERDA) and then into the Department of Energy (DOE), the testing structure remained essentially the same within DMA through the last test in 1992.

The two earliest post-war operations, CROSSROADS AND SANDSTONE, were in the Pacific. President Truman and his senior advisors were reluctant to test in the continental United States. They were concerned about both the political repercussions stemming from conducting these types of weapons experiments in the lower forty-eight and the assurance of safety from radioactive fallout associated with the detonations, particularly of higher yield devices. Eventually, however, it became clear that there were compelling arguments for a continental test site.

In August 1949, the Soviets startled the world by detonating a nuclear device. This news, along with the development of missiles and long-range aviation, meant that the U.S. would become increasingly vulnerable to a devastating nuclear attack. The political repercussions were serious. The U.S. wanted an appropriate response to what was considered the challenge of the Soviet nuclear detonation. Two options were explored. One was to launch a crash program to develop a hydrogen bomb, if indeed it were possible to develop such a weapon. The second was to substantially increase nuclear weapons production, which necessitated designs that used scarce fissionable material more efficiently. Ultimately the U.S. did both. It soon became clear that increased nuclear testing would be needed for the development of new weapons, regardless of the specific response to the Soviet test.

The subsequent history of nuclear weapon testing is inextricably tied to the history of the Cold War. Nuclear-capable military forces provided the foundations for the deterrence posture through the period of confrontation between the Soviet Block and the West. Throughout the 1950s, nuclear weapons technology advanced rapidly and was embraced enthusiastically by the defense community. During the 1960s, and beyond, the major advances were made in the ever more sophisticated delivery platforms. It was usually necessary to develop new nuclear weapons to fit into the modernized reentry vehicles, missiles, or bombs. In addition, safety and security technology evolved throughout the whole period of the Cold War. All of the advanced developments from the 1940s onward required nuclear testing.

The geopolitical situation became gravely worse in June 1950 when the North Koreans attacked South Korea. This was to have a major impact on the military's ability, and desire, to support nuclear testing in the Pacific.

When the Korean War broke out, Los Alamos was planning a test series, code named Operation GREENHOUSE, to be performed at Enewetak during the spring of 1951. High on the priority list were several new fission weapons and an experiment to explore the fundamentals of thermonuclear design. Two major concerns surfaced during the summer of 1950. The first was that the defense establishment was worried about both the logistics for GREENHOUSE and about the security of the remote operation. The second concern, focused at Los Alamos, was about the predicted behavior of key fission devices to be tested in the Pacific. Los Alamos wanted one, or a very few, preliminary experiments before committing to the extensive GREENHOUSE tests.

It was these concerns that spawned a serious, and time urgent, reconsideration of a return to continental testing that had been abandoned after TRINITY.

However, the idea of continental testing had been percolating at a low level in Los Alamos and Washington since 1947. There were many advantages to testing in the continental United States. The logistical problems would be much less than for operations in the Pacific, it would be more convenient for those involved, and the drain on the military' support requirements would be much less. On the other hand, the public relations issue was thought to be a major problem.

There were significant advantages to testing in the Pacific; most notably the ability to test very high yields, but there were also troubling features associated with what was called, at that time, the Pacific Proving Ground (Bikini and Enewetak atolls). These included concerns with the difficulties of adequate weather prediction, particularly the wind direction and velocity at high altitudes. Such concerns had been raised during and after the SANDSTONE series and would lead to a number of surveys of potential alternative sites. No particular urgency was felt, however, until the Korean War broke out in June 1950. The perspective in Washington changed quickly. On December 18, 1950, President Truman made the selection of a nuclear weapons test site within the Nellis Bombing and Gunnery Range in southern Nevada. Also, in support of Operation GREENHOUSE, the fielding of a test series, called Operation RANGER, was approved for January and February 1951.

The next six chapters of Part I examine the transition period, 1945 to 1950, from the first nuclear test in New Mexico to the establishment of the Nevada Test Site.

CHAPTER 1. TRINITY

The world's first full scale nuclear explosion was the Trinity test in south central New Mexico on July 16, 1945. The lessons learned from the selection of the Trinity test site and the fielding of the test laid the foundations for the selection and operation of the Nevada Test Site when the United States decided to resume continental testing in 1951.

Background material regarding some of the basic concepts of nuclear weapons design is given in Appendix A. Appendix B supplements this chapter on Trinity by providing sidebars regarding organization and approach to the technical work that existed at Los Alamos as well as at the Trinity site, along with some items of "personal interest".

EARLY CONCEPTS FOR TESTING BEFORE DEPLOYMENT

In Spring 1943, under the leadership of J. Robert (Oppie) Oppenheimer, the initial efforts at Los Alamos were focused on gun assembly of the fissionable components. Two guns, one with uranium components and one with plutonium components, were being studied, see Appendix A. There was confidence that a reasonably efficient gun assembly could be built.

However, as a fallback, Oppenheimer established a small research effort in a group under Seth Neddermeyer to explore implosion assembly. Neddermeyer, a physicist from the National Bureau of Standards and a former Cal Tech student of Oppenheimer's, was an early champion of the implosion concept. The concept of obtaining a super-critical configuration of fissionable material by implosion was not new. At a 1942 conference at the University of California, Berkeley, a small number of physicists met to explore the feasibility of nuclear weapons; and during this conference Richard C. Tolman* suggested the implosion concept. (Serber 1992: 59) In Spring 1943, Neddermeyer's group was E-5, a part of E Division, Ordnance and Engineering, which was led by Navy Captain William S. (Deak**) Parsons. (Hoddeson 2004: 67, 87) [*Footnote: Tolman was an original member of the National Research Council, which was formed in June 1940 by President Roosevelt and chaired by Vannevar Bush, to search for new opportunities to apply science to the war effort. [Hewlett and Anderson 1962: 25] In 1943, Tolman became General Groves' Scientific Advisor. [Hewlett and Duncan 1969:15]] [**Footnote: The spelling *Deak* is used here, but some authors use Deek]

In late 1943, or very early 1944, Parsons, presented a colloquium at Los Alamos where he discussed the need to test a nuclear weapon. Parsons had been responsible to Vannevar Bush for successfully field-testing the naval anti-aircraft proximity fuze. He was familiar with the failure of the then-new Mark 6 torpedo exploders in the early days of World War II and was sensitive to the need for proof testing of a new weapon system prior to military deployment. Following his stint at Naval Ordnance, Parsons was assigned to Los Alamos Laboratory, also called Project Y, to assist with the weaponization of the nuclear device. If a gun assembly were to be dropped in a military strike, and failed for some reason, but remained intact, the enemy would be presented with a substantial amount of fissionable material and would, no doubt, gain technical information about its construction. Initially, a plutonium gun was being considered for a test. (Ben Diven private communication, October 17, 2001)

The idea of a test was well received by the Los Alamos scientists who felt that there were too many unknowns in the performance of such a new device to chance military deployment without at least one full-scale test. The disadvantage of doing a test, of course, was the expenditure of a large fraction of the then existing U.S. stockpile of fissionable material.

Work was progressing on the implosion design. In March 1944 General Leslie Groves, the commanding general of the Manhattan Project, decided to permit scientists at Los Alamos to study the issues associated with a full-scale nuclear test of an implosion design. Groves was not particularly enthusiastic about a test and viewed one as a waste of precious plutonium worth about a billion 1944 dollars. (Hoddeson 1993: 174-175) In 1944 the production of plutonium was very slow, and it was thought that the use of even one weapon for a proof test might well delay the production of a weapon that could significantly benefit the war effort. Groves finally agreed to a test on the condition that the fissionable material would be recovered in the event of a dud or misfire.

Several ideas were explored for recovering the plutonium, but it was rather quickly decided that a large containment vessel would be the most effective solution. A 214-ton cylinder was designed by a team lead by Robert W. Henderson to contain the high explosive and plutonium in the case of a failure to produce a significant nuclear yield. The containment cylinder was named Jumbo.

Within E Division, Parsons formed a group, E-9 (High Explosives Development), under Harvard physicist Kenneth Bainbridge, to look into all aspects of the feasibility of a nuclear test. (Hoddeson et al 1993:174). In the spring of 1944 a Site Selection Committee was established to pick a test location. The members were: Oppenheimer, Bainbridge, Henderson, Major Wilbur A. Stevens (who was responsible for the construction of the Trinity base camp (Norris 2002: 12)), and Major Peer de Silva (head of security at Los Alamos (Hoddeson 1993: 310; Norris 2002: 268-269)).

From a scientific perspective the site had to be flat in order to minimize extraneous effects of the blast; there should be a minimum of haze, dust, and inclement weather to interfere with the optical experiments; and the site should be as convenient to Los Alamos as possible.

From a military standpoint the main issues were security; a low population density and a minimum displacement of people; isolation of activities from association with Los Alamos; and logistics and infrastructure.

Possible locations that were identified were the Tularosa Valley in New Mexico; the region known as the Journada del Muerto (Journey of the Deadman -- named by the early Spanish to reflect the very harsh environment) Valley in New Mexico; an Army training area near Rice, California; San Nicolas Island off the coast of Southern California; the lava beds south of Grants, New Mexico; the area southwest of Cuba, New Mexico, and north of Thoreau, New Mexico; Sand bars which form the coast of South Texas, located about 10 miles from the main coast; and the San Luis Valley region near the Great Sand Dunes National Monument in southern Colorado.

The site options were quickly reduced to the Journada del Muerto location, in New Mexico, and the desert training area near Rice, California. The Army's southern California training area near Rice was under the control of General George Patton. General Groves preferred not to approach Patton on the use of this site by the Manhattan project. Consequently the choice fell to a site located in that part of the Jornada del Muerto in south central New Mexico that is included in what is now the White Sands Missile Range, see Figures 1-1.1 and 1-1.2. The name of the military reservation during the war was the Alamogordo Bombing Range.* [*Footnote: The test site is frequently referred to as the Alamogordo site, after the town of that name, 55 miles distant. The town of Socorro is closer, however, being less than 40 miles away.] In September 1944 General Groves obtained the approval of army General Uzal G. Ent, the commanding general of the Second Air Force, which governed the Alamogordo Bombing Range, to use the area as a test site.

REORGANIZATION TO EXPEDITE IMPLOSION WORK

Also, in the Summer of 1944, it became painfully clear that a plutonium gun would not work. Emilio Segre, and his Los Alamos group, discovered that there was entirely too much Pu240 with reactor produced Pu239. Neutrons from spontaneous fission of the Pu240 would cause a premature chain reaction in a gun-type weapon resulting in a fizzle yield. By mid July Oppenheimer was convinced that the lab should focus on an assembly that was not dependent upon a low neutron background. This meant a much faster assembly was required for a plutonium weapon than for a uranium weapon. Implosion was the solution. (Hoddeson 1993: 228-245)

Implosion would be much faster than gun assembly, but the details of the final configuration could only be known approximately. A proof test of an implosion system would be even more important than a proof test of a gun-assembled device.

Oppenheimer felt that Los Alamos' progress was too slow in implosion technology and that this area had to be beefed up. Los Alamos was desperate. If the plutonium produced by the Hanford reactors was not to be a total loss, the implosion technique had to be made to work. Oppenheimer felt that Neddermeyer could not manage the hundreds of personnel who would be working on implosion. He would replace Neddermeyer with Norris Bradbury, a Navy Commander who had been a physics professor at Stanford before the war.

In August 1944, Los Alamos was reorganized in order to expedite work on the implosion weapon. "The reorganization transferred technical control over implosion from Parsons to Kistiakowsky and Bacher ---"who would head two new divisions devoted to implosion, X (Explosives) and G (Gadget).

The new X-Division was tasked with: the design of explosive components for the implosion device; development of detonation techniques; improvement of the quality of explosive castings and lens systems; and supplying experimental assemblies. George B. Kistiakowsky, Harvard professor and explosives expert, was named X-Division Leader. Most of the new X-Division groups came from the old E-Division. (Hoddeson 1993: 245)

In the new X-Division organization, Neddermeyer's E-5 became X-1 under Bradbury. Bainbridge's group, E-9, High Explosives Development, became X-2, Development, Engineering and Tests. This group would eventually evolve into a division at Los Alamos and would have the responsibility for the world's first full scale nuclear explosion, Trinity. Section leaders within X-2 included Robert Henderson for Engineering, Wilber Schaffer for High Explosives and Lewis Fussell for Test measurements (Roger Meade, Private Communication November 2001). Appendix B provides a description of the organization and some of the field activities associated with Trinity.

The new G-Division would provide experimental support for explosives development. Robert F. Bacher (from the Radiation Laboratory of the Massachusetts Institute of Technology) was named Division Leader, Darol Froman and Marshall Holloway were appointed Associate Division Leaders. (Hoddeson 1993: 245-246) Neddermeyer became a scientific advisor for implosion research and a joint group leader with Donald Kerst (University of Illinois and inventor of the betatron) of G-5, the betatron group. (Hoddeson 1993:274)

Parsons was named Associate Director with oversight of interdivisional questions and broad policy questions. He was also the leader of the new O(Ordnance)-Division which had responsibility for weaponization of both the gun and the implosion gadgets: ordnance, assembly, delivery, and engineering of nuclear weapons*. [*Footnote: Parsons later served as Officer-in-charge of Project Alberta which included preparations for overseas operations against the enemy and was the weaponeer aboard the Enola Gay when it struck Hiroshima on August 6, 1945. (AFSWP 1954 Vol. 1:4.2.2-3)] O-Division was composed of the parts of the old E-Division that did not go into X-Division. The Deputy Division Leader of O-Division was Edwin McMillian who retained technical control of the gun work that now took place in O-Division under Lt. Comdr. A. Francis Birch. (Hoddeson 1993:245,249) This gun work focused on streamlining the uranium gun.

Enrico Fermi*, who would arrive at Los Alamos in September 1944, was named an Associate Director with oversight responsibilities for the "research and theoretical divisions and all nuclear physics problems". Oppenheimer recognized Fermi's "senior standing and talents" and also made him the head of F (Fermi) Division which "---consolidated several miscellaneous research projects." (Hoddeson 1993: 245-6) [*Footnote: When Fermi received the 1938 Nobel Prize in physics, he left Italy for the United States. He was Professor of Physics at Columbia prior to his arrival at Los Alamos.]

PREPARATION FOR TRINITY

Oppenheimer named the first nuclear test Trinity; and the location of the shot was known as the Trinity site. Trinity would be a test of the plutonium implosion concept. The design progressed rapidly through the summer and fall of 1944 and was frozen on February 28, 1945 at a meeting in Oppenheimer's office at Los Alamos. Oppenheimer, Groves, Tolman, Kistiakowsky, James Conant (former President of Harvard), Hans Bethe (Theoretical Physics Division Leader at Los Alamos and physics professor at Cornell University), and Charles Lauritsen (physics professor from Cal. Tech.) were in attendance at the meeting. (Hoddeson 1993: 312)

To assure that the development and production of the implosion components were proceeding on a schedule that would allow for a field test of the weapon within a few months, Oppenheimer formed a group on March 1, 1945, called the Cowpuncher Committee. This group, chaired by Samual Allison (University of Chicago professor) was to "'ride herd' on the implosion development." The committee included: Oppenheimer, Bethe, Kistiakowsky, Parsons, Bacher, and Cyril Smith (who had come from the National Research Council (NRC) to head Los Alamos' metallurgy efforts in CM Division), and Bainbridge. (Hoddeson 2004: 316) Bainbridge also remained the Group Leader of X-2 as well as the overall leader of the test operations for Trinity.

By May 1945 there was sufficient confidence in the design, and enough plutonium available, that Oppenheimer decided to dispense with the containment vessel, Jumbo*. [*Footnote: Not all senior Presidential advisors were as confident. Admiral Leahy, Chief of Staff to President Truman, remarked in reference to the atomic bomb "This is the biggest damn fool thing we have ever done. The bomb will never go off, and I speak as an expert on explosives." Also, Gen. Groves commented: "If our gadget proves to be a dud, I and all of the principal Army officers of the project ... will spend the rest of our lives so far back in a Fort Leavenworth dungeon that they'll have to pipe sunlight into us." (Norris 2002: 175)] Firing the test shot inside a heavy steel container would complicate the analysis of the data and the additional fallout from the activated, vaporized, material would be undesirable. Trinity would be fired atop a 100-foot tower. Robert W. Henderson was in charge of the tower operations for the test.

On May 7, 1945 a high explosive test, known as the "100-ton test" was conducted at the Trinity site, about 800 yards SE of the Ground Zero (GZ) for the test of the nuclear gadget. (Hoddeson 2004: 361) The purposes of this test were: "(1) To provide a full dress rehearsal in preparation for the later gadget test; (2) to provide calibration of blast and earth shock equipment." One hundred and eight actual tons of TNT high explosive (HE) contained in wooden boxes were carried up a wooden platform and stacked. The center of gravity of the TNT stack was 28 feet above the ground. This height was chosen as an approximate scaled height* that was planned for the nuclear gadget whose yield was approximated at 4,000 - 5,000 tons on the 100 feet height of the tower.[*Footnote: 100/28 ~ $(4500/100)^{1/3}$] (Bainbridge 1945: 15) "An irradiated slug from the Hanford pile was dissolved and poured into flexible tubing threaded through the HE." (Hoddeson 2004: 361) It was hoped that post-test distribution of the 1,000 curies of radioactive material would provide clues for the distribution of radioactive material resulting from the nuclear test. (Szasz 1984:64) The 100-ton test was "much larger than any previously measured explosion" (Hoddeson 2004: 360); and Bainbridge states: "The test appears to me to have been successful as a trial run". (Bainbridge 1945: 28)

In late May, Frank Oppenheimer, brother of "Oppie" and also a physicist, arrived at Trinity to troubleshoot the test. In this assignment, he was working for his brother while serving as an administrative aide to Bainbridge. On his arrival, he noted: "people were feverishly setting up wires all over the desert, building the tower, building little huts in which to put cameras and house people at the time of the explosion." (Rhodes 1986: 654)

THE TRINITY TEST

On June 30, 1945, Oppenheimer and his colleagues selected July 16 as the test date. Firing Trinity was time urgent not only for military reasons, but also because President Truman had announced that he would meet with Churchill and Stalin at Potsdam beginning on July 17, 1945. He wanted very much to use the occasion to notify Stalin of the U. S. development of nuclear weapons. (Roger Meade, private communication)

A yield estimate of 4kt was made on July 10th. For safety purposes, a maximum yield of approximately 100kt was adopted for personnel planning. Bunkers for the scientists and their equipment were placed to the north, west and south at 10,000 yards, see Figure 1-1.2.

A team composed of Lewis Fussell (from MIT - Roger Meade private comm.), Phillip Moon (a senior British physicist), Bernard Waldman (from the University of Chicago Metallurgical Laboratory - RM private communication), and Victor Weisskopf (from the University of Rochester) reviewed proposals for experiments on Trinity. (Hoddeson et. al. 1993: 351)

The experimental program was quite extensive for this first nuclear shot. All told about 250 technical people were involved. Many of the experimental techniques continued to be used for decades after Trinity. Perhaps the most noteworthy, from the standpoint of the historical importance, was the measurement of alpha (α) by Bruno Rossi and his collaborators. One over alpha ($1/\alpha$) is the time for the neutron population in the device to increase by a factor of e (base of the system of natural logarithms having the approximate numerical value 2.71828). It is a measure of how fast the fission neutrons were multiplying, see Appendix A. Actually, the measured observable was gamma-rays, which are proportional to the neutron population over the times of interest.

There were two approaches for the Trinity alpha measurements. Robert R. Wilson (from Princeton) had the responsibility for measuring alpha. He elected to do this with photomultiplier tubes* which directly measured the brightness of the gamma-rays. [*Footnote: the term *photomultiplier tube* is sometimes used interchangeably with the term *scintillation detector*.]

The Cowpuncher Committee had cut off additional experiments; but Bruno Rossi (from Cornell) proposed to Wilson that he, Rossi, measure alpha as well with ionization chambers*.(Rossi 1990: 93-96) [*Footnote: An ionization chamber consists of a material (usually gas) between two conducting electrodes (a + charged anode and a - charged diode). When gamma-rays ionize the material, the + charged ions and - charged electrons move to the electrodes of opposite polarity thereby creating a current that is measured.] Wilson graciously accepted the offer and provided Rossi with the space and support that he needed for his experiment.

The data were to be recorded on oscilloscopes, but the difficult part was to trigger the sweep at just the right time to have the display on the scope. Rossi ingeniously used a continuous sine wave in place of a conventional oscilloscope sweep while the signal from the ionization chambers was applied to the vertical axis. The sine wave provided the timing information for the data presented on the 'scope. Alpha was subsequently calculated from the observed trace*. [*Footnote: Ben Diven was a member of Rossi's team. He was a graduate student from U.C. Berkeley during the Manhattan period and received a Ph.D. in physics from the University of Illinois after the war. In later years Diven would play a significant role at the Nevada Test Site, making basic physics measurements using bomb neutron sources. He would also be a Scientific Advisor to the manager of NVOO.]

At 05:10 am, on the morning of July 16, a panel composed of J. R. Oppenheimer, Brig. General Farrell (Deputy to Gen. Groves), meteorologist J.M. Hubbard*, [*Footnote: Col. Ben G. Holzman, an army air force meteorologist, was brought in to advise the inexperienced Hubbard. Holzman was later a senior meteorologist on the 1948 SANDSTONE test series (Norris 2002: 402)] and K. T. Bainbridge gave permission to fire with a 05:30 am zero time (Mountain War Time, which was the same as daylight saving time and used all year during the war).

Enrico Fermi, who was at Station 10,000 during the test, began tearing up paper into small pieces during the final seconds before the detonation. After he saw the flash, he dropped the paper from his hand; and when the blast hit, measured how far it carried them. Then, he pulled out his slide rule and calculated the yield as 10 kt. (Szasz 1984:83) The final reported yield was 21 kt.

The bomb performance and the associated diagnostic experiments were very successful. The energy released by the device was deduced from several experimental observations: measurements of the gamma-ray flux; measurements of the neutron fluence; measurements from the fireball dimensions as a function of time; blast; and radiochemistry.

RESULTS

The most important result of the Trinity test was that the implosion design worked. The yield and size of the fireball allowed the delivery group to fix the explosion height for the Hiroshima and Nagasaki bombs at 1850 feet. (Hoddeson 2004:374) For the alpha measurements, both groups were able to show that alpha was consistent with theoretical calculations and that the alpha value implied a more efficient explosion than originally anticipated. The successful test of the implosion concept gave the scientists confidence in their predictions for the performance of the gun-assembled weapon; and it was decided to deploy this device, without a test, in the first strike on Japan, August 6, 1945, three weeks away.

One immediate result of the Trinity test was that Oppenheimer raised with Groves the possibility of using the gun assembled (i.e. Little Boy) weapon's uranium with reduced quantities of plutonium in so-called composite cores, or pits, to produce more implosion-assembled weapons than would otherwise be available. On July 19, just 3 days after Trinity, Oppenheimer sent a teletype to Groves outlining the possibilities. In the teletype, Oppenheimer mentioned the number of weapons that would be available by November first, the planned date of the start of Operation Olympic, the first phase of the anticipated invasion of the Japanese home islands.

Groves immediately discussed the options that Oppenheimer raised with his Washington associates. He responded the same day to Oppenheimer: ... "Factors beyond our control prevent us from considering any decision other than to proceed according to existing schedules for the time being. It is necessary to drop the first Little Boy and the first Fat Man and probably a second one in accordance with our original plan. It may be that as many as three of the latter in their best present condition may have to be dropped to conform with planned strategic operations."(Los Alamos Archives: files 471.6 Little Boy)

This redesign and fabrication of the more efficient nuclear weapons was not pursued in 1945 because of the enormous time pressure to deploy the weapons then available. Composite cores were, however, included in the next weapon designs to enter the stockpile in the post-war years. The first opportunity to test the new concepts would come during Operation SANDSTONE in the spring of 1948.

The United States made nuclear strikes on Hiroshima on August 6, 1945 ("Little Boy") and on Nagasaki on August 9 ("Fat Man"). Japan surrendered on August 15.

CHAPTER 2. OPERATION CROSSROADS AND CREATION OF THE ATOMIC ENERGY COMMISSION (AEC)

OPERATION CROSSROADS

On August 28, 1945, two weeks after the Japanese surrender, Vice Admiral E. L, Cochrane proposed a test series to study the effects of nuclear weapons on warships. Lt. Gen. B. M. Giles seconded this when he proposed on September 14 that at least two atomic bombs be used to test their destructiveness on naval vessels. Adm. Ernest King, on October 16, recommended that the Joint Chiefs of Staff control the operation and that the Navy, with cooperation of other military branches, carry out the tests.

The operation, named CROSSROADS, was formally initiated on January 11, 1946 with the formation of Joint Task Force One, JTF-1, under the command of Vice Adm. W. H. P. Blandy. The tasking orders stated that both air and underwater detonations should be included if feasible.

Bradbury, who succeeded Oppenheimer as Director of Los Alamos in October 1945, asked John Williams (a senior physicist at war-time Los Alamos) to return to Los Alamos from the University of Minnesota to prepare a tentative plan for the tests. In addition, several other veterans of the Manhattan Project were persuaded to remain at Los Alamos through CROSSROADS (see Chapter 3).

Initially CROSSROADS was thought by Los Alamos to be of little use to the Laboratory, and was considered by Bradbury to be largely a drain on scarce resources. The manpower level at the lab was at low ebb. Most of the senior scientists and engineers left Los Alamos after the surrender of Japan to resume their pre-war careers. Many of the younger scientists departed to finish graduate school. This left large gaps in the capability of the laboratory to design and field nuclear weapons tests. In fact, there was some concern at Los Alamos about whether or not the Laboratory could even field the devices for the Navy. To some extent the successful support of the military was beneficial in bolstering the confidence of Los Alamos in their ability to carry on without those who left after the war (Betty Perkins, private communication).

On January 15, 1946 Bradbury formed B Division, with Marshall Holloway as the division leader, to support CROSSROADS. B-Division was dissolved after CROSSROADS work was completed. Holloway was the senior Los Alamos person on the test series. In addition, Z-Division (see Chapter 4), which had been formed in 1945, with Roger Warner now its division leader, was assigned to B-Division with ordnance-related responsibilities in support of the Navy. In this organization plan, Warner had the title of B-Division Associate Division leader.

B-Division had 16 groups. The group numbers, title, and group leaders were as follows:
<u>B-1,</u> Firing - W. O. McCord;
<u>B-2, Fuzing</u> - George A. Koester;

B-3, Assembly - Arthur Machen (this group became Z-3 by December 1946);
B-4, Engineering - Robert W. Henderson;
B-5, Pit Assembly - Raemer E. Schreiber;
B-6, Logistics/Supply - Harlow W. Russ;
B-7, Air Coordination and Air Force Liaison - Glenn A. Fowler;
B-8, Air Collection - T. V. Davis;
B-9 Photography - Berlyn Brixner;
B-10, Fast Neutron - G. A. Linenburger;
B-11, Gamma Timing - Norris Nereson;
B-12, Electronics (Timing and Firing) - Jerome Wiesner;
B-13, Airborne Blast Damage - James Wieboldt;
B-14, Radiochemistry - William Rubison;
B-15, Phenomenology - Joseph O. Hirschfelder; and
B-16, Radiography - James L. Tuck.
This organization for B-Division, as well as subsequent test organizations, had many similarities to the organization used for Trinity, see Appendix B.

A number of the British contingent to wartime Los Alamos returned to assist with this first Pacific test series. Sir William Penny, who later would be the director of Aldermaston, the United Kingdom's Atomic Weapons Research Establishment, was an advisor to Los Alamos and a consultant on blast damage and theory. Earnest Titterton, from Australia, was in charge of the laboratory's technical electronics work. James Tuck, who would eventually return to Los Alamos permanently, was the B-16 group leader responsible for radiography. Klaus Fuchs, who would become infamous as the master spy for the Soviet Union during WWII, helped predict the weapons effects such as the bubble formation from the underwater burst, for group B-15.

Ralph Sawyer, of the Naval Research Laboratory (NRL), was appointed to be the technical director for Operation CROSSROADS reporting directly to Admiral Blandy. In later years, Sawyer's son, George, would spend his career at Los Alamos as a physicist in the controlled fusion program.

CROSSROADS which was conducted at Bikini Atoll*, see Figures 1-2.1 and 1-2.2, was an enormous operation. [*Footnote: The US used the Bikini and Enewetak atolls of the Marshall Islands as the Pacific Proving Grounds during 1946-1958. These atolls were not US territory but were a part of the Trust Territory of the Pacific Islands which also included the Mariana Islands, Micronesia, and Palau. This trust territory was a trusteeship of the United Nations over the former Japanese possessions in the Pacific and was granted following World War II. The trusteeship was administered by the United States.] It was by far the largest operation that the United States has fielded. There were over 42,000 people, over 200 vessels, and more than 150 airplanes involved. This was an international show, with representatives of the press, governments, and military from around the world. The U.S. government clearly wanted to impress the observers with the enormous potential of nuclear weapons. Two shots were fired, both identical to the plutonium implosion device used on the Nagasaki strike. The first shot, Able, at 9:00 am on

July 1, 1946, was an airdrop with a detonation altitude of 520 feet. It missed the target of anchored ships by about 700 yards and apparently wasn't particularly impressive to the observers who were miles away on board support ships. The second shot, Baker, was detonated on July 23 underwater at a depth of 90 feet. Two million tons of water was contained in the ensuing eruption. This dramatic event made up for the Able shot and certainly impressed those observers still around. Unfortunately, many had left after the rather disappointing first event.

Los Alamos provided the two nuclear weapons, detonated the underwater bomb, provided radiochemical yield data and made other diagnostic measurements to assess weapon performance and to measure the blast wave. They also supplied additional support in the form of timing signals and technical advice. About 150 scientists, engineers and technicians and 50 GIs from Los Alamos were assigned to Bikini for the tests. This made up about 40 percent of the Navy technical director's personnel.

A number of lessons for Los Alamos came out of the CROSSROADS experience. One dealt with the reliability and reproducibility of the Fat Man implosion weapon. After the successful CROSSROADS tests, a total of four almost identical implosion devices had been detonated; and within the experimental uncertainty all gave the same yield. This, of course, was comforting to the laboratory and gave them confidence in their ability to move forward with new designs. CROSSROADS also stimulated the military to request additional, and more advanced, nuclear weapons. The result of the increase in demand for weapons was to strengthen the rationale for a continuation of the laboratory's research and development role. Surprising as it may seem, after more than five decades, this was an issue that was by no means clear in 1946. CROSSROADS, while not popular with many at Los Alamos, did in fact cement the laboratory's future as a design and development facility rather than as a production plant.

Operation CROSSROADS also served as a precedent for the Joint Task Force organization, which would be used successfully in the Pacific test series through the Tightrope event above Johnston Island, southwest of the Hawaiian Islands, in November 1962.

In 1945 and 1946, Los Alamos engineers and scientists handcrafted all of the atomic weapons (which in 1946 were the Mark III, i.e. the Fat Man design). This was a slow process requiring a group of about 30 people approximately a month to assemble a bomb from the basic components. At the time of Operation CROSSROADS, there were only 9 weapons in stockpile. Partly because President Truman wanted to carefully husband these valuable resources, he cut CROSSROADS from the planned 3 tests to 2. By the end of CROSSROADS, in August 1946, there was a very seriously limited capability available for assembling nuclear weapons. (DTRA 2002: 13-14)

CREATION OF THE ATOMIC ENERGY COMMISSION (AEC)

The organization of the nuclear weapons program and the leadership responsibilities were matters of some concern immediately after the Second World War. During the war, the Army ran the Manhattan Project under the direction of Gen. Leslie Groves; but after VJ-Day, there was a growing feeling that nuclear weapons development should be under civilian control. The Manhattan Project scientists in particular were chaffing under what they considered oppressive security constraints.

After several months of acrimonious debate Senator Brien McMahon (D. Connecticut) introduced S. 1717 on December 20, 1945 to address the issues surrounding civilian control. The debate continued through the winter of 1945 and into1946. The vital issue was: "Are we going to have military domination or civilian control of atomic energy in this country?" (Hewlett and Anderson 1962: 488-492). The major stumbling block was the definition of the role to be played by the military. Clearly, they would be the ones to employ nuclear weapons in a conflict; and it was recognized that they should have some influence on the nuclear development program. At issue was how this influence should be exercised. Senator Arthur H. Vandenberg, chairman of the Foreign Relations Committee, proposed a military liaison board, but McMahon felt that this would undercut civilian control.

Eventually a compromise was reached that called for a five member Atomic Energy Commission (AEC) and a general manager appointed by the President to provide overall direction and day-by-day management of nuclear weapons development. The Commission would appoint the directors of the four divisions of research, production, engineering, and military applications. The compromise position included the establishment of a Military Liaison Committee (MLC) between the military and the Commission; a General Advisory Committee (GAC) for the Commission; and a Joint Committee on Atomic Energy in Congress (JCAE). The establishment of the MLC addressed the problem of input by the military to those in the AEC who were charged with the responsibilities for nuclear weapons research, development and production. The MLC chairmanship was a statutory position, where the President appointed the Chair. Usually the Chair was dual hatted and was also the Assistant to the Secretary of Defense for Atomic Energy (ATSD/AE). Both civilian and military personnel staffed the MLC.

Section 6.a(2) of the enabling legislation provided that "The Commission is authorized and directed to have custody of all assembled or unassembled atomic bombs, bomb parts, or other atomic military weapons, - except that upon the express finding of the President that such action is required in the interests of national defense, the Commission shall deliver such quantities of weapons to the armed forces as the President may specify." (Hewlett and Anderson 1962: 717)

S.1717 passed the Senate on June 1, 1946 and on July 26 both houses accepted the compromises. President Truman signed the bill creating the Atomic Energy Commission on August 1, 1946. It went into effect on January 1, 1947. David Lilienthal was appointed the first Chairman of the AEC on November 1, 1946; and he remained in that position until February 15, 1950. J. Robert Oppenheimer was appointed Chairman of the GAC on December 12, 1946, a position he held until August 8, 1952. Lt. Gen. Lewis H. Brereton, United States Army Air Force (USAAF), was appointed chairman of the MLC on August 1, 1946. He served in this capacity until April 5, 1948.

The Director of the Division of Military Applications (DMA) was specified as an admiral or general officer on active duty. The first appointee was Brig. Gen. James McCormack, Jr. USAAF who was selected on February 1, 1947. He served until August 19, 1951.

On July 2, 1947 Carroll Tyler arrived in Los Alamos as the Manager of the Santa Fe Operations Office (SFOO) of the AEC, which was located in Los Alamos at what is now the southwest corner of 15th and Central. This office served the official liaison functions between the AEC and Los Alamos. At that time the Santa Fe Office oversaw the operation of approximately 1200 employees at Los Alamos and an additional 320 in Z-Division at Albuquerque.

Nuclear weapon testing was under the purview of the DMA and would remain there through the various reorganizations of the parent organization as long as the United States continued to test.

CHAPTER 3. LOS ALAMOS LABORATORY, ORGANIZATION AND LEADERSHIP: 1945-1950

DEVELOPMENT OF PEACETIME LOS ALAMOS

After the war ended, the long months of intense effort were followed by a crushing letdown and a mass exodus from Los Alamos. The status of the Laboratory itself was faltering. There was a lack of national policy to determine the use and control of atomic energy; and little direction was being provided for its future course. (Furman 1990:125)

Oppenheimer left Los Alamos during the early fall of 1945; and Navy Commander Norris Bradbury was appointed in October by Groves, with a recommendation from Oppenheimer, to serve as director. Bradbury was discharged from the Navy just prior to his appointment as Director of Los Alamos in October 1945. He had the immediate task of forging a cohesive organization out of the general confusion. The job would extend into a twenty-five year commitment. The wisdom of selecting Bradbury was quickly proven. "To end the doldrums that seemed to hang like a pall over the Laboratory, he decisively delivered an ultimatum to Project Y personnel. 'You've got three months to get off the Hill or go to work.' 'That shook up a few', Bob Henderson recalled, 'but those that decided to stay went to work, and the bickering and bitching stopped.'" (Furman 90:132)

High on Bradbury's priority list was to find out what the government expected the laboratory to do. He wrestled with a definition of the proper role of Los Alamos during the winter and spring of 1946. It was during this period that Bradbury came to grips with the real responsibility of Los Alamos - to design and develop new atomic weapons. In August he submitted to the AEC Commission an untitled report on a proposed program of work for the laboratory (Los Alamos Archives: Roger Meade, A-84-019; 37-5)

In this August report, Bradbury proposed disassociating the laboratory from bomb testing: "not part of the job of a research laboratory. —If it cannot avoid the responsibility now accepted, Los Alamos will have to undertake to test the Fat Man modifications upon design completion. But unless new investigations are proposed, the measurements made should be the minimum required to check the success of the model. Probably yield, alpha, gamma ray timing, and radiochemistry are needed." (Roger Meade, private communication)

By November 14, 1946 in a letter to the AEC, Bradbury wrote (Meade unpublished manuscript: Sandstone): "It is further suggested that Los Alamos retain the responsibility for testing the nuclear reactions for new atomic weapons, but that such tests as have purely military significance be carried out by the armed forces. The distinction, which is intended is that of separating a test of the 'Alamogordo' type* from a test of the CROSSROADS type." [*Footnote: i.e. Trinity] Thus, between August and November 1946, Bradbury had completely changed his mind about laboratory involvement in test activities. Never the less, Bradbury's philosophy remained remarkably consistent over the years that Los Alamos stay as "pure science" as possible, even to the exclusion of military staffers on assignment in the weapons systems area of the Laboratory (military personnel were, however, assigned to the test division).

On January 1, 1947 the Los Alamos Laboratory, also known as Project Y during the war, officially became the Los Alamos Scientific Laboratory (LASL). This was clearly a reflection of Bradbury's view of the Laboratory's mission.

Bradbury, in a January 13 letter to AEC general manager Carroll Wilson, asked the AEC Commission to "indicate in a general way the extent of effort which the laboratory should spend on (1) basic research applicable to weapons [15% - the numbers in square brackets indicate the percentage spent by the lab at that time], (2) productions of weapons components [25%], (3) short range weapon development [35%], (4) long range weapon development [2%], (5) ordnance engineering [16%], (6) power or propulsion applications of nuclear energy [1%], and (7) basic scientific research not obviously applicable to atomic weapons [6%]." Bradbury went on to argue for weapons tests and for a continental test location that could be used with proper precautions (Meade, Sandstone, unpublished and private communication). The January 1947 letter to Wilson is the first reference in print to a continental test site.

At the end of March or early April 1947 Norris Bradbury sent to Washington a proposed directive, to be from the AEC to Los Alamos, articulating the Laboratory's mission. This was done at the request of the AEC Director of Research, James B. Fisk, and Director of the DMA, Brig. Gen. McCormack. In it Bradbury suggested, "The primary purpose of the Los Alamos Laboratory is to conduct research, development, and such production as is necessary on atomic weapons. The secondary objective of the laboratory should be to conduct such scientific, technical, and administrative operations as are necessary to insure a vigorous and progressive scientific laboratory." (Meade unpublished manuscript: Sandstone. "Proposed directive for the Los Alamos Laboratory," B-9, 310.1, AEC)

This mission statement is remarkably similar to the mission of the Laboratory over 50 years later. It reflected the view that Bradbury had that Los Alamos remain mainly a scientific laboratory, doing the nuclear part of nuclear weapons R&D. Others should do ordnance and military effects activities. (Harold Agnew, who succeeded Norris Bradbury in 1970, established closer, more collaborative, ties to the military. In particular, Agnew encouraged the services to assign military staff members to the laboratory in order to improve liaison on weapons systems and nuclear policy.)

In April 1947 Norris Bradbury asked Joseph O. Hirschfelder, who in 1945 made fallout predictions for the Trinity test, for help in assessing the safety problems associated with a continental test site (Los Alamos Archives: B9 353.4 1/47 - 4/66). Bradbury noted that, "We must prove in advance, and as conclusively as possible, that such a test can be conducted safely." (Betty Perkins, Private communication) Shortly thereafter Bradbury, in an informal memorandum to Roger S. Warner, then the Director of the AEC Engineering Division in Washington, presented his views on the importance of a comprehensive test program and discussed potential sites, both continental and overseas. (Betty Perkins, private communication)

Obviously, testing in the Pacific was complicated and very expensive, and Los Alamos proposed that the AEC activate a continental site to make it easier and faster to test the new concepts.

FIELD TEST DIVISION

Bradbury had formed B-Division which existed for nine months in 1946 to support Operation CROSSROADS. He assembled J-Division (for Joint, as in Joint Task Force) in October 1947 to conduct Los Alamos' technical portions of Operation SANDSTONE at Enewetak in the spring of 1948. Like B-Division for CROSSROADS, J-Division was set up for SANDSTONE as a temporary division.(Ogle 1985: 62-63) People from other divisions would be lent to J-Division for the operation, then it was intended that they would return to their respective divisions after the operation. The initial J-Division group structure, cited below, again had many similarities with that used on Trinity, see Appendix B. (Roger Meade, Private Communication and Los Alamos Archives)

<u>J-1 Theoretical Physics.</u> Group Leader: Fred Reines (who would later win the Nobel Prize in Physics for his experimental observation of neutrinos. This work was done in collaboration with Clyde Cowen, who died before the Nobel Prize was awarded), Alternate Group Leader: Joseph Mullaney
<u>J-2 Radiochemistry.</u> Group leader: Roderick W. Spence, Alternate Group Leader: Melvin G. Bowman
<u>J-3 Measurement of Fast Neutrons.</u> Group Leader: G. A. Linenberger, Alternate Group leader: William Ogle (who would become the third J-Division Leader in July 1965 upon the death of Alvin Graves)
<u>J-4 Measurement of Alpha.</u> Group leader: Richard Taschek (who would become the Physics Division Leader and later the associate director for research). Ernest Krause, and his Naval Research Laboratory (NRL) team, reported to Taschek
<u>J-5 Gamma Spectrum.</u> Group Leader: L. D. P. King
<u>J-6 Transit Time.</u> Group Leader: Norris Nereson. Herbert Grier, of EG&G, reported programmatically to Nereson
<u>J-7 Technical Photography.</u> Group Leader: Berlyn Brixner
<u>J-8 Blast and Jets.</u> Group Leader: John C. (Jack) Clark (who would later be a test director at Nevada)
<u>J-9 Assembly.</u> Group Leader: Arthur B. Machen. Section Leader: William Bright. Neil Davis, who spent his career in the Los Alamos weapons program, was in this group
<u>J-10 Engineering.</u> Group Leader: Charles Runyan
<u>J-11 Communication.</u> Group Leader: Louis A. Hopkins
<u>J-12 Timing and Firing.</u> Group Leader: William McCord. Herbert Grier and Bernard O'Keefe, both from EG&G, were programmatically assigned part-time to this group (see also Group J-6 above). Also, Lewis Fussell, who participated in a very visible

way in the Manhattan Project, was in this group. (See, for example, Critical Assembly by Hoddeson, Henriksen, Meade, and Westfall)
J-13 Construction. Group Leader: Roy W. Carlson
J-14 Logistics. Group Leader: Harry Allen
J-15 Administration. Group Leader: Armand Kelly
J-16 Circuit Diagrams and Maps. Group Leader: Carl Hedberg
J-17 Gamma Ray Exposure. Group leader: James F. Nolan
J-18 Safety. Unknown
J-19 Classification. Group Leader: Herbert. I. Miller
Meteorology. On staff of JTF-7* (SANDSTONE), Col. B. G. Holzman
[*Footnote: The Pacific operation prior to SANDSTONE was CROSSROADS, JTF-1. The Pacific operation subsequent to SANDSTONE was GREENHOUSE, JTF-3. However, SANDSTONE was JTF-7. (Berkhouse. L.H.1983: 1)]

The first J-Division Leader was Darol Froman, who served from October 1947 until October 1948. Alvin C. Graves served the same period as the Associate Division leader. Froman and Graves had already been named (nominated by Bradbury to the AEC who then forwarded the nomination to the task force command for approval) as Technical Director and Deputy Technical Director respectively for SANDSTONE. Robert Henderson was an Assistant Division Leader from October 1947 until August 1948, and John C. (Jack) Clark was Assistant Division leader from October 1947 until October 1948.

The Division was reorganized in October 1948 when Graves became J-Division's second Division Leader. Henderson returned to Z-Division in Albuquerque in August 1948.

J-Division was made a permanent division in 1949. It would continue to field technical experiments on operations as well as study technical issues and problems related to those experiments throughout US nuclear testing. (Ogle 1985:62-63)

CHAPTER 4. SANDIA LABORATORY 1945-1950

In 1942, the Secretary of War had appropriated 1,100 acres of land on Albuquerque's East Mesa, including Oxnard Field, the old Albuquerque municipal airport, named for James G. Oxnard who had developed it. This acreage became the Albuquerque Army Air Field for training aircraft mechanics and repairmen; but also in 1942, its name was changed to honor Colonel Roy C. Kirtland, an Army pioneer military pilot. Then, during this same early period, the unofficial term "Sandia Base,"[^] [^Footnote: Sandia is Spanish for watermelon.] for the mountains east of Albuquerque, began to be used by the construction engineers active in the erection of facilities. In 1944, the base became a convalescent center for wounded airmen; and by 1945, it had become a dismantlement center for surplus military aircraft. (Furman 1990: 122-125, Johnson 1997: 14-15)

Today, Sandia Base is incorporated into Kirtland AFB. The area that was Sandia on the east, the airport, and the old western area are now all part of the Kirtland AFB. The Base is under one Command, the 377th Air Base Wing. Today, there are over 100 other organizations on the Base which are referred to as "Associate Units". Included in the 100 associate units are: Sandia Laboratory; The Air Force Research Laboratory; Air National Guard; the Defense Threat Reduction Agency (DTRA) which is the successor to the Armed Forces Special Weapons Project (AFSWP); and the Albuquerque airport. (Verbal communication with Lt. Rose Richeson, January 28, 2004)

1945

In March 1945, a Special Weapons Committee was created at Los Alamos that reported to "Deak" Parsons (see Chapter 1) and was responsible for "all phases of work peculiar to combat delivery". This committee soon became a part of Project Alberta* which was also formed in March 1945 and had responsibility for movement overseas. Some of the members of Project Alberta would become a part of the first contingent from Los Alamos to be stationed at Sandia Base.(Furman 1990: 84) [*Footnote: Project Alberta was activated in March 1945 to conduct such activities as: planning and establishing the advance base (on Tinian island) where bombs were assembled and field tests with non-active bombs were conducted. Project Alberta personnel also conducted the assembly and loading of the nuclear bombs and tested and armed the bombs in flight. Furman 1990:89,125]

In the spring of 1945, Oppenheimer wanted to relocate ordnance activities to a site that would be more accessible to rail and air transport, but would still be relatively convenient to Los Alamos. Groves and Oppenheimer agreed that much of the ordnance work including assembly, testing, and production could be performed elsewhere. Additional factors at that time were the crowded conditions that existed on the Hill, which was described as "bursting at the seams", and the limited water supply. (Furman 1990: 119-125)

Kirtland was selected as the preferred site. It had been an important staging ground for the transportation needs associated with Trinity and Project Alberta. Toward the end of July, 1945, Kirtland Field was transferred from the Army Air Force to the Army's Manhattan Engineering District. (Furman 89: 123)

> Also, in July 1945, the Los Alamos Ordnance Division, O-Division, began to metamorphose into Z-Division, named after its first leader, Jerrold R. Zacharias, who had been a part of the MIT radar group before going to Los Alamos. The formal announcement of the new organization was made in late August 1945, and it was the nucleus of what would eventually become the Sandia National Laboratories. Most of Z-Division would transfer to the newly acquired base in Albuquerque between March and July 1946.

In September, 1945, Z-Division was composed of the following groups:

Z-1 - Experimental Systems - Commander Norris E. Bradbury
Z-1A - Airborne Testing - Dale R. Corson; Glenn A. Fowler (Alternate)
Z-1B - Informers (Telemetry) - Jerome B, Wiesner; B. Wright (Alternate)
Z-1C - Coordination with Using Services - Glenn A. Fowler
Z-2 - Assembly Factory (originally Production) - Colonel Lyle E. Seeman Roger S. Warner, Jr., Deputy
Z-2A - Procurement, Storage, and Shipment - Colonel Robert W. Lockridge; Major Parker (Alternate)
Z-2B - Production Schedules, Manuals, Roger S. Warner, Jr.
Z-3 - Firing Circuits, Lewis Fussell, Jr - Commander Stevenson, Earl Thomas (Alternates)
Z-4 - Mechanical Engineering - Robert W. Henderson; Richard A. Brice (Alternate), Frank Oppenheimer (Coordinator for Redesign)
Z-5 - Electronic Engineering, Robert B. Brode - E. B. Doll (Alternate)

Zacharias didn't stay long. He returned to MIT in the fall of 1945. Roger S. Warner was selected as Division Leader on October 17, 1945. In future years, Warner and a number of other initial Z-Division individuals would serve prominent roles in the weapons testing community. Commander Norris Bradbury became Director of Los Alamos in October 1945. Glenn Fowler, one of the principals in Z-1, would play a leading role at Sandia for many years. Jerome B. Wiesner, another senior person in Z-1, would eventually become the President's Science Advisor in Washington and the President of MIT. Also, Richard A. Bice, Lewis Fussell, Donald Hornig, Robert W. Henderson, and Harlow Russ all served the nuclear weapons testing community for many years in a wide variety of roles.

The new Z-Division would be responsible for specifying ordnance engineering details, establishing production lines at various sites, and with assistance from the armed forces, for setting up routine methods for assembling, testing, and maintaining weapons in a ready state.

In November, 1946, when the Z-Division Leader, Roger S. Warner, left Los Alamos, Bradbury asked Robert W. Henderson to assume leadership of Z-Division on an acting basis. Henderson lead the division from Los Alamos until Group Z-4, Mechanical Engineering, moved to Albuquerque in February 1947.

Ever since the summer of 1945, the ordnance activities had grown at Sandia Base; and the work there eventually assumed the role of a laboratory separate from Los Alamos. Robert Henderson served as interim director of Sandia until Bradbury appointed Paul Larsen director on December 4, 1947. Larsen had just completed production engineering of a proximity fuze at the Johns Hopkins Applied Physics Laboratory, and Henderson considered him "to be the perfect man for the time".

The Sandia Laboratory's main mission remained weaponization - the casings to house the explosive systems as well as the design of fuzing and firing systems to detonate a weapon at a particular time and altitude. (Furman 1990: 252) The approximate division of effort between Los Alamos and Sandia was as follows: Los Alamos was responsible for the "physics package", and Sandia was responsible for everything else associated with the weapon. For example, Los Alamos had responsibility for nuclear, explosive, and firing unit components. Sandia had responsibility for power supplies, fuzing systems, and ballistics. The dividing line was admittedly arbitrary, and was from time-to-time renegotiated between Los Alamos and Sandia, and later between Livermore and Sandia as well.

The activities and volume of work at Sandia expanded rapidly, and Sandia's participation in Operation SANDSTONE in April and May 1948 added to the work load. Larsen soon learned that the work environment at Sandia Base was one of crash programs and crisis management. (Furman 1990: 258) To address this, he split his operational responsibilities by placing research, development, engineering design, and field activities under Henderson. (Furman 1990: 252-259)

By April 1948, Z Division was officially declared a separate branch of the Los Alamos Scientific Laboratory. Bradbury explained to University of California President R. G. Sproul: "Responsibilities of Z Division have grown to such an extent that divisional status became completely inappropriate." Five days later, Larsen reorganized the Sandia Laboratory into nine departments: Engineering, Applied Physics, Field Test, Road, Surveillance, Development Fabrication, Administration, Documents, and Procurement and Supply. (Furman 1990: 256-8) In July 1948, Larsen's title was abbreviated from "Associate Director, Sandia Laboratory, Branch of the Los Alamos Scientific Laboratory" to simply "Director, Sandia Laboratory". Larsen's tenure as director witnessed a major increase in manpower and corresponding increases in funding and building programs at Sandia Base. (Furman 1990: 258)

In 1947, the University of California, contractor to the AEC for the management of Los Alamos, was also managing the Sandia Laboratory. The Z-Division operation at Sandia Base had emerged gradually out of necessity, without a formal statement of its relationship to Los Alamos. "Robert M. Underhill, in charge of business affairs at the University of California wrote to Bradbury in June, 1947, that in his opinion the University never contemplated operations anywhere but at Los Alamos. He considered Sandia a shoestring operation covered neither by government contract nor by insurance." (Hewlett and Duncan 1969:135)

FORMATION AND EARLY YEARS OF THE SANDIA CORPORATION

In December 1948, the University requested that the AEC release them from the oversight responsibilities for the Sandia Laboratory. On July 1, 1949, Leroy Wilson, President of American Telephone and Telegraph (AT&T), agreed to accept responsibility for managing Sandia for the government on a no-profit, no-fee basis. The agreement, whereby Western Electric became the contractor for the Laboratory went into effect November 1, 1949. (Furman 1990: 343-360)

The contract provided for the loan of management and technical personnel from the Bell System to Sandia. If, for example, Sandia needed a technical, financial, purchasing, or accounting expert, or a general manager for overall corporate administration, the Bell system (primarily Western Electric and Bell Laboratories) would provide appropriate people from their nationwide operations. (Furman 1990: 360)

George A. Landry, a Western Electric employee, became the first director under the new arrangement. Landry had many years of experience in manufacturing operations. Since 1945, he was Operating Manager of Western Electric's nationwide installation services organization. Landry was to tighten up the operation of Sandia Laboratory: instilling a production atmosphere; introducing the Bell System personnel benefits, and continuing the construction efforts started by Larsen. (Furman 1990: 357)

Landry's management style of "go-by-the-book" with an emphasis on stratification and protocol, and the "tightening up of things", met with some resistance from the young and loosely structured laboratory. The average age of male employees was 33 and women 29. The informality and pragmatic atmosphere prior to Landry's arrival, which also existed at Los Alamos and later at Livermore was described by a Sandia employee:

> "There were some people who came in with very stilted opinions of themselves and were dismayed to find out that there really wasn't any hierarchy of educational level. The assignment, and who could get the job done ... was the real criterion of a man's worth. Everybody was pretty flippant with their conversations; we even greeted the janitors as 'Hi, Doc.'" (Furman 1990: 369)

The structured environment of the eastern manufacturing world had not prepared the Bell arrivals for the lack of formality, the first-name basis, and the ethos of cooperative interaction and "getting the job done". There was not even an executive washroom; and although Landry designated one, it became the source of such humor that it didn't last long. (Furman 1990: 369) While the Landry style would be softened over the years the Sandia environment would always appear more formal than that at the University of California labs.

On November 1, 1949, Sandia had about 1,700 employees. The number of Bell people transferred or loaned to Sandia was 14. In the November 1949 Landry organization, a new level of management was inserted between Landry and the divisions. There was no such level separating Larsen from the divisions. This new level of management was populated mostly by the newly arrived Bell people. (Furman 1990:374-5)

In the late 1940s, many employees of Sandia and the AEC lived on base, which was about six miles from downtown Albuquerque. There was a definite need for a social and recreational center, and the employees took the initiative to develop one. As a result the Laboratory operated a dining room while a social club offered additional benefits at a facility called the Coronado Club, located at 3210 West Sandia Drive. It included a swimming pool, game rooms for ping pong and pool, a small party room, and bowling alleys in the basement. Approximately 2,500 people attended the dinner dance celebrating the Club's grand opening on June 9, 1950. (Furman 1990: 261-2)

CHAPTER 5. ARMED FORCES SPECIAL WEAPONS PROJECT, ORGANIZATION AND LEADERSHIP: 1946-1950

1946 - FORMATION OF THE 2761st ENGINEERING BATTALION (SPECIAL)

Due to the loss of personnel from Los Alamos, by July 1946, the nuclear weapon assembly capability was seriously limited. To address the problem General Groves formed a new Manhattan Engineering District (MED) unit in August, 1946, that would take over the assembly function that Los Alamos appeared to want to give to the services. The new unit was the 2761st Engineer Battalion (Special) under Col. Gilbert Dorland. (DTRA 2002: 1-2)

To staff the new group, General Groves sought the army's best and the brightest. By September young men started arriving in Albuquerque and Col. Dorland formed Company B, a technical group that comprised 40 officers and 60 enlisted men, enough to form 3 bomb assembly teams. As Groves suspected, Los Alamos was skeptical about the army's capability to handle the complex assembly procedures. From the perspective of the services, at least, Los Alamos remained less than committed; and for the first few months the Army felt that their instructors received inadequate training from Los Alamos. (DTRA 2002: 22-25) The shortfalls in training, however, were soon addressed by the Los Alamos Z-Division in Albuquerque, and the assembly teams served as the nucleus for the Armed Forces Special Weapons Project's (AFSWP's) Field Operations Command.

After their initial training at Sandia Base on bomb assembly and non-nuclear component testing, some of the military personnel were selected for assignment to the AEC where they served in a dual capacity as AEC-Military inspectors and certifiers of weapons for the stockpile. They would also use their training in the development of capabilities for the manufacture and assembly of specific weapon components; see Attachment I "Early Working Relationships Between Laboratory, Military, and AEC Personnel", by Robert Duff. Such working relationships often resulted in life-long friendships.

FORMATION OF AFSWP

Executive Order 9816 abolished the army organizations of the Manhattan Project on December 31, 1946. By this order, property and facilities were transferred to the AEC. Personnel were reassigned either to the AEC or to another army organization. Dorland's group in Albuquerque was one of those reassigned to the AEC.

David Lilienthal had been appointed Chairman of the AEC Commission on November 1, 1946, even before the abolishment of the Manhattan Project. However, it took the military a bit longer to react to the abolishment of the Manhattan Project. A memorandum issued on January 29, 1947, by the Secretaries of War (Robert P. Patterson) and the Navy (James V. Forrestal), established "... effective midnight December 31, 1946, a joint Army-Navy atomic energy organization which will discharge all military Service functions relating to atomic energy and will be known as the Armed Forces Special Weapons Project (AFSWP)." (DTRA 2002: 29)

It took even longer for the military to appoint a head for AFSWP and to develop a charter. Groves was of course the primary candidate to become head of AFSWP. However, he was not immediately appointed. Part of the reason for this may have been that a charter was not yet formulated. Also, Groves and Lilienthal had clashed during the war when Lilienthal was the director of the Tennessee Valley Authority (TVA). Groves had called Lilienthal in 1942 "to discuss the provision of electric power for Oak Ridge". Lilienthal had refused to see Groves outside office hours. Groves was miffed and left all further contact with the TVA to Brig. Gen G. Kenneth Nichols*. [*Footnote: Nichols served under Groves during the Manhattan Project, as District Engineer, where he supervised the construction and operation of the uranium production facilities at Oak Ridge, Tennessee.] Groves let it be known that he was not pleased with Lilienthal's appointment as Chairman of the AEC, and Lilienthal struck a hostile attitude toward Groves. Lilienthal all but refused to consult directly with Groves during the six-month transition period of the MED to the AEC. Groves balked at providing the Commission with a new and detailed inventory of project holdings, and he continued to fight to keep custody of atomic weapons and weapon facilities in the hands of the military (Lawren 1988: 264)

Groves was appointed an Army representative to the MLC on January 31, 1947. However, he was still not appointed as Chief AFSWP. Finally, on February 28, 1947, two months after the abolishment of the Manhattan Project, General Eisenhower and Admiral Nimitz appointed Major General Leslie R. Groves as Chief, AFSWP. There was no officially appointed Chief of the AFSWP between January 1 and February 28, 1947, but Colonel S. V. Hasbrouck, who was Groves' aid, assumed command as the senior officer of the group assigned to establish AFSWP.

A short charter for AFSWP was finally issued with a date of March 18, 1947 as Memorandum No. 850-25-8, by order of the Secretary of War. This memo made the establishment of AFSWP effective 31 December 1946. The charter specified that the head of AFSWP was to be appointed by and report directly to the Chief of Staff of the Army and the Chief of Naval Operations. The Army and Navy Chiefs at this time were Dwight Eisenhower and Chester Nimitz respectively. Although the original charter does not specify, it was understood that the head of the new organization would be a flag officer from one of the services. Throughout the years, some people judged the current prestige of the organization by the number of stars held by the current head of the agency. (DTRA 2002: 403)

The two chiefs would also select an AFSWP deputy from the opposite Service, and both the head of AFSWP and his deputy would serve as members of the Military Liaison Committee. The head of the AFSWP (Chief AFSWP) would assume responsibility for ".All military service functions of the Manhattan Project as are retained under the control of the Armed Forces." This included training of special personnel, coordination with the AEC, technical training of bomb commanders and weaponeers, and participation with other agencies in developing joint radiological safety measures. (DTRA 2002: 32)

Groves only served about a year as Chief AFSWP. He was promoted to the temporary rank of Lieutenant General and retired from the Army on February 29, 1948. (AFSWP 1954: Vol. 1:4.2.1-3)

In March, 1947, Groves appointed Rear Admiral William S. "Deak" Parsons who had earlier served in important roles at Los Alamos (see Chapter 1) as Deputy Chief AFSWP. At that time a total of 19 officers, (18 Army and 1 Navy) and 13 civilians were on duty at the AFSWP in Washington. By the end of 1948 there were 165 persons. Between 1951 and 1956, the total hovered around 260-270. The number reached 330 by 1958. Roughly 2/3 of the personnel were military and the rest civilian. This ratio existed for the duration of AFSWP. The total manpower levels at Headquarters were always just a few percent of those in Albuquerque. Both Headquarters and Sandia Base experienced rapid growth. At the end of December, 1950, the manpower at AFSWP headquarters in Washington was 175, while at Sandia it was over 6000. (AFSWP 1954 vol 3:3.1.4-5)

The first Washington offices of AFSWP were located in the New War Department Building at 21st Street and Virginia Ave. NW, Washington, where the Manhattan District offices were located. In mid April, 1947, AFSWP moved to the Pentagon, where they remained until after DASA was formed in May 1959.

G. Kenneth Nichols, now a Major General, became Director of AFSWP on February 29, 1948, upon the retirement of General Groves. By 1949, the organization of Headquarters AFSWP consisted of the Executive Offices of the Chief and Deputy Chiefs and 7 Divisions: Operations and Training; Fiscal and Logistics; Radiological Defense; Development; Special Projects; Security; and Personnel and Administration. Earlier, a CROSSROADS Division had been established to complete "unfinished business" from the Operation, such as publication and distribution of technical and scientific reports. The division was disbanded on December 22, 1948, but AFWSP and its successors continued such archival work for all subsequent operations. (AFSWP 1954 vol 2:3.1.3-3.1.5)

NATIONAL MILITARY ESTABLISHMENT AND THE DoD

The National Securities Act of 1947 (61 Stat. 465) was enacted July 26, 1947. By this act, the armed forces organizations that then existed in the War Department, including AFSWP, were subsumed into the National Military Establishment (NME). This act also created the independent Department of the Air Force. The US Air Force (USAF) was founded from the US Army Air Force (USAAF) on September 18, 1947. There were now three independent military services whose personnel would serve in AFSWP. Also in July 1947, AFSWP received a revised charter that made it responsible for storage and surveillance of weapons in the custody of the armed forces. (DTRA 2002:37)

In 1949, the National Security Act Amendments of 1949 (63 Stat. 578 August 10, 1949) formed the Department of Defense from the National Military Establishment. The same year, the Joint Chiefs of Staff broadened AFSWP's charter to include the collection, review, and dissemination of data on the effects of atomic weapons. Since only the AEC (in practice only Los Alamos and later also Livermore) could legally detonate nuclear devices in peacetime, there was a natural collaboration between the DoD and the AEC in the weapons effects program.

AFSWP FIELD OPERATIONS

When the AEC took control of the Manhattan Engineering District's properties on January 1, 1947, the Sandia Base remained under the control of the War Department; and it came under the control of AFSWP. The first commander of AFSWP's field operations was Major General G. Robert Montague, who was appointed July 15, 1947 and reported to General Groves.

However, Colonel Dorland and his 2761st Engineering Battalion, who had come to Sandia Base in 1946 with the responsibility for the assembly of atomic weapons, reported directly to AEC General Manager Carroll L. Wilson, rather than to Montague. Others on the Base, such as those who came from Los Alamos's Z-Division and worked with nuclear weapons assembly also reported to the AEC. (Hewlett and Duncan 1969: 136, 666)

The issue of civilian versus military control of atomic weapons was a major consideration in the development and passage of the legislation that formed the AEC. But the issue was not yet resolved in practice. Hewlett and Duncan (1969:136) summarize this by stating: "The Commission had established the principle in December, 1946, that it would assume custody of all atomic weapons and fissionable material, but how did this square with the fact that custody of such materials at Sandia remained with a military officer (Dorland)?" This issue, of civilian versus military control, was a continuing problem for the nuclear weapons community and was debated for years. On the local working level at Sandia Base, such issues required considerable managerial skill by both civilians and the military.

Colonel William Canterbury who was on Montague's staff greatly aided in the interfacing of military personnel with Los Alamos' Z-Division civilians. A number of the officers from the 2761st battalion had even been assigned to Z-Division, including Lieutenant Colonel Ellis E. Wilhoyt, who served as the alternate and then acting Z-Division leader in 1946. (DTRA 2002: 31)

Many challenges were presented to the AFSWP personnel during the late 1940s: changes in the organizational structure; primitive working conditions and housing facilities (particularly at Sandia Base); providing support to the AEC; and the development and fielding of new types of effects measurements for the nuclear test operations. Despite these challenges, AFSWP was fulfilling its assignment for collecting, reviewing, and disseminating data on nuclear weapons effects.

AFSWP 1950
Important changes were made in the divisional organization at AFSWP headquarters in 1950. The Radiological Defense and the Development Divisions were abolished. All technical responsibilities and activities of these two divisions were distributed among three newly named divisions under a new "Director, Technical Divisions": 1) Weapons Development Division, 2) Weapons Defense Division, and 3) Weapons Effects Division.

The Weapons Development Division had responsibility for all DoD work in connection with the development of atomic weapons and ancillary hardware. The Weapons Defense Division had responsibility for work in radiological defense. (AFSWP 1954: Vol 3:3.5.20; 3.5.29)

The Weapons Effects Division was the most involved with nuclear weapons testing. (AFSWP 1954: Vol. 3: 3.1.6) It was responsible for all AFSWP work in connection with:

- The effects of existing and proposed atomic weapons.
- Thermal and nuclear radiation measurements.
- The AFSWP budget for effects work.
- Dissemination, maintenance, and classification of records, data, and photographs. (AFSWP 1954: Vol 3:3.5.69)

At its inception, the Weapons Effects Division, under Air Force Col. R.N. Isabell was organized into four branches:

- Radiation Branch - Herbert Scoville Jr. - Effects of ionizing radiation and thermal radiation (excluding medical research), and special studies. The division also provided scientific advisors to the Chief AFSWP, and Scoville was also a technical and scientific advisor to the Director of the Technical Divisions.
- Blast Branch - Lt. Col. G. F. Blunda USAF - Military effects other than radiation and thermal, base surge investigations, underwater and underground explosions and effects.
- Test Branch - Lt. Col. Max S. George, USA, Responsibilities pertaining to weapons tests. The Free Air Pressure Group became a part of Test Branch. (AFSWP 1954: Vol 3:3.5.69-92)
- Technical Library Branch - Maj. J.B. Gulley USA - Records, data, photography. (AFSWP 1954: Vol 3: 3.1.6, 3.5.68)

In 1950, AFSWP prepared, jointly with the AEC, the report, "The Effects of Atomic Weapons", edited by Samuel Glasstone. (DTRA 2002: 7) Subsequent updated editions titled, "The Effects of Nuclear Weapons", also edited by Glasstone, were printed in 1957, 1962, 1964, and 1977. These books, which were unclassified and publicly available, became the definitive sourcebooks on nuclear weapons effects. Every technical person's office had one along with a slide rule, then later an electric desk top computer.

CHAPTER 6. OPERATION SANDSTONE AND ITS INFLUENCE ON THE FUTURE OF TESTING

Operation SANDSTONE was the first weapons development test series after the 1945 Trinity event in New Mexico.

In terms of the test objectives for SANDSTONE it should be noted that the Los Alamos weapons scientists, working under the direction of Hans Bethe, the leader of the war-time Theoretical Division, had several new designs for fission weapons by the end of World War II. The main objective of the new designs was to develop weapons that used the scarce plutonium and uranium more efficiently. In addition, during the 1946 -1948 period, there was strong pressure to test designs that would soon enter stockpile. Since the gun-assembled weapon used fissionable material very inefficiently, the design emphasis continued to be on new and better implosion systems and pit (fissionable material) configurations including the use of uranium in implosion weapons.

The top levels of government were keenly interested in the location for SANDSTONE. On June 12, 1947 President Truman, met in his office with AEC Chairman David Lilienthal, Secretary of State George Marshall, the members of the Joint Chiefs of Staff, including the Army Chief of Staff General Dwight Eisenhower, the Service Secretaries, and the Secretary of War Robert Patterson to discuss the proposed SANDSTONE series and its location. Marshall and Patterson favored a continental test site while Eisenhower and Lilienthal strongly argued for a location somewhere in the Pacific. Lilienthal told Marshall that the tests would have to take place outside the United States because "the alternative within the U.S. would create considerable fears and concern, possible damage suits, and general public relations problems." (Lilienthal 1964: 196-198) The decision was made to go with a site outside the continental United States. This turned out to be the Enewetak Atoll. Lieutenant General John E. Hull, U.S. Army, was appointed Commander of Joint Task Force 7 for the execution of SANDSTONE. One of General Hull's two JTF-7 Deputies was Rear Admiral William S. Parsons, the W.W.II Navy Captain who had been assigned to Los Alamos and was now Deputy Chief AFSWP. Navy Captain James S. Russell was appointed Test Director. Major Frank Camm, who would later become a general officer and the Director of the AEC's Division of Military Applications, was administrative officer for the newly formed AFSWP (see Chapter 5).

Formal approval for Operation SANDSTONE was received by the AEC Santa Fe Operations Office (SFOO) Manager Carroll Tyler, from AEC General Manager Carroll Wilson on July 23, 1947. Wilson's memo was the formal authorization for Bradbury to begin formal preparations for SANDSTONE.

J-Division had the responsibility for scientific measurements and for nuclear device assembly and firing. EG&G, a civilian contractor, played a major role, as they would throughout the years of nuclear testing, in fielding experiments, providing timing signals, and in technical support. A Naval Research Laboratory team, under the direction of Ernest H. Krause, fielded a number of the gamma ray diagnostic experiments. Los Alamos shared responsibility with the AEC and the military for logistics, supply and property, health and safety, and infrastructure. Infrastructure included housing, transportation, food and water, power, sanitation, and communications. Security was provided by the military or by the AEC. Support, surveillance, and radioactive sampling aircraft and aircraft operations were provided by the military. Civilian and military scientists supplied meteorological assistance. The overall operational leadership was the responsibility of the military commander, but the coordination and liaison were responsibilities shared by all of the participating organizations.

During the planning stage for SANDSTONE, the international situation was critical. (AEC Chairman Lewis L.) "Strauss warned Forrestall against permitting all the weapon assembly teams to go to Enewetak for the forthcoming SANDSTONE weapon test series." Strauss was concerned about the possibility of a Pearl Harbor-like surprise enemy strike. (Hewlett and Duncan 1969: 159)

Despite these concerns, the SANDSTONE test series on Enewetak, see Figure 1-6.1, went off with hardly a hitch. Three devices were tested in a carefully designed program. All three shots were fired on 200-foot towers. The first was X-Ray at 37 kt on April 14, 1948 GMT. Yoke, at 49 kt, was fired on April 30, and Zebra at 18 kt was fired on May 14. The locations of these shots on the Enewetak atoll are shown on Figure 1-6.1. All were considered very successful and provided important data for the weapons designers.

Yield was derived from radiochemistry, fireball growth, and gamma-ray flux. Alpha was measured with gamma-ray detectors employing many of the same techniques that Bruno Rossi developed for Trinity.

One week after X-Ray was fired Col. B. G, Holzman, JTF-7 staff meteorologist, sent a memo to Admiral ("Deak") Parsons, Deputy Commander JTF-7, expressing concern about weather forecasting problems at Enewetak. He was specifically disturbed by the difficulties of forecasting rain, clouds, and the details of the complex upper level wind structure. These concerns were echoed by Col. James P. Cooney, radiological safety officer for SANDSTONE, who pointed out to Parsons the fallout prediction problems encountered in the Marshall Islands. These communications turned out to be particularly significant because they started the ball rolling to find an alternate test location. As it turned out, an alternate site was eventually adopted; but the Marshalls continued to be used for nuclear testing until curtailed by the 1958 moratorium.

In mid-May, just before Zebra was fired, Admiral Parsons sent General Hull, the JTF-7 Commander, a memo recommending that a survey be made to identify a continental test site. Parsons touched on weather forecasting and safety and security as issues to examine. Clearly the military was beginning to take seriously the questions surrounding a continental test site.

During peak manning periods, the Army provided approximately 2,100 personnel, the Navy 5,850, the Air Force 1,800 for services such as security, operations, and logistic support on SANDSTONE. The joint task force was divided into functional and service-branch oriented task groups. (Berkhouse 1983: 30, 32)

The AFSWP collaborated with LASL early in the planning stage for SANDSTONE, and provided 116 military personnel that were integrated into the AEC Proving Ground Task Group. The roles that AFSWP undertook on SANDSTONE were typical of the roles they would undertake in subsequent AEC tests in the Pacific. AFSWP assisted the AEC in weapon diagnostics experiments that measured nuclear device outputs and supervised the extensive effects experiments by the military services. They included: measuring air blast, testing specially constructed structures to ground motion and blast, exposing aircraft to blast, and a wide variety of tests associated with radiation exposure and instruments for measuring radiation. In addition to AFSWP, the military laboratories that participated in these effects experiments were: Naval Ordnance Laboratory, Chief of Engineers (Army), Navy Bureau of Yards and Docks, Navy Bureau of Medicine and Surgery, and the Army Chemical Corps. (Berkhouse 1983:89-101)

Los Alamos, the AEC, the contractors and the military learned a number of lessons from SANDSTONE. The benefits of the formation of a test division, with complete scientific authority and responsibility for technical operations far from home, were to be realized over the years as tests were fielded in the Pacific, in Nevada, on Amchitka Island in the Aleutians, and elsewhere.

Since EG&G played such a major role in the history of testing it would be appropriate to digress briefly to describe their background. EG&G, are the initials of the three founders Harold Edgerton, Kenneth Germeshausen and Herbert E. Grier. A fourth key individual in the EG&G hierarchy was Bernard (Barney) J. O'Keefe. EG&G played a major role in support of Los Alamos (and eventually Sandia, Livermore, the AEC and DoD) during almost the entire history of the nuclear test program. The Edgerton, Germeshausen and Grier collaboration grew out of the practical applications of pre-WWII research at MIT and was formalized as a corporation in November 1947. Some years later the name was officially changed, from the complete names of the founders, to the briefer EG&G. (Peter Zavataro, private communication)

In some sense the EG&G nuclear involvement goes back to the WWII period when O'Keefe was a Navy ensign assigned to Los Alamos. Most notably he participated in the preparation of the nuclear strike bombs while on Tinian Island in the Pacific. During SANDSTONE all four of the EG&G principals, plus a number of their colleagues, were involved in the timing and firing of the devices and in photography. They also participated in several experimental programs to measure the reaction history and gamma-ray flux.

The complex scientific measurements that were the basis for the tests provided severe challenges for the technical teams working in the corrosive tropical salt air atmosphere of the Pacific Proving Ground. Despite the inherent difficulties, the long hours and the absence from homes and families, the technical teams generally enjoyed their experiences and emerged with feelings of accomplishment in their participation in the field tests.

The SANDSTONE Test Director's report has an interesting, but rather petulant, observation about the scientists and engineers engaged in the tests. It was probably written by Darol Froman, of Los Alamos, the scientific director of SANDSTONE: "It is a striking psychological fact that scientists and organizations associated with atomic weapon tests tend to lose their sense of judgment as to the value of performing certain experiments in connection with nuclear explosions. This is particularly true if the personnel concerned have originated the idea of the experiment, if the performance of the experiment is their only reason for witnessing or taking part in the tests, or if they think their organization will receive a certain amount of kudos for such activity." (Test Director, Joint Task Force Seven. Report to The U. S. Atomic Energy Commission on Operation SANDSTONE. Part 1, Volume I p. 103,104) It should be noted that as the testing technology progressed over the years, the number and complexity of the diagnostic experiments increased enormously. The stimulation for new and ever more complex experiments could not be attributed solely to the experimentalists. The theoretical designers also were always on the lookout for more and better quantitative observations of nuclear explosion phenomena.

CHAPTER 7. PUSH FOR A CONTINENTAL TEST SITE

INTRODUCTION

A variety of key: organizations, individuals, objectives/motives, and political issues influenced the final push for a continental test site. Among these were:

Organizations - AEC, DoD, Los Alamos National Laboratory, US Congress, and White House.
Key Individuals - President Harry Truman, Norris Bradbury, Alvin Graves, John Clark (Los Alamos Lab); William Webster, R Adm. William Parsons, Robert LeBaron, Maj. Gen. Kenneth Nichols, Brig. Gen. Herbert Loper (AEC MLC); Lewis Strauss (AEC Commission); Brian McMahon (US Senate); Navy Captain Howard B. Hutchinson (AFSWP).
Objectives/Motives -

weapons - Need for quickly addressing new designs and diagnostics experiments.

features offered by a continental test site - like: simplified management (no large Task Force), less expense, simplified logistics, easier to acquire & maintain work force, more realistic venue for the effects of nuclear weapons on continental environments, better weather prediction than possible at the Pacific Proving Ground.

promotion of thermonuclear weapons - by Edward Teller.

Operation GREENHOUSE - Tests of new implosion weapons with uncertainties & an experiment to explore fundamental design and physics of fusion weapons.

Political Issues -

*August 29, 1949, Soviet Union conducts its first nuclear test.

*1950, Klaus Fuchs, British citizen who had worked at Los Alamos admits passing information to Soviet Union from 1943 until after the war.

* DoD interested in earth penetrator weapon, Operation WINDSTORM.

*June 25, 1950, North Korean military forces attack South Korea.

*July 5, 1950, US troops go into battle in Korea.

*October 25, 1950, Chinese forces joined North Korea in the war.

*Military concerned about supporting GREENHOUSE due to war.

*President Truman steps in

The time sequence and details of how these interacting elements "played out" to result in the AEC acquiring a continental Nevada Test Site are described in this chapter.

POST SANDSTONE: SUMMER 1948 - JANUARY 1949

By the summer of 1948 the timing of a follow-on test series after SANDSTONE had slid from 1950 to 1951. The Los Alamos weapons leaders felt that more time was required to develop new designs and to prepare diagnostic experiments. The objectives were to further refine fission weapon designs for more efficient use of

SNM (Special Nuclear Material) and to develop a family of weapons with a choice of yields.

On July 22, 1948 Alvin Graves wrote Bradbury requesting consideration of a continental test site for the 1951 test series. His reasons included easier management, less expense, and fewer problems if there were to be a war. (Los Alamos archives: A. C. Graves to Norris Bradbury, Subject: *Continental Site for Atomic Weapons Tests*, July 28, 1948 7/22/48-5/18/51). It is interesting to note the influence of the unsettled, and tense, international situation.

The attractions of a continental test site also continued to percolate within the DoD. In late July the Joint Chiefs of Staff asked the Military Liaison Committee (MLC) to get input from the AEC on their views regarding such a facility.

In response, in early August 1948, the AEC Division of Military Applications (DMA) presented to the Commission their views on the "Location of Proving Ground for Atomic Weapons." They highlighted, as an advantage of a continental test site, the ease of access. They also quoted Los Alamos to the effect "that continental operations would make possible greater participation of scientific and technical personnel." Los Alamos was also credited with confirming the views of Colonels Holzman and Cooney that the wind structure at Enewetak, with its reversal of direction aloft, caused some concern and that a wind structure consistent in direction to high levels would reduce the radiological hazard. A continental test site would also greatly reduce logistical problems. The disadvantages that were identified by the DMA included the radiological safety outside the test site proper, physical security, and the public reaction to testing within the continental U.S. Their conclusions were that continental operations would pose difficult domestic and possibly international relations problems, and that any decision on a continental site must be made at the highest levels of government.

On September 3, 1948 Bradbury issued J-Division Directive #1, to Alvin C. Graves, where he outlined the role of J-Division in the laboratory. "The programmatic research responsibility of J-Division in the Los Alamos Scientific laboratory is, in its broadest phases, directed to accomplishment of the research, development and operations necessary for the testing of weapons designed and fabricated by the laboratory. ... It is the expressed policy of the laboratory that the J-Division Leader should be designated as the Scientific Director of a proving ground operation and have sole authority with respect to the technical operations or experiments to be done and the manner of accomplishment."

In October J-Division was given permanent status with responsibility for full-scale nuclear tests. Alvin Graves was appointed Division Leader; John C. Clark and Herbert I. Miller, Assistant Division Leaders;

At the beginning of fall, 1948, William Webster an engineer and New England utilities executive (Hewlett and Duncan 1969: 87) was appointed chair of the MLC. His first order of business was to send a memo to the AEC suggesting that the national military establishment, with the assistance of the AEC, conduct a preliminary study of potential continental test sites. He felt that this would result in a more extensive evaluation of the relevant factors, and assist in an evaluation of future action.

Rear Admiral "Deak" Parsons, who was then a member of the MLC, asked the Chief of the Armed Forces Special Weapons Project (AFSWP) to do a study on possible continental test sites. This resulted in Project Nutmeg, directed by Navy Captain Howard B. Hutchinson.

At a meeting held on October 14 with representatives from both the AEC and Los Alamos, Graves presented his perspective on Proving Ground Operational procedures. (Los Alamos archives: A-99-019 310.1 62-22 J-Division 11/18/1947 - 7/2/1953) "It was understood by those present that the Santa Fe Operations Office would act for the Atomic Energy Commission in all matters concerning Proving Ground Operations, except those which required policy approval on the Commission level, or matters of concern to other government departments, all of which would be handled by the Division of Military Applications, USAEC, Washington, D.C. Under such circumstances, it was agreed by those present, that economy of effort would best be served by designating the Division leader of J-Division as a representative of the Atomic Energy Commission for such purposes as may be necessary in carrying forward Proving Ground activities." Joint appointments between the laboratories and the AEC would be successfully employed over the years, and would not be restricted to just the J-Division Leader. For example, the Test Group Director, usually an Assistant J-Division Leader in the Los Alamos structure, would have a joint appointment between the AEC and the Laboratory for shot execution.

On November 10, 1948 Hutchinson reported to Admiral Parsons that Project Nutmeg was being coordinated with the MLC, Los Alamos and other interested parties. Five days later an interim report was made to the MLC. (AFSWP 1948)

1949

The final report for Project Nutmeg was completed in late January 1949 and concluded "tests conducted within the continental United States at properly engineered test sites, under proper meteorological conditions, will result in no harm to population, economy or industry." It went on to say that the preferred locations were in the desert regions of the west. (See Pruess and Russell 1949)

Interestingly, on March 8 the AEC informed the MLC that a continental test site was not at that time considered necessary, but that Project Nutmeg would be valuable for emergency test planning. This seems to be the first time that the concept of an emergency test site appeared. Apparently Los Alamos had lost interest in a continental site for the proposed 1951 test series, but it did want to keep a back-up site in case the Pacific Proving Ground became unavailable for any reason. The concept of an emergency site was to come into sharper focus in the very near future.

The world situation, at least from the perspective of the West, took a turn for the worse on August 29, 1949 when the Soviet Union fired their first nuclear shot, called Joe-1 by the United States and First Lightning by the Soviets*. The Russians report the yield as 22 kt. (Mikhailov 1999: 14) The United States intelligence community was surprised, because the official estimate for the first capability by the Soviets to have a bomb was "two years." The fact that the estimate had been two years for the previous several years seems to have escaped attention. The forecast was stalled and had not been updated. (York 1995: 6) [*Footnote: This first nuclear test was detonated at the Soviet test site, Semipalatinsk. Appendix C provides a brief description, translated from Russian, of the characteristics of the site. There are many similarities between the criteria used by the Soviets for site selection and those described for the US in this chapter.]

Washington wanted an appropriate response to what was seen as a threat of the new Soviet capability. With the rapid pace of development of missile technology and long-range aircraft, it was clear that a nuclear-armed Soviet Union could soon threaten the United States in a new and terrifying way. At this time the U.S. government, and the advisors on nuclear matters, were debating the future of a thermonuclear weapons program. In fact it was by no means clear that such weapons were even feasible. Edward Teller was chaffing at what he perceived as foot dragging by the AEC General Advisory Committee and by the senior leadership at Los Alamos. Teller and his allies, mainly E. O. Lawrence at Berkeley, felt that a crash hydrogen bomb program should be the cornerstone of a proper response to Joe-1. Others felt that expanding fission weapon development and production was the most reasonable and appropriate response, partly because it would enhance the immediate U.S. military posture. Ultimately the U.S. followed both tracks and ended up with multiple requirements for an expanded testing capability.

The secret debates over the hydrogen bomb development program were highly acrimonious. Teller, becoming ever more alienated from Los Alamos and from Norris Bradbury, campaigned for an additional nuclear weapons design laboratory with broad responsibilities in competition with Los Alamos.

1950

Teller was at least partially successful. With the help of E. O. Lawrence and Luis Alvarez at Berkeley, and powerful allies in Washington (Lewis Strauss and Brien McMahon were particularly influential Washington-based supporters of the Super), the faction supporting vigorous hydrogen bomb research succeeded in convincing President Truman to order, on January 31, 1950, a crash program to develop the thermonuclear weapon. (York 1976: 69) This directive fostered discussions of a testing capability that would not limit the theoretical design process.

Klaus Fuchs, a German-born physicist, became a British citizen in 1942; and in 1943 he was sent to Los Alamos where he had worked until after the war. After being arrested in 1950, he admitted passing information to the Soviet Union since 1943. (Encyclopedia Britannica) In early 1950, Robert LeBaron who was then the Deputy to the Secretary of Defense for atomic energy and chairman of the MLC, asked the Chief of AFSWP, Brig. Gen. Kenneth Nichols, and Brig. Gen. Herbert Loper (who would succeed Nichols as Chief of AFSWP in 1951), both of whom were then members of the MLC, to estimate the damage done by Fuchs' disclosures. Their alarming and sensational conclusion was that the Soviets might be much further advanced in nuclear weaponry than the Americans believed. They concluded that the Russians might even have a thermonuclear weapon in production. The Nichols/Loper report moved up the command chain to the president. After digesting the report, the military wanted a crash program to develop a super bomb, even if it meant cutting back on the existing atomic bomb program. (DTRA 2002: 80)

In mid-1950, Los Alamos was planning a test series, code named Operation GREENHOUSE, to be performed at Enewetak during the spring of 1951. High on the priority list for testing were several new fission weapons and an experiment to explore the fundamentals of thermonuclear design. However, Los Alamos was concerned about the predicted behavior of the fission devices that were planned for testing on GREENHOUSE. Los Alamos wanted one, or a very few, preliminary tests before committing to the extensive GREENHOUSE series. These tests would be used to validate the reliability of the yield calculations of new implosion designs that were slated for testing during Operation GREENHOUSE. These pre-GREENHOUSE tests would have to be conducted in early 1951 in order to meet the spring schedule for the Pacific operation. As it turned out, this test series would be code-named Operation RANGER. In addition, by November 1950, Los Alamos was planning another series of tests for the fall of 1951 that would become known as Operation BUSTER. The BUSTER tests were designed to evaluate new device designs developed by Los Alamos and to obtain data on the basic phenomena associated with nuclear weapons. (Ponton 1982c:20)

In addition to weapons design tests, which were of particular interest to Los Alamos and the A.E.C., the military services were eager to continue a study of the effects of nuclear weapons as a follow-on to CROSSROADS. The DoD was becoming interested in an earth-penetrating bomb and had questions concerning the effects of surface and shallowly buried nuclear detonations on land*. (Ponton 1982c: 20-1) [*Footnote: The geology and the properties of the geologic materials in the Pacific atolls are very different from those on a continental land mass. Therefore ground shock from a nuclear explosion is very different in atolls and continents.] Bradbury was reluctant to have Los Alamos more than peripherally involved in such tests. He felt that Los Alamos had its hands full with weapons development and that the effects of nuclear weapons were the proper purview of the DoD. Consequently the military initiated their own effects test program. The AEC agreed to support such experiments under DoD auspices.

The DoD was initially planning on three tests to address the effects of detonations on land: one at the surface and two underground. The plan evolved into two detonations and was code named WINDSTORM: a surface and an underground, but near surface, at about a 50 foot depth. AFSWP planned to request that the AEC fire two 20 kt devices in Operation WINDSTORM.

AFSWP hosted a meeting at the Pentagon, on April 10-11, 1950, to discuss a land-based site for WINDSTORM. Presidential approval came for WINDSTORM in mid-1950 with the time and place to be determined. After a wide-ranging search AFSWP settled on Amchitka Island, in the Aleutians. It was far from centers of population, had a good World War II runway, a fair harbor, and belonged to the United States. A major disadvantage was that it had a wretched climate. The White House approved the Amchitka operation in November 1950. (Hacker 1994: 60-61)

Up to November 1950, the only yields that had been tested, by the United States, were 18, 21, 37, and 49 kt. The military wanted a known yield in order to effectively plan its effects tests and therefore was planning on about 20 kt for WINDSTORM. Lower yields could be used to address the military's effects issues, but such devices had not yet been tested.

As described later in this chapter, a site in Nevada would be in the hands of the AEC before WINDSTORM could be conducted. As described in Part 2, after the successful 1 kt tests on Operation RANGER, WINDSTORM would evolve into Operation JANGLE conducted in Nevada. Eventually Amchitka was used for nuclear tests, but not until the DoD-sponsored the Long Shot event in 1965.

KOREA

On June 25, 1950, North Korean military forces poured across the border into South Korea. With United Nations approval, the United States and its allies came to the defense of South Korea. This conflict was a major crisis in the international arena. As a consequence of the attack, the United States increased its commitments to the defense of East Asia in order to deter communist aggression, increased defense spending, strengthened NATO, and campaigned to rearm West Germany. (The American Heritage Encyclopedia of American History, Henry Holt 1998: 499-500)

IMPACT OF KOREA ON TESTING AND GREENHOUSE

The Korean War was heating up rapidly. U.S. troops went into battle on July 5th, ten days after the initiation of hostilities. The military services, the Navy in particular, had their hands full and were concerned about the support of Operation GREENHOUSE scheduled for the spring of 1951. A Pacific test series required dozens of ships and airplanes, and thousands of servicemen in support of the several hundred scientists, engineers and technicians. In addition, the government was even more concerned about the security in the Marshall Islands than they were during the SANDSTONE operation of 1948.

Meanwhile, Gordon E. Dean, who had been a member of the AEC since May 1949, was appointed AEC Chairman on July 11, 1950, replacing David Lilienthal whose tenure ended in mid-February. On Dean's first day as Chair, the Commission received a draft memorandum from DMA which it proposed to send to Robert LeBaron, the Pentagon's chair of the MLC. The title of the memo was "Location of Proving Ground for Atomic Weapons" (the same title as the August 1948 DMA memorandum to the Commission). The AEC/DMA felt that a national emergency was possible and that at least one continental test site should be identified.

Two days later Dean proposed to LeBaron that the DoD and the AEC collaborate "to locate at least one site recommended for emergency atomic test use and possibly one alternate site. ... We now feel that a national emergency is, at least possible. While we do not wish to seem unduly pessimistic, we believe it would be wise to reexamine the question of a continental site with the objective of having available a definite and specific site which could be recommended for use if needed." Dean also wrote to President Truman who directed that a study to identify an emergency test site be undertaken with a "high degree of security."

Brig. Gen. James McCormack, Jr., the Director of DMA, in a July 20, 1950, letter to Carroll Tyler, AEC/SFOO, pointed out that because of the Korean War there might be problems with the scheduling of GREENHOUSE at Enewetak. McCormack noted "the Commission faces a basic problem with regard to GREENHOUSE in light of the immediate shortage of shipping and particularly air transport in the Pacific and in light of uncertainties in predicting the situation which may prevail at the scheduled time of the tests." McCormack also noted that earlier surveys to find a continental test site considered: Trinity (White Sands Missile Range); Tonopah Bombing Range; Dugway (in Utah); the North Carolina coast; and the Texas coast. McCormack requested Tyler's (i.e. Los Alamos's) views on the following:

1. "Order of desirability of continental sites.
2. Major delays, which might be introduced by conducting at a continental site, the tests now scheduled for the spring of 1951.
3. The requirement for retention of the Enewetak proving ground.
4. The technical advantages of a continental test site as a supplement to Enewetak.
5. The requirement for continued support by the DoD for tests conducted in the continental United States.
6. Radiological safety problems introduced by tests conducted in the continental United States."

(McCormack to Tyler. Subject: GREENHOUSE 6/30/50 - 1/7/58)

BRADBURY WRITES TO DMA

It didn't take Los Alamos long to respond. The next day Bradbury, probably with input from Graves, wrote to McCormack, through Tyler, emphasizing the importance of GREENHOUSE to "the stockpile position of the United States (and the GREENHOUSE tests are) ... expected to make a direct contribution to the rapid understanding of the technical and economic feasibility of a thermonuclear weapon program." Bradbury went on to argue for conducting GREENHOUSE on schedule. Enewetak must be retained in order to preserve the capability to test at high yields. "The following comments are numbered in the same order as given in the reference document and are based on technical considerations only."

1. "We consider the suggested sites to have the following order of desirability: Tonopah bombing range, Trinity, Texas coast, North Carolina coast, and Dugway. Tonopah is given preference over Trinity in spite of convenience and prior use of Trinity because of the lower population density.
2. If an extremely early decision is reached in determining the location of a site, it is believed that the same time scale as that proposed for Operation GREENHOUSE can be met.
3. The Los Alamos Scientific Laboratory believes it imperative that the Enewetak area be retained as a potential proving ground. ... Moreover, in times of greatly reduced international tension, the Enewetak area may prove a generally more practical area for the conduct of nuclear tests from the point of public psychology and practical security than a continental site.
4. A continental test site offers a number of technical advantages over Enewetak: The logistics problem; the problem of obtaining capable construction firms in sufficient numbers to insure economy, efficiency, and speed of construction; the problem of obtaining competent technical personnel willing to participate in a test; the problem of obtaining sufficient area, appropriate foundations, and suitable soil conditions for many experiments; and the problem of transporting and assembling nuclear devices, technical equipment, and experimental set-ups are obviously much simplified at a continental test site. The Los Alamos Scientific Laboratory can only foresee a continued program of practical nuclear explosion tests as a supplement to its research and development program. This has been true from 1945 until the present time. And there seems equal reason, and from the point of view of thermonuclear reactions, even more reason, to predict the continuance of this requirement. The requirement becomes even greater in times of tension when less time can be spent in theory and laboratory experiment, and more reliance must be given to early test programs. The occurrence of tension in international relations makes an extra-continental site less available, or even actually unavailable, and an alternative site must be immediately ready for use.
5. The basic structure of the Joint Task Force THREE is required if the GREENHOUSE tests are to be conducted on the present time scale. Some units of the Joint Task Force could and should be dispensed with but we would prefer to postpone such discussions until our plans for the use of a continental United States test site are more nearly complete.
6. Based upon a preliminary analysis of the factors of accuracy of meteorological predictions, prevalence and stability of specific wind patterns, population densities in potentially affected areas, the radiological results of the Alamogordo test in 1945 (Trinity), and the technical plans for Operation GREENHOUSE, it is believed that such an operation could be conducted at the continental site recommended in Nevada with a degree of public radiological safety which would considerably <u>exceed</u> that of the Alamogordo

operation. While the precise nature of test planning will be affected by meteorological considerations, to which much further attention must be given, we see no reason to anticipate that such a goal cannot be met." (Underlining in original)

This was almost exactly two years after Graves had suggested that the GREENHOUSE shots be moved to a continental site. One major change was the crash program to develop a thermonuclear weapon. The George shot on GREENHOUSE, a critical event in that program, was predicted to have a higher yield than would be acceptable in the atmosphere at any potential U.S. test site. (George went at 225 kt in May 1951 at Enewetak. It was about three times the yield of Hood, the highest yield atmospheric shot fired in Nevada)

COSTS AND SCHEDULES INFLUENCE CONTINUING DEBATE
Gordon Dean told Secretary of Defense Johnson that GREENHOUSE was necessary for improving the nuclear stockpile and for obtaining new data on the effects of nuclear weapons. (Fehner 2002: 43) Johnson responded that the Joint Chiefs of Staff (JCS) had requested a review of the costs and schedule of Operation GREENHOUSE. With the information from the review, the JCS would make their decisions regarding the "necessity for postponement" of GREENHOUSE in light of the "necessity for reallocation of both shipping and personnel from the tests, as originally scheduled, to the support of operations in the Far East." The JCS would also examine the possibility of limiting the logistical costs by a "reduction in scope of the tests." (Fehner 2002: 43)

Bradbury and his colleagues at Los Alamos thought this incredible. The international situation was very tense. The Korean conflict could escalate at any time, and the Soviet Union had, just a year earlier, detonated a nuclear weapon. Now the JCS was talking about reducing their support of the critical nuclear weapons tests scheduled for GREENHOUSE.

It is interesting to speculate on how, and who, would review the technical details of the nuclear weapons test program for the JCS if it were not Los Alamos. At that time there were few nuclear weapons experts outside the AEC complex.

BRADBURY WRITES AGAIN
Bradbury responded to the issues raised by Defense Secretary Johnson by writing to the AEC/DMA on August 22. He went into some detail on the rationale for testing and the relationship to the weapons development program and to the proposed 1951 GREENHOUSE series. The following are quotations relevant to the establishment of the Nevada Proving Ground.

i. "To Los Alamos Scientific Laboratory it appears almost fantastic that, precisely at a time in international relations, when the most rapid progress should be made in this field, that such a denouement can be contemplated without the gravest concern.

ii. If we cannot use Enewetak in the spring of 1951 what then can we do? The following possibilities occur to us:
    a. The date is postponed, possibly indefinitely, until the international circumstances again permit the use of Enewetak.
    b. A continental United States site is selected and appropriate planning conducted at a rate, which would permit its use by the summer of 1951.
    c. A continental or extra continental site is selected and the most rapid planning carried out so as to permit its use at the earliest date possible - probably sometime in 1952 or 1953.
    d. Preparations are made as rapidly as possible to permit the testing of simple atomic weapons by air observations of airplane dropped bombs, presumably over open ocean.

iii. Possibility (b) has been given particular attention in the last few weeks. The Nevada location appears to be such that atomic weapons - almost certainly up to 50 kt in yield - can be tested there with much greater public safety than the original 1945 shot at Alamogordo. If vigorous planning were carried out in the near future, it could be ready for use in the summer-fall of 1951. If Enewetak had been unavailable in the spring, this area could be used for several new weapons designs. Shots would have to be fired in increasing order of yield and the termination of the program might be indicated by observation of fallout phenomena other than those predicted...

iv. In any event the Los Alamos Scientific Laboratory sees an immediate, practical, and important program, which can be jeopardized by being unable to use Enewetak, almost at will, during 1951. We see no obvious way to maintain this flexibility other than to be able to retire to a United States site on short notice, if Enewetak is unusable. We, therefore, recommend as strongly as possible that a United States site be selected and given initial preparation as soon as possible and that such a site be proposed for use if (but only if) Enewetak is unusable during the course of the year."

The frustration felt by Los Alamos clearly came through. It would probably be an understatement to say that the nuclear weapons program during the summer of 1950 was in a state of flux. The laboratory was in the midst of a crash program, mandated by President Truman, to develop a thermonuclear weapon, and no one had a clear idea of how to do it. In addition, new fission weapons were soaking up valuable design resources; but these were just the weapons that would be needed by the nation, in the immediate future, in the event of a critical nuclear confrontation.

There was to be an experiment on GREENHOUSE to examine some aspects of thermonuclear burn, but so far the theoretical foundation for the experiment was lacking. This was a very trying time at Los Alamos, and the staff felt under the gun to produce. It was imperative, they felt, that the ability to test not be the limiting factor.

SAFETY

Radiation Concerns

The principal concern of both the AEC and Los Alamos with a continental test site was safety in general and radiation safety in particular. By this time it was clear that the front-runner for a site in the U.S., outside of Alaska, was somewhere on the Tonopah Bombing and Gunnery Range in Nevada. The other sites all had serious disadvantages, such as sizeable populations down wind, or that the site acquisition involved the displacement of large numbers of people. On the first of August 1950, Los Alamos hosted a meeting on the radiological hazards associated with testing in the atmosphere. (Reines 1950) The group focused on the Tonopah Range. Attendees included Norris Bradbury, John Clark, Enrico Fermi, Alvin Graves, William Ogle, Fred Reines, and Edward Teller. The discussions, which included both Cooney and Holzman, who had called attention during SANDSTONE to the weather prediction problems at the Pacific test sites, resulted in a conclusion that a tower burst having a yield of 25 kt or less "could be detonated without exceeding allowed emergency tolerance dose of 6 - 12 R outside a 180 degree test area sector 100 miles in radius." Bradbury noted that the Nevada south site provided "much greater safety than the original 1945 shot at Alamogordo." (Hacker 1994: 42)

Maximum Yields

It is worthwhile to digress briefly to discuss perceptions of the maximum yield limit for the south site in Nevada. At the August 1, 1950, Los Alamos conference 25 kt was tentatively offered as a plausible maximum, but that was not a hard and fast number. The actual upper limit would ultimately be set by experience. Lower yield shots would be fired and the radiological hazards assessed. If the radiological hazard was deemed well within the accepted guidelines then, and only then, could the yield be increased. Other values for a possible upper yield limit appeared during the summer and fall of 1950, but most were around 25 to 50 kt. The highest yield atmospheric shot fired in Nevada, at 74 kt, was Hood, suspended from a balloon 1500 feet above the surface, and detonated on July 5, 1957 during Operation PLUMBBOB.

As a technical aside, after 1950, it was found theoretically, and validated by tests, that there is a height of burst at which early (or local) fallout ceases to be a serious problem. This height, H, is very roughly related to the weapon yield, W, by:

$$H = 180W^{0.4},$$

Where H is in feet, and W is in kilotons. This height of burst is greater than the radius of the fireball at *breakaway*, which is when the air shock breaks out of the fireball. At heights of H and greater, the extremely hot early time fireball does not interact with the ground; and the air shock that emerges from the fireball has cooled considerably before it interacts with the ground. Thus the amount of ground material that becomes vaporized, lofted, and subsequently becomes fallout is significantly reduced. (Glasstone 1962:73-79) This is the reason why almost all of the aboveground tests at the Nevada Proving Grounds (NPG) were conducted on towers, as airdrops, or suspended from balloons. The devices usually had a height of burst that exceeded the above equation.

Surface bursts are the worst configuration for producing fallout. Of the 28 surface tests conducted by the U.S., 10 were conducted at Bikini and Enewetak. One safety and four storage-transportation tests, all with yields of "zero", were conducted on the Nellis Air Force Range (NAFR); adjacent to the NTS. Eleven of the remaining 13 surface shots were conducted at the NTS during the 1951-1958 operations described in Part II. Two, were conducted after the moratorium, in 1962 and 1963.

THE HOLMES AND NARVER REPORT

Two areas within the Tonopah Range, one in the northern part and one in the southern section, were candidates for the new test site, see Figure 1-7.1. The engineering firm of Holmes and Narver was given a contract, on July 22, 1950, by the AEC Santa Fe Operations Office, to undertake a preliminary feasibility study and to make a recommendation regarding site selection. On August 14 they delivered their conclusions: "The South Site ... has certain advantages that definitely point to the selection of that site in preference to the North Site." The advantages included: terrain relative to the prevailing winds; visual security; the close proximity of the Air Force's Indian Springs camp facilities and aircraft runway; and finally the near-by town of Las Vegas, with a population of about 50,000 in 1950, would make support logistics less expensive than for the north site.

It should be noted that the military range in Nevada went by a variety of names: Las Vegas Bombing Range; the Las Vegas Bombing and Gunnery Range; the Las Vegas Aerial Gunnery Range; and the Nellis Bombing and Gunnery Range, after the Nellis Air Force Base that ran the range. The potential nuclear testing sites went by these names as well as by Tonopah site (after the Nevada town near the north end). These names were often used interchangeably, but it may be that some referred to the north site as the Tonopah site. It is also apparent that some references to the Tonopah site could refer to either the north or south sites, or both. By this time, Mercury was an additional name for the Nevada site. It was often called Site Mercury, or just Mercury, after an old mine in the region. During the early days of the search for a continental site, the term Mercury was so classified that it couldn't be used in writing.

By mid to late summer of 1950, it was clear that Los Alamos and the AEC had zeroed in on the Tonopah south site in Nevada. Of the sites considered there really wasn't much choice. Southern Nevada was obviously the most attractive from almost any perspective. Usually sites were eliminated either because the down-wind population was too high or the number of people that would be displaced to make room for the site was excessive. As it turned out, when atmospheric explosions were banned by treaty and testing went underground in the early 1960s, the geology of the Nevada site turned out to make it uniquely the ideal location.

AEC AND PENTAGON BATTLE OVER OPERATION GREENHOUSE

Through the end of August and the first half of September the battle raged at the highest levels of the AEC and the Pentagon over whether Operation GREENHOUSE was to receive the necessary support of the services. During this period AEC Chairman Dean decided that the Commission should definitely identify an appropriate proving ground and test atomic weapons in the United States, regardless of the fate of GREENHOUSE. An additional argument for a U.S proving ground was the likelihood that some of the tests planned for GREENHOUSE would be suitable experiments for the continental site. Bradbury, and probably Graves as well, had previously alerted Dean of the possibility of needing one or a few shots prior to GREENHOUSE.

The AEC was serious about the need for a continental site, but reluctant to declare a national emergency. On the other hand it was equally clear that with the war in Korea the Department of Defense had their hands full. "Dean fought for GREENHOUSE because the tests were vital to the thermonuclear program, and he quickly won Deputy Secretary of Defense Stephen Early to his cause. He eventually triumphed, for in mid-September the Joint Chiefs of Staff decided they could spare the resources for the tests."(*Forging The Atomic Shield*, excerpts from the office diary of Gordon E. Dean, edited by Roger M. Anders. p65) Deputy Secretary Early not only supported GREENHOUSE, but became an advocate for a continental test site as a necessary adjunct.

Continental testing, however, was still viewed by many of the players as an emergency measure necessitated by the crisis brought on by the Korean War. Carroll Tyler received a letter from AEC/DMA, dated September 15, 1950 that expressed the concept that the Nevada site would "... be restricted to purely emergency development ..."

i. "I wish to express to you my feeling concerning the probable limited scope of development to be planned for a U.S. site."
ii. "We all recognize that there is some approximate upper limit to the usefulness of Tonopah (or any other U.S. site) on the basis of acceptable radiological safety. Regardless whether this might be 100 or 200 kt, it is not believed any U.S. site can handle anticipated explosions upward of 1000 kt. For this reason alone Tonopah could only be considered a temporary or partial answer to our problem."
iii. "... It seems obvious to us, therefore, that our thinking in this site should be restricted to a purely emergency development of a type quite different and much less expensive than was contemplated in the Holmes and Narver preliminary report on Tonopah."

(Tyler to distribution. "Documentation of Establishment of Continental Test Site." Report. September 14, 1953)

The quoted yields are interesting in that the "100 to 200 kt" is far higher than those attending the August 1 meeting at Los Alamos would subscribe to. Of course the 1 Mt yield (i.e. 1000 kt) was clearly beyond the pale for explosions in the atmosphere over the U.S. As will be seen, these limits would safely be exceeded in Nevada and

Alaska when nuclear testing went underground in the 1960s. One underground shot with a yield 1.3 Mt, code-named Boxcar, was fired at the NTS in 1968. Another, with a yield of less than 5 Mt, code-named Cannikin, would be fired on Amchitka Island, Alaska, in 1971.

Los Alamos and the AEC/DMA were correct to anticipate the need for a location to test the high yield weapons that would result from the thermonuclear crash program. Except for a brief interlude during the moratorium of the late 1950s, the Pacific continued to be used by the United States for atmospheric testing for more than a decade.

As an aside, the French began testing at Mururoa, in Polynesia, in the atmosphere in 1966 and moved underground, at Fangataufa (adjacent to Mururoa), in 1975. The last atmospheric shot in the Pacific was the French Verseau test on a balloon. It went at about 330 kt on September 14, 1974 at Mururoa.

Meanwhile, at Los Alamos the Committee for Weapons Development (established in July 1948 with Edward Teller as chair) met on September 9, 1950 to review the GREENHOUSE device proposals. Those present included G. H. Best, H. A. Bethe, G. Breit, F. de Hoffman, E. Fermi, G. Gamow, R. W. Goranson, A. C. Graves, M. G. Holloway, E. R. Jette, R. M. Page, F. Reines, A. R. Sayer, R. E. Schreiber, T. B. Taylor, E. Teller, J. L. Tuck, S. Ulam, and R. D. Richtmyer. Concerns were expressed about the GREENHOUSE designs. Teller asked Graves whether it was possible to activate either a Nevada site or the Trinity site for a low yield test prior to going to the Pacific. A test at Enewetak or Bikini was much more costly, and thought to be more visible to the world, than a test in the U.S.; and there was a natural reluctance to take a major gamble on an experiment in the Pacific with the uncertainty that prevailed in September. (Betty Perkins, private communication)

PRESIDENT TRUMAN STEPS IN

Chinese forces joined North Korea in the war on October 25, 1950. Immediately after this escalation in the hostilities on the Korean Peninsula, President Truman directed the AEC to find a continental test site. The President appeared to favor continental testing even before his National Security Council formally considered the issue. (Anders 1987: 68) The next day the DMA requested the Corps of Engineers to make a more detailed study of the Tonopah Test Range than that by Holmes and Narver during the summer.

Then on November 14 the White House sent a memo entitled "Additional Test Site" to the Chairman of the AEC, the Secretary of Defense and the Secretary of State. This directed the AEC, with the assistance of the DoD to survey suitable sites and to recommend one for prompt development. It stressed consideration of security, accessibility, safety and a minimization of requirements for military logistical and security support. Two days later the AEC assigned staff responsibility for the survey to the DMA.

In less than a week DMA received two reports from Los Alamos: the first entitled "Desirability of an Area in the Las Vegas Bombing Range to be used as a Continental Proving Ground for Atomic Weapons" (Lab-J-1609. November 12, 1950); the second entitled "A Discussion of Radiological Hazards Associated with a Continental Test Site for Atomic Bombs". (LAMS-1173, September 1, 1950) No authors are listed for the first report, but Graves, Ogle, Clark and Reines probably wrote it. Frederick Reines authored the second report, which was based upon the notes of the August 1 radiological hazards meeting in Los Alamos. These reports summarized the advantages of the south site in the Las Vegas Bombing Range and recommended that it be made available for the tests planned for the fall of 1951.

The LASL report (Lab-J-1609) summarized the situation in the following way: "The Frenchman Flat area, remotely located in Southern Nevada, is relatively free from radiation hazards, has a minimum of operational limitations, and offers many operational facilities for an atomic proving ground in the continental limits of the United States. Although present knowledge of radiation hazards appears to limit the size of test weapons to 50 kt, it is expected that knowledge gained from small yield weapons may extend the allowable maximum yield. However, in the event this is not true, there are indeed small-sized weapons (10-30 kt) to be developed and tested, and this area is very well situated with respect to the laboratory for such tests."

No mention was made in the Los Alamos letter to DMA of pre-GREENHOUSE tests. There was still much uncertainty in the Los Alamos weapons program during this period, and it wasn't clear what they would test, prior to going to the Pacific for GREENHOUSE, even if they could field one or a few shots in Nevada.

Although in mid-November, 1950, it may not have been clear how they could mount a test series in early 1951, Los Alamos, with the strong support of the AEC and the Air Force, ended up doing just that.

By the end of November, J-Division responded to the September 9 request of Teller's Committee for Weapons Development for a feasibility study of doing a pre-GREENHOUSE test in the continental U.S. Graves recommended that any low-yield tests be done in Nevada. This was less than a week after the report released by Los Alamos, as J-1609, requested the preparation of a continental site for a fall 1951 test series.

To underscore an increase in the emphasis on this program Bradbury, on November 21, 1950, established the Fission Weapons Committee to oversee implosion weapon development. This group assumed the fission weapons responsibilities of Teller's Committee for Weapons Development. The initial members were: Froman, Carson Mark, Holloway, Schreiber, Eric Jette, Teller, Graves, Goranson, Eugene Eyster, and Duncan MacDougall as chairman.

On November 24, DMA received a report from AFSWP on potential testing sites in the continental U.S. (T/S OA1-1342-1A) The report summarizes the studies regarding five areas "--- which may be suitable for the conduct of air burst atomic tests." These are:

i. The Alamogordo - White Sands Guided Missile Range in New Mexico
ii. The Dugway Proving Ground - Wendover Bombing Range in Utah
iii. The Tonopah Bombing Range and Las Vegas Aerial Gunnery Range in Nevada
iv. An area in Nevada about fifty miles wide and extending from Fallon to Eureka.
v. The Pamlico Sound area in North Carolina. (AFSWP November 24, 1950: 2)

These were not, of course, new areas. They had been studied extensively before and, except for i and iii had been rejected.

The AFSWP report recommended:

1. Selection be limited to - "those sites now under the control of the Department of Defense" (i, ii, and iv)
2. The site be considered in the following priority:
   (a) Tonapah Bombing Range
   (b) Alamorgordo
   (c) Dugway.
3. That in selection of a continental site, consideration be given to the possible use of the site for future atomic tests. (AFSWP November 24, 1950: 8-9)

DECEMBER 1950

On the first of December Carroll Tyler reiterated to Brig. Gen. James McCormack, the director of DMA, Grave's request for an early decision on a test site in order to be able to field a series in September of 1951.

While Bradbury was in contact with AEC Chairman Gordon Dean about the possibility of an early 1951 test series in preparation for the next Enewetak experiments, Los Alamos did not want to make a formal application to DMA until after all parties had agreed upon a location for a continental test site. As a result, within Los Alamos, device design and development went forward without a clear picture of a feasible test program. (Betty Perkins private communication)

During the first week of December 1950, it looked more and more as if the Nevada south site would indeed become the continental proving ground. Teller was encouraged and proposed, at the December 6 meeting of the Fission Weapons Committee, that the lab plan to test one or more designs in Nevada before going to the Pacific in the spring. Bradbury went further and proposed a series of airdrops, which would not require the preparation of tower shots. This meeting yielded the definitive directives for the pre-GREENHOUSE test series. The Committee continued with a discussion of the test sequence and device designs five days later.

It took about a month for the AEC to respond to the White House's November 14, 1950, request for a continental test site recommendation. On December 12, the

AEC approved the DMA recommendation included in the report entitled "Selection of a Continental Test Site" which was based on Los Alamos' (J-1609 and LAMS-1173) and AFSWP's (OA1-1342) reports described above. The next day the AEC Chairman, Gordon Dean, submitted the recommendation included in the DMA report to the special committee of the NSC for Atomic Energy matters. Dean stated that while the southern Nevada site was recommended it would not be suitable for very high yield tests and that there was some urgency to find a secure site alternate to Enewetak or Amchitka for use in an emergency. The NSC forwarded Dean's recommendation, with their endorsement, to President Truman on the 18th.

The Fission Weapons Committee met on December 18. MacDougall, the chairman, noted, "it appears essential that firm decisions be reached concerning the details of "... the gadget designs for the January/February 1951 tests." The Committee was able to settle on a bill of material for the Nevada shots. Clark was queried on his views regarding the tempo that J-Division could sustain in the field and how fast they could fire the series of shots. He said "J-Division is preparing to fire shots on successive days. It plans to have the data from a morning shot analyzed by late afternoon of the same day, so that the committee may designate the shot for the following morning." This proved to be a close estimate.

During the final weeks of December the primary design groups, T-Division under Carson Mark and W-4 under Arthur Sayer, refined predictions and prepared a handbook compiling data felt to be useful in the conduct of an operation. Fred Reines edited the handbook, which included contributions by G. Felt, P. Flor Cruz, P. Galentine, R. Goranson, R. Newman, R. Patten, F. Porzel, F. Reines, L. Seeley, H. Stewart, B. Suydam, T. Taylor, J. Whitener, and E. Zadina.

Diagnostic experiments were defined and final equipment orders placed. Weapons effects measurements were planned with the AEC and with the DoD. Safety procedures were established, monitoring personnel trained, and monitoring equipment procured and calibrated.

A week before Christmas, on December 18, President Truman signed a memorandum by James S. Lay, Executive Secretary of the National Security Council on Atomic Energy, which "---- recommends approval by the President of the development of a portion of the Las Vegas Bombing and Gunnery Range as an atomic weapons test site." This document is considered to be the "birth certificate" for the Nevada Test Site.

On the day that the Presidential approval was received officials from the State Department, the DoD, and the AEC met to plan a public relations strategy for a resumption of testing in the United States. The participants at the meeting felt that the public should be told that nuclear testing was a routine activity and that radiological safety was well in hand. They also felt that the Nevada atomic weapons testing site should be viewed as analogous to the Army's Aberdeen Proving Ground, which had the understanding and support of the public. The Washington decision makers probably overestimated the public's apprehension. In hindsight some of the comments that were made at the meeting appear ludicrous: "The most important angle to get across, they concluded, was the 'idea of making the public feel at home with neutrons trotting around.'" (Fehner 2002: 52) It would be comforting to believe that such a statement appeared ludicrous at the time.

The pre-GREENHOUSE test plans had been on the fast track since mid-December. Graves was the test director, and Clark was his deputy. Clark, with assistance from Gaelen Felt and others, had the day-to-day responsibility for the continental operations while Stanley W. Burriss and William Ogle focused mainly on the operations in the Pacific. Graves monitored both programs and provided most of the liaison with the AEC in Washington and with the field offices.

On December 19, Bradbury sent a letter to DMA director McCormack outlining the rationale for the upcoming test series. Briefly, there were concerns about the reliability of the yield calculations and the yield reproducibility of the new implosion designs that were slated for stockpile and for testing on Operation GREENHOUSE in the Pacific. Bradbury told McCormack "it is proposed to carry out at Site Mercury* a series of drop tests of up to but not more than five atomic bomb explosions during the first part of February 1951". The diagnostic measurements required were alpha and yield. [*Footnote: Note that Bradbury refers to the site as "Site Mercury". A number of names were used for the site prior to its official naming. It wasn't until July 8, 1951, that the AEC settled on the Nevada Test Site as the official name. It was changed to the Nevada Proving Ground in February 1952, and was changed back to the Nevada Test Site on December 31, 1954. It retained that name to the end of US nuclear testing in 1992.]

The Air Force, AEC, and the Los Alamos J-Division met in Las Vegas on December 21 to negotiate final agreements for the joint use of the range. The summary of the agreements focused on two issues: (1) use as a permanent site for AEC tests, and (2) arrangements for the first test series. The final Summary of Agreements included the following items:

a. "The AF will surrender its lease to an area of the Las Vegas Gunnery Range consisting of a rectangular tract approximately 12X30 miles, pending approval of an accurate survey." The 30 mile sides of the rectangle would be oriented north and south. The southern boundary of the area would be contiguous with the southern boundary of the gunnery range. The existing 4 lanes of the Training Command gunnery range would be to the east of the AEC land, the existing SAC gunnery range would be to the west, and the existing SAC and ATRC bombing range to the north.
b. "The AF will permit joint occupancy with the AEC of Indian Springs encampment for a period from January 1, to March 1, 1951. Specifically, barracks and mess buildings for 200 and possibly 250 will be made available with the understanding that all services and equipment must be provided by the AEC."
c. "The AF will provide, on a temporary basis only, space at Nellis AFB for a communications center for the AEC radiological safety activity," and

d. “AEC will be responsible for providing all services required by observers invited by the AEC to be present on test days.” (Rutledge 1950:1-3)

Figure 1-7.2 shows the Nevada Test Site with the approximate boundaries described above, approximately 12x30, 360 square miles. The southern boundary of the bombing and gunnery range did not run in a true E-W direction, rather it slanted, from NE to SW by approximately 4 degrees.(USGS 1952:Mercury Quadrangle) Note that Mercury was outside of the initial boundaries. The boundaries of the NTS would increase with time, and these increases are discussed as they occurred, as well as summarized in Appendix D. However, the SE corner of the test site would remain fixed. Appendix E briefly describes the environmental conditions at the site. Appendix D provides a brief history of the acquisition of real estate for the site.

CHAPTER 8. JANUARY 1951 - PREPARATION FOR THE FIRST NEVADA OPERATION - RANGER

Largely because Bradbury and Graves had kept the AEC informed of the Los Alamos test objectives during the late summer and fall of 1950, the proposal for a test series before GREENHOUSE, received a rapid and favorable response in Washington. On January 3, 1951 the MLC gave their approval. The same day, McCormack told Bradbury that the code name for the five shot series would be RANGER. FAUST was an early acronym for the series and stood for "First Airdrop United States Test."(Perkins 1992:25) The next day, January 4, the AEC submitted a description, and rationale for the tests, to the Special Committee of the NSC.

The General Advisory Committee of the AEC, chaired by Oppenheimer, enthusiastically endorsed the proposed test program for RANGER on January 6, 1951. Approval by the NSC* was more difficult. They were very concerned about two issues. One was the possible radiological hazard of the thirty to forty kiloton design yield of the proposed "F" test: Fox. The other issue was the wording of the AEC's planned press release. (Fehner 2002: 53-54) With regard to shot F, the AEC explained that the first explosions would have lower yields and that assessments of the radiological hazards would be made after each shot. If the experience with the lower yield shots indicated that shot F could not be done safely, then it would be cancelled for Nevada, and perhaps moved to a region over the open ocean in the Pacific. (Fission Weapons Committee 6th meeting, January 9, 1951(SRD)) [*Footnote: The National Security Council (NSC) was established by the National Security Act of 1947; and in 1949, the NSC was placed in the Executive Office of the President. It is chaired by the President and attended by the Vice President and Secretaries of Defense, State, Treasury, and the Assistant to the President for National Security Affairs. The function of the NSC is to advise and assist the President on national security and foreign policies. It is the principal arm for coordinating these policies among the various government agencies. Presidential approval is obtained for all nuclear tests. A NSC committee reviews and presents to the President their recommendations regarding proposed nuclear tests. While the NSC is not formally in the approval chain for nuclear testing, their views and concurrence regarding a test (or operation) particularly in the areas of public safety and international relations are of the utmost importance for securing Presidential approval.]

Secretary of Defense Marshall wanted a meeting with Secretary of State, Dean Acheson, and Gordon Dean to discuss a press release. Marshall was concerned with a public statement that implied we had small nuclear weapons. He felt that this would be detrimental considering the tense world situation of the time. Presumably there would appear to be a greater incentive to use low yield weapons than high yield weapons in a crisis. This could result in the nuclear threshold being crossed without appropriate consideration of the escalatory consequences, the so-called "fire break" issue. With the wrangling and the short time frame Gordon Dean began to worry about not having any press release. He was very concerned about the public relations implications and argued convincingly for some sort of announcement.

Ultimately a rewritten draft of a press release proved acceptable to Marshall and the NSC. It was then forwarded to President Truman along with the suggested test program, including shot F. Truman approved the program for RANGER and the press release on January 11, 1951.

Meanwhile Dean and others in the AEC, along with Bradbury and Graves from Los Alamos, were making the rounds to brief federal and Nevada and California state officials on the impending test series and the radiological safety implications. In general the public relations activities were successful and there was little, if any, objection to the upcoming test series in either southern Nevada or California. (Fehner 2002: 58) In fact the Las Vegas Review-Journal was quite pleased when southern Nevada was chosen for the Nation's atomic test site, while Alamogordo, NM, was disappointed that the Trinity location was not selected.

Carroll Tyler, manager of the AEC Santa Fe Operations Office (SFOO) was named "the responsible officer of the USAEC for the conduct of Operation RANGER." (M.W. Boyer, General manager AEC to Tyler, January 15, 1951) The leadership and management for Operation RANGER was a partnership between the AEC and Los Alamos. The AEC had the overall governmental responsibility for the operation, but the Laboratory had oversight of the technical execution of the tests. Los Alamos had established J-Division to field nuclear tests at the Pacific Proving Ground, and with Operation RANGER the Division began its long history with continental testing as well as with off-continent testing. Alvin Graves, as J-Division Leader, was responsible for all of the Laboratory's nuclear test related activities. For RANGER he was Chief, Test Group, with responsibility for the overall technical direction of the test series. John C. (Jack) Clark, the associate J-Division Leader, was the Deputy Chief, Test Group.

An immediate question, within the Federal government, was the role of the military in RANGER. One of the main reasons for going to a continental test site in the first place was to reduce the burden on the services. In the July 21, 1950 message to McCormack, Bradbury had postponed addressing the question of the military role at a continental test site that had been raised in McCormack's July 20 message to Tyler. However, at the beginning of January 1951, Air Force Lt. Gen. Elwood R. Quesada, commander of JTF-3 for Operation GREENHOUSE, suggested that the test series should be the responsibility of his task force. "The Atomic Energy Commission disagreed. Brig. Gen. George Schlatter* argued that the task force was 'neither necessary nor sufficiently flexible' for the purpose of the test series. McCormack stated that this was a responsibility that the Commission could not 'appropriately share' through the mechanism of a task force. In the end, with the relative proximity to Los Alamos and much reduced logistical and security requirements, task force support was not needed; and the task force played no role in RANGER." (Fehner. 2002: 49-50) This set a significant precedent for the tests conducted by the laboratories that carried through until testing stopped in 1992. For the military's weapons effects tests, the services assumed responsibility for the construction and the measurement of nuclear weapon effects; the laboratories provided the weapons, the firing, and the yield certification by conducting diagnostics experiments. [*Footnote: Brigadier General George F. Schlatter was in attendance at the 21 December 1950 meeting between the AF, AEC, and Los Alamos. At the time, he was with the AEC Military Applications Division of which he became the deputy director. He played an important part in the coordination of public relations necessary to make possible the first series of atomic tests in Nevada and the later participation of troops and civil defense agencies in these tests.(Air Force 2005: Biographies)]

While the AEC, Los Alamos, Sandia, and soon-to-be Livermore, enjoyed close and fruitful collaborations with the services during the atmospheric testing operations in the Pacific, the management structure adopted for Nevada proved best for the on-continent operations. In addition, the DoD continued to contribute to nuclear testing in Nevada though air support and base operations at Indian Springs throughout the whole period of testing. The Air Force was particularly generous in providing office space, housing and vehicles at Nellis Air Force Base, Las Vegas, during RANGER. In addition, they provided housing and mess facilities at Indian Springs, just south of the proving ground.

There was not an AFSWP test group on Operation RANGER, because of the rapid development and deployment of the operation. There was, however, an AFSWP contingent of six men who were placed in the Scientific Tests Section under the supervision of Los Alamos scientists and AEC officials. (DTRA 2002: 82-83)

Over 350 service men participated on RANGER, mostly in Air Force weapons delivery, weather, and cloud sampling programs. (DTRA 2002: 82-83) The Air Force flight crews from the 4925th Special Weapons Group at Kirtland A.F. Base, in Albuquerque, NM, had the very significant responsibility of dropping the nuclear devices in the right place at the right time. This was accomplished in a skillful and professional manner.

The stage was set for Operation Ranger at Nevada. The first shot, Able, would be fired January 27, 1951, with a yield of about 1 kt. It would be the first nuclear explosion in Nevada, the seventh test for the United States (not counting the two weapons used in strikes over Japan, which were not tests), and the eighth in the world.

In retrospect, the compressed time scale and successes of Operation RANGER were most impressive. The people from Los Alamos, Sandia, the AEC, EG&G, the Air Force, the Naval Research Laboratory, the McKee Company, REECo, and many others accomplished an incredible feat in successfully fielding five shots in little more than a month after the AEC formally assumed ownership of the site. Much of the credit goes to Carroll Tyler, Alvin Graves and John Clark, for their superb leadership.

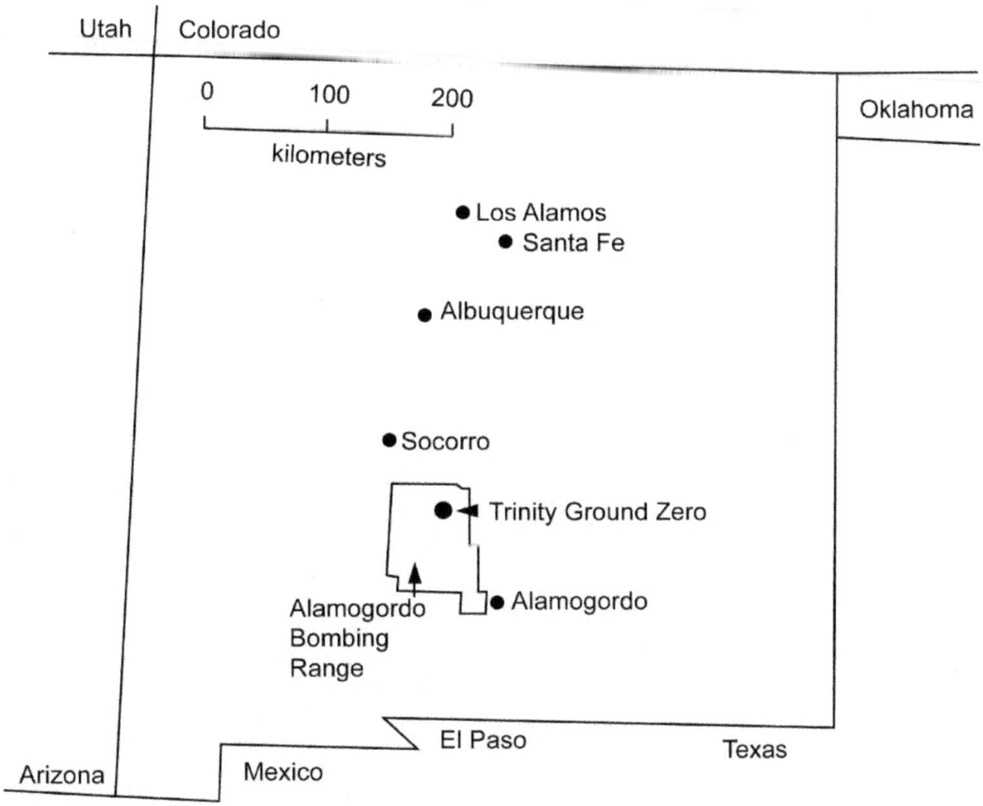

Figure 1-1.1 Location of the Trinity test site within New Mexico.

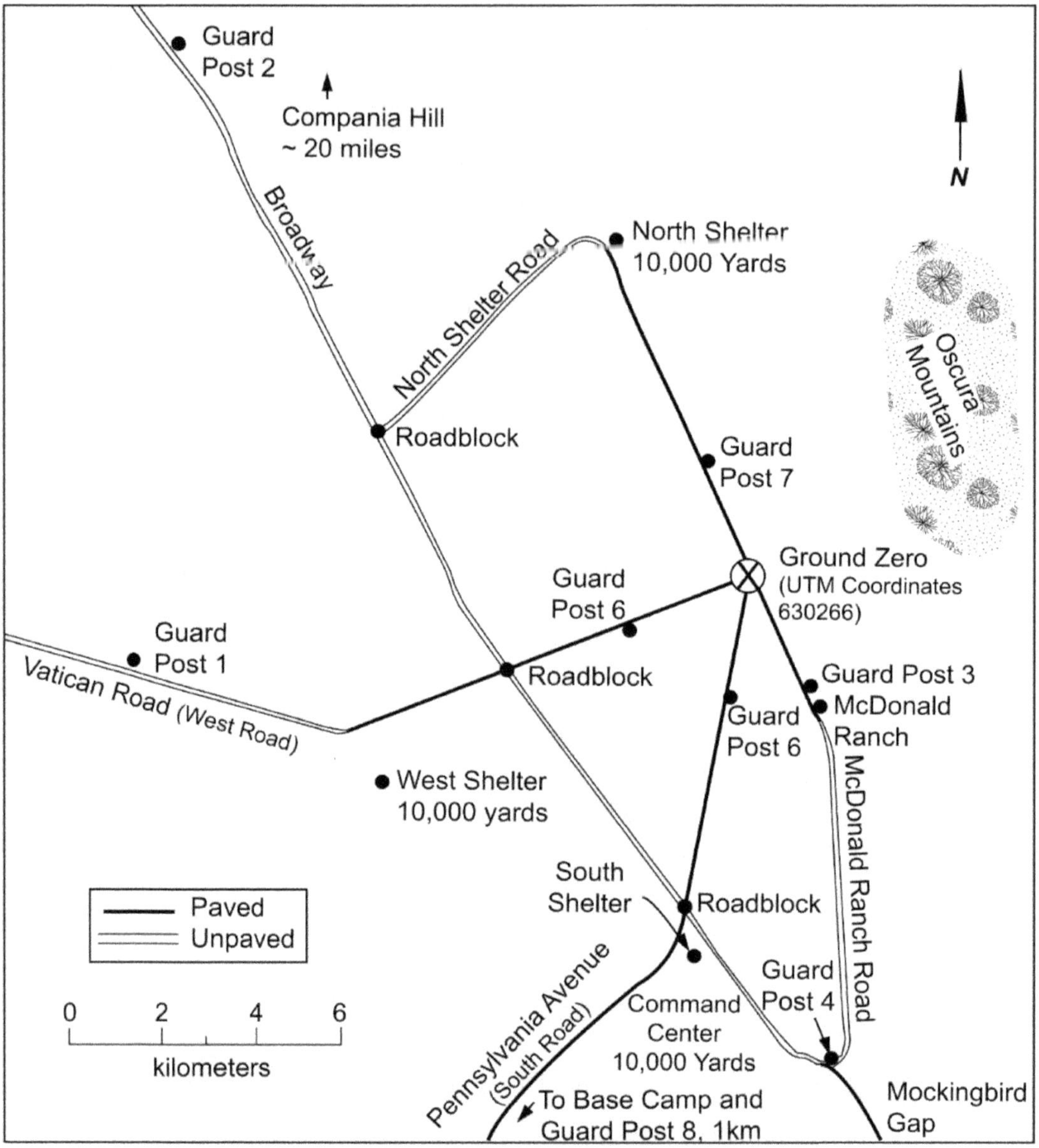

Figure 1-1.2 Trinity Site.

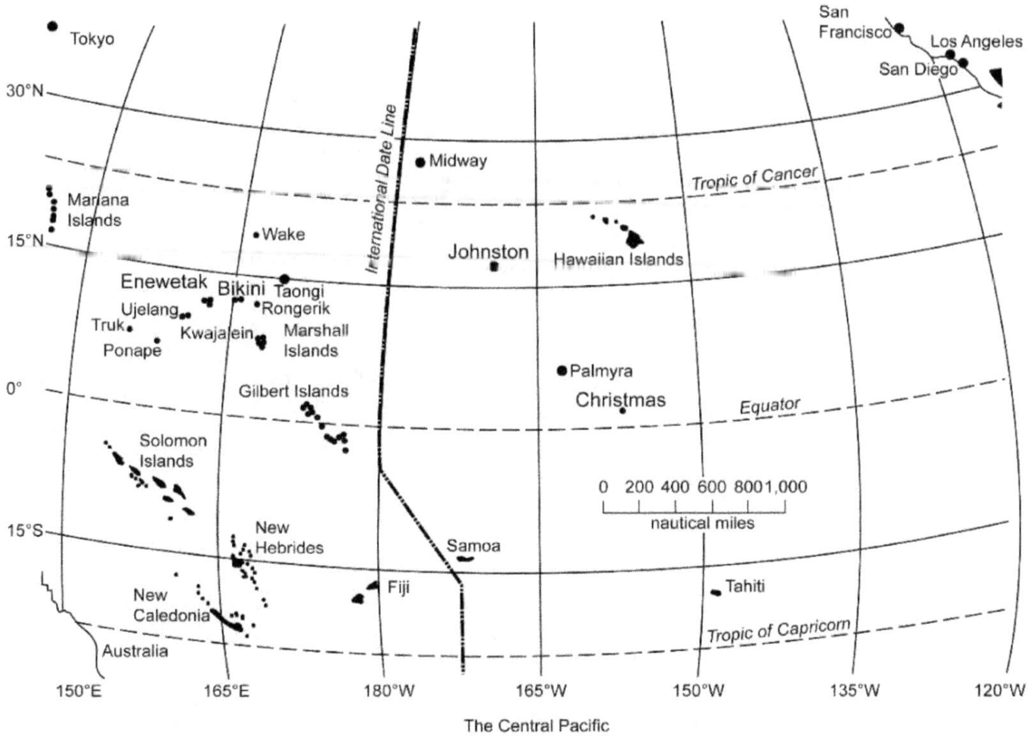

Figure 1-2.1 Central Pacific - Marshall Islands with Bikini and Enewetak highlighted.

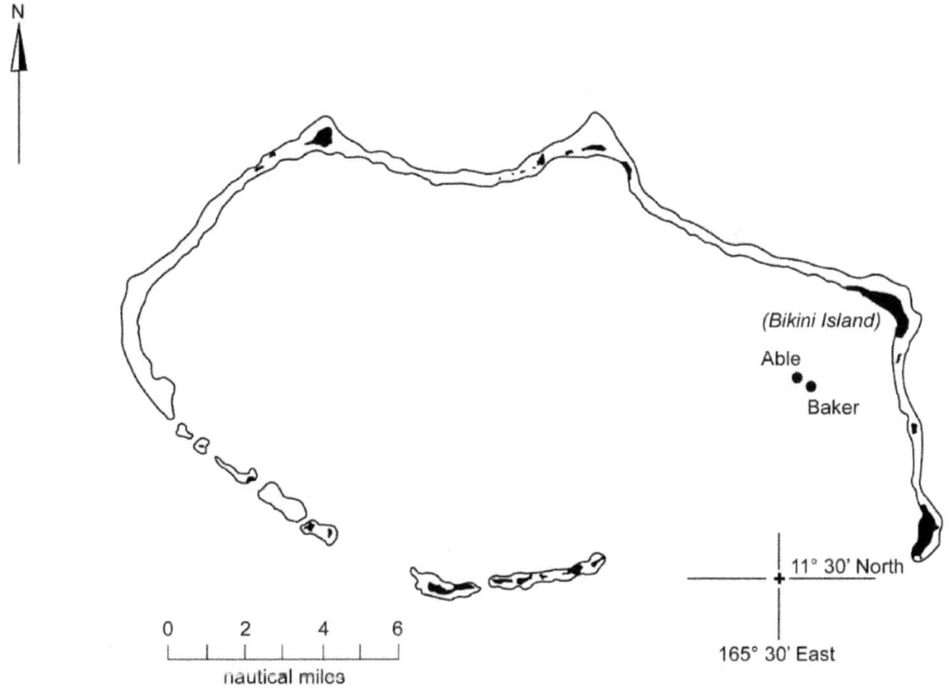

Figure 1-2.2 Bikini atoll 1946, with location of CROSSROADS shot.

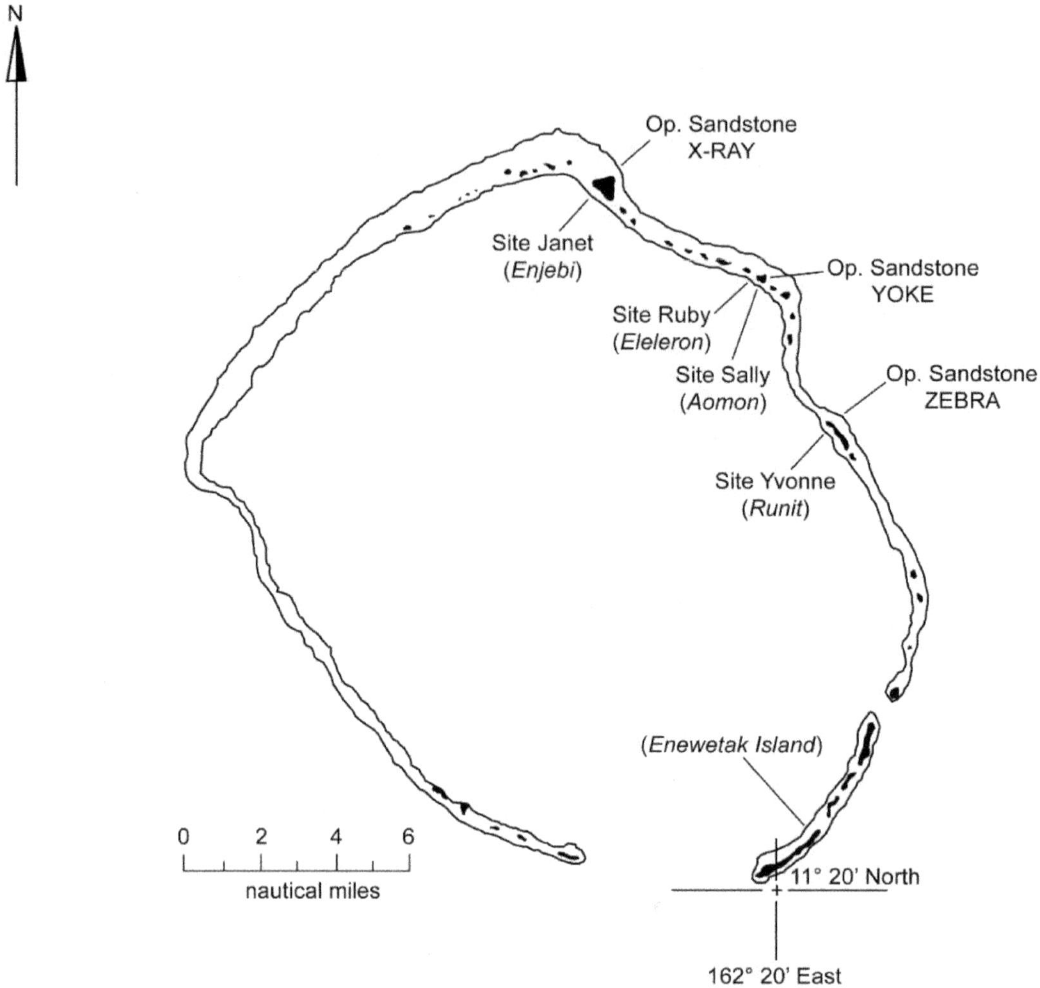

Figure 1-6.1 Enewetak atoll 1948, with location of SANDSTONE shots.

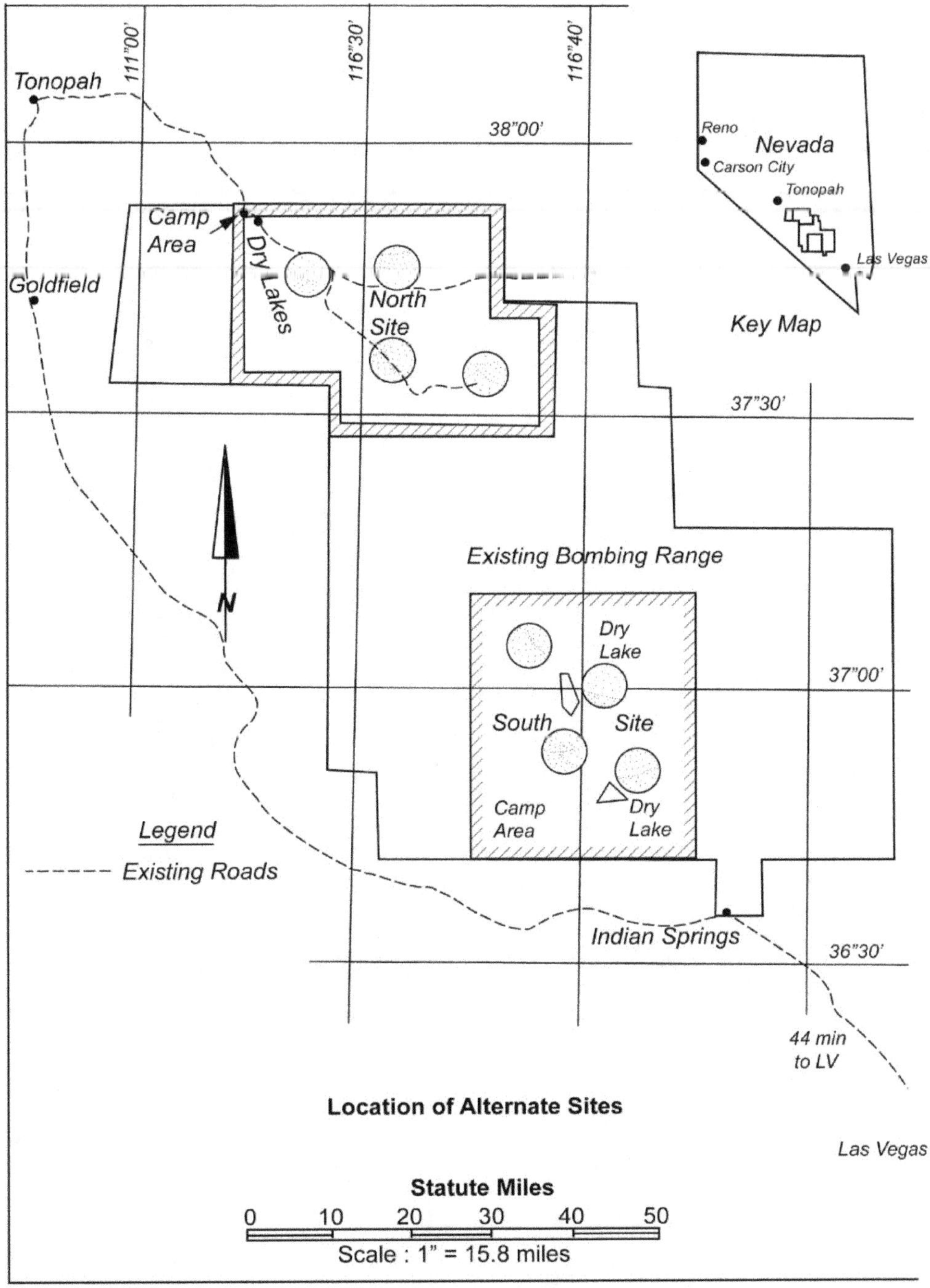

Figure 1-7.1 Nevada with location of Bombing Range and the North and South sites.

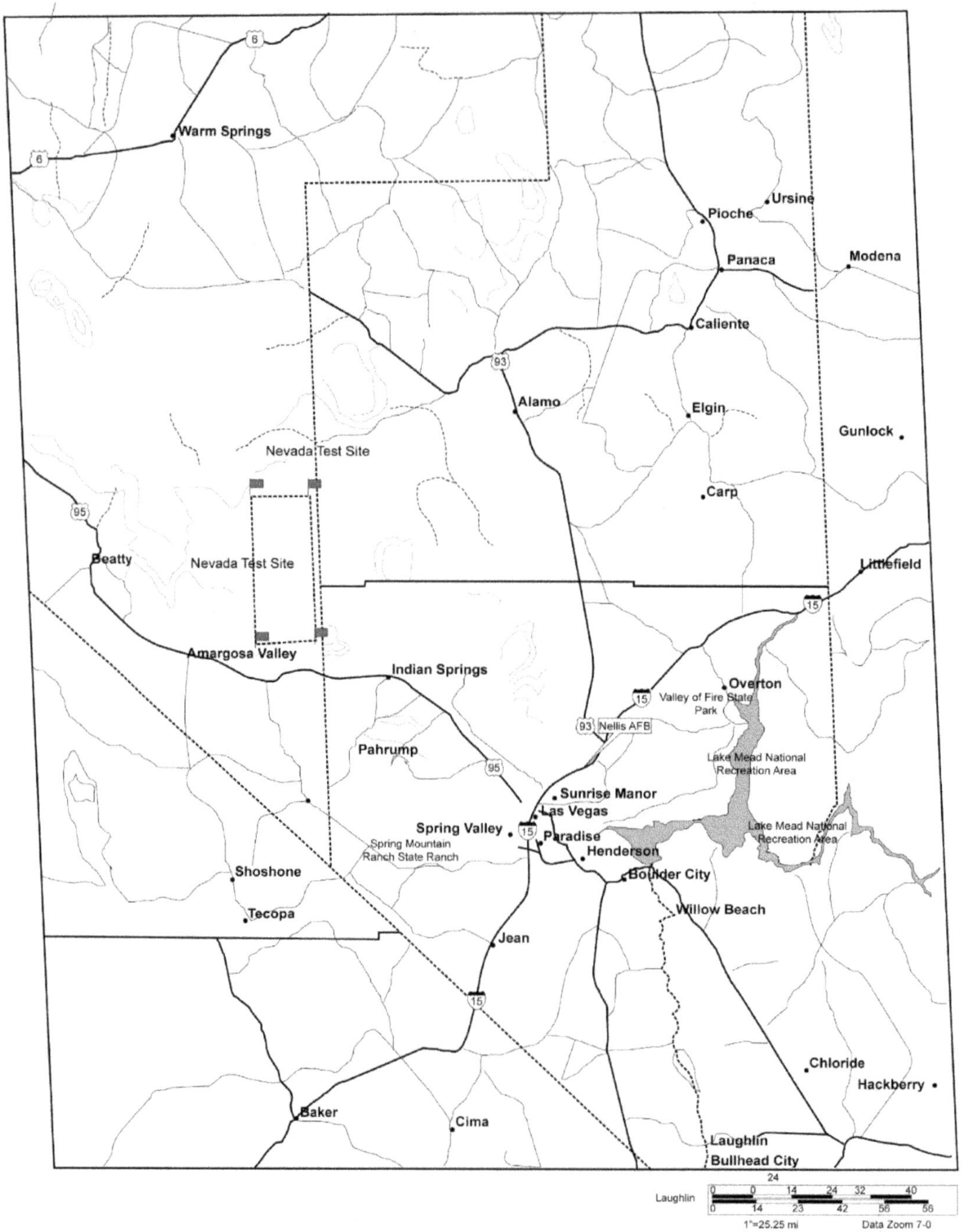

Figure 1-7.2 Approximate initial boundaries for the site of 12 X 30 miles.

J. Robert Oppenheimer and Major General Leslie R. Groves at Trinity site, postshot.

William S. "Deak" Parsons

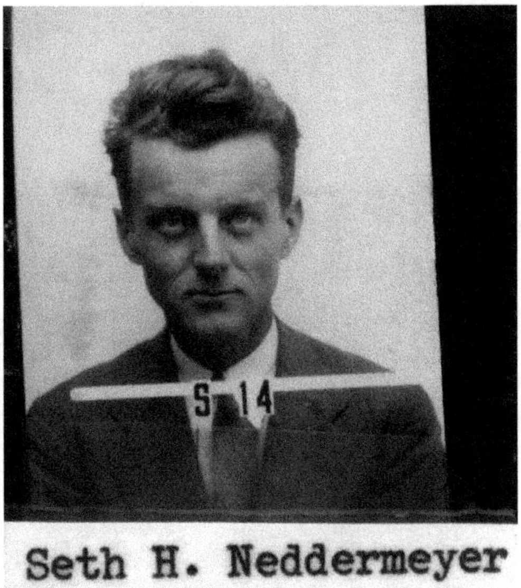

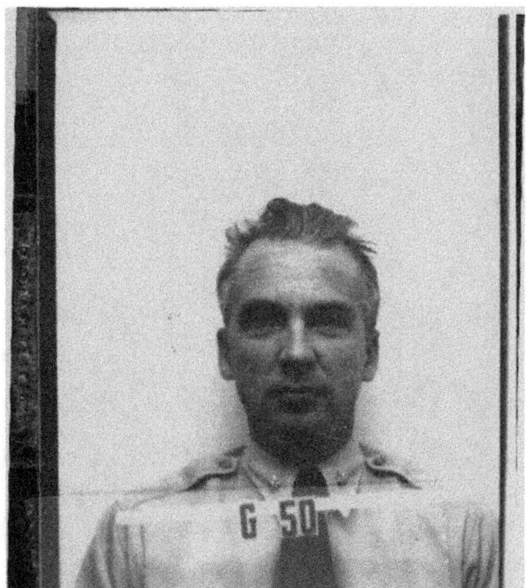

George B. Kistiakowsky

Hans Bethe

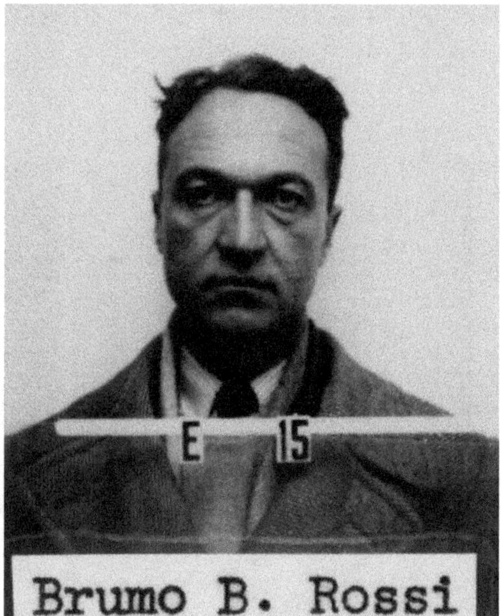

Harold Agnew, 1955

William Penny

Carroll Tyler

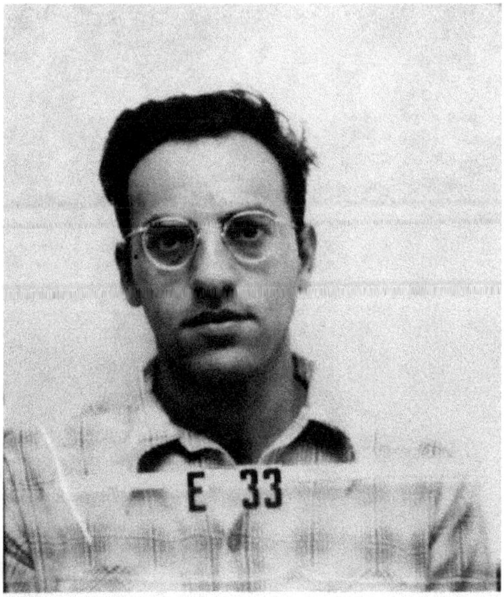

Frederick Reines

Rod Spence

Bill Ogle

John Clark

Paul J.Larsen

Major General Kenneth D. Nichols

Major General Herbert B. Loper

Duncan MacDougal

Gaelen Felt

Eugene Eyster

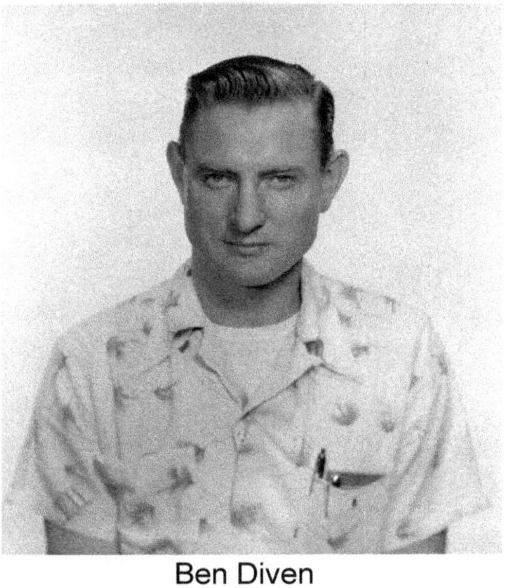

Ben Diven

Caroll Wilson

Samuel Glasstone

PART II. THE FIRST DECADE

INTRODUCTION

Continental testing offered the opportunity for organizations to more realistically assess the effects of nuclear weapons on a wide variety of targets. This was very important to the DoD and to those responsible for Civil Defense. In addition, military maneuvers in an environment simulating a nuclear battlefield were feasible at a continental test site, but were impossible at a Pacific atoll where there is very limited real estate.

Finally, it was vastly easier, and less expensive, to field a test series in Nevada than it was at the Bikini or Enewetak Atolls. Los Alamos, and later Livermore, viewed Nevada as a "back yard" range that provided the opportunity to rapidly test new ideas or answer time-urgent questions prior to committing the resources necessary for a high yield detonation at the Pacific Proving Grounds.

WEAPONS DEVELOPMENT

Immediately following World War II the United States recognized that it was not feasible to match the Soviet Union's conventional land forces division for division. Fortunately the U.S. had a fledgling nuclear weapons capability to counterbalance the Soviet's formidable conventional capability. The United States military services were enthusiastic supporters of both strategic and tactical nuclear armed forces. With the enormous success of the thermonuclear designs, first tested on Mike in 1952 and in weaponized forms in 1954, the military had a whole new war-making paradigm to consider. The first reaction was to go for ever larger yields with a wide assortment of weapons systems.

Three technological developments contributed to a dramatic change in the United States' defense posture during the 1950s, particularly late in the decade. The first was the development of naval nuclear propulsion. The second was the development of much more accurate, and militarily useful, weapons delivery platforms. For example, the development of solid fuel missiles that were vastly more accurate than their predecessors was extremely important. The third was the development of thermonuclear weapons of a size that could easily be delivered with feasible weapons platforms.

The enthusiasm for ever larger weapons was dampened by the delayed recognition that collateral damage was usually a negative rather than a positive feature. This observation gained currency in the United States before doing so in the Soviet Union, but by the mid 1960s it was clear to all nuclear states that there was precious little military or political benefit from multi-megaton weapons. It also became clear, in the United States, by the end of the decade of the 1950s that a wide assortment of tactical nuclear weapons had little or no practical value. Some tactical nuclear weapons were, indeed, potentially useful, but a nuclear counterpart to virtually every conventional system made little sense. The reasoning is complex, but the essence is that nuclear weapons are very specialized tools of war and are not just extensions of conventional weapons with enhanced explosive power.

The first decade of the Nevada Test Site witnessed three major nuclear weapons development thrusts: Smaller, more efficient fission weapons; the design of primaries for thermonuclear (TN) weapons; and the exploration of new TN concepts utilizing low yield mock-ups.

At the beginning of the 1950s the AEC complex designed and built nuclear weapons, and the services designed weapons systems around them. By the 1960s it was the other way around. The military designed weapons systems, or delivery platforms, and the AEC designed and built nuclear weapons to fit. Ultimately this resulted in the requirement for more nuclear testing because more devices were tailor-made for new weapons systems.

Enormous advancement was made during the atmospheric test days of the 1950s. At the close of the 1940s the United States nuclear arsenal was a modest extrapolation of the weapon designs on hand at the end of World War II. Within a decade the arsenal included a complex mix of tactical and strategic weapons spanning a yield range of more than ten thousand. Size was also reduced and delivery systems improved. The WWII arsenal required a huge B-29 and its crew for the delivery of 20 kt. By the end of the 1950s, a megaton could be delivered by a single pilot from an aircraft carrier.

WEAPONS EFFECTS

In the immediate post-war period Norris Bradbury, the director of Los Alamos, defined the nuclear weapons role for Los Alamos, and subsequently by precedent, for Livermore as well. That role was research and development of new nuclear weapons. It did not include weapons production, routine stockpile maintenance, or research on the effects of weapons. While the dividing lines were occasionally fuzzy, and shifted from time to time, the general responsibilities stayed remarkably fixed until the end of nuclear testing in 1992.

Bradbury recognized the need to study weapons effects; he just didn't think that it was an appropriate AEC responsibility. As a result the job of finding out what nuclear weapons did in a military environment fell to the DoD; and to address that responsibility, the DoD established the Armed Forces Special Weapons Project (AFSWP).

Los Alamos scientists made some estimates of the damage that might be expected prior to the military strikes on Japan in 1945. Of course, such estimates were made without the benefit of observational data. As the AFSWP scientists and engineers started seriously looking into this area it quickly became clear that the effects of nuclear weapons were much more complicated than anyone imagined during the war. It was not possible to rely solely on theoretical calculations. An experimental program was absolutely necessary. This began initially with Operation

CROSSROADS, during the summer of 1946, and continued through the Hunters Trophy event on September 18, 1992. That, incidentally, was the next to the last shot that the United States fired prior to the 1992 moratorium.

The weapons effects experiments were generally labor-intensive activities involving many military organizations and civilian scientific and engineering firms. In the process the United States, over the years, developed a large, sophisticated, technical family of weapons effects experts.

DESERT ROCK - MILITARY EXERCISES

The DoD was naturally very interested in using nuclear tests as targets of opportunity to indoctrinate combat troops in the nature of a nuclear war-fighting environment. Thus the Desert Rock exercises were born. Tens of thousands of military personnel were involved as observers or maneuver troops throughout all of the NTS test operations of the 1950s except the first and last, RANGER and HARDTACK II. These exercises included subjecting military hardware and fortifications to nuclear blasts and the analysis of the resulting damage. From a military standpoint this was extremely valuable training for an unfamiliar environment during the early years of the Cold War. These exercises had repercussions decades later when some of the military personnel, who had been involved in Desert Rock, began to notice medical problems that they attributed to radiation exposures in Nevada.

CIVIL DEFENSE

Federal and state organizations responsible for civil defense matters were also major participants at the nuclear tests of the 1950s. The civil defense organizations focused on the exposure of many facets of civilian life to nuclear environments and the orientation of officials with civil defense responsibilities to nuclear explosions. They also developed proposals to the AEC for so-called open shots to familiarize government officials and the media with atomic bomb blasts. In collaboration with the AEC, the federal civil defense organizations invited the NATO allies to participate in the exposure of non-military structures, such as bomb shelters, to nuclear environments. Also, industrial partners were invited to participate, at their own expense, in civil defense related weapons tests. Many of the structures still visible in Frenchman Flat are left over from the civil defense studies of the 1950s.

PLOWSHARE - PEACEFUL USES OF NUCLEAR EXPLOSIVES

During the 1950s Los Alamos focused almost exclusively on nuclear weapons R&D while Livermore was more adventurous and searched out potential new civilian roles for nuclear energy in general and nuclear explosives in particular. Thus was born the Plowshare program. This was focused on exploring peaceful applications for the enormous explosive energy of nuclear devices. The Soviet Union was also taken with this potential and for many years pursued an active peaceful nuclear explosive (PNE) program. Cynics, however, suspected that at least some of the interest was only a cover for military-related R&D programs. This was particularly relevant when negotiations were in progress that would limit weapons programs but permit peaceful nuclear activities. Ultimately the Plowshare applications proved not to be politically feasible, largely because of the fear of the residual radioactivity that accompanies all nuclear explosions.

PUBLIC IMPACT OF NUCLEAR TESTING

Initially, there was relatively little public concern with radioactive fallout. The Federal government in general and the AEC in particular initially enjoyed a great deal of trust and confidence on the part of the population that surrounded the Nevada Test Site. The people, in fact, were proud of their community's contribution to national defense and the role that southern Nevada was playing in nuclear weapons tests.

Public concern over fallout increased over the years as requests to stay indoors during radioactive cloud passage, or injury to livestock, occasionally arose. Also, the down-winders, those with the most exposure, or potential exposure, to fallout to the east and northeast of the test site, started to look for, and find, what they viewed as increases in cancers and other diseases associated with nuclear testing in the atmosphere. This is an issue that is still with us today.

For their part, the laboratories and the AEC looked for ways to reduce the local fallout by going ever higher, above the surface, with the nuclear tests. The objective was to keep the fireball well away from the ground. Towers got taller and taller, and eventually the test community went to balloons for device suspension. An alternative approach was to go underground, either to reduce the fallout or to eliminate it altogether. Livermore took the lead in this and fired the first fully contained shot, Rainier, in a tunnel, in mid-September 1957. Los Alamos elected to use shafts rather than tunnels. At first the shafts were not stemmed, or backfilled, and the resultant fallout was substantially reduced, but not totally eliminated.

TEST BAN DISCUSSIONS

Following the 1954 Bravo shot in the Pacific that produced the deadly fallout on the Japanese fishing boat, the Lucky Dragon, there was an international clamor to end nuclear testing altogether. The Soviet Union and the United States were roundly criticized for exposing the world's population to what some considered a dangerous environment whose long term effects were, at best, poorly understood. The major powers entered into talks, under the auspices of the United Nations, to curb nuclear testing. This was in an environment of suspicion and distrust between the eastern and western blocks, and the success of any such discussions turned on the issue of verification.

UNDERGROUND TESTING

Edward Teller and his colleagues at Livermore lead the efforts to argue that underground testing addressed the fallout and public health concerns. On the other hand Bradbury and his Los Alamos colleagues (mostly Carson Mark, the leader of the theoretical design division) argued that there were both fallout and arms control concerns, and that underground testing only addressed the fallout component. Also, Los Alamos was not convinced that underground shots could be diagnosed adequately, i.e. that measurements of the number of neutrons or gammas and the yield could not be made accurately.

At the same time the new technology for underground testing raised the specter of evasion opportunities. Could the Soviet Union, for example, evade a test ban, at some low threshold level, by testing underground in a large room and effectively decoupling the ground shock of the shot? The resulting seismic signals might or could be greatly diminished compared to those emanating from an explosion where the device was detonated in a relatively confined space. As a result of these concerns, the VELA program was developed in the late 1950s to explore the verification capabilities of current technology.

Eventually both Livermore and Los Alamos went to stemmed shafts for development shots while the DoD and Sandia Laboratory went to tunnels for the weapons effects tests. Ultimately the technology achieved a level where no radioactive release was the norm. In addition to the public health benefits, underground shots had the added advantage of substantially reducing weather delays, which was a major scheduling factor during the 1950s. Also, diagnostic techniques and instrumentation became even more accurate and were fabricated for tunnel and downhole emplacement.

MORATORIUM

In 1958 the Conference of Experts in Geneva recommended a moratorium on nuclear testing. The governments of the United States, United Kingdom, and the Soviet Union agreed and set a date of November 1 for the cessation of all tests. The U.S. stopped testing on October 31, and the Soviet Union followed suit a few days later. The moratorium lasted almost three years. France tested in February, April and December 1960 and April, 1961, at their Reggane, Algeria test site. The Soviets used the late 1960 declaration by President Eisenhower that the U.S. would no longer be bound by the moratorium and the French shots as excuses to resume testing in September 1961. The Soviet rationale was that if the West tested, and France was viewed as included in the West, then it was appropriate for the Soviet Union to test. Clearly the Soviets had been planning to resume for some time. They fired twenty-eight shots the first month.

Part II begins with Operation RANGER and concludes with the 1958 to 1961 nuclear test moratorium. In between, the other six major test operations (BUSTER-JANGLE, TUMBLER-SNAPPER, UPSHOT-KNOTHOLE, TEAPOT, PLUMBBOB, AND HARDTACK PHASE II) and the four subsidiary projects (Project 56, 57, 58 and 58A) are discussed. There are also chapters that discuss such related topics as: growth of the testing community, issues associated with the test site, and testing procedures.

CHAPTER 1. RANGER JANUARY 27 - FEBRUARY 6, 1951

INTRODUCTION

On January 27, 1951, a B-50D aircraft with a crew of 11 from the 4925th Air Force Special Weapons Command and Gaelen Felt, a Los Alamos scientist, took off from Albuquerque to make the first nuclear weapons airdrop over the continental United States as the initial test in Operation RANGER. (Maag 1982:51-55)

RANGER was to be executed prior to and in support of the GREENHOUSE test series planned for the spring of 1951 in the Pacific. GREENHOUSE would test new fission weapons and the fundamentals of thermonuclear design. It was discovered, however, in the fall of 1950 that minor variations in the hydrodynamics of the implosion process could produce significant changes in the yields. The designers desperately wanted a few rather simple shots prior to GREENHOUSE to verify their calculations and assumptions. This was the motivating factor for doing some time urgent experiments closer to home.

Uncertainties in the calculational models were identified; and a series of tests planned to explore, as separately as possible, each of the major issues. This resulted in four nuclear device designs, where experiments to determine yield and alpha could be expected to give insights into the nuclear processes. Five devices were fired, with one design fired twice to explore performance reproducibility. (Clark 1953: 22-23)

The most noteworthy characteristics of RANGER were the incredibly fast response time and the outstanding success of the operation. President Truman approved the use of a part of the Air Force's Las Vegas Bombing and Gunnery Range for nuclear testing on December 18, 1950, see Figure 1-7.2; and the first nuclear device was detonated there six weeks later. Los Alamos did four more shots in the next ten days, completing the operation with a 22 kt airdrop on February 6th.

NEVADA TEST SITE ORGANIZATION (NTSO) FOR OPERATION RANGER

The organizational structure used for the Nevada Test Operations evolved over the years, as experience was acquired and as the requirements for test operations changed. A brief description of the key positions in the NTSO and the persons who held these positions during RANGER is discussed here and shown in Figure 2-1.1. While the beginning and ending dates for the test operations are generally given as the dates of the first and last nuclear tests, NTSO was active prior to the first test and for some time after the last test. RANGER was conducted on such a short time scale that it did not include a significant level of participation by the military in effects measurements and military exercises that existed during the subsequent atmospheric operations at the NTS.

The AEC and Los Alamos, and their contractors, with significant military support, supplied the infrastructure for RANGER. The General Manager of the AEC, Marion W. Boyer, appointed Carroll L. Tyler the Test Manager for RANGER on January 15,

1951. Tyler, manager of the AEC's Santa Fe Operations Office (SFOO), was the senior federal official in the test organization. He had overall responsibility for test operations, but not for the content of the technical programs. (AEC/Washington TWX to AEC/SFOO January 11, 1951)

The Test Manager selected a group of experts, called the Consulting Committee, for technical advice during RANGER. Alvin Graves, from the Los Alamos test division, was the chairman of this group, which was convened at the call of the Test Manager or the Consulting Committee Chairman, or Alternate Chairman. Meetings were concerned primarily with evaluating forecasts presented by representatives of the Blast, Fallout, and Weather Prediction Units. The committee made recommendations to the Test Manager to proceed with a scheduled event or to defer detonation for a specified number of hours. When a test was in its final day of preparation, the committee met every few hours to assess if weather conditions were favorable or unfavorable for the detonation. As a result of their deliberations the committee made a recommendation to the Test Manager to either continue preparations or to delay. If a 24 hour (or more) delay was called, they would reconvene on the next day (or days) and again begin the frequent meetings until zero time. The Consulting Committee also ascertained the readiness of participating scientific, technical, and support personnel for a specific detonation. (Tyler 1952: 18)

The Consulting Committee was basically what would later be called the Advisory Panel, and much later, the Test Controller's Panel. Committee members, who came from the AEC, the laboratories, and the military, were selected on the basis of their knowledge of nuclear testing. They were respected technical experts in their fields; and like Graves, many served on the advisory panels for many years.

The Consulting Committee for RANGER, in addition to Graves, included Brig. Gen. James P. Cooney and Shields Warren from the AEC; Col. George F. Taylor and Col. Benjamin G. Holzman from the USAF; Capt. Howard L. Andrews from the U.S. Public Health Service (USPHS); and Darol K. Froman and Thomas Shipman of Los Alamos.

Appendix F outlines the AEC and DoD senior leadership, advisory panels, and planning boards for RANGER and the subsequent operations thru HARDTACK II. Figure 2-1.1 shows the NTSO for RANGER and similar figures are given for the subsequent operations.

Tyler appointed SFOO Director of Personnel and Organization, John W. Macy, Jr., to serve as his Executive Officer, see Figure 2-1.1. (Tyler 1952:14-18) Macy spent more time at the site than Tyler and served as the senior federal official in Tyler's absence. Lt. Col. W. R. Sturges, from the DMA in Washington, was Operations Planning Officer, and Richard G. Elliott of SFOO was named Public Information Officer. Other Tyler appointments included: William J. McElwreath as Chief of the Security Group and Richard L. Kennedy as Chief of the Communications Group. (Tyler 1952:4)

Ralph P. Johnson was appointed Chief of the Administrative Services Group. He was borrowed from the SFOO Engineering and Construction Division and managed the newly formed AEC Las Vegas Field Office located at 817 South Main Street. (SFOO 1951: Press release; Tyler 1952: 14, 18-20)

Tyler also appointed Graves to serve as Chief of the Test Group (i.e. Test Director). Graves appointed John C. (Jack) Clark, J-Division Associate Division Leader, Deputy Chief, (i.e. Deputy Test director) on January 19, 1951. Operations, planning and execution of the technical programs were under the direction of Clark who was assisted by William E. (Bill) Ogle, who would soon become Los Alamos's Assistant J-Division Leader. In addition to his duties as Deputy Test Director, Clark lead a team that provided construction planning services to the test groups and acted as the liaison between Los Alamos and the SFOO offices of Engineering, Construction and Communications. Clark and Ogle were at the test site "Zero Control" (i.e., the Control Point or CP) for each test. (Graves 1951a: Memo)

During the nuclear tests, Tyler and Graves and their immediate staffs were usually at the test headquarters at Nellis Air Force Base, which at the time was just outside Las Vegas, Nevada.

Graves named Frederick Reines the Technical Deputy of the Test Group. Reines was at Los Alamos during the tests, to oversee reception of the experimental data and their rapid analysis in order to facilitate any potential modifications to the test sequence that the data might suggest. Reines kept Bradbury, MacDougall, and the Fission Weapons Committee informed of the results. (Graves 1951a: Memo; Tyler 1952: 22)

On January 19, two additional appointments were made by Graves. Gaelen Felt of LASL's J-Division was appointed Kirtland Field Representative of the Test Group. He served as LASL's liaison with the Air Force and flew on the drop missions in the strike aircraft. Thomas L. Shipman, who was head of LASL's H (Health Sciences) Division, was appointed Chief of the Radiological Safety Section of the Test Group.

Shipman's section directed the test site radiation safety organization, which provided on-site Rad-Safe services and off-site monitoring out to approximately 200 miles from the nuclear detonation. Radiation monitors were recruited from the following divisions at Los Alamos: H (Health), CMR (Chemistry and Metallurgical Research), P (Physics), and J (Test)-Divisions. In addition monitors came from AEC Area Offices and from the U.S. Corps of Engineers*. (LASL Memo: Graves to Shipman, *Delegation of Authorities and Responsibilities*, January 24, 1951) [*Footnote: Additional information on radiation levels and monitoring can be found in Barton C. Hacker's book *Elements of Controversy - The Atomic Energy Commission and Radiation Safety in Nuclear Weapons Testing 1947 - 1974* University of California Press, Berkeley /Los Angeles/London, 1994]

Monitors accompanied the cloud sampling aircraft and the scientific teams that recovered data from the experimental bunkers in the vicinity of ground zero following a shot. The B-29 debris collection aircraft operated out of, and were usually decontaminated at, Nellis AFB.

A. W. Kelly, of Los Alamos's J-Division, provided administrative services for Los Alamos, including travel and housing, support with fiscal matters, security clearances, and visitor programming. Harry Allen, of LASL's Supply and Property Department (SP), provided logistical support.

The Indian Springs AFB hosted the vehicle maintenance facilities as well as other facilities for personnel working in the forward area such as housing, messing, and office space.

The U.S. Air Force Air Weather Service (AWS), under the command of Brig. Gen. W. O. Senter, provided meteorological support. Major D. H. Russell served as the on-site Air Weather Officer for the nuclear tests.(Russell 1951:15-49)

The Air Force Special Weapons Command (SWC) at Kirtland AFB in Albuquerque, under the leadership of Brig. Gen. J. S. Mills*, provided the airdrop services, cloud tracking within 200 miles, and airlift of personnel and equipment between Albuquerque and the test site. [*Footnote: The SWC efforts, which were led by Mills for RANGER, represented a significant and successful involvement by the military. This would result in SWC and Mills being appointed to lead the military efforts on the next operation, BUSTER-JANGLE.] Captain E. Miller, reporting to Col. O. J. Ritland at Kirtland, was the SWC test-site representative. The Air Force Office for Atomic Energy under the leadership of Lt. Col. J. Cody provided cloud-tracking services beyond 200 miles.

LOCATIONS OF RANGER ACTIVITIES

Figure 2-1.2 shows the 12X30 mile area of the test site during RANGER and in addition identifies five separate and distinct locations that were occupied or used in support of RANGER: (1) locations within the city of Las Vegas, (2) Nellis AFB, (3) Indian Springs AFB, (4) the Control Point (CP) in the test area, and (5) the blockhouse at ground zero (GZ), which was not occupied at shot time.

Col. George Schlatter, who was assigned with the AEC, noted the following distances and 1951 car travel times between these locations:

a. "Las Vegas to Nellis - 8 miles northeast (from Las Vegas), 12-15 minutes by staff car.
b. Las Vegas to Indian Springs - 45 miles on U.S. black-top highway - 50 minutes by staff car.
c. Indian Springs to Turn Off (to the site) - 15 miles black-top - 20 minutes. Turn off to control point via reasonably good graded road 15 miles - 25 minutes.

d. Control Point to (Ground) Zero - 8.9 miles direct, 11-12 miles by graded road - 25 minutes."

(Schlatter 1951:1)

Las Vegas

Early in January the McKee Company leased a vacant garage building at 817 South Main Street in Las Vegas to serve as city headquarters. This building had a switchboard and space for Tyler, Johnson and other AEC security and service personnel. (Schlatter 1951: 2) These facilities were not large enough to include the Public Information Office, which occupied a room in the El Cortez Hotel on Fremont Street when the tests started.(Tyler 1952: 24)

Nellis Air Force Base

Without the active assistance and cooperation of the Commanding Officer at Nellis AFB, the test program would have been difficult, if not impossible, to accomplish prior to GREENHOUSE. The operation's headquarters, building 926, was located at Nellis, a large H type building that housed: briefing rooms, a communications center, operations for aircraft trackers, a cloud track plotting center, and the headquarters for the off-site radiological safety teams. Also, there were facilities for the weather detachment and additional motor vehicles. (Tyler 1952: 24) There was "plenty of room for administrative staff, telephone switchboard, latrine facilities and space for 20-30 bunks for emergency use if needed." (Fehner 2002: 65; Schlatter 1951: 2)

Indian Springs

Indian Springs AFB, a satellite station to Nellis, was utilized for housing and feeding of construction personnel and as a warehousing and maintenance center. After the construction phase these facilities were used by the operational personnel who worked at the Control Point (CP) or Ground Zero (GZ). Indian Springs was also the site for maintenance of vehicles used in the operation. (Tyler 1952: 24) "Housing consists of one-story wooden barracks in reasonably good condition with two mess halls, sufficient for maximum of 600 or more." (Schlatter 1951: 2)

Control Point

The CP for RANGER, was located a short distance north of the saddle between Mercury and Frenchman Flat, see Figure 2-1.3. It was 8.9 miles south of the GZs and about 4.5 miles west of the eastern boundary of the test site. It was about 8.5 miles north of the southern boundary of the test site, which was just north of what would become the Mercury area.* [*Footnote: In 1951, the test site did not include Areas 22 or 23, present day Mercury and Desert Rock]. The CP was located at an elevation of 3960', and provided a good view of target Ground Zero (GZ) at an elevation of 3135' and of the Frenchman Lake at 3080'.

Except for the blockhouse at ground zero, existing facilities were adapted for this first test operation. The CP was a surplus Los Alamos frame building that housed the control room, the hardware for starting the sequence timer, and the camera controls for the theodolite measurements of the bomb trajectory and burst location.

As a precautionary measure the CP building was shored up prior to the first blast. (Schlatter 1951: 2)

In the Control Room, Deputy Test Director Clark had contact with the drop aircraft from the time of take-off from Albuquerque. Subsequently, Clark had direct operational control of the aircraft when they entered the test area. (Schlatter 1951: 2)

The CP also housed Clark's office, the medical doctor and first aid station, a Rad-Safe briefing room, and an office for AEC on-site security. Conditions were primitive at best. The building had no corridors. To go from one room to another a person had to walk through the intervening rooms. There were portable toilets, but no hand-washing facilities. Hot dinners were trucked from the Indian Springs mess hall to the CP on D minus 1 evenings. Mid-day meals were box lunches from Indian Springs. The road from the main highway (U.S. 95, from Las Vegas to Reno) and all roads throughout the test site were gravel. They were occasionally watered to reduce the worst of the dust.

Water was trucked from Indian Springs. Electricity for the CP area was supplied by two 30 kw diesel-driven generators. Telephone service for the CP tapped into lines running parallel to highway 95. Radio communications operated with an antenna located atop a nearby mountain peak about a half-mile from the CP.

Ground Zero

On the test site, the only significant construction for RANGER was at GZ where a blockhouse was constructed that had been designed to withstand 10 kt 1050 feet overhead. It housed recording equipment for the alpha detectors. (Fission Weapons Committee January 15, 1951(SRD)) Although sturdy, the GZ blockhouse was designed as simply as possible so that the construction could be completed in time for the experimenters to install their equipment before the first shot. This, in fact, was accomplished; and the blockhouse and its electronics were operational for a dry-run airdrop on January 25, 1951.

There were photo stations NE and SE two miles from GZ. Additional diagnostic stations were to the west and north. Diesel generators, for electricity, were stationed two miles south of GZ.

TRANSPORTATION TO LAS VEGAS AND ITS GROWTH

Occasionally there would be chartered direct flights between Los Alamos or Albuquerque and the test site, but the vast majority of travel went by way of Las Vegas. It was a place where travelers would occasionally spend their first and/or last night of a trip to the site. When one had a long stay in the field, it was also the place to go for R&R or for personal supplies. Ultimately, the Las Vegas experience became an integral part of most NTS veteran's memories of nuclear testing.

Originally, Las Vegas, which means "the meadows" in Spanish, was a way station on the Old Spanish Trail. It is located in the center of one of the more spacious valleys of southern Nevada and reportedly possessed free-flowing perennial springs and extensive meadows. In 1855, Mormon settlers came to proselytize the local Indians and to raise crops to provision travelers. Until after the turn of the century, Las Vegas Valley contained little more than a few scattered ranches. In 1905, the San Pedro, Los Angeles and Salt Lake railroad came. Because of its location about midway between Los Angeles and Salt Lake City and because of the water supply, the railroad established maintenance facilities, laid out the town of Las Vegas, and auctioned off lots. In 1931, Nevada legalized gambling; and the same year construction of the Boulder (later Hoover) Dam, which would then be the largest dam in the world, began. The dam brought significant federal funding to Las Vegas. Even more federal funding came during the Second World War when the Army Air Corps established a gunnery and training base and the government built a giant magnesium plant south of town. After the war, the resort industry became the primary economic driver. (Fehner 2000: 17-20)

During the 1930s, casinos had been limited largely to Fremont Street in the downtown area. In 1941, the El Rancho Vegas opened on what became known as the Las Vegas Strip. The Flamingo Hotel was built next by Benjamin "Bugsy" Siegel, a member of the Meyer-Lansky crime organization. (Fehner 2000:20)
By 1950, the post war affluence, the growth of the resort industry, and "-- the creation of a reliable highway link with southern California, had pushed the population of the valley to almost 50,000." The 1950s would witness the growth of both the test site and of Las Vegas. "Las Vegas would become the primary bedroom community for workers who daily commuted to the site." (Fehner 2000:20)

Travel connections to Las Vegas were not particularly good in the early 1950s. Although Las Vegas was a "railroad town", Schlatter (1951: 3) describes connections as being "poor for our purposes".

Charter flights from Los Alamos or Albuquerque were the fastest, but they were not regular or frequent. When they did run they were generally conducted by Carco Charter and used Bonanza aircraft from Los Alamos, which had to refuel at Prescott, AZ, or 14-place Lockheed or Twin Beech aircraft from Albuquerque. TWA had one DC-3 flight per day each way between Albuquerque and Las Vegas. United and Western had connections through Salt Lake City. By air, a trip between Los Alamos and the site would take the better part of a day, while a traveler from Washington could count on killing at least a whole day. (Schlatter 1951: 3)

The AEC transferred several sedans, mostly from Los Alamos, to Las Vegas. They augmented these with Hertz "You-Drive-It" (as the 1950s commercials would say) vehicles when necessary. Trucks and utility vehicles were largely Los Alamos transfers. In addition, the Air Force, at Nellis AFB, loaned the AEC vehicles from time to time. (Schlatter 1951:3)

SUPPORT ACTIVITIES AND THE REYNOLDS ELECTRIC & ENGINEERING COMPANY (REECo*)

[*Footnote: REECo is an acronym for Reynolds Electric and Engineering Company which was also referred to simply as Reynolds.]

Lou Reynolds, the founder of Reynolds Electric & Engineering Company (REECo) started out as an electrical contractor in El Paso, TX, in 1923. The company was a subcontractor under Robert E. McKee, the prime contractor at Los Alamos from the very earliest days, and it was only natural that they were asked in late 1950 to become involved as the prime contractor in the preparation of the continental site for the first tests.

Harold D. Cunningham, the long-time REECo President and General Manager recalled that "Reynolds was in a joint venture agreement with Robert E. McKee and Brown-Olds, mechanical and Industrial contractors in which all of the craftsmen and supervisory personnel were brought from Santa Fe and Los Alamos areas because of the clearances that these persons had (from working in Los Alamos) and the types of capabilities that existed in the three joint venture organizations for doing the mechanical, electrical and structural work. The people were based at the Indian Springs Air Force base and the work was done in Frenchman Flat." Joe Lopez, an electrical engineer with Reynolds, was the General Manager of the joint venture group. (REECorder 1982: Vol. 14 No.2)

Over the years, REECo played a major role at the test site, responsible for construction and engineering as well as for housing, messing, and related personnel services. They were involved in all facets of engineering and construction in support of the nuclear testing program. The history of the site is, to a large extent, the history of REECo.

TECHNICAL ACTIVITIES AND TESTS

Technical teams from Los Alamos and EG&G started installing diagnostic equipment at the GZ blockhouse and at the other experimental stations in early January 1951. Clark and his staff arrived at the site on January 10 and remained through the roll-up in February.

On January 19, Graves assembled a committee to assess the advisability, from a radiological safety perspective, of doing airdrops. (Graves 1951a: Memo) This was one of many advisory panels that were formed during the years to advise the test site leadership on specific technical matters. Some of these panels were short lived for a specific issue; while others, such as the Weather Panel, continued in one form or another throughout the entire period of nuclear testing. These panels were important to the management, and the people who served on them often played important roles over many years in various areas of nuclear weapons testing.

Parts for the nuclear devices were trucked from Los Alamos to the Sandia Laboratory, in Albuquerque, where the final assembly was completed. The drop case was delivered to the aircraft loading area at Kirtland AFB approximately an hour before departure. Los Alamos group W-1, under the leadership of W. C. Bright, was responsible for the nuclear components. A Sandia team under the direction of Walt Treibel incorporated the nuclear package into the drop case and installed the remainder of the non-nuclear components. Gaelen Felt, the Los Alamos representative at Kirtland, was responsible for overseeing the loading of the proper device and the arming of the weapon aboard the drop aircraft while it was in flight. (Perkins 1992: 105)

Favorable weather predictions for shot time were required for takeoff from Albuquerque. The flight plan called for arrival near Indian Springs approximately two hours before drop time. The aircraft then proceeded at 10,000 feet to a region north of the drop point where Gaelen Felt armed the device for the shot. The pilot then took the aircraft to the altitude designated for the test, 22,840 feet MSL, for the first four shots and 30,000 feet MSL for the fifth event. The drop aircraft then conducted practice runs until notified that the shot had been approved. Shot times were planned for just before sunrise. Darkness was desirable for photographic reasons; but sunlight was needed so that the resulting cloud could be tracked. After shot approval, the crew began the final run for bomb release by the bombardier using an optical bombsight. The bombardier who made the five test drops during RANGER achieved a maximum horizontal deviation of 490 feet. The observations were designed to allow for a deviation as great as 1000 feet. (Fission Weapons Committee January 15, 1951(SRD); Perkins 1992: 106)

In his report on RANGER, Carroll Tyler stated, "The aircraft bombing crew furnished for the job conducted every drop and turned in a splendid performance." (Tyler 1951: 57) A B-50 with photographic equipment accompanied the drop airplane. In addition, a C-47 with a disaster team of 10 and a crew of 4, also from the 4925th Special Weapons Group, followed the drop aircraft to the Las Vegas area and flew holding patterns during the events. After completing a mission, the aircraft returned to Albuquerque. (Maag et. al. 1982: 51-55)

The drop cases had radar fuzes called Archies. Gaelen Felt asked Donald Cotter*, from Sandia, in collaboration with J-Division, to investigate the ability of the Archies to consistently detonate the devices at an altitude within 50 feet of 1050 feet above the terrain. [*Footnote - Cotter would later fill various senior positions in the AEC and DoD, where he was Assistant to the Secretary of Defense for Atomic Energy - ATSD/AE - from 1973 to 1978.) Ground zero was approximately 3140 feet above sea level, and the detonations were at 4190 feet above sea level. (Fission Weapons Committee January 15, 1951(SRD)] For the 5th and last shot, Fox, the planned detonation altitude was increased to 1350 feet above the terrain in order to compensate for the higher yield and to reduce the damage to the blockhouse at ground zero.

The target for the bomb drops was identified as a cross of lights extending 100, 300, and 500 feet from ground zero in NE to SW and NW to SE lines. Ground Zero (GZ) was also identified with a red light.

The test execution was conducted from the Control Room, located in the Control Point (CP), with Deputy Test Director Jack Clark in charge. W. E. Ogle, in addition to being the leader of the Scientific Test Group, was a technical advisor to Clark. H. Grier and B. J. O'Keefe, of EG&G, oversaw the operation of the timing signal equipment located in the room adjacent to the control room.

At the CP, Captain E. Miller served as the liaison between Clark and the SWC and the SAC aircraft in the operation. Captain Robert Smith provided liaison with the aircraft. Sandia's E. A. Aas operated the release tone radio receivers. The bomb release initiated a radio signal ("release tone") from the drop airplane that was received in the control room. This radio signal actuated a relay, which set the sequence timer mechanism in operation, and ultimately produced the time signals required by the diagnostic experiments. Sandians, E. J. Klink, and an assistant operated the theodolite camera control equipment for the determination of the burst location.

On the morning of the day before the shot, ten vehicles with two-man rad-safe monitoring crews were dispatched to the small communities in the region that might be exposed to the debris cloud. At 1:00 pm of the day before the shot the Test Manager and the Test Director held a weather briefing at Nellis AFB. If the weather was favorable, Test Director Graves notified Clark at the CP; and he in turn notified the Weapons Assembly Group at Sandia and all other operational and experimental units to proceed as scheduled. Another weather briefing was held the same evening at 8:00 pm to review the weather situation. If the weather still looked favorable the Test Manager made the decision to proceed. The Test Director notified the Deputy Test Director, and the "execute" order was put into effect. If the weather was questionable, the execute procedure was continued, but another weather meeting was held at Nellis at 3:00 am on the day of the shot.

Two hours prior to shot time, AEC security guards at the forward area check posts determined that the area was clear of all personnel by means of access lists and reported this information to the Deputy Test Director at the CP. At this time all test personnel who may have been working at their experimental stations in the target area were to be out of the area. At 4:00 am, which was usually an hour and a half before shot time, the CAA closed the airspace in the vicinity of the test site.

During the period from two hours before shot time to shot time, rad-safe briefings of the recovery teams were taking place at the CP. As soon as the detonation occurred, the Rad-Safe operations started. Upon agreement between the Rad-Safe Officer and the Deputy Test Director, monitors and recovery teams moved into the target area. This generally got started at approximately plus 20 min and continued through D-day. Simultaneously, the cloud-tracking and -sampling operations started. After completing the mission the sampling aircraft landed at Nellis AFB and the samples, along with neutron foil samples from the site, were dispatched to Los Alamos via courier aircraft supplied by SWC.

The Data Analysis Group for Operation RANGER was in J-Division at LASL, under the direction of Frederick Reines. Data were transmitted to this group from the test site as soon as they were available. Since the order of firing was dependent upon results of previous detonations, the analysis for each shot had to be completed for presentation to the Fission Weapons Committee by 7:00 PM on the day before the next shot. The Fission Weapons Committee then recommended to Norris Bradbury which of the remaining devices should be fired next. (Clark 1953, 57-58)

A complete dry run was executed on January 25, 1951; two days before the first nuclear shot. The strike aircraft, manned by the crew that was scheduled to fly the nuclear missions, dropped a bomb with inert nuclear components. The dry run proved to be quite useful to the test teams. The Rad-Safe group in particular uncovered a number of deficiencies that were at least partially addressed prior to the actual events. The main problem was in the communications area. The quickly installed communications system relied on line-of-sight, which is subject to geographical limitations. For the preparation of this first test series, there just was not time available to install a more robust communication system.

Table 2-1.1 Nuclear tests conducted during BUSTER-JANGLE.

TEST	DATE-1951	TYPE: (*ft)	AREA	YIELD (kt)
Able	Jan 27	Airdrop 1060 ft	5	1
Baker	Jan 28	Airdrop 1080 ft	5	8
Easy	Feb 1	Airdrop 1080 ft	5	1
Baker-2	Feb 2	Airdrop 1100 ft	5	8
Fox	Feb 6	Airdrop 1435 ft	5	22

* feet above surface
(Lewis 1977:16; DOE 2000: 2-3)

The first nuclear shot, Able, was detonated at 0545 PST, January 27, 1951, at 1060 ft above GZ, within 100 feet of the intended target. (Fission Weapons Committee January 27, 1951(SRD)) The shot went very smoothly, and there were no operational or experimental problems. The yield was about 1 kt and the ground zero radiation level at plus one hour was deemed sufficiently low to permit the experimental teams to quickly set up for a test on the following day. The radioactive cloud rose to an altitude of about 17,000 feet above sea level and traveled on an easterly trajectory.

Baker was detonated on January 28th at 0552 PST at an altitude of 1080 ft. The yield was approximately 8 kt. The radioactive cloud rose to 35,000 feet on a generally easterly trajectory, with upper level shear. Unfortunately the high level of radiation discouraged work in the vicinity of GZ. This, along with the grueling schedule over the previous week, convinced the test leadership to call for a two-day break to rest and regroup.

The third shot was scheduled for January 31, but weather delayed the event by one day. The prediction was favorable at the 8 PM January 31 weather briefing and permission to proceed was granted for a February 1 shot date. Easy was fired at

0546 at an altitude of 1080 ft. The yield was approximately 1 kt. The radioactive cloud rose to an altitude of 16,000 ft over a generally southeasterly trajectory. There was some low-level wind shear that took a component of the cloud over Death Valley and southern California.

The fourth shot, Baker-2, was fired February 2nd at 0549 PST. The detonation point was 1100 ft above the terrain. The yield was 8 kt. The cloud rose to 25,000 ft on a southeasterly trajectory with substantial wind shear. There was some destruction at ground zero from this shot. The entrance to the alpha station was damaged, and a bulldozer was required to clear a path to the entrance after the shot. This, coupled with the fatigue experienced by the test teams, convinced Jack Clark and his colleagues to call for a three-day delay before the fifth shot. It was scheduled for February 5. The weather was favorable, but the strike aircraft experienced an engine oil leak about 16 miles west of Albuquerque, and the mission aborted until the next day.

Fox was fired at 0546 PST on February 6th, at an altitude of 1435 ft above the terrain. The yield was 22 kt. The cloud rose to about 43,000 ft on a southeasterly trajectory.

Because of the damage from Baker-2, the ground zero for RANGER Fox had been moved 500 feet due west of the alpha blockhouse. It was felt that this move would reduce additional damage to the blockhouse at ground zero without degrading the diagnostics. Bill Ogle had an interesting anecdote regarding the target move. (Ogle 1985: 64)

> As the operation went on at approximately one shot per day, the (alpha) detectors were gradually destroyed and for the last shot could not be replaced without overdosing personnel appreciably. To solve this problem, the field team at Frenchman Flat, namely John (Jack) Clark and William Ogle, simply moved the lighted target array to the one set of detectors that was still operating properly, in order to increase the probability of getting a signal. Since the bombers were bombing on the lighted array, it did not occur to the field team that anyone else could possibly care about this movement, so no notice was given to the Air Force or the Test Manager and Scientific Advisor, who for Ranger were in Las Vegas. ... However, some three days after ... (shot F) the reporters, in a normal press briefing, inquired of Alvin Graves, who was the Scientific Advisor, as to whether the target had been moved. He commented 'No,' and a few hours later asked Ogle the reason for the question. The answer was that it had been moved but notice of that fact had not been considered important. Graves was extremely embarrassed and from then on, rejected the philosophy that the Test Manager and Scientific Advisor could be physically separated from the rest of the technical organization in conducting an operation.

TEST DIAGNOSTICS

Appendix G entitled "Diagnostics Experiments" provides descriptions of the array of technical experiments fielded during the atmospheric era. The main experiments conducted on RANGER were: yield, alpha, transit time, and detonation location. (Reines 1952:Vol.2; LASL 1952:Vol.4; Perkins 1992: 114-129)

Yield

Yield was primarily measured in three ways: radiochemistry, fireball growth, and time to first minimum in the light intensity (Bhangmeter Experiment).

Los Alamos radiochemists from group J-2 under, the direction of R. W. Spence, conducted yield analyses on samples obtained from cloud samples. (Perkins 1992: 120) The United States Air Force took samples of the material in the radioactive cloud as part of their program for detection, analysis and interpretation of bomb debris. They sent their samples to the McClellan Laboratory of Tracerlab, Inc. The British Long Range Detection Group also analyzed radioactive cloud samples.

The rate of growth of the fireball was measured photographically by EG&G. The yield deduced from rate of fireball growth was expected to be accurate to better than ten percent. (LASL Fission Weapons Committee January 15, 1951(SRD); Houghton 1951: 58-73)

The yield measurement from the time of first light of the explosion to the first optical intensity minimum (the Bhangmeter measurement) was performed by EG&G under the direction of Capt. R. A. Houghton. Also, B. Brixner using a FASTAX camera measured the time to the first minimum. Ogle explained the origin of the name of this diagnostic technique:

> An intense afternoon was spent by the entire Group J-7, with its Group Leader Fred Reines, early in 1950, picking a name for this world-shaking device that was going to produce simple, cheap, and easy yield measurements. At the end of the afternoon, Reines picked a name which we all knew would be misinterpreted for the rest of history. Bhangmeter is not synonymous with bangmeter. Bhang is a variation of Indian hemp, the leaves and seed capsules of which are chewed or smoked, and which then produce the same euphoria as other variations of hashish. The now obvious connotation is that we were off our rockers to think that this thing would ever be particularly useful and anyone else who ever believed it must also have a little something wrong with them. (Ogle 1985, 67)

In addition, yield was measured by foil activation, using gold and arsenic. It was also deduced from photographic film measurements of the gamma-rays as a function of distance.

Alpha
A Los Alamos team led by Robert B. Patten measured alpha with Rossi ionization chambers. The Naval Research Laboratory (NRL) measured alpha with scintillation detectors. See Part I-Chapter 1 for a description of these measurement techniques.

Gamma-Rays
Two groups made gamma-ray measurements at different distances from the device. Film experiments were used by the first group composed of individuals from different parts of the LASL, including military personnel who were on assignment to Los Alamos. The second group, Los Alamos Group J-1, under the direction of Ellery Storm, used detectors.

Neutron Measurements
Los Alamos Group J-3 under Bill Ogle and Clyde Cowan measured neutrons.

Transit Time
Don Schuster of Sandia and Newell Smith of Los Alamos measured the transit time, which is approximately the time between the firing pulse to a detonator and the emergence of the first gamma rays from the device.* [*Footnote: Transit time had different meanings depending upon the context. Technically it was defined as the time between the signal which was sent to the detonator and 50 neutron generations. Some calculations were required to relate the measured time (firing pulse to first gamma rays) to the technically defined transit time.] (Reines 1952: 48) Shuster, supervisor of Sandia's Special Project group, developed a new technique for the measurement --- the use of radio telemetry. The crew installed the radio telemetry in the weapon's case and the monitoring system in the rear end of the drop aircraft. The monitoring system, which consisted of high-speed oscilloscopes, recorded the time interval between the firing pulse and the detection of the first gamma rays. (Furman 1990, 575-576)

Detonation Location
An EG&G team, under the direction of E. J. Klink and R. I. Liebman, tracked the bomb drop with phototheodolites*.[*Footnote: Two cameras are located at exactly the same altitude and take pictures. From differences in the two pictures taken at the same time, locations in three dimensions can be determined.]

Another EG&G team, led by Herbert Grier, determined the actual burst location. This was done optically, with an accuracy of about 20 feet, using two photo trucks. (Reines 1952: 37)

Airblast
The Los Alamos GMX-9 group fielded gauges to measure pressure in the air as a function of time. The gauges were located in the vicinity of the control point, about 8.9 miles from the detonation. Two records were obtained, one from the Baker-2 test and the other from the Fox test. In addition, there were several informal observations made by spectators near the CP of multiple blasts. There were also personal observations of multiple blasts in Las Vegas, about 70 miles away. A shop

window was broken, and the Desert Inn hotel reported cracking of windows and plaster from Baker-2. Interestingly, the first Baker caused no damage. Sandia was assigned to study what was known as the “skip-zone phenomenon”. (Furman 1990, 603-4) These anomalous blast effects would be studied in considerable detail by Sandia and others in order to provide guidance about meteorological conditions that would minimize the effects of the blast on Las Vegas.

Other Experiments and Measurements

Enrico Fermi and Richard Garwin raised an interesting issue regarding the possibility of a pre-detonation in a hypothetical scenario: “The detonation of two or more fission bombs within a few minutes is subject to restrictions imposed by serious pre-detonation of succeeding bombs by ... the delayed neutrons from the first.” David and Jane Hall, along with W. D. Schafer, confirmed that this could indeed be an issue of concern. They measured neutrons minutes after a detonation and found that it was large enough to cause a pre-detonation of a second near-by fission bomb. (Garwin 1950: 13; Hall and Hall 1952: 183-200; Perkins 1992: 132)

Louis Rosen, from Los Alamos, used the RANGER series as an opportunity to test neutron collimators in preparation for Operation GREENHOUSE. (Perkins 1992:1333)

DOD NUCLEAR WEAPONS EFFECTS (NWE) MEASUREMENTS, OPERATION HOT ROD

RANGER was conducted with such haste that the DoD did not have time to organize programs with more complexity than film badges and exposure samples. However, such relatively simple programs provided important and useful information and continued in some form on essentially all of the subsequent atmospheric test operations at the site.

A DoD army group, assisted by Sandia Lab personnel under the direction of Lt. Colonel Merwin Forbes, measured the radiation in foxholes with film badges. The film badges were placed at various depths in the foxholes. Sandians also assisted the military by setting film badges in foxholes at various distances between 0 and 6000 ft from GZ. They measured both initial gamma radiation and residual radiation. The badges for initial radiation were placed in “mousetrap gadgets”, which had thick lead walls and doors that closed after the initial exposure. (Furman 1990, 576-577)

About six army and navy agencies conducted measurements of thermal effects on a variety of materials that were placed on panels at various distances from GZ. (Steadman 1952: 194)

There was general interest in knowing how much protection a car might provide during a nuclear attack. To address this issue, the AEC Division of Biology and Medicine, under the direction of Walter Claus and Joe Deal, conducted a program codenamed Operation HOT ROD on Fox whose yield was 22 kt. Five sedans of 1936-1939 vintage were fielded at various distances from GZ:

½ mile	Buick 4 door sedan	front windshield toward blast
1 mile	Oldsmobile 4 door sedan	45° to blast with front-side window opposite the blast open
1 ½ miles	Chevrolet 2 door sedan	60° to blast
2 miles	Lafayette 4 door sedan	windshield facing blast
2 ½ miles	Plymouth 4 door sedan	side toward blast

Except for the Oldsmobile, all sedans had their windows closed. The results were:

- At ½ mile, individuals would probably be "killed twice" by radiation and by a combination of blast and fire.
- At 1 mile, there was a danger of injury from radiation or a possible fire. Unless fires were started, individuals would probably survive.
- At 2 miles or more, the chances of survival without injury were very good. (Claus and Deal 1952, 249-258)

Personnel from the Desert Game Range observed that there were no effects on cattle 8 miles from GZ. (Perkins 1992: 134)

FALLOUT

The Oak Ridge National Laboratory (ORNL) was asked by the AEC Division of Biology and Medicine on January 17 to participate in measuring fallout from RANGER. An extensive program was conducted with:

- Air filtration and precipitation collection stations at 5 sites from Kansas through the southeastern United States. (Davis 1951: 39)
- Hanford air-filter monitors located in CO, NM, and UT, and
- UCLA had monitors located in the western U.S.

Also, Dr. A. K. Chapman of the Eastman Kodak Company reported "relatively high activities in the snowfall at Rochester, N.Y. on January 29, 1951." (Betty Perkins, private communication) The Canadian Chalk River Laboratory also found fallout in snow near Ottawa.(Hacker 1994: 43-52)

AIRDROP TECHNOLOGY

There were plans to develop the technology associated with airdrops for weapons R&D tests on Operation GREENHOUSE. Since airdrops had been successfully demonstrated on RANGER, that part of the GREENHOUSE program was cancelled; and all four GREENHOUSE shots were fired on towers.
(Perkins 1992: 161)

Airdrops were easier, and hence faster, to field than tower shots. However, there are serious downsides to airdrops. The first is safety. A device conceivably could be dropped in the wrong place. Shot Able, the airdrop on Operation CROSSROADS, missed the target by over 700 yards. The Cherokee airdrop, during the 1956 Operation REDWING at Bikini, missed the intended target by miles. Conservative safety protocols, however, ensured that no one was injured.

Another concern regarding the safety of airdrops focused on the reliability of the fuzing. If the fuze failed and the bomb detonated on contact with the ground the fallout of bomb debris and entrained dirt would be much worse than that planned from the airdrop. The critical point here was confidence in the reliability of the fuzing system*. [*Footnote: In the Pacific, the unintended detonation of a large nuclear explosion on or below the surface of the open ocean could cause a serious tsunami problem.]

An additional problem with airdrops is that the experimenters did not know the exact detonation point prior to the test. This, of course, “complicated” the diagnostic and effects measurements. In some cases, the whole project could be lost on a shot; in others the amount of good data could be seriously lessened. Fortunately, the Air Force did a superb job of delivering the bombs to the right spots on RANGER. With the Air Force state-of-the-art bombsite and radar all RANGER devices were dropped very close to the intended target. (Clark 1953: 58-65; Perkins 1992: 106)* [*Footnote: Froman and Bradbury in an October 2, 1951, letter to Col. Kenner F. Hertford, DMA, emphasized the importance of an accurate airdrop capability to the military. Accurate bombing techniques with low yield weapons provide point target damage comparable to that obtained from near misses with high-yield nuclear weapons. (Clark 1951: 162)]

CONCLUSIONS AND LESSONS LEARNED FROM OPERATION RANGER

The RANGER test series was a resounding success, both operationally and experimentally. It went particularly smoothly considering the short time available for preparation. Almost everyone who commented seemed pleased. “A story from the Associated Press stated, ‘The Nevada test explosions will have the effect of putting Russia on notice that the United States is confident of its atomic weapon lead.’ The story noted that the tests could have high political as well as technical value.” (LANL archives February 1951, A-92-016)

Technically RANGER was extremely important. Substantial progress was made in improving weapon designs that made more efficient use of special nuclear materials (SNM, i.e. uranium or plutonium). More significantly, the uncertainties associated with the Operation GREENHOUSE test design predictions were largely resolved. One specific accomplishment was the verification of the performance of low-yield nuclear explosives. The 1 kt design first tested in RANGER would be employed in the surface and subsurface weapons effects tests being planned by the DoD for Nevada in the fall of 1951.

The test series satisfied the public safety requirements of the day, was cost-effective in terms of time of the technical and support personnel, and required a minimum of military resources. Also, the planning and execution time was very short compared with that required for a test series in the Pacific. From a weapons design perspective, having the capability to test in the continental United States often meant that the testing would not be the limiting factor in the development of new weapons and that more rapid progress could be made than would be possible if the Nation were dependent upon testing only at Bikini or Enewetak.

Although the yield limitation of the Nevada Test Site was thought to be in the 50- to 75 kt range, it was clear that Los Alamos viewed the site as having more potential than just as the emergency test facility suggested by the September 15, 1950 AEC/DMA communication to Tyler. This issue was revisited when the role and future of testing in Nevada was reviewed in 1953.

The feature that caused some concern was the larger than expected airblast pressure and the multiple blasts observed in Las Vegas. Clark, in his post-operation report, said, "These anomalous blast patterns were attributed to refraction and reflection of the blast wave, refraction due to upper wind shear, and reflection due to temperature inversion in the air structure between the target and Las Vegas." (Clark 1953, 65)

Jack Clark also observed, "Operation RANGER was not exactly an experience that the test personnel would like to repeat once or twice each year. The fine working spirit of the test personnel was able to survive the handicaps of minimal working conditions and the condensed shot schedule only because of intense interest both in the test results and in exploring the possibility of establishing a 'backyard' testing laboratory within the continental limits."

Clark went on to conclude that RANGER did demonstrate that on-continent testing was feasible and that the main question now was the extent of future development of the infrastructure at the test site. Clark offered four recommendations: (Clark 1953, 68-69)

1. The test schedule should provide for a four- to six day preparation and rest period between detonations. Only for the very simplest of tests can the test and operational personnel efficiently perform their duties on a schedule such as was used for Operation RANGER; namely, five shots in ten days.
2. Permanent facilities should be established at the test area: Several test areas, including underground and shockproof instrument stations, with permanent power, signal, and communications facilities. These test areas should be in Yucca Flat, and preliminary surveys indicate that as many as eight such areas can be properly located, although for the next test series perhaps not more than three areas need be established.

   An operational control building which would serve as a test Control Point, a communication center, a weather center, and a Rad-Safe operation center which would have facilities for a technical machine shop and technical-data-reducing laboratories and offices.

   A radiological safety building located near the control building, provided with the facilities required to carry out all the Rad-Safe operations.

A permanent housing facility for test personnel which would include dormitories, messing, recreation, PX, first aid, etc. The number of test personnel to provide for is highly questionable, depending entirely upon the participation of the DoD, the Federal Civil Defense Agency, and personnel other than weapon-development-test personnel. This housing facility should be located within the test-site boundaries in order to minimize travel of test personnel.

A permanent set of utilities for the test site, including electrical power, water, and communications.

A blacktop road system throughout the test site, connecting the test areas, control and Rad-Safe buildings, and the living area.

Permanent facilities for assembling nuclear devices within the test-site boundaries.

3. It is recommended that as many as possible of the services which are required by the test organization be provided by an agency of the SFOO, probably a field office established at Nevada. In general, such services include transportation within the test area; living facilities; construction; security; communications including radio, telephone, telegraph; and handling of official visitors to witness tests.
4. During Operation RANGER, women participated in the test operations as secretaries, technical assistants, scientists, and telephone operators, and their services proved very valuable. It is therefore recommended that provision be made in all permanent facilities for the presence of women at the Nevada Test Site.

Conclusions (Clark 1953, 69):

1. The scintillation detectors for alpha measurements were 'proved in' during this test series, simplifying considerably the interpretation of neutron multiplication rate data as compared with the Rossi chamber technique previously employed.
2. The use of manned aircraft rather than drones for cloud sampling greatly simplified air operations.
3. The devices detonated in RANGER were well suited for the examination of fallout and blast effects.
4. The fact that all the devices tested were airdropped, along with the streamlined experimental program, made it possible to detonate the five devices in a minimum overall time.

OBSERVATIONS ON THE SIGNIFICANCE OF OPERATION RANGER

From an operational perspective RANGER was extraordinary. The test phase lasted only ten days. In comparison to the following operations, RANGER was, in retrospect, a very modest effort.

On the other hand, there are several very significant features of this test series. The first is the very rapid response. RANGER was fielded less than six weeks after the barren site was transferred to the AEC. The plans and devices for the tests were developed over a period of a few months, in the fall of 1950.

The second observation is that Jack Clark, the deputy test director, was incredibly successful in his anticipation of the requirements for future operations in Nevada. Essentially all of his recommendations were followed; and the site, including the support camp, Mercury, developed just the way that Clark visualized in 1951.

The third observation was that the very short lead time for this first operation precluded much participation by the DoD, or any others who would eventually engage in the exploration of the effects of nuclear weapons. The effects community made up for this in later operations where they outnumbered those working on weapons development programs. Also, military maneuver exercises were possible in Nevada, but not at the Pacific Proving Grounds. These were the Desert Rock exercises of subsequent operations and involved thousands of troops. Clearly, the Nevada site was a most valuable facility for DoD activities.

Finally, the test site in Nevada proved to be very successful. The Los Alamos concept of a "back yard" test area proved to be immensely important over the next four decades. So successful, in fact, that shortly after GREENHOUSE (in April and May 1951) the testing community geared up for a follow-on Nevada test series named BUSTER - JANGLE for the fall of 1951. Los Alamos was soon to learn that this "back yard" would not be private.

Nevada Test Site Organization (NTSO) for Operation Ranger

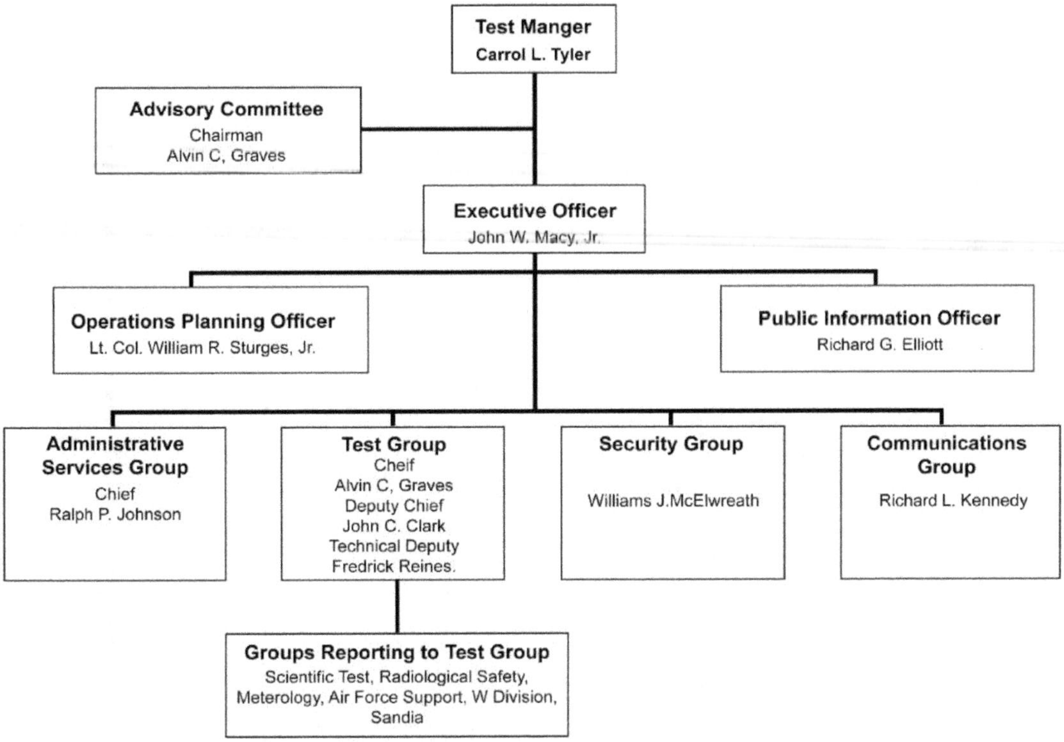

Figure 2-1.1 Nevada Test Site Organization (NTSO) for Operation RANGER

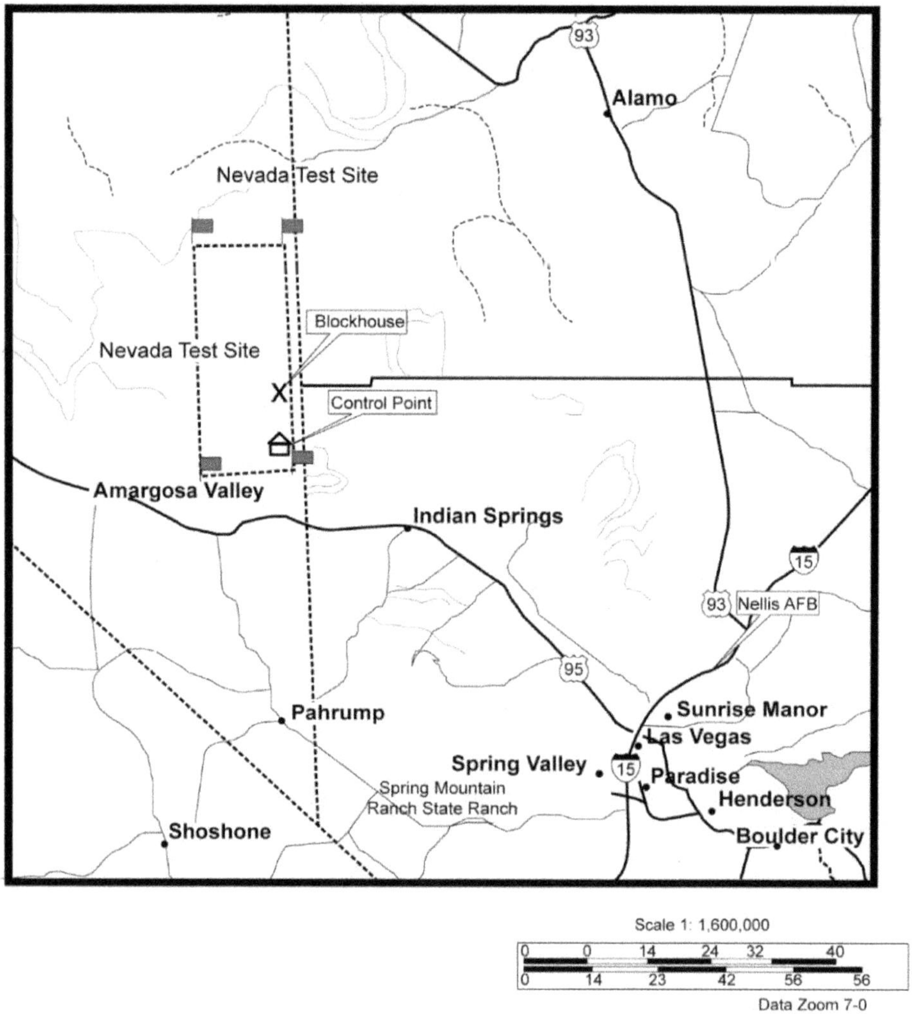

Figure 2-1.2 Location of RANGER at the site.

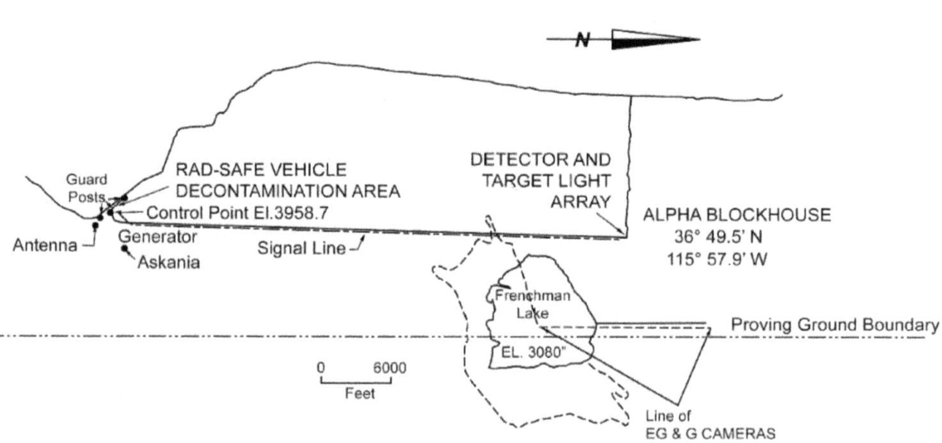

Figure 2-1.3 Location of RANGER tests, Control Point, and cameras.

Block House near Ground Zero where recordings were made.

Control Point (CP) building, moved from Los Alamos.

Instrument room in Block House.

Buick four-door sedan placed at one-half mile from ground zero, with windshield oriented toward the blast. All windows were blown out, as was the rear of the car. The doors away from the blast were blown off their hinges, and the hood was blown some 50 to 100 yards from the car. Burning of the automobile was extensive. The rear tires were burned, and the car sank into the ground to the axel level. The front tires were undamaged and still inflated. The motor appeared to be undamaged. Source: Los Alamos National Laboratory.

Oldsmobile four-door sedan placed at one mile from ground zero, oriented at about a 45-degree angle to the blast. The windows on the blast side were broken. One was blown in and the other badly crushed. The windshield was cracked. The paint and tires on the blast side were charred, but the tires remained inflated. The side facing the blast was bashed in. The hood was lifted but not blown off. Apparently the door on the blast side had been left open, because there was a sharp line of demarcation of charred area visible on the upholstery. The motor seemed undamaged, as was the battery, given that the horn still operated. Source: Los Alamos National Laboratory.

Chevrolet two-door sedan placed one-and-a-half miles from ground zero still burning four hours after the shot. Oriented at about 60 degrees from the blast, the car was completely burned. The glass was destroyed as a result of the fire. The headlights were not broken, and the chrome was not charred. The top was warped. The front tires remained inflated and intact. Source: Los Alamos National Laboratory.

OPERATION HOT ROD

CHAPTER 2. OPERATION BUSTER-JANGLE: OCTOBER 22 - NOVEMBER 29, 1951

INTRODUCTION

The availability of a continental test site in Nevada permitted a tremendous change in the how and why of nuclear testing. "The RANGER series ... represented a new philosophy, essentially, in the conduct of laboratory work, since this was the first occasion on which nuclear tests were carried out chiefly to study the properties of atomic warheads and were not simply proof tests of systems which had been exhaustively studied in other ways."(Schrieber 1953: 3)

The changes in thoughts and needs after RANGER resulted in the Los Alamos Test Division (J-Division) becoming a permanent organization.

LOS ALAMOS' FIELD TEST (J-DIVISION) BECOMES PERMANENT

Before the summer of 1951 LASL's test Division, J-Division, largely consisted of the leadership, administrative and engineering cadre of the nuclear test program. The scientific groups were drawn from the other divisions at the Laboratory, and also from Sandia, for the express purpose of executing a particular test series. After the operation the groups were disbanded and the group members returned to their original divisions. With the availability of the Nevada Test Site, it was clear that testing was a full-time program requiring a full-time staff. In July 1951 J-Division was reorganized and groups were formed on a permanent basis, rather than on a campaign basis. The test series, however, continued to be run as campaigns for another decade.

RANGER TO BUSTER-JANGLE

Immediately following RANGER, which centered on Frenchman Flat and relied upon the hospitality of the Air Forces installations at Indian Springs and Nellis AFB, the AEC decided to establish permanent facilities at the site for the whole operational crew, including space for the Air Operations Center and the Communications Center. Plans were formulated and set in motion to develop four regions of the test site, see Figure 2-2.1:

1) Yucca Flat - north of Frenchman Flat and north of the usually dry Yucca Lake bed, was chosen as the area for most of the testing and was called Target Area 7.
2) A Control Point (CP) was constructed at its present location, which is the saddle between Frenchman Flat and Yucca Flat.
3) Housing and warehousing areas were built at what is now Mercury, which was just south of the site.
4) The development of Camp Desert Rock was started, also outside of the southern boundary of the site, near the present air strip. Camp Desert Rock housed the troops that participated in military exercises.

PERMANENTIZATION
The first BUSTER-JANGLE (B-J) shot was scheduled for October 3, 1951 (later changed to October 17). Soon after RANGER, planning and construction was started for: Target Area 7 at Yucca Flat, roads and utilities, the Control Point building, the rad-safe building, weapons assembly facilities and a camp for billeting personnel. After the layout plans were completed, barely six months remained for carrying out the large construction program. (Clark 1951: 60-61)

The building program at the site between RANGER and BUSTER-JANGLE and between BUSTER-JANGLE and TUMBLER-SNAPPER was called "permanentization." Examples of the work conducted can be found in press releases from that time. These releases were advance notifications of upcoming requests for bids by the AEC, the results of the bids, and identification of the successful bidders.*
[*Footnote: Some examples of construction that were mentioned in the press releases regarding the permanentization work in 1951 are:
- 17 miles of black top asphalt roads and 35 miles of additional roads.
- 125 miles of fencing.
- Production of concrete aggregate.
- Production of paving mix.
- A concrete reservoir, two pump houses and about 15 miles of water line; 10 miles of power line and about 10 miles of signal line.
- A group of one-story wood frame buildings and other facilities, including housing, messing, and administrative facilities for the 400 to 600 person support camp.
- 3 one-story reinforced concrete buildings having about 14,000 square ft of area.
- A power house.]

A September 19 press release noted: "For the past several months between 300 and 400 construction workers have been employed, although a brief peak period being reached this month will account for 1,200 workers. All work has been done by private firms under contracts with the Commission." It further mentioned that although much of the planned construction remained to be completed, construction had progressed sufficiently by mid-August for the initiation of a continuing series of conventional high explosive blast wave experiments connected with meteorological studies. In late August, conventional high explosive for cratering and for BUSTER-JANGLE were also begun.(Opennet 322845)

REAL ESTATE
As described in Appendix D, the specific dates associated with the process of land acquisition by the AEC for the NTS were often somewhat ambiguous. The previous 2 sections describe the permanetization development conducted by the AEC in early 1951 in areas beyond the 12 X 30 mile area agreed upon with the Air Force on December 21, 1950.

After Operation RANGER, the AEC decided that they needed more testing space and an additional buffer and distance from Las Vegas. Subsequently, the AEC presented the Air Force with a request for a total parcel of approximately 16X40 miles. In a letter dated June 8, 1951 (AEC141/12) the Air Force agreed to make the 16 X 40 mile area available. (AEC DMA 1952: 1)

Subsequently, on October 25, three days after the start of operation BUSTER, Gordon Dean, Chairman of the AEC, wrote to Oscar Chapman, Secretary of the Interior. Dean specified the longitude and latitude of the real estate parcel under consideration for the 16X40 mile area, see Appendix D. With the additional land the new site was nearly double the size of the original 12X30 mile area. Dean requested that ... "an appropriate land order be issued withdrawing the land ... from all forms of appropriation ..., including the mining and mineral leasing laws, and reserving it for the use of the Atomic Energy Commission. It is also requested that the Atomic Energy Commission be permitted to continue its occupancy and use of the land pending the issuance of such an order." The last paragraph of Dean's letter makes an interesting commitment: "The Atomic Energy Commission agrees to assume any obligation which the Air Force may have had to restore the land described ... to a condition similar to that existing at the date of initial occupancy of the land by the Air Force; provided, however, that this obligation will not be performed by the Atomic Energy Commission until such time as it no longer has a requirement for this land." (Dean 1951)

On November 2, 1951, Dean again wrote to Chapman requesting: "an additional 2,000 acres of public domain land, approximately, for use by the AEC as a camp site." (Dean 1951b) This is an approximately 3 square mile rectangular area that includes Mercury and lies south of the southern boundary of the Bombing and Gunnery range. By the time of this November 2 letter, the AEC had already undertaken considerable construction to permanetize the camp at Mercury, which at this time was also being called Camp 3. (AEC DMA 1952: 3) Camp Desert Rock and the air strip were still both outside of the areas designated by Dean in his two letters to Chapman.

SANTA FE OPERATIONS OFFICE (SFOO) MOVES TO ALBUQUERQUE

Between July and October 1951 the AEC's Santa Fe Operations Office (SFOO), that had been located in Los Alamos, was moved to Albuquerque. Carroll Tyler continued as manager. This move recognized the realities of the increased work load associated with Sandia, the military services, and Nevada. Although located in Albuquerque, the office would still be called the Santa Fe Operations Office until mid 1955. SFOO retained a much-reduced Area Office at Los Alamos to continue to provide liaison with the laboratory.

SITE NAMED NEVADA TEST SITE, JULY 8, 1951

The test site was going by a variety of names in early 1951. It is interesting to note that at this time, Los Alamos often referred to the Nevada test site as Tonopah in their minutes of meetings.* [*Footnote: See, for example, Fission Weapons Committee Seventeenth and Eighteenth meeting announcement 19 February 1951(SRD) from D. P. MacDougall p. 1, and Minutes of the 20th Fission Weapons Committee, 29 March 1951] It was also called Site Mercury** [**Footnote: See, for example, Minutes of the 21st meeting of the Fission Weapons Committee, 11, April 1951]

It finally got an official designation when on July 8, 1951 the United States continental test area was officially named the Nevada Test Site (NTS) by the AEC. This was the official name during BUSTER-JANGLE. (Hacker 1994, 73)

PREPARATION FOR JANGLE

Studies of the underwater explosion on Operation CROSSROADS, in 1946, led to questions concerning the effects of surface and shallowly buried nuclear detonations on land. (Ponton et. al 1982c, 20-22) As a result, the DoD was developing an interest in the effectiveness of an earth penetrating weapon that was expected to penetrate about 50 feet in dry soil. (Lewis 1997, 18) Two tests using the same device design, one detonated at the surface and the other detonated in a shallow shaft, could address the relevant effects.

The Defense Department was planning Operation WINDSTORM with two shots of about 20 kt at Amchitka Island in the Aleutians. The plan was to conduct the tests between September 15 and November 15, 1951. The Navy would be responsible for administering the test series with support from the AEC. The military organized Joint Task Force 131, which was similar to the task forces that were used for the test operations at Enewetak and Bikini (CROSSROADS, SANDSTONE, and GREENHOUSE). A. F. Spilhaus* of the University of Minnesota, an AFSWP consultant, was designated the Scientific Director of the Task Force. (AFSWP 1954: Vol 4: 3.5.96 and AFSWP 154: Vol 5: 3.10.5) Lt. Col. M.S. George*, US Army, chief of the Test Branch of the Weapons Effects Division at Headquarters AFSWP, was also a key member of the Task Force. [*Footnote: These two individuals would also participate in the Nevada Test Site management organization of Operation BUSTER-JANGLE.] On November 30, 1950 President Truman approved the plan for Operation WINDSTORM.

The 20 kt shots intended for WINDSTORM were larger than could safely be accommodated in Nevada on the surface or in shallow holes. From the weapons effects perspective, it wasn't necessary to have such high yields; but prior to RANGER, the country had only tested yields* of 18, 21, 37, and 49 kt. [*Footnote: The military generally wanted devices for their programs with a high probability of producing the planned yield. The effects measurements were generally quite sensitive to the actual yield. For instance, the type of instruments used and their location with respect to GZ were yield dependent. Although many of the military measurement systems were purposefully designed with redundancy for possible yield variations, there was a limit to the range of redundancy that could feasibly and economically be fielded.] Fortunately, the RANGER Able and Easy shots, at approximately 1 kt each, demonstrated the reliability of producing a low yield.

On March 28, 1951 representatives from AFSWP, the AEC, and the Joint Chiefs of Staff met to reconsider the location of the site for Operation WINDSTORM. In addition to logistic problems, Amchitka had a wretched climate. Also, the geologic strata (fairly hard rocks) were not suitable to produce the results expected in dry soils, where the earth penetrator would most likely be deployed. (Hatlem 1951, 8) The meeting concluded that it made sense to use the new 1 kt nuclear device designs and to move the operation to Nevada. Under those circumstances it was more appropriate to put the operation under the Air Force rather than the Navy, and the executive agent for the tests was changed from the Chief of Naval Operations to the Chief of Staff, USAF. (Ponton 1982c, 20) On May 9, 1951, the JCS postponed the Amchitka-based Operation WINDSTORM, which in effect amounted to its cancellation. (Hatlem 1951: 8)

The Chief of Staff had the Air Force Office of Atomic Tests (AFOAT) draft a directive to the Air Force Special Weapons Command (AFSWC). This directive, dated May 15, 1951, ..." broadened the responsibilities of the Command to include 'coordination of military participation' instead of (just) 'US Air Force Participation'". (Hatlem 1951: 5, 18) Thus, the Commanding General of AFSWC, Brig. Gen. John S. Mills, became the responsible agent for all military participation for atomic tests at the NTS. The directive continued by describing roles for AFSWP and AFSWC* [*Footnote: "Chief, AFSWP will be responsible for preliminary planning and budgets for the military phases of atomic tests. The Chief of Staff, USAF has been directed to establish an appropriate, permanent test group, jointly staffed, for the purpose of coordinating military participation and assistance as desired by the AEC in the conduct of the tests. Individuals and organizations from the three Services and from AFSWP will be assigned to this Joint Test Group, as required.

Because of its close association with the field offices of the AEC, its inherent knowledge of the problems involved in atomic weapons testing and its direct contact and/or liaison with the Army, Navy, AFSWP, and major commands of the Air Force in the conduct of normal operational activities, the Special Weapons Command has been designated the Air Force Agency primarily responsible for coordinating military participation and assistance desired by the AEC for future atomic tests conducted at the Las Vegas site.

Accordingly, the Commanding General, Special Weapons Command is directed to develop and establish under the SWC the necessary organization with which to carry out those responsibilities accepted by the Chief of Staff, Air Force, with regard to the conduct of future atomic tests at the Las Vegas site. The permanent Joint Test Group is visualized by this Headquarters as a small, permanent, jointly staffed organization, designed primarily for the continuous planning of future atomic tests. This organization may be temporarily augmented to insure satisfactory implementation of the actual test programs." (Hatlem 1951:11, 12)]

“The Joint Test Group was organized at Kirtland AFB, SWC Regulation 20-1, dated 16 August 1951.” It “did not play an important part in BUSTER-JANGLE mainly because the planning was nearly complete when the Group was organized and also because certain individuals, slated to become members of the Group, were already actively engaged in the tests.” (Hatlem 1951:12)

General Mills appointed an officer, Lt. Col. Earl W. Kesling, whose ---“duties were to represent the Commanding General and his staff on matters pertaining to joint SWC-AEC test activities as directed.” (Hatlem 1951:26) Kesling served on the NTSO and was present at the site after October 10, 1951.]

The March 28, 1951 meeting also added a deeply buried test to the new Nevada plan. Relevant data on shots at or very near the surface of the ground did not exist. The AEC wanted, as a preliminary test: a “shot, set off far enough below the surface to minimize surface rupture, intended primarily to help assess the potential radiological hazards of subsequent shots”. If all went well, then “a second shot would be detonated at a lesser depth.” Finally, “a third test would be made, this time at the surface.” (Hacker 1994: 61)

The DoD shots at a shallow depth and on the ground surface could well pose a fallout problem. All of the players in the nuclear testing arena, including the AEC, the DoD, and the laboratories, were keenly aware of the paramount importance of safety and the public health concerns. Specifically they appreciated the sensitivity on the part of the general public to the real and imagined dangers of radioactivity, and to the fact that the AEC was responsible for the safe and professional execution of full scale nuclear tests within the continental United States. Clearly if testing were to continue the AEC, the DoD, and the labs would have to do their best to operate in a safe and responsible fashion and to make it known to the public that they were doing so.

To this end, Shields Warren of the AEC’s Division of Biology and Medicine, organized a committee of experts who met at Los Alamos on May 21 and 22 to consider the feasibility and conditions for the deeply-buried preliminary radiologic safety shot for JANGLE. This group of more than 20 experts was called the Jangle Feasibility Committee. They unanimously concluded, “a test involving the explosion of a 1 kiloton ... bomb ... can be carried out without undue hazard. The Committee recommends that the test be made”. (Hacker.1994, 61-62,313)

In June, Bradbury still had some misgivings about firing nuclear weapons on the surface of the ground or shallowly buried. He thought that the deeply-buried preliminary test demanded more than a few remarks about purported lessons from Trinity and Enewetak. Bradbury was concerned that if there were to be trouble on the first underground shot, the AEC would be called upon to explain “how come.” He wanted to see some credible calculations of the downwind particulate size and radiation levels.

Bradbury asked Los Alamos scientists to look into the problem further. A theoretical model derived from Trinity data seemed the most useful. It was the only experience even remotely similar, even though Trinity was 20 kt on a 100 ft tower. Problems revolved around uncertainties in the phenomenology; and the phenomenology of the buried shot was significantly more uncertain than that for the surface shot. The scientists even concluded that it was possible that the deeply buried shot could have more serious problems than the shallower shot. In effect, the Los Alamos scientists reversed the preferred order of the three tests that had been decided upon on March 28. As a result of this study, Bradbury urged that the surface shot be fired first.

On July 13 the Jangle Feasibility Committee met in Washington to address Bradbury's concerns and Los Alamos's calculational results and uncertainties. They ended up recommending that the preliminary, relatively deeply buried, shot be scrubbed and that the surface shot be fired first, followed by the shallowly buried shot if the fallout hazard found on the surface shot was within acceptable limits. (Hacker 1994, 63)

The AEC was still concerned about possible evacuation of the downwind population in the event of an emergency. They asked the then-new Civil Defense Administration, which had been toying with possible roles in nuclear testing, to take charge of evacuation in an emergency. (Hacker 1994, 65) Fortunately evacuation was not necessary. Both the surface and slightly buried underground shots were conducted, in that order, successfully and safely.

PREPARATION FOR BUSTER

In November 1950, about a month before the AEC acquired the Nevada site, the AEC notified the DoD that they were planning a test series, subsequently code named BUSTER, for the fall of 1951 in Nevada, if indeed that site were to be selected as the new continental test location. (George 1952: 20) On February 12, 1951, six days after the last RANGER shot, AFSWP outlined for the JCS a proposal for DoD participation on BUSTER; and on March 8 AFSWP solicited from the Services suggestions for specific measurements on BUSTER.

Meanwhile, Los Alamos was somewhat tardy compared with the DoD in planning for the fall series in Nevada. The same day AFSWP was outlining its proposals for participation on BUSTER, Bradbury asked Duncan MacDougall and the Fission Weapons Committee to begin thinking about the operation. (Fission Weapons Committee February 12, 1951(SRD)) This was indeed pursued, but with Operation GREENHOUSE slated to begin in the Pacific in less than a month it was hard to find anyone at Los Alamos who could give a fall continental test series much thought.

MacDougall, however, did ask Clark for his opinion on the optimum number of nuclear tests per series. Clark replied that there was an "economical number." He went on to say that three tower and three airdrops would be satisfactory. The interval between shots was of prime concern to the diagnostic scientific and engineering groups. They wanted at least four or five days between shots. (Fission Weapons Committee February 12, 1951(SRD)) As the experiments became ever more complex over the years, the time between shots increased enormously. To reduce this time between shots, the laboratories and EG&G formed crews that would be assigned to one or a few shots rather than to every test. For instance, different crews would simultaneously conduct field work on alpha experiments for different upcoming shots.

GREENHOUSE did go off as planned, starting on April 7, 1951, at Enewetak Atoll, with four shots. This was a particularly significant series. The George shot, the third one in the sequence, with a yield of 225 kt, and was the first test of a thermonuclear concept. The last shot, Item on May 24th, was the first boosted device (using a deuterium and tritium gas mix to enhance the fission yield of an atomic bomb) at 45 kt. (DOE/NV 2000: 2-3)

During the GREENHOUSE operation it was even more difficult to get people to think about testing in a forthcoming series in Nevada. MacDougall put it this way: "During the past few weeks, it has seemed impossible to get together a reasonable fraction of the membership of the Fission Weapons committee because so many people have been vacationing in the Pacific." (Fission Weapons Committee March 1951)

After GREENHOUSE, Los Alamos was able to devote serious attention to the BUSTER series, and planning proceeded rapidly. In mid-July Graves sent SFOO a preliminary plan for BUSTER-JANGLE. BUSTER was to have three to six shots with five being the most probable. Of the five, Graves proposed 4 airdrops and one tower shot. JANGLE, the DoD series, was to have one surface and one shallowly buried shot at a depth of 17 feet. (Hacker 1994, 65) Also in mid-July, the MLC requested AEC approval of troop exercises on BUSTER-JANGLE, which would be code-named Desert Rock.

NEVADA TEST SITE ORGANIZATION (NTSO) FOR OPERATION BUSTER-JANGLE

The main organizational differences between RANGER and BUSTER-JANGLE were stimulated by the presence of the DoD in two important activities that were present on BUSTER-JANGLE, but not on RANGER: 1) A large number of well planned and fielded Effects Measurements and 2) Desert Rock Exercises. After RANGER, Effects Measurements were conducted on all NTS atmospheric operations prior to the 1958 moratorium. The Desert Rock Exercises would take place on all future atmospheric operations at the NTS except for HARDTACK Phase II*. [*Footnote: The Desert Rock exercises involved many thousands of military personnel on each operation. These exercises were conducted on the laboratory's weapons development tests (like BUSTER) as well as on the DoD's effects tests (like JANGLE). The Effects Measurements involved far fewer military and civilian personnel, only about 500 to 2000 on the atmospheric operations at NTS.]

As was the case for RANGER, the authority and responsibility for all tests at the NTS, as well as in the Pacific, rested with the President of the United States; and he approved each operation and test.* [*Footnote: Technically, for each test, the president approved the expenditure of fissionable material.]

As shown in Figure 2-2.1, both the Secretary of Defense and the AEC Commissioners reported to the president who was assisted in nuclear testing issues, as well as other national security matters, by the National Security Council. The Military Liaison Committee, described in Part I, provided liaison between the AEC and DoD. The AEC Commissioners delegated test authorities to the Director of the AEC's Division of Military Applications (DMA), Brig. Gen. Kenneth E. Fields. Fields in turn delegated to Carroll L. Tyler, the Manager of the AEC SFOO. Tyler also assumed the role of the Test Manager who headed the Nevada Test Site Organization (NTSO). The delegated authority for the detonation of nuclear devices resided with the AEC Test Manager, not the military.

On the military side, the Secretary of Defense delegated authority to the Joint Chiefs of Staff. During atmospheric testing at NTS, the military's interests resided in two branches. One branch was responsible for the Effects Measurements, which was the development and execution of programs exploring the effects of nuclear weapons on military structures and equipment. This branch recognized the requirements for effects data from all of the military services. They also provided military support (mostly aircraft and weather information) to the test operation. The other branch was involved with the Desert Rock Exercises, which exercised military troops from all of the military services in nuclear environments. The interests of both the technical and exercise branches were represented by one military officer, and his staff, who served in a liaison role between the AEC Test Manager and the DoD.

The military liaison role for BUSTER-JANGLE was conducted by Lt. Col. E. W. Kesling, who had been appointed by the Commanding General of AFSWC, Brig. Gen. John S. Mills, to represent him on matters pertaining to joint DoD-AEC test activities. Kesling was named "Special Projects Officer" (Ponton 1982c: 30-33, DeMeers 2003, Hatlem 1951: 26-27) for BUSTER-JANGLE. Kesling was the individual in charge of overall military activities at the site, including the Desert Rock Exercises and the support services provided by SWC, as well as AFSWP's effects measurements. (Ponton 1982: 32 and Hatlem 1951:27)

There was a single organizational structure at the NTS for the combined LASL and DoD BUSTER-JANGLE joint test series, see Figure 2-2.2. The Test Manager had the ultimate responsibility for safety of the activities conducted at the site prior to and during the operations. This included the responsibility for authorizing detonation at a time when weather conditions and other factors were considered to be acceptable. Tyler's Advisory Panel, see Appendix F, was again chaired by Alvin Graves from Los Alamos.

For BUSTER-JANGLE, Tyler did not have an Executive Officer as he did on RANGER. Rather, he appointed Ralph P. Johnson the Field Manager. The Support Groups reporting to Johnson conducted many of the activities that had been directly under Tyler and Macy during RANGER. The Field Manager and his staff were largely AEC personnel, with contractors performing most of the services.

Tyler again appointed Graves Test Director, which was essentially the same position he held as Chief, Test Group on RANGER. Graves and his deputy, Jack Clark, managed the operations for the tests and were responsible for all decisions and activities at the site regarding the programmatic aspects of the operation. They oversaw the scientific experiments and measurements, and were responsible to the AEC for the safety of those operations performed by the laboratories and the user agencies.

Graves was in charge of both the Weapons Development Test Unit, headed by W.E. Ogle, LASL, and the Weapons Effects Test Unit, headed by Col. Max George, AFSWP. (Hartlem 1951:24) A. F. Spilhaus was on Grave's staff to provide liaison between the weapons development and the weapons effects programs. Clark, Ogle, and George were usually the senior NTSO people at the site during the operation.

The Weapons Development Test Unit conducted the diagnostic experiments on the tests and was mainly staffed by Los Alamos, EG&G, NRL, and Sandia personnel. The DoD contributed to the Weapons Development Test Unit by providing a variety of technical and other support functions on all of the shots.

The Weapons Effects Test Unit was responsible for the DoD weapons effects programs. Although the Weapons Effects Test Unit reported to Graves, Graves did not review and was not responsible for the technical merits or rationale of the programs conducted by the military. He was, however, responsible for the administrative and functional aspects of the unit. On BUSTER-JANGLE, this unit was mainly staffed by military personnel or civilians from DoD laboratories.

The DoD operated on a non-interference basis with respect to the Laboratory's programs during BUSTER. On the DoD's JANGLE tests, LASL performed basic experiments to verify the proper performance of the nuclear device, but otherwise tried not to interfere with the DoD operations. After having a total of no more than about 500 people issued clearances for RANGER, the 2500 people at the site associated with the DoD effects programs stretched the facilities and logistics almost to the breaking point. (This did not include the 6,500 military persons that participated in Desert Rock troop exercises.) Consequently, close, continuous, and reliable liaison and communications between the AEC, Los Alamos and DoD was essential. As mentioned earlier, DoD input was provided at two main points: 1) at the Test Manager's level - by Lt. Col. E. W. Kesling* and 2) in the Test Director's organization - by the Deputy Test Director for Effects Tests, A.F. Spilhaus*, and by the Director of Weapons Effects Tests, Col. Max George*. [*Footnote: Kesling was from the Air Force's SWC which was the responsible military organization for

BUSTER-JANGLE. Spilhaus and George were from AFSWP, but they may have been assigned to the SWC Joint Test Group for BUSTER-JANGLE.]

There was a Los Alamos group called Plans and Operations, J-3, under Duncan Curry, Jr. that was responsible for the full range of administrative support.* [*Footnote: J-3 was responsible for ... "coordination of experimental programs and projects; an information bulletin; determination of operational requirements by means of questionnaires and interviews; establishment of a system of status reports; procurement of planes and vehicles; allocation of vehicles; scheduling and meeting special plane and helicopter flights; preparation and dissemination of an Administrative Order and Operation Orders and Annexes as well as general supervision of their execution; liaison with the Civil Aeronautics Administration; coordination with the Field Manager's security section; preparation of access lists; coordination of operations with rad-safe requirements; and certain messing and dormitory arrangements at the Control Point." (Clark 1951: 16)]

Los Alamos group J-6, under the leadership of Newell Smith, was responsible for engineering and construction liaison until November 1951. Robert H. Campbell* took over from Smith on November 29, 1951. [*Footnote: Robert H. Campbell remained group leader of J-6 until August 1,1957 when he transferred to the Division Office with the title of Assistant Division Leader and Test Director. He retired from Los Alamos National Laboratory on February 5, 1982. Campbell played such an important role over the years as an advisor, mentor, and counselor that he became known to the test community as "father Campbell."]

Clark described the J-6 role as follows: "In the earliest phase the Test Director gave construction requirements to Holmes and Narver, architect engineers. Later, a member of LASL Group J-6 was given authority to establish an AEC engineering office at the site, where drawings were prepared for later construction in Operation BUSTER and all of Operation JANGLE. Haddock Engineers performed this work on a cost-plus-fixed-fee contract." The Haddock Engineering Co. performed last minute changes on both operations according to work orders with the approval of J-Division. (Clark 1951: 18-19)

Personnel control in the test area was the responsibility of Los Alamos group J-1 (Personnel and Administration) under group leader Armand Kelly and alternate group leader S. R. Whitaker. Beginning in the morning of D - 1, the day before a shot, a muster station was established at the entrance to the danger, or forward, area. All entrances and departures from the area were monitored. Roving patrols were also instituted to ensure that no one would be left behind at shot time in the test area. AEC Security also manned a station adjacent to the J-1 station to limit access to those with the correct badge authorizations. Clark recommended that in the future these functions be combined and handled by the AEC Security Service. (Clark 1951:14-15)

BUSTER-JANGLE TESTS

The first shot was originally scheduled for October 1, 1951, but construction problems caused a delay until October 15. This delay was not particularly bothersome since Presidential approval was not received until October 9th, and the test site leadership was reluctant to embark on the final preparations until that approval was in hand.

Little was done in the way of overt public relations for BUSTER-JANGLE. Most of what the AEC did in this arena centered on keeping news related to the test series low-key. On August 28 the press notified the public that testing would be resumed soon, but that there was no danger. The DoD, in their announcement, conveyed a brief outline of their program and emphasized troop safety. (Hacker 1994: 69-70)

LASL group J-8, under the leadership of Sandian Walt E. Treibel (until May 1952), was responsible for the preparation and delivery of the devices both on-site and to the strike aircraft in Albuquerque. The responsibility for the preparation of one of the airdrop weapons was shared with Sandia's R. W. Henderson and his colleagues C. F. Robinson and J. Benish. The Able device assembly and handling were accomplished by military personnel under the supervision of J. R. Heaston and E. L. Jenkins, attached to Sandia Division 1632, working with LASL's J-8.

Weapons assembly in Nevada was hampered by incomplete facilities, but Clark acknowledged the contributions of T. Roehl and J. Lopez of the AEC, and the Haddock Engineers superintendents and foremen for their work at the three static sites (i.e. non-airdrop). "Without their complete cooperation, shot dates could not have been met."(Clark 1951: 41)

The Operation BUSTER weapons development tests were generally designed to explore comparisons between theoretical calculations and observed device performance for several new fissionable material geometries. The weapons effects tests, Operation JANGLE, were focused on the phenomenology of surface and shallow subsurface nuclear explosions. The tests conducted during BUSTER-JANGLE are listed in Table 2-2.1, and their locations in Area 7 at the Nevada Test Site are shown in Figure 2-2.3.

Table 2-2.1 Nuclear tests conducted during BUSTER-JANGLE.

TEST	DATE-1951	TYPE	AREA	YIELD (kt)
BUSTER - LASL, WEAPONS RELATED				
Able	Oct 22	Tower 100 ft	7, Station 5	< 0.1
Baker	Oct 28	Airdrop 1118 ft	7, Station 3	3.5
Charlie	Oct 30	Airdrop 1132 ft	7, Station 3	14
Dog	Nov 1	Airdrop 1417 ft	7, Station 3	21
Easy	Nov 5	Airdrop 1314 ft	7, Station 1	31
JANGLE - DoD/LASL, EFFECTS				
Sugar	Nov 19	Surface, 3.5 ft	9	1.2
Uncle	Nov 29	Crater, -17 ft	10	1.2

(DOE/NV 2000, 2-3; Lewis 1977, 18)

Able was the first tower shot conducted in Nevada. However, as described by Walt Treibel*, things did not go as planned on October 19. [*Footnote: Walt Treibel joined Los Alamos in 1944 and moved to Sandia as an early member of Z Division. He had participated in development of the timing system at Trinity and was on the four-man arming team at Operation GREENHOUSE.] After the arming party, composed of Treibel, Jack Clark, Barney O'Keefe, and Joe Dawson of Sandia, returned to the CP,

> "... the countdown began. At minus 10 seconds, the automatic sequence timer took over 10, 9, 8, 7, each tick seemed an eternity. Then 'zero'. And silence. The PA system came to life. 'Misfire. There's been a misfire' I wondered - 'Is it alive? Had an electrical relay stuck that might give way any second?'
>
> We reviewed emergency procedures, but the next step was inescapable. The device had to be disarmed. Al Graves said, 'Treibel, you put that thing up there. You go get it down.'
>
> We (the four man arming party) immediately began jotting down an outline of what needed to be done and what we needed to look for in the way of evidence. When the time came, we nodded to our friends. We weren't sure we'd see them again. We got in a car and drove toward the tower. We stopped at the switching station two miles from the tower to disconnect electrical circuits. When we arrived at the tower, we started up the ladder. I don't recall the order. It didn't matter anyway. We were a team, and we'd agreed right up front that nothing would occur without the concurrence of all.
>
> We stopped a lot on the ladder to catch our breaths because of the anxiety and because of our awareness that each step could be our last. Finally we reached the cab containing the device, which was housed in a 60 inch-diameter cylindrical can that rested on legs about two feet off the cab floor.
>
> We were careful not to disturb anything - just looked at the electrical relay rack voltmeters. No voltage was going into the unit. But it could still be charged. We disconnected the cable between the power supply and the device. We all rested a lot easier when that cable came out.
>
> No electrical power had even reached the device. In spite of a keyway designed to prevent the symmetrical connector insert from rotating in its mating connector, one insert had shrunk, and rotated 90 degrees, causing a fuze to blow."

After repair and checkout, a shot date of the 21st was selected. There was a one-day weather delay, and Able was fired at 0600 PST October 22. Its yield was less than 100 tons. The low yield raised concerns within the AEC about a possible public perception that the nuclear device was either a dud or a fizzle.

Actually it was neither. Shot Able was a physics experiment, and the results were neither particularly surprising nor disappointing to the Los Alamos scientists. (Fission Weapons Committee, July 7, 1952(SRD))

The next three shots were airdrops using a B-50D aircraft flying out of Kirtland AFB at Albuquerque. Easy, the last BUSTER test, was dropped by a B-45C jet also from Kirtland. The specific number and types of support aircraft depended upon the particular device being tested. A typical complement of aircraft was as follows: Two B-29 and two T-33 cloud sampler aircraft operating from Nellis Air Force Base at Las Vegas; two cloud trackers; two or three terrain survey (for mapping ground contamination) C-47s operating from Indian Springs, and one helicopter stationed at the Control Point, at the site. In addition, two aircraft (P2V-2 and B-17) were used for RADIAC (RAdiation Detection, Indication, And Computation); and one cloud tracking aircraft operating out of Kirtland AFB might be used. (Clark 1951, 45) An additional B-29 and T-33 aircraft were available for cloud sampling as was one additional C-47 for terrain survey. (Clark 1951: 50-2)

A dry run for Baker was executed on October 24. Icing conditions and poor visibility at the site forced a two-day delay of the nuclear shot, which was eventually fired at 0720 PST on October 28, with a yield of 3.5 kt. Charlie was fired October 30 at 0700 PST. Its yield was 14 kt. Dog was fired on November 1 at 0730 PST with a yield of 21 kt. These three airdrops used the same target, Station 3. The actual Surface GZ (SGZ) locations of these three detonations with respect to Station 3 were: Baker, 140'N, 13'W; Charlie, 162'N, 99' W; and Dog: 56'N, 37'E. (Corsbie 1952: 8)

Easy was fired November 5 at 0830 PST with a yield of 31 kt. This was the first live nuclear weapon dropped from a jet aircraft. Station 1, slightly NW of Station 3, was the target.

There were a number of electrical problems that caused the loss of data on both Baker and Dog.

The DoD shots on Operation JANGLE were named Sugar (for **S**urface) and Uncle (for **U**nderground). Sugar was actually fired at a 3.5 foot height-of-burst (HOB). That is, the center line of the device and its surrounding canister were 3.5 feet above the surface of the ground. After a four-day weather delay, it was detonated on November 19 at 0900 PST with a yield of 1.2 kt.

For Uncle, a shaft 20' deep with 8.4' diameter was excavated. A ¼" thick metal corrugated tunnel liner with 8.4' OD and 8.1' ID was placed inside of the shaft for lining. The device, the same as Sugar's, was placed with its center of gravity about 3 feet off the floor at a 17 foot burial depth. The dirt backfilled was placed around the device and for 12' on top of it. (George 1979:112)

Uncle was fired on November 29 at 1200 PST, also with a yield of 1.2 kt. (Lewis 1997, 23) The DoD was interested in earth penetrating bombs, and the 1.2 kt yield at 17 feet scaled to about 30 kt at a depth of about 50 feet, assuming cube root scaling.* [*Footnote: $(30/1.2)^{1/3} \sim (50/17)$]

The average dimensions of the craters formed by these two tests were:

Table 2-2.2 Average crater dimensions of Sugar and Uncle.

Parameter	Sugar	Uncle
Volume (ft^3)	5.18×10^4	1.04×10^6
Maximum Depth (ft)	16.3	50.9
Diameter (ft)	90	254.4

(Schuster 2001: 68, 77 and Bishop 1952:12)

WEAPONS DIAGNOSTICS PROGRAM

Bradbury and Ogle describe the overall rationale for the tests in this way: "The general philosophy and practice of supplementing part of the theoretical and experimental work on fission bombs at Los Alamos by means of test detonations using fissionable materials has been well established. Although use has been made of this approach ever since the stockpile of fissionable materials became great enough to afford it, it was only with the advent of the continental test site, Mercury, that the exploitation of this method became very effective. The high effectiveness of this method of study combined with the present heavy development load has indicated the value of an extensive program of test shots, both for this fall and from time to time in the future."(Ogle 1951: 9)

There were approximately 130 people directly involved with weapons diagnostics on BUSTER-JANGLE. The experiments, program numbers, approximate numbers of people involved in each at the NTS, and the operating agencies are presented in Table 2-2.3. (Clark 1951: 50-52)

LASL diagnostics programs were numbered 10.1 through 10.11, but in succeeding operations the programs were usually numbered 10 through 19. All of the program numbers were not necessarily used on every operation. (LASL DO-555 No Date: 23)

Table 2-2.3 Diagnostics measurements associated program numbers for BUSTER-JANGLE.

DIAGNOSTICS MEASUREMENTS	# PERSONNEL	AGENCY
10.1 Alpha, BUSTER	15	LASL
10.1 Alpha, JANGLE	20	NRL
10.2 Transit Time, BUSTER	6	LASL and Sandia
10.2 Transit Time JANGLE	2	NRL
10.3, 10.4 Yield	9	LASL and EG&G
10.3, 10.9 Thermal Radiation	10	LASL and NRL
10.5, 10.6 Gamma Rays	5	LASL
10.7 EMP Signals	11	LASL and Sandia
10.8 Neutron Measurements	8	LASL
10.10 Blast	30	LASL and Sandia

10.11 Growth of Fireball Underground	6	LASL
10.3 Other Technical Photography	6	EG&G

The weapons development experimental program was similar to that on Operation RANGER. Measurements of yield and alpha were the principle objectives, with additional experiments designed to enhance the understanding of the device behavior details. Appendix G provides descriptions of most of the diagnostic experiments.

Alpha, transit-time, and yield were conducted on all seven shots. On the six above ground tests, yield was determined by radiochemistry, fireball growth, and bhangmeter. The optical techniques for fireball growth and the bhangmeter were, of course, not used on the below surface Uncle shot. This was the first radiochemistry yield determined for an underground test.

In LASL's Measurement 10.11, "Two types of switches (called pins)" were used to measure "the time-of-arrival (TOA) of the first hydrodynamic signal*." (Baker et.al. 1951, 71-77) [*Footnote: Baker et.al. have an interesting footnote which indicates the technical "state of the art" regarding underground explosions in 1951: "The term 'first hydrodynamic' signal is used rather than the specific terms such as shock wave, plastic wave, etc. because of the uncertainty as to the precise nature of the phenomena."] The first type, "the pressure closure pin, can complete an electrical circuit when a pressure is applied to it. The second, the ionization pin, will close an electrical circuit when it is immersed in a region sufficiently ionized to short it." The pressure pins measured the TOA of the shock front, and the ionization pins the TOA of the expanding ionized fireball.

Pins were used on Sugar as a "wet run" for Uncle. A "wet run" is a more complete dress rehearsal than a "dry run." For Uncle, two lines of pins were located horizontally at the 17 foot depth of the device location. These two lines ran from the wall of the emplacement caisson (4 ft from bomb zero) to 38 ½ feet with pins located at 18 inch intervals. Another two lines were run at a depth of 5 foot from the surface. These two lines started at about 20 feet from the wall of the emplacement caisson and also ran to 38 ½ feet.

This is the first instance where yield was measured at the NTS by using TOA of the ground shock. In later years, when testing went underground, the techniques for obtaining yield from ground shock TOAs would be greatly improved and frequently used. Such techniques, known as CORRTEX and SLIFER, which are cables that crushed on arrival of the ground shock, were reliably used in the Joint Verification Exercises conducted with the Russians in the 1980s and 1990s at NTS, Semipalatinsk, and Novaya Zemlya.

A preshot hand book would usually be developed for an operation. These handbooks were valuable documents and were extensively used by the experimenters as well as by NTSO.

In the BUSTER-JANGLE preshot handbook, Conrad Longmire and Ted Taylor presented descriptions of the calculations of the performance of the BUSTER-JANGLE devices. Specifically, they calculated the alpha, yield, and transit time for each shot. Transit time is defined in two ways, both of which were calculated by Longmire and Taylor: 1) the time from a signal from the X-unit to the time of criticality; and 2) the time to a certain number of fission generations. Also in the handbook Clyde L. Cowen described the neutron measurements; L. Seely the thermal radiation measurements; and E. J. Zadina the blast measurements. (Ogle 1951: all)

DoD PARTICIPATION ON OPERATION BUSTER-JANGLE

For BUSTER-JANGLE, the military did what it does so well: organize, mobilize, and conduct large formations in action. The magnitude of the DoD activities almost overwhelmed the weapons development personnel. Both TRINITY and RANGER had been manned nearly entirely by Los Alamos people. RANGER had been done in such a hurry that the DoD had not mounted much of an effort, but they made up for it on BUSTER-JANGLE.

Approximately 9,000 military and civilian DoD personnel (6,500 for Desert Rock, 2,500 for effects measurements) participated in the operation. (Ponton et.al.1982c, 27) While 9,000 is a large number, it was about the size of a typical Pacific test operation, except for CROSSROADS, which involved about 42,000 people.

The presence of large numbers of un-cleared military personnel at the site presented a substantial challenge for the AEC security people. The issues were significant; but they were solved in a cooperative way between the DoD, the AEC, Sandia and Los Alamos. The areas of concern centered, in most cases, on what un-cleared personnel could see and what might be inferred about classified details from the visual clues.

Interestingly, several enterprising newsmen set up shop high on the slopes of Mt. Charleston, northwest of Las Vegas, with telephoto lens-equipped cameras to photograph nuclear shots and pre-shot preparations. They clearly identified the Uncle shot as an underground blast, but the AEC declined the opportunity to confirm or deny the press's conclusions. Clark viewed the AEC's position as rather foolish since the facts really were obvious to the reporters. (Clark 1951: 10)

The DoD's participation on BUSTER-JANGLE was in three main areas: 1) Desert Rock Exercise I on Dog, Desert Rock II on Sugar, and Desert Rock III on Uncle; 2) Effects Measurements Programs; and 3) Support of Los Alamos' Weapons Development Test Programs.

DESERT ROCK OPERATIONS: I (DOG), II (SUGAR), and III (UNCLE)
During the summer of 1951, the Chairman of the MLC sent an Army proposal for troop maneuvers to be fielded on BUSTER-JANGLE to the AEC. The idea was to permit military people to observe a nuclear explosion and to participate, to the extent that was safe and feasible, in military activities in an environment like that encountered in nuclear warfare. These activities were approved by the AEC, and similar programs would be conducted on all subsequent NTS atmospheric operations except HARDTACK Phase II. These activities were known as the Desert Rock Operations or Desert Rock Exercises. Camp Desert Rock, near the present day Mercury airport but outside of the 1951 boundaries of the NTS, was constructed to house the military participants. With about 100 semi-permanent buildings and more than 500 tents, it accommodated more than 6,000 troops at a time. (Ponton 1982c: 27)

The Desert Rock organization functioned separately from the test organization, with liaison between the two groups to ensure that Desert Rock training programs did not interfere with the weapons development and weapons effects programs. (ibid., 29). Desert Rock Operations were staffed and administered by the Army with frequent participation by the Navy, Marine Corps and Air Force. More than 150 different military units and armored battalions, paratroopers, transportation companies, engineers, and a veterinary detachment were present on BUSTER-JANGLE. (DTRA 2002: 88; Ponton 1982c: 29)

Approximately 70 percent of the BUSTER-JANGLE DoD participants, about 6,500 of the 9,000 total participants, took part in the Desert Rock Operations, which consisted of three exercises. (ibid., 27, 46)

The three Desert Rock exercises were developed to test tactics and protective measures for use during a nuclear conflict. Their "objectives were to:

- Study the military uses of nuclear weapons
- Train military personnel in the tactical use of nuclear weapons
- Study the psychological reactions of military participants to the detonation of a nuclear weapon
- Test the effects of a nuclear detonation on animals and military equipment
- Determine the effects of a nuclear detonation on field fortifications and defensive structures
- Determine appropriate measures for radiation protection and instruct participants in those measures."

(Ponton 1982c, 46-47)

The participants in the Desert Rock Operations were divided into two groups: 1) Camp Desert Rock Support Troops and 2) Desert Rock Exercise Troops. The Support Troops numbered about 2,500 soldiers at the beginning of Desert Rock I. Some of them were stationed at Camp Desert Rock throughout all three exercises, but many returned to their home stations after the Dog shot on November 1, 1951. They provided "support functions for the camp, such as administration, transportation, construction, communications, security, food, and laundry." Some participants worked in the "forward testing areas of Yucca Flat to help prepare for specific Desert Rock activities, to assist in operations during test events, or to help ensure safe post shot recovery operations. Three units involved in shot-day operations were the Control Group, the Radiological Safety Unit, and the AFSWP Advisory Group." The support troops provided military police and medical support as well as field fortification construction for the Dog, Sugar, and Uncle shots. (ibid., 47-49)

Approximately 3,700 Exercise Troops participated as observers, in maneuvers. These soldiers, plus a few DoD civilians, were stationed at Camp Desert Rock for short periods ranging from several days to about two weeks. (ibid., 49-50)

The observer program was designed to acquaint DoD personnel with nuclear detonations. The program consisted of preshot lectures and films, observations of nuclear detonations in the forward area of the test site, and post shot tours of equipment display areas. (ibid., 50-51)

Approximately 2,800 Army, Navy and Marine Corps personnel were engaged in Desert Rock I, with most being from the Army. There were less on Desert Rock II and III, but the activities were basically the same.

The observers got to their assigned areas about an hour before shot time. They were then briefed on the scheduled detonation and safety procedures by AFSWP instructors. "Shortly before the shot, the instructors directed observers to sit on the ground with their backs toward ground zero. After the initial flash of light from the detonation ... the observers were directed to turn and view the fireball and cloud." After the post-shot radiological conditions permitted safe reentry the observers were permitted to inspect the equipment and facilities around ground zero. (ibid., 51)

"The troop maneuvers were designed to train participants in the tactical use of nuclear weapons and to demonstrate nuclear effects to the participants." A troop maneuver exercise with 883 men was conducted on the Dog event as part of Desert Rock I. Troop maneuvers were not fielded on Sugar and Uncle. (Ponton et.al.1982c, 49-50, 54) The participants on Dog were Army units that comprised a Battalion Combat Team (BCT) for this particular exercise.

During the two weeks prior to Dog, the BCT dug foxholes and built gun emplacements and bunkers in a Tactical Defensive Position (TDP) SW of ground zero, see Figure 2-2.4. The position, which had film badges in foxholes, was not occupied at shot time but was developed to test the effects of nuclear weapons on the structures and emplacements. (Ponton 1982c, 54)

Several hours before the Dog detonation, the BCT and observers went to an observation position about 7 miles south of ground zero. After the detonation, the troops moved to the Tactical Defense Position that they had built to see first- hand what a 21-kt nuclear weapon did. On order, the BCT moved forward in an attack formation to an objective SW of ground zero. At the closest point, the objective was slightly more than a quarter of a mile from GZ. The troops were accompanied by radiological safety monitors and were preceded by radiation survey teams who determined the limits of safe advance. After reaching the objective, the BCT toured display positions at roughly 0.5, 0.75 and 3.0 miles south of GZ. After being monitored at the decontamination station at Yucca Pass the troops and observers returned to Camp Desert Rock. (Ponton et.al.1982c, 55)

The Human Resources Research Office (HumRRO), a civilian agency under contract to the Department of the Army, investigated the psychological reactions of the maneuver troops. HumRRO was particularly interested in troop behavior during the maneuver and the changes in troop attitudes about nuclear weapons before and after participation in the test. The data collected by HumRRO presumably assisted the Army in determining the probable behavior of troops involved in nuclear warfare. (Ponton et.al. 1982c, 55)

Evaluation teams of up to10 men each were assembled from the Chemical, Signal, Engineering, Medical, Ordnance, and Quartermaster units for the Desert Rock Exercises. These teams were responsible for putting together equipment at the display areas and film badges in the fox holes prior to the shots. After the detonations, in the company of rad-safe monitors, the teams recovered equipment while a combat battalion recovered film badges. The information was assembled in after-action reports detailing the exercises and measurements. (ibid., 57)

DoD NWE PROGRAMS AND PROJECTS

During the 1950s at the NTS (during operations BUSTER-JANGLE, TUMBLER-SNAPPER, UPSHOT-KNOTHOLE, TEAPOT, PLUMBBOB, and HARDTACK II), the DoD conducted 324 projects that addressed NWEs. These 324 projects as well as 158 Civil Defense projects conducted between 1953 and 1958 are all identified and very briefly described in Killian 2011. Three special points about the DoD NWEs are:

1) 324 is a lot of projects - an average of over 50 per operation. Nearly all of the projects were conducted on more than one shot during an operation.
2) The work done by the DoD and its contractors for the projects was staggering. In terms of numbers of people and the amount of construction and materials used, the DoD NWE technical activities at NTS dwarfed those conducted for nuclear weapons development by Los Alamos, and later by Livermore.
3) The projects often involved cutting edge technology, as did those in the weapons development arena. Also, as with the weapons development experimentation, the work in the field was a small fraction of the total effort; most of which took place “back home” in laboratories and offices. The formal project reports often include additional topics such as background, previous research, rationale for the field activity, analyses of resulting data, computations, successes and failures, and general results. We focused our

attention mainly on the field work in Nevada, concentrating on the following information:

- The nuclear shots that were involved.
- Location with respect to ground zero.
- Objectives of the projects.
- The approximate numbers of people involved in the field operations.
- The nature of fielded objects and significant construction, equipment, or materials.

Synopses of the activities that were undertaken during field operations at the NTS for several of the major categories of NWE projects are presented in Appendix H. These synopses are intended to provide a general overview only of what people did in the field to conduct these categories of projects: Exposure; Biomedical Exposure Animals; Airblast, Ground Motion, and Thermal Measurements; Nuclear Radiations; and Structures.

Herein we try to convey a sense of the objective, magnitude, and complexity of the field work conducted on only some of the NWE projects conducted. The selection of projects discussed in the following was based upon consideration of: its importance at the time of the operation; the magnitude of the construction or field activities; general human interest; and/or "firsts" at the NTS of measurements that were also fielded on subsequent NTS operations. As might be expected, the "firsts" were greater on BUSTER and JANGLE than on subsequent operations.

A number of projects that addressed a similar technical aspect of NWEs were usually grouped together, managed, and funded under a program number. For instance, on BUSTER, with its 5 tests, there were 7 DoD programs with a total of 27 projects.

OPERATION	PROGRAM	# PROJECTS
BUSTER	2) Thermal and Nuclear Radiation	7
	3) Blast Effects on Structures and Equipment	3
5 Tests	4) Biomedical	4
	6) Test of Service Equipment and Operations	4
	7) Long-range Detection	5
	8) Supporting Measurements	2
	9) Personnel Shelter Evaluation	2
	Total	**27**

Each of the projects within a program was assigned a number*. Thus, for a given operation, each project was identified by:

Program #.Project #.

For instance Project 2.3 would be Project #3 in Program #2. Projects usually had long titles, and simply referring to Project 2.3 must have aided numerous forms of communication. Herein, and in practice at NTS, projects are generally referred to simply by their number. [*Footnote: There may be missing program numbers in an operation. However, AFSWP's programs were fairly consistent throughout the operations. For example: Program 3 usually included projects that addressed structures and their vulnerability to airblast and/or ground shock. Project numbers might also not be consecutive with numbers missing or have an additional designation of a, b,-1, etc. This is probably due to the addition or deletion of projects from a program at some time after an initial number designation was made.]

In the following discussion of projects, the "Program #.Project #" is cited along with the project description. These numbers will aid in locating a particular project in other literature or search systems.

BUSTER NWE PROGRAMS AND PROJECTS

Approximately 30 percent of the DoD participants on BUSTER-JANGLE, about 2,500 people, participated as part of the test organization, and most of them were in the Weapons Effects Test unit engaged in DoD's NWE projects. (Ponton et.al.1982c, 58) Los Alamos' Weapons Development Test unit remained about the same size throughout the 1950s as on TRINITY and RANGER, a couple to a few hundred persons. Personnel working on DoD NWE projects were always much more numerous than Los Alamos personnel - but for BUSTER-JANGLE, the factor was about 10.

While most of the AEC and DoD work went smoothly, there were growing pains during BUSTER-JANGLE. After, RANGER, it was a shock to many Los Alamos folks to witness the number of DoD personnel at the NTS. Los Alamos people had worked with the military in the Pacific where operations were organized as military Task Forces; however, the NTS operations were not Task Forces. So, the relationships between AEC and DoD were new, and responsibilities were not yet well defined. The participants were on a steep learning curve with respect to both operational procedures and technical advances; and both operational procedures and technical advances could cause changes or glitches in field activities on a project. In retrospect, it is surprising that the early operations, with their diverse objectives and numerous participants, went off as well as they did. Clearly, the individuals who were involved, on all sides, deserve an enormous amount of credit for developing a system that succeeded and resulted in numerous accomplishments. It is said that a key part of the "system" was the general feeling that getting the job done well was much more important than who could (or did) do what to whom.

The continental site, with its large land area and relatively easy access for US participants, enabled many types of NWE projects that would not have been feasible on the Pacific atolls. Most of the military's interests in nuclear weapons were associated with nuclear detonations over, on, or within landmasses; and the small Pacific atoll environment did not provide a representative venue.

The valuable technical experience on weapons effects that was gained in the Pacific (like that on airblast and the large structures program on GREENHOUSE

(Northrop 1951a and 1951b; Kirkpatrick 1951) was of course applied to the work in Nevada. Therefore, many of the weapons effects projects represented a continuation of, or a similarity with, measurements made in the Pacific and were not necessarily "firsts" of their kind.

While the DoD did not conduct airblast or ground shock programs on BUSTER both Los Alamos and Sandia did, and these activities are very briefly described in this chapter in the section titled WEAPONS DIAGNOSTIC PROGRAM.

Thermal and Nuclear Radiation

On project 2.4-1he DoD fielded a "thermal line", which ran between 2,000 feet and 12,000 feet from the intended ground zero (IGZ) of Able and Easy. Calorimeters and other sensors were placed along this line for the measurements of thermal intensity vs. time and total thermal outputs. Shots Able and Easy had 5 stations along this line, and the other shots had 6 stations. A motion picture film of the thermal line showed that "... large quantities of smoke or dust appear long before the arrival of the shock front ..." (Broido 1952: xi)

Operation BUSTER marked the start of numerous exposure projects conducted by the DoD at the NTS that used panels mounted on racks to expose different samples. Racks that held the objects being exposed were being developed and built at the Army's Engineer Research and Development Laboratories, ERDL. The racks were like saw horses, 5' x 10' with two 3' x 5' panel sections that could be placed at any angle from 90° to 180°. "They are inexpensive, light, but sturdy racks anchored by steel stakes and require only 6 to 8 man hours to erect in the field." The racks were placed at 4 and 5 different distances from the Baker and Dog Intended Ground Zeros (IGZs) respectively. BUSTER project 2.4b used samples of 15 different plastics, 7 coated fabrics, 10 packing materials, and paints.

The design was modified after BUSTER such that the racks would drop down with shields covering the exposed panels following the thermal flux but prior to the arrival of the blast wave. This improved design was used extensively throughout atmospheric testing for numerous types of exposure samples. (Miller 1952: ix, 4, 27-37) Project 2.4a exposed fabrics and uniforms in a stand-alone configuration that appeared almost ghoulish. (Davies 1952: ix)

The protection provided to a soldier by a foxhole was important information which was sought by project 2.6. On BUSTER, and future operations, foxholes were instrumented for the measurement of gamma-rays and neutrons,. (Walsh 1952a: ix, 6, 12-15)

Blast Effects

The Army used non-detonating Universal Indicator Mines to determine the probability of detonation of anti-tank mines that were subjected to a nuclear explosion. Project 3.5 was the first of four minefield clearance projects that were conducted at the NTS. These were extensive undertakings that used numerous mines in different configurations. Later operations used live mines in addition to the indicator mines. The results of this first minefield clearance project demonstrated that the response of mines to atomic detonations did not obey the scaling laws that had been established using HE tests. (Thurston 1952: ix, 1-2, 43-4)

Two aircraft, an F-47 fighter and a B-17 bomber, were both located on the ground 4,250 feet from the Dog GZ for project 3.8. The fighter had its tail toward GZ, and the bomber had its left side facing GZ. These two aircraft were relocated for Easy where the fighter again had its tail toward the blast at a range of 2,675 feet, while the bomber had its nose toward GZ at 5,847 feet. Aircraft were also exposed on subsequent operations, often with instrumentation to measure the levels of airblast, ground motion, thermal and nuclear radiations. (Gilroy 1952: xiii)

On Easy, four US Army 3,000 gallon rubber-coated nylon-fabric water tanks filled with drinking water were left uncovered and exposed at 4 distances for project 3.9. The tanks were essentially undamaged, and the water was not contaminated. Canned samples of sea water in various dilutions were also exposed to Easy at 2 distances. These sea water samples showed considerable induced activity. (Lindsten 1952a: vii)

Biomedical

Project 4.2, provides a lesson on how a field project rapidly got complicated. Dogs had been used extensively in laboratory studies of burns by radiations. On BUSTER, they were used for this purpose as well as to correlate laboratory and field studies. The dogs wore protective jackets with holes on the side of the jacket facing GZ. Two dogs were to be exposed on a trial run during Baker, which was scheduled for detonation at 0700. At 0630, the shot was postponed. Project personnel immediately proceeded to remove the animals only to find that they had both succumbed to the elements. The army then took more elaborate protective measures against the cold.

- Two chemical heating pads were included in each dog jacket.
- All animal surfaces except for the exposure apertures were wrapped with aluminum foil.
- A 500-watt infrared lamp was focused on the exposure apertures of each animal.

Six dogs (3 at 7000 feet and 3 at 9000 feet from GZ) were exposed along the thermal line on (of course) shot Dog. (Brooks 1952: 5-14)

The first flash blindness project at NTS was project 4.3. It was conducted to determine the size, depth and duration of scotoma* occurring after the exposure of the human eye to the light from a nuclear detonation. [*Footnote: a partial loss of vision or blind spot in an otherwise normal vision field.] Human test subjects were trained to chart their scotoma on portable instruments. The participants orbited at an altitude of 15,000 feet in an Air Force C-54 approximately 9 miles from the Baker, Charlie, and Dog detonations. They observed the flash, then immediately began recording their visual acuity. Data recording continued until pretest acuity was regained. Some participants wore different types of goggles, others were unprotected, facing the detonation or facing away from it. Unfortunately, useful data were only obtained on the Charlie shot. Baker's low yield produced a low flash, and Dog had an inaccurate detonation position. (Byrnes 1952: vii, 1-3)

Service Equipment and Operations
The disruption of the ionosphere due to a nuclear detonation was studied throughout the period of atmospheric testing. Project 6.9 obtained data to explore the effects of an atomic explosion on radio propagation at all frequencies. In one measurement transmissions were reflected from the ionosphere at a point nearly over GZ and received and recorded at Beatty, NV.(Stanford 1952: x, 1-4)

Several projects focused on hardware development. RAdiation Detection Indication And Computation (RADIAC) equipment included various types of electronics (portable hand held devices or larger systems for installation in aircraft or land vehicles) that were used to measure nuclear radiations for: 1) decontamination work, 2) health monitoring, and 3) surveys for determining location and level of contamination.

In project 6.4, both the Navy and Air Force developed and tested airborne RADIAC equipment to detect atomic clouds at sufficient range for the aircraft to take evasive action. (Terry 1952: v, 1-4)

The USAF had a requirement for the development of an all-weather Indirect Bomb Damage Assessment (IBDA) system that would evaluate the effects of atomic weapons on enemy installations by the correlation of data that the strike aircraft could collect. In a conflict scenario this capability would assist the planning of follow-on strikes and eliminate the requirement for immediate reconnaissance missions. It was also thought that some of these types of techniques might also be used on the ground for the determination of GZ, height of burst, and yield.

The all-weather requirement rendered radar techniques attractive. Some measurements had been made in the Pacific, but the ability to detect radar returns from atomic explosions over inland terrain was not in hand prior to BUSTER. IBDA radar sets were tested in two B-50D and one B-29 aircraft for project 6.5. (James 1952: ix, 1-3)

Long-Range Detection
During the early years of nuclear testing, little was known about the various possible methods of detecting nuclear detonations from remote locations. On Program 7, Long-Range Detection, 4 different detection methods were examined:

RADIOACTIVE DEBRIS
The first method consisted of two projects involving radioactive debris: One project, 7.1, determined: the initial cloud dimensions; the movement of the clouds within a few hundred miles of the site; the cloud width and the concentration of debris, primarily over the eastern part of the US; and the concentration of fallout at the surface. This project used aircraft and filters for tracking the cloud and specialized

in meteorological analyses and predictions. On occasion additional ground samples were obtained. (Allen 1952:1, 5-9) Project, 7.3, the Air Force conducted radiochemical analyses of the radioactive debris samples that were collected by aircraft with filters at locations close-in and out to about 1200 miles from the detonation point. Radioactive particles and decay processes were identified in the samples that were collected. (Singlevich 1952: v, 1-3)

LIGHT
Project 7.2 involved stations that were set up at Las Vegas, NV, Flagstaff, AZ, and Albuquerque, NM to detect light from the detonations. The stations were also equipped to obtain time to minimum light intensity (like a bhangmeter measurement) and to obtain data on the attenuation of light over paths of hundreds of miles. While this was the only light detection project conducted by AFSWP, the AEC laboratories did pursue similar experiments in later years. (Colson 1952: xi, 13-16)

SEISMIC WAVES
Project 7.5 involved the measurements of seismic waves generated by detonations. At less than 20 km from GZ, an assortment of 28 accelerometers, displacement meters, and tiltmeters were deployed. Between Reno, NV and Prescott, AZ, 74 displacement meters were fielded; and 26 velocity meters and a long-period displacement meter were placed between 900 and 2700 km. In addition, data from existing seismic observatories were also used in the analyses.
(Crocker 1952: ix, 1-4)

SOUND
Project 7.6 explored the feasibility of low-frequency sound detection from the detonations on BUSTER. They were made at 10 locations in several directions and distances. Baker, Charlie, and Easy were detected at 2200 miles and Dog was detected at 2600 miles, in Oahu.(Olmsted 1952: viii, 1-6)

Supporting Measurements
The Air Force's Air Weather Service provided meteorological assistance during RANGER; and in March 1951, Brig. Gen James McCormack AEC/DMA requested their support for BUSTER-JANGLE. This consisted of meteorological support for all tests as well as advice to the Test Director during the planning and operational stages. At a meeting at Los Alamos in July, Jack Clark "stressed the need for a completely operational weather station at the CP, with sufficient outlying upper air observing stations" to obtain wind values up to 25,000 feet. Cloud dispersion due to wind shear factors from 10,000 to 25,000 feet was important, and the proper general wind flow from 20,000 to 25,000 feet would assure the dispersion of the cloud. It was also desirable to have winds from the SW in order to move the cloud off into the sparsely populated NE quadrant. Clark "also stressed the need for cross-sectional analysis of pressure, temperature, humidity, and wind" velocity to allow proper evaluation of the effects of blast-reflection, which had been a problem on RANGER. This information would be obtained before each shot from the surface at GZ to an altitude of 1,500 feet. It ultimately required an instrumented

balloon anchored near GZ that was capable of being raised and lowered by remote control from the CP. (Karstens 1951: 1-5){8.2}

Federal Civil Defense Agency (FCDA) Collaboration With DoD On BUSTER
In the spring 1951, the Federal Civil Defense Agency (FCDA) was the organization responsible for civil defense (CD) in the US; and it sent observers to GREENHOUSE. Civil Defense organizations in the United States from WWII to 1951 had a torturous history, see Appendix I. The frequent changes that occurred in the organization and the fluctuating governmental funding for civil defense work are in stark contrast to the relative stability of the AEC and DoD over the same period. On BUSTER, the FCDA had a limited project to test family shelters with the DoD; and, in collaboration with the AEC, participated in another project for a larger communal shelter. The DoD conducted these two projects as their nuclear weapons effects projects 9.1a and 9.1b.

By UPSHOT-KNOTHOLE, the FCDA had its own test group at NTS that conducted a large sub-operation titled DOORSTEP. However, throughout the years of atmospheric testing, AFSWP assisted and collaborated with the FCDA. Today, the documents by the FCDA and their contractor's reports can be found in the DoD DTRIAC archives.

Project 9.1a was conducted to determine the effects of atomic explosions on small shelters that could be built by the average householder with available materials. Data developed by Lehigh University Institute of Research served as a guide to selecting four types of shelters.

A total of 29 shelter structures of four types were constructed:

- Type A - 18 Covered Trenches
- Type B - 5 Metal arches
- Type C - 4 Wood-arches
- Type D - 2 Basement Lean-tos.

Several configurations were tested. The Type A covered-trench shelters were prefabricated of wood and were small enough to be moved by truck and lowered into its site by an A-frame. It appears that a workman could neither fully stand nor fully lie down in one of these structures. The Type B, C, and D shelters were constructed in situ. The sites were located 25 feet apart along an arc 1200 feet from the Station 3 target point. (Flynn 1952:5-20)

Project 9.1b was the study of nuclear explosion effects on a 48-person communal shelter that had been designed by the AEC in liaison with the FCDA. Such shelters were intended for installation at what was considered to be the AEC's prime targets, like Hanford. The communal shelter was built 800 feet from Station 3, which was the air-drop target for shots Baker, Charlie, and Dog. The shelter was a horizontal cylinder, 48 feet long with a 7.5-foot inside diameter, perpendicular to the radial from GZ. Half was pre-cast concrete pipe, and the other half was iron pipe. Two ramps were poured-in-place at the ends of the pipe. Three feet and 3 2/3 feet of earth were placed over the concrete and metal sections of pipe respectively. (Corsbie 1952: 1, 3, 8)

JANGLE NWE PROGRAMS AND PROJECTS

JANGLE's seven NWE programs, with 56 projects, focused on blast and ground shock, cratering, structures, and thermal and nuclear radiation.

OPERATION	PROGRAM	# PROJECTS
JANGLE	1) Blast and Shock	15
	2) Radiological Phenomena	19
2 Tests	3) Blast Effects on Structures	5
	4) Special Phenomena	5
	6) Test of Service Equipment and Operations	7
	7) Long-range Detection	4
	8) Supporting Measurements	1
	Total	**56**

Scientists found that their measurements of air blast on Operation GREENHOUSE did not correspond to "ideal" wave forms predicted by air blast theory, i.e., characterized by a rapid rise to a single peak followed by a single decay. Double peaks occurred at ranges where the overpressure was about 15 psi, with the second peak higher than the first. At a given distance from GZ, structures were damaged less than they would have been had the airblast been the single higher peak predicted by the theoretical models available at the time. As a result, scientists were just starting to recognize some of the environmental conditions that could affect airblast generated by a nuclear detonation. (Lewis 1997:16)

In 1951, the understanding of the transmission of energy from a detonation through the ground was even less well understood than airblast. Ground motion created by an earth penetrating weapon was a key objective of JANGLE. A report by The Ohio River Division Laboratories states: "No information has been found of any previous attempts to determine permanent displacements at significant depths below the ground surface." (Ohio River Division Lab. 1952: 1).

In preparation for the nuclear airblast and ground motion measurements on JANGLE, the DoD conducted ten high explosive (HE) blast and shock-related projects at the NTS between August 28 and September 9, 1951 to explore:

- geologic, hydrologic, and thermal features of the test site
- seismic refraction characteristics of Nye County
- the accuracy of four theoretical studies regarding airblast and ground shock*
- scaling, cloud formation, craters, and base surge** analyses through a series of high explosive (HE) tests.

These tests were used along with scaling concepts to provide airblast and ground motion predictions for JANGLE. They were also used for the development of instrumentation to be employed on JANGLE. (Ponton 1982c: 79) [*Footnote: ground motion, the phenomena of energy transmitted through the ground, is generally referred to as ground shock, even when the velocity of transmission is less than supersonic.] [**Footnote: "Base surge is a cloud which rolls outward from the bottom of the column produced by a subsurface explosion." "For subsurface land bursts, the surge is made up of small solid particles but it behaves like a fluid. A soft earth medium favors base surge formation in an underground burst." (Glasstone and Dolan 1977: 630)]

The Uncle shot was located about 5 km N of Sugar, and the HE shots fired prior to BUSTER were located about half way between the 2 nuclear tests. Morris comments that having the HE shot between Sugar and Uncle "proved to be of immense value ... and is highly recommended for future operations." Diagnostic trailers were located about 8,000 feet from each GZ. (Morris 1952: xxi, 14-20)

Blast and Shock

Five projects focused on airblast phenomenology. In Project 1.2a-1, the BRL placed blast switches and microphones along two blast lines to measure the shock arrival times and peak pressures.(Ponton 1982c:80; Jackson 1993:8-9) In 1.3a, a partially successful attempt was made to measure the free-air velocity with sensors suspended from balloons.(Rankowitz 1952: ix, 1-10) In 1.3b, the velocity of the shock front was successfully obtained by using smoke rockets and photography.(Moulton 1952: v, 1-7) Canisters containing pressure measuring instruments were dropped from two aircraft in 1.3c. But, the drop accuracy was poor, and the data disappointing.(Haskell 1952: vii, 5, 14). Pressure versus time was successfully measured along the major blast lines in 1.4. (Howard 1952: 1-4)

Two ground motion projects are described in some detail in Killian 2011 which illustrate the state-of-the-art of the emerging technology in ground shock during 1951. In project 1.6, fifteen 45' long shafts were emplaced vertically underground to determine displacements.(Ohio River Division Laboratories., OCE 1952: vii, 1-5, 10, 20) Project 1.7 used a Japanese technique that was researched as early as 1899 in which shock pins were used. It was based on the principle that: "The magnitude of acceleration required to overturn a given size pin is dependent not only on the diameter/height ratio of the pin, but also upon the actual height of the pin, and the frequency of the ground motion". (Hansen 1952a: 2, 6)

Three projects involved extensive electronic gage measurements of accelerations along the blast lines. A large NOL project, 1.1, successfully measured horizontal, vertical, and transverse ground accelerations with gages installed at 12 ranges at depths of 10, 20, and 30 feet.(Morris 1952: xxi, 14-20) As a backup, in 1.2a-2, BRL installed 17 self-recording accelerometers at 10-foot depths at many of the same NOL stations.(Andrews 1952: 11, 17) For project 1(9)-a, horizontal and vertical accelerations, as well as earth pressure, and air blast measurements were successfully made by SRI at 20 ranges and depths between 5 and 68 feet. (Stanford Research Institute 1952: ix, 1-11)

Ground shock arrival times (TOAs) were successfully obtained on two projects. On 1.2b, the Navy used electronic switches at depths of about 17 feet in 31 holes on Uncle located between 5 and 33 feet from GZ. (Gannon 1952: 1-23) On 1.5b, the David Taylor Model Basin (DTMB) obtained TOA data using a seismic detector that sent a signal to an underground flash lamp when first motion was detected. A mirror reflected the flash from the lamps to remote surface cameras. (Cook 1952: 1-7)

The NOL used photographic techniques on 1(9)-b to measure cloud phenomena on Sugar and the base surge, column, jet, and smoke crown on Uncle. They then developed scaling laws based upon comparisons with the conventional high explosive results. (Young 1952: Abstract)

Radiological Phenomena

Program 2 focused on radiological phenomena, only some of which are described here. The radioactivity associated with the two JANGLE craters was of considerable interest, and two innovative methods were developed to obtain early time samples from the two craters.

Four remotely controlled "weasels" were developed for 2.6a and successfully used to obtain surface and core samples from around the lip and crater areas. The weasels had gamma ray detectors and television cameras to guide and observe the operations. (Forbes 1952a: ix, 1-23)

The second method, 2.6c-3, used a rocket-born sampling head and a towed line to retrieve the sample. The sampling head was driven into the ground on impact. After sample acquisition, the rocket was towed away from the highly radioactive area. This rocket system was truck mounted and had a range of about 1100 feet with a probably impact area of approximately 75 feet in diameter. Sampling was done at D+2 on both craters. (Maxwell 1952a: vii, 1-6)

Five projects measured gamma-ray activity. Project, 2.1a, fielded by Evans Signal Laboratory (ESL) and National Bureau of Standards (NBS), involved a large field effort with an extensive detector array that measured dose rates as a function of time. The dose rate versus time data curves were later used to obtain total dose levels. (Costrell 1952: 1-4, 42-44, 46, 51)

Three of the five projects used photographic film detectors. To record initial and residual total gamma dosage in project 2.3-1, ESL placed about 200 photographic films in lines on the surface as well as in/on foxholes, tanks, structures and animal cages. (Forbes 1952b: vii, 19, 25)

Three instrumented C-47s flew at about 600 feet over the area downwind from ground zero four hours after each shot to locate fallout and to evaluate the hardware. The aircraft surveyed about six thousand square miles in approximately four hours of flight time for 2.1c-1. (Harlan 1952: vii, 1-7)

Blast Effects On Structures

The most important, and by far the most extensive, DoD program on JANGLE focused on blast effects on structures. For Program 3, the DoD built a panoply of defense-related structures at the test site in order to make the most and best use of the nuclear shots in the hopes of learning the most in the shortest period of time.

The effort was indeed enormous, but unfortunately the structures were essentially untouched by Uncle "because the coupling of energy into ground shock had been grossly overestimated." (Lewis 1997: 18) The construction of extensive structures on JANGLE was a major gamble on the part of the DoD. It failed in its original purpose of obtaining significant damage to the structures; but never-the-less, the weapons effects community learned a lot about ground coupling in the Nevada alluvium. It also acquired important information about nuclear cratering. Figure 2-2.5 shows the contours of the Uncle crater.

Program 3, Blast Effects On Structures, consisted of five projects on Uncle. The 5 projects were: 3.1 Navy (Hazzard 1953: iii, 6, 118); 3.2 Army (Hansen 1952b: 4-30); 3.3 Air Force (Armour 1952: 7); 3.28 Sandia's instrumentation of the structures; and 3.29 Navy's soil mechanics tests.

The construction in support of Projects 3.1, 3.2, and 3.3 was staggering! There were 65 major structures or structural elements, categorized by 26 different designs. Sketches of the designs are given in Appendix H. (George 1979:128-33)

Sandia fielded the instrumentation of all test structures on JANGLE. "The location and type of instruments and the expected magnitudes of loadings were specified by the structure research groups ---". Sandia procured and installed the necessary instruments, operated them during the test and reduced the field records to a form usable by the structure research groups --." (Lenander 1952: 1, 3-8)

The Navy's Civil Engineering Research and Engineering Laboratory (CEREL) analyzed soil in the vicinity of the structures tests in order to provide information that was necessary for assessing the response of structures to ground shock. (Jackson 1993: 8-12)

Special Phenomena

In 1950, JANGLE (formerly code-named WINDSTORM), was being planned for Amchitka Island in the Aleutians where the near surface geologic material contains rocks and boulders. The issue of damage produced by missiles* generated by an underground explosion became of interest. [*Footnote: missiles here refers to objects present in the environment or debris generated by the destruction of objects in the environment which the detonation causes to become ballistic projectiles.] High explosive tests in 1951 at the Dugway Proving Grounds, showed that explosive generated missiles had large ranges. As a result a project was added to JANGLE to further explore the phenomena. Its objective was: "to obtain data on the range, size, and source location of potentially damaging missiles produced from a typical concrete highway or landing strip, and a typical concrete wall of a type that might be used in a small factory building of several stories."

Construction for project 4.5, was considerable. It consisted of: A) Highway strips of concrete and B) Walls and Foundation. The strips and walls were to become the source of missiles. The Highway strips were located between 15 and 300 feet of GZ. The Walls and Foundations were between 18 and 54 feet.

The walls and foundation were within the postshot crater as were most of the highway targets. In order to permit determination of the source location of missiles, the walls and highway strips were in small sections with different pigment and aggregate sections.

The Uncle results indicated that a 25 kt penetrator weapon, which was then of interest to the Army, detonated at the same scaled depth beneath a concrete runway, would generate missiles that could damage or destroy buildings out to about 1000 feet or airplanes on the ground out to 3000 feet. However, airblast from the same explosion would damage buildings out to about 2200 feet and airplanes on the ground out to about 6000 feet. Therefore, it was recommended that "further study of the missile problem is not justified."* (Vaile 1952:1, 9,10,13,38) [*Footnote: While the DoD did not conduct further missile studies, the civil defense community pursued a number of programs to assess potential missile damage in the civilian sector.]

Service Equipment and Operations
This program's 7 projects were extensive and diverse. A wide range of decontamination activities were conducted as the single Project, 6.2. The decontamination focused on: land reclamation; structures with a wide variety of types of construction materials; vehicles; paved areas; and the study of basic phenomenology. Numerous exposures were made, and a variety of decontamination techniques were tried on the various classes of contaminated objects. (Earle 1952: full text) A second project, 6.7, investigated various washing techniques for clothing and used a WWII standard field laundry unit in a trailer van. (Hughes 1952: 1-12)

Two medium tanks and one personnel carrier were used in 6.3-2 to assess the inhalation hazard to crews of armored vehicles, both during and following a detonation. The vehicles were exposed at 2000 feet in downwind from Uncle. One tank was head-on to GZ with hatches open, one side-on with hatches closed, and the Personnel Carrier was head-on with the commander's and driver's hatches open. " -- at H+50 hours, after decontamination, the vehicles were operated with one tank leading with hatches open, and the other vehicles following with hatches closed." They proceeded "--up to and beyond the crater lip and return." Fortunately, the H+50 hours decontamination had occurred. The combined exposure of the vehicles during Uncle and through the contaminated area after the shot resulted in airborne activity that exceeded "by a large degree the maximum allowable concentrations established by the Department of Defense and the U. S. Atomic Energy commission for lifetime exposure." (Engquist 1952: ix)

Long-Range Detection and Supporting Measurements
Program 7, Long-Range Detection, was essentially the same as BUSTER's. Program 8, Supporting Measurements, had only one project, technical photography for IBDA.

LIVING AND WORKING CONDITIONS AT THE SITE DURING BUSTER-JANGLE
Comments and observations in various documents indicate that the leadership at the site viewed the workforce as the most important nuclear testing resource. This view, however, did not always translate into desirable living and working conditions.

Jack Clark noted that incomplete facilities offered the major challenges for people at the time of BUSTER-JANGLE. These included everything from delays in occupancy of data recording stations, which limited proper system checkout, to inadequate signal cable installation. There were not enough cars or trucks. The communication systems were inadequate. Laundry, housing, and messing facilities were often insufficient to serve the test site population. All of these problems were addressed over time, but during the fall of 1951, conditions at the site were still quite primitive. (Clark 1952: 60)

Clark, while describing sleeping accommodations during BUSTER-JANGLE, mentioned that people worked such long hours that they were often unable to return to Mercury to rest. Perhaps this was merciful, since the on-site housing consisted of crude hutments that could accommodate 8 people each with 4 double-bunk beds. The alternatives included sleeping at the recording stations or at the CP. Either one could be better than the hutments. Even minor annoyances made it into Clark's post-operation report. For example, Clark recommended that the sheets for the beds be larger, to tuck in all the way around the mattress, in order to enhance comfortable sleeping. He also notes that there was no hot water available in the living quarters in the camp until after operation BUSTER. (ibid., 15)

Just after BUSTER-JANGLE, on November 30, 1951, the Director of Military Applications wrote a report to the AEC Commission regarding the "Expansion of the Permanent Camp". The report states: "The initial plans for the permanent AEC Camp were based on a normal capacity for housing, messing and administration of 400 men, expandable under short-term operating conditions to 600 by double-bunk beds." ... "No capacity was included for housing military and other people engaged in effects programs, or for their office and laboratory space." (AEC 1951: 1) Table 2-2.4 gives the population of the Permanent Camp during the BUSTER-JANGLE period.

Table 2-2.4 Average population housed in the camp during the period October 6, 1951 to November 17, 1951. Dates are for the end of the week over which the averages were calculated:

Date	SFO Admin	Scientific Program	Logistics Support	Effects Programs	Other Staff	Total
10/13	123	165	49	292	0	629
10/20	138	290	84	366	0	878
10/27	138	261	87	431	45	962
11/3	139	290	86	500	60	1075
11/10	129	256	84	510	47	1026
11/17	129	290	84	557	46	1106

(ibid., 5)

As justification for the expansion, the overcrowded living conditions during test series were described:

> Under current conditions for the BUSTER-JANGLE tests, personnel are living for prolonged periods (up to two months) with eight men to the room in double-decked bunks. The latrine facilities are designed for half this number. There are no recreational facilities of any kind, nor any areas for relaxation. The intensive and varied work schedules of the personnel make reasonable rest and relaxation well-nigh impossible in the barracks. Sanitation in the crowded mess-halls is difficult to maintain under overload conditions and the living spaces, crowded as they are, constitute unsatisfactory housing. It is emphasized that the present housing is inadequate for men who must spend from one to three months on-site away from their families. (ibid., 6)

The Camp Desert Rock troops were billeted in even more primitive conditions than those at Mercury. During BUSTER-JANGLE, it was mainly tents without electricity. There were attempts to keep the DoD personnel involved with the effects measurements separated from those participating in Desert Rock activities because Desert Rock personnel generally did not have security clearances. (Byron Ristvet 2003: Personal Communication)

In addition, the DoD had a building program near Camp 3 (which eventually became Mercury). This construction consisted of permanent and temporary buildings for housing, warehouses, etc. They built 28 Quonset huts, two shower-latrines, two Butler warehouses, and fencing. "The temporary structures located in and around Camp #3 consisted of approximately two hundred 8 to 10 men buildings. These, as well as the permanent structures were erected by Haddock Engineers, using various funds and at present it is difficult to place exact ownership and actual cost of each structure." (Hatlem 1951: 29)

The SWC made arrangements to be tenants at Indian Springs with initial plans for the accommodation of about 30 officers and 125 airmen. Ultimately, however, there were about 350. This probably included at least some of the personnel engaged in support flights. (Hatlem 1951: 11)

Over the years, the living conditions at the site improved enormously in terms of creature comforts, such as air conditioning, private rooms with attached bathrooms, and television. But, these improvements came at the cost of the communication and camaraderie that the close quarters of the early days engendered.

The work week at the site was set at 48 hours by the AEC and by Los Alamos, but the Laboratory soon realized that much longer hours were, in fact, expected and being worked by their employees. As a result Los Alamos adopted a special compensation package with a 30 percent increase above and beyond the 48-hour workweek for all time spent at the NTS by exempt employees. LASL non-exempt personnel were paid for the hours actually worked. Clark, commenting on the lack of uniformity in the policies of the various employers, urged that this be addressed prior to the next test series. (Clark 1952: 13)

Meals at the site cost three dollars per day. “Housing” (“bunking” might be more accurate) was fifty cents. A subsistence allowance of $6.00 per day was authorized for those off-site. The mess policy at the site was an issue that would be revisited over and over again throughout the years. During the BUSTER-JANGLE era, after much top-level attention, the meal rate was established at a flat $1.00 per meal. Eventually a turnstile, accepting silver dollars, was installed at the entrance to the Mercury cafeteria.

The Pacific operations were well known for the quality and quantity of the fare. This tradition was carried over to the NTS. Meal times provided not only nutrition but also an opportunity to socialize and to discuss work. These times were important and enjoyed, and they made being away from home a little bit easier.

The health and welfare of those at the NTS has always been a responsibility of both the Test Director and the Test Manager. Clark was also concerned with providing leisure time services to the personnel who were now spending three to eight months a year at the site. Specifically, he proposed that before the next operation in Nevada, TUMBLER-SNAPPER, Mercury have a movie theater, expanded PX facilities, and space available for a library and game room.

The amount of time that people spent at the site was highly variable and depended on the jobs they were doing. Those who were a part of the management organization with on-site responsibilities might spend the entire duration of the operation at the site plus some time prior to the first test and after the last shot. Few of these people had home bases in the Las Vegas area. They might be able to return home for a weekend or on rare occasions perhaps for a full week. A trip to Las Vegas was considered good “R&R”, and sometimes spouses would come for a brief holiday in Nevada. Those in the management organization without specific on-site responsibilities would come to the site either as needed or as their management style prescribed.

The Haddock Engineering Company was the support contractor for Operation BUSTER-JANGLE. Most of their crafts and support people lived in Las Vegas and commuted daily to the site. Usually these people began work some time before the first test of an operation and would terminate shortly after the last test. Individuals often participated in test operations over many years and became old hands at the site.

The amount of time that the scientists and engineers spent at the NTS was highly variable and depended on the specifics of the work. Their stay might be just a couple of days; or in some cases, it could be for the duration of the operation. Few of the technical people lived in Las Vegas during the 1950s. Although the number of scientists and engineers with a Las Vegas residence grew during the years, it always remained a small fraction of the total.

While many at Los Alamos and some at AFSWP had participated on RANGER, BUSTER-JANGLE was a new experience, and was much more complicated and involved. People in all areas of work faced new administrative and technical challenges, and the solutions to these challenges set precedents. In retrospect, for many, these were the most difficult of times as well as the best of times that occurred during their careers in nuclear testing.

BUSTER-JANGLE EPILOGUE

Test Director Clark had some observations regarding the facilities at the site:

> The BUSTER operation presented a construction program which was in itself of considerable scope.
>
> The situation was further complicated because the Control Point building was not completed in time for the operation. This factor considerably increased the Test Director's operational problems.
>
> ... (T)est personnel cannot be expected to live and work in partially finished facilities. No hot water was available in the living quarters in the camp until after Operation BUSTER, and permanent power was not available in the experimental area until two days before the first scheduled dry run. These conditions must be rectified before the next series of tests.
>
> The test organization for continental tests was assembled largely by the Test Director, although the Test Manager (Carroll Tyler) established an organization for over-all operations. Many of the Test Manager's personnel were newly assigned and had no previous test operation experience. With the experience gained in this operation, these personnel will undoubtedly provide for much more efficient planning of future tests. It is recommended that the Test Manager provide an organization for conducting continental tests, with responsibilities of all personnel clearly defined.

> The philosophy of requiring that the Test Director 'accept' the Military Effects Test program without screening on his part and solely on an operational non-interference basis was introduced in this operation with the JANGLE program. If this concept is to prevail in the future, suitable organizational changes must be made to clearly delineate responsibilities and authorities. Furthermore it is quite clear that with the addition of large military test programs considerable additions must be made to living and operating facilities at the Nevada Proving grounds. (Clark 1952: 60-62)

The recommendations described in the last section, which found their way into the AEC DMA report "Expansion of the Permanent Camp," were implemented in two increments. The first started immediately upon Commission and Bureau of the Budget approval. It included building: four barracks for men and one for women, a recreation facility, an administration building, a mess hall seating 365 and a warehouse. It also included upgrades of communications facilities, utilities, streets, and walkways. The second increment began in March 1952. It included: six more barracks for men; the expansion of the mess hall; additional utilities; 26 Quonsets including the installation of heat, water, and coolers; and the rehabilitation of Butler buildings. (LASL Archives: AEC DMA 1951: 2-4)

The report raises an interesting issue, the DoD was not asked to contribute to the cost of the proposed expansion even though a significant portion of the expansion was for housing military personnel. The report comments: ---"it is definitely desirable from an operation standpoint to maintain the unity of control of the Nevada Test Site by the Atomic Energy Commission. Since this site was acquired by the Commission for the purposes of furthering its weapons development program, it is believed that full control of the site should at all times remain with the Commission." (ibid., 6-7)

THE SITE: 1951-1952

Real Estate

On February 12, 1952, Public Land Order No. 805 withdrew from the public domain 417,459 acres (652 square miles) for the approximately 16X40 square mile tract plus a 2000 acre tract of Camp Mercury, both of which were requested by AEC Chairman Gordon Dean in November 1951. This area did not contain the Army's Camp Desert Rock, see Appendix D.

Site Named Nevada Proving Grounds, February 25, 1952

The AEC changed the name of the Nevada Test Site (NTS) to the Nevada Proving Ground (NPG) on February 25, 1952, prior to Operation TUMBLER-SNAPPER. This was to cause considerable consternation among some at Los Alamos who believed that the site was for testing, not proving. Norris Bradbury and Darol Froman were particularly irritated by the name, and they didn't pass up many opportunities to tell the AEC what they thought of it.

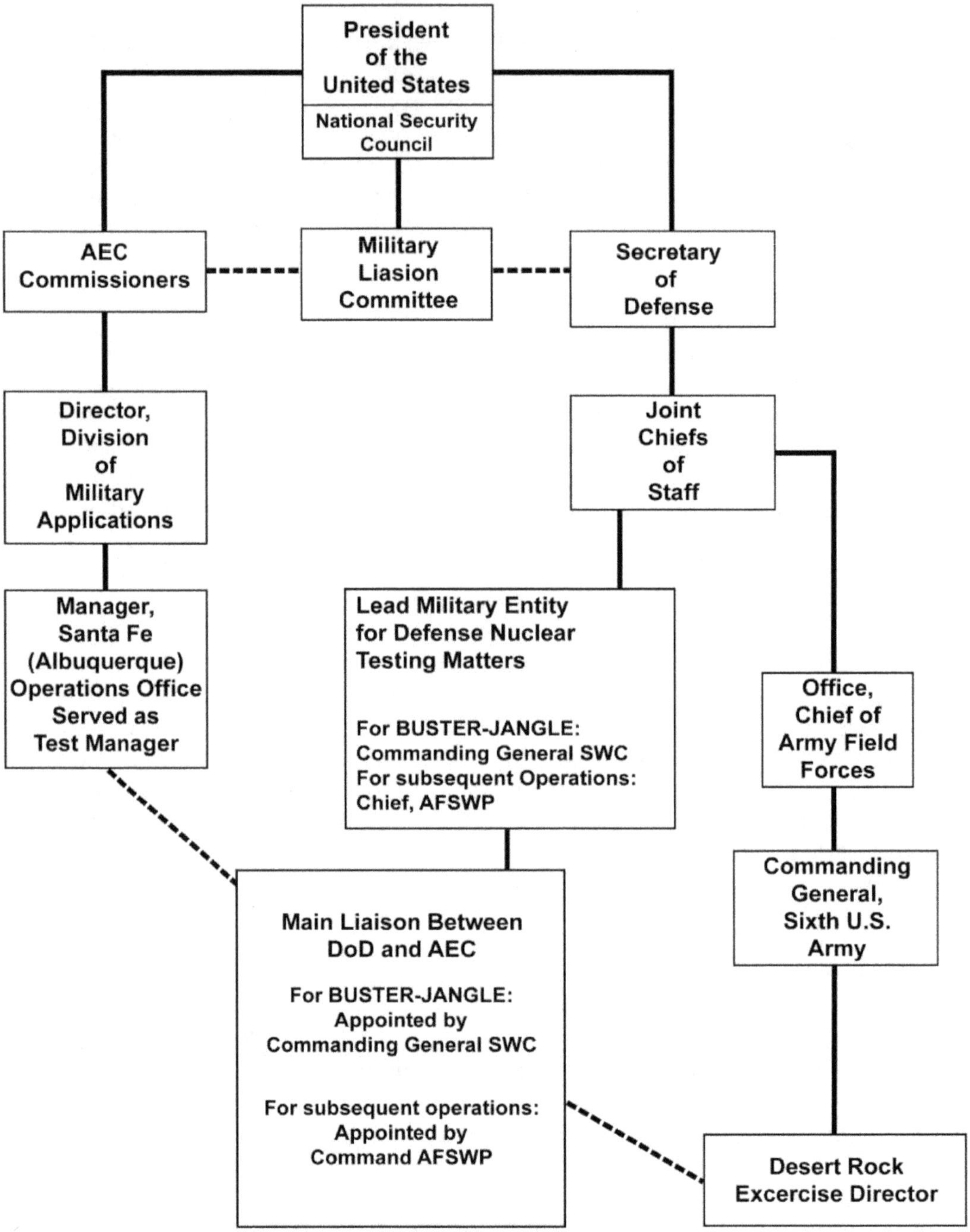

Figure 2-2.1 Upper Level Management During Operation BUSTER-JANGLE.

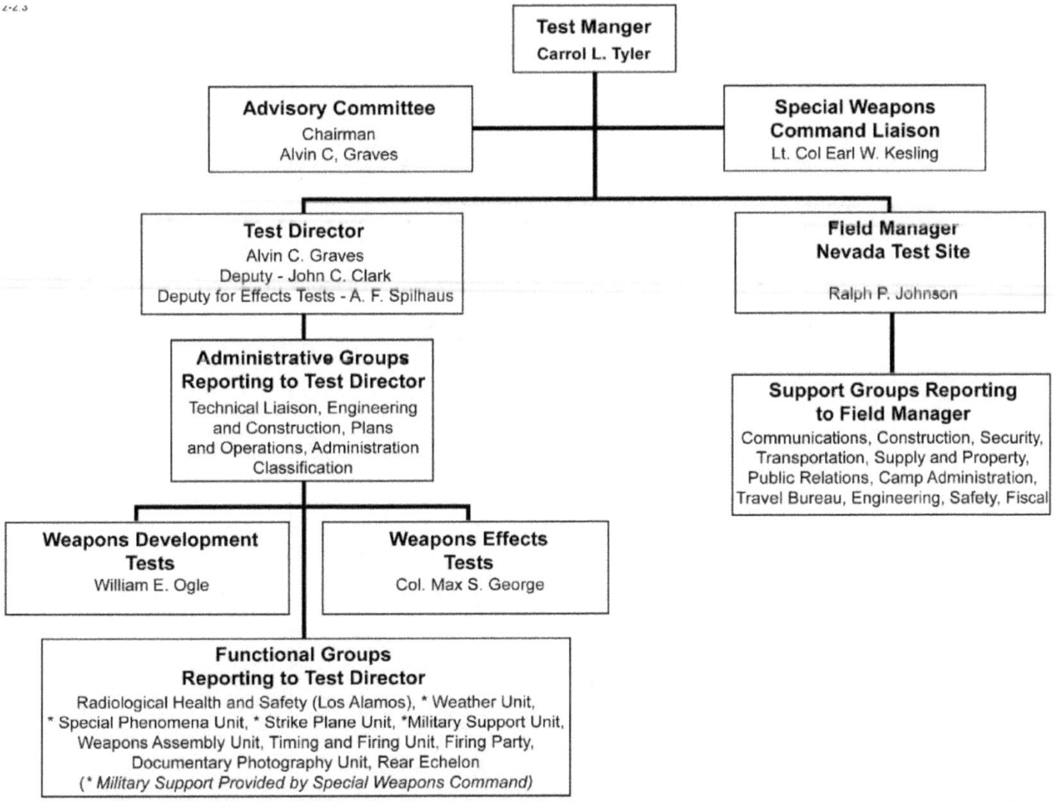

Figure 2-2.2 Nevada Test Site Organization (NTSO) for Operation BUSTER-JANGLE

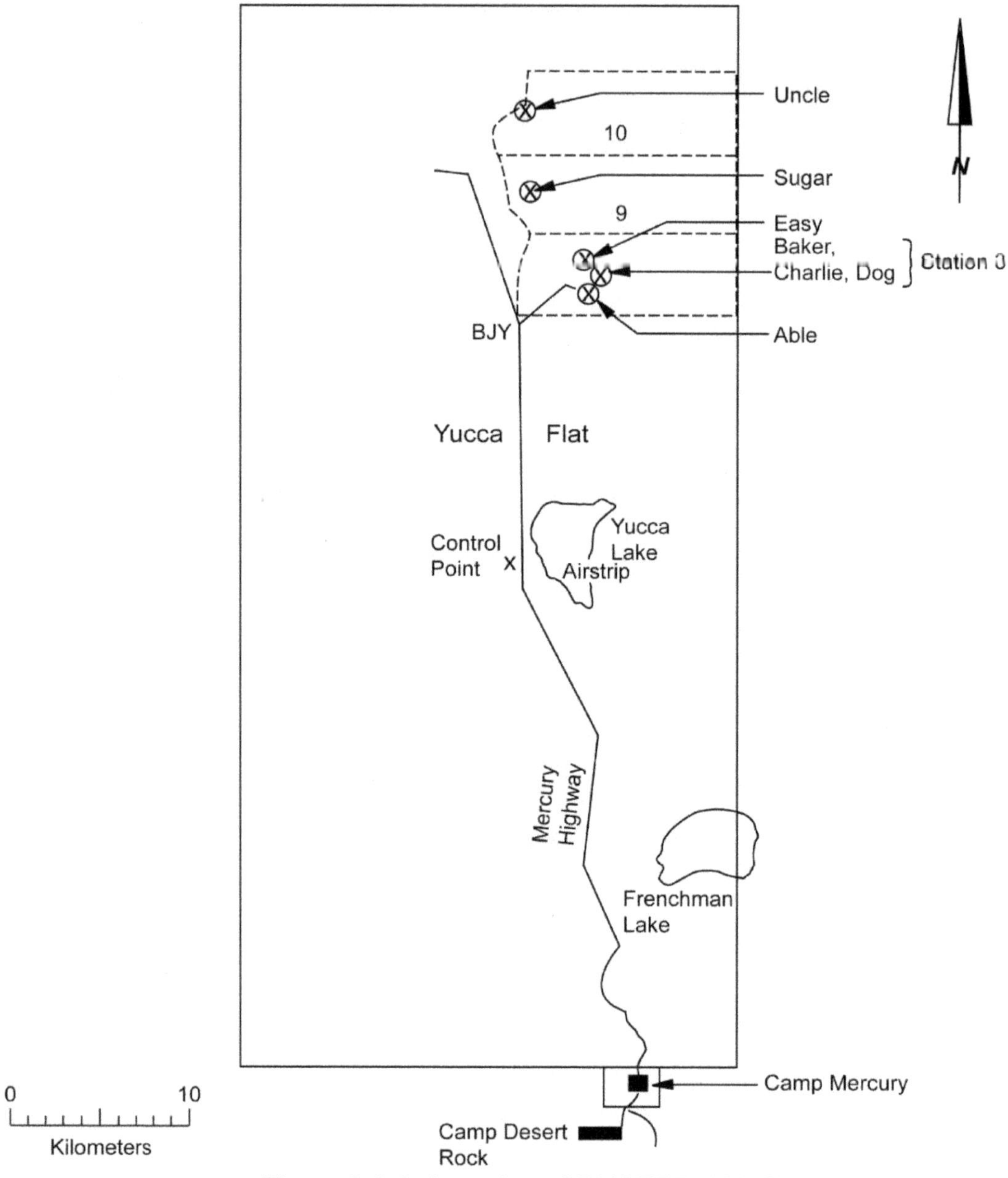

Figure 2-2.3 Location of BUSTER-JANGLE Tests.

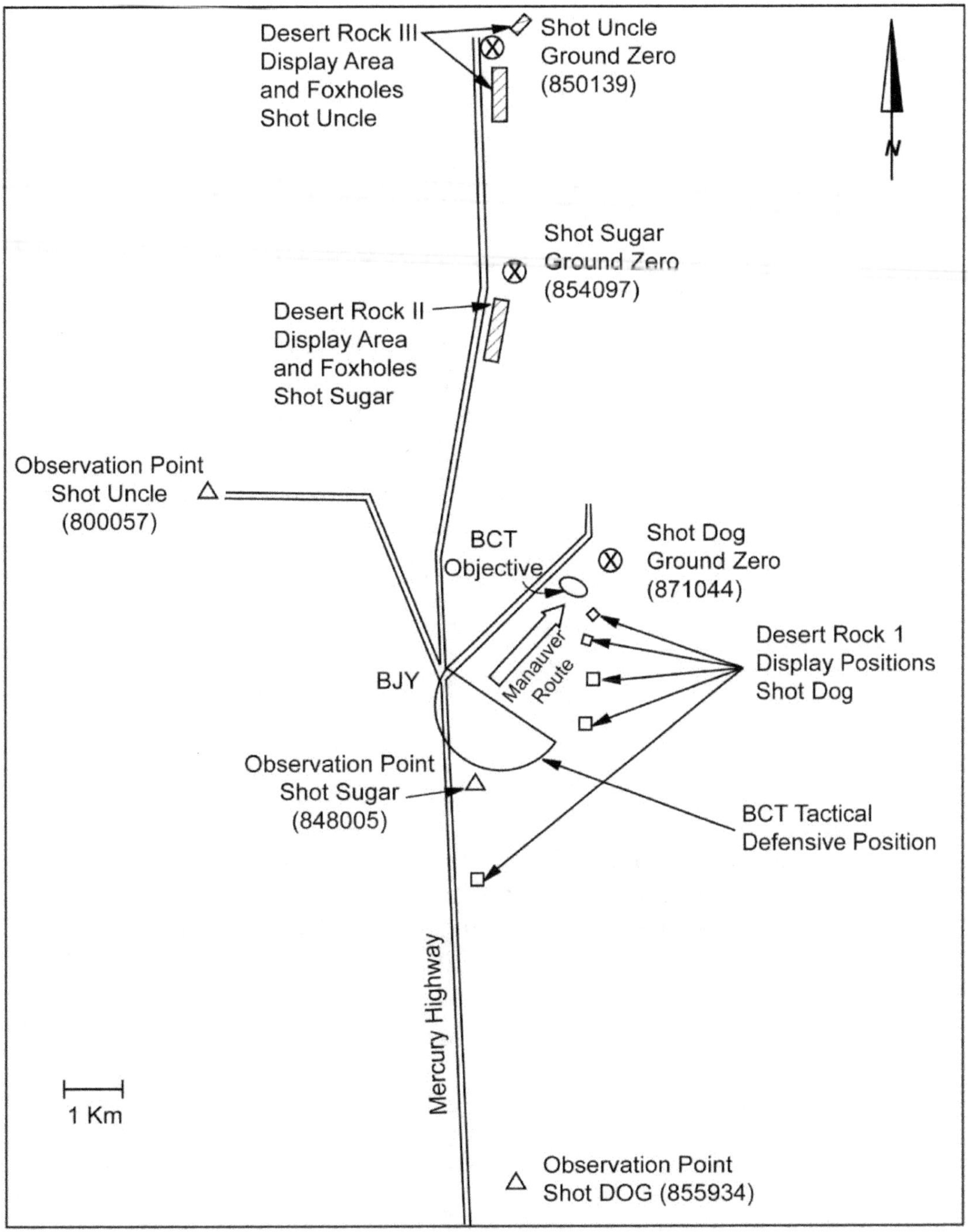

Figure 2-2.4 Desert Rock Exercises I (Dog), II (Sugar), and III (Uncle), during BUSTER-JANGLE.

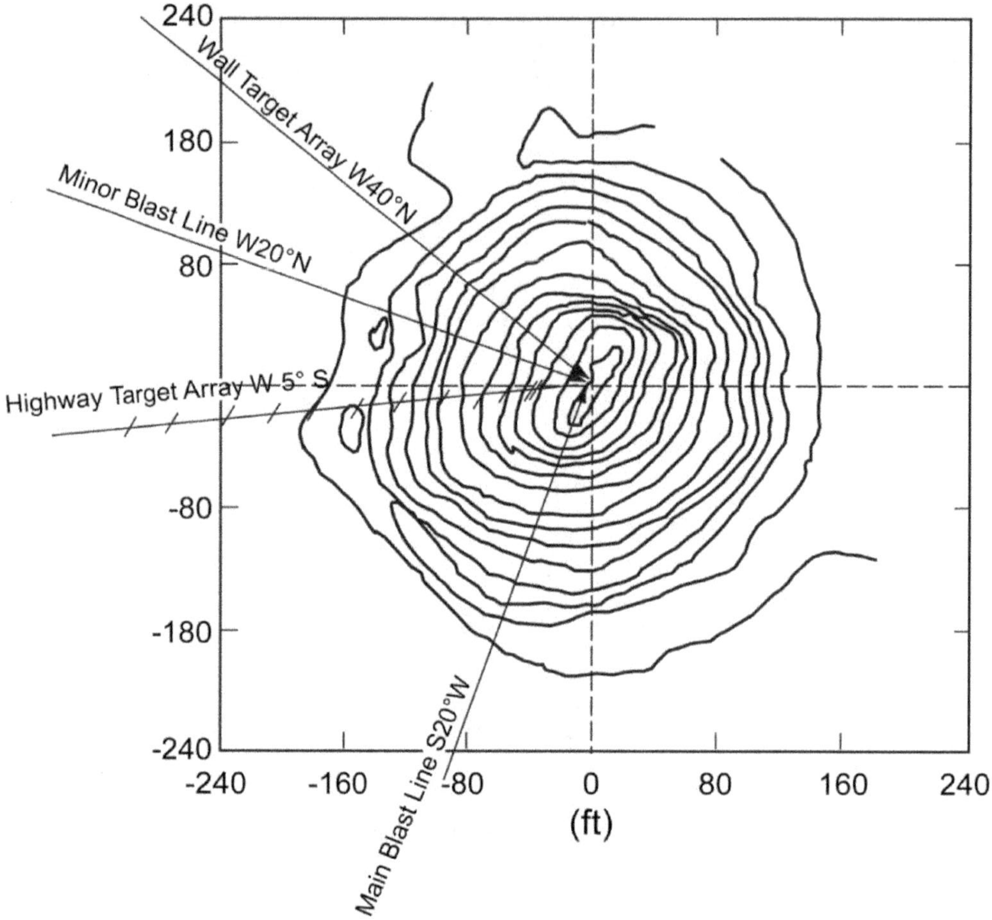

Figure 2-2.5 Uncle, 5 foot depth contours of crater with pre-shot SGZ at 1009.3 feet and bottom of crater at 956.7 feet. Pre-shot locations of Main and Minor Blast Lines. Lines for Wall target array (out to 54 feet) and Highway Target array (out to 300 feet).

Maj. Gen. John S. Mills

Lt. Col. E. W. Kesling 40-11PL12-2

John Clark

Robert "Father" Campbell

Control Point with Yucca "Lake" and BUSTER areas in background

Control room preparing to set off bomb. Seated (L to R): Col Max George, Maj Sam Connelly, and Jack Clark. Standing (L to R) Barney O'Keefe and Armstrong.

Troops waiting for blast

Troops watching Dog

Control Point for BCT troops

Chemical Corps protective clothing prepared for exposure (BUSTER 2.4a)

Personal Shelters, BUSTER {9.1a}, about 6 ft on a side, before being placed underground

Communal shelter with steel construction, BUSTER {9.1b}

Laying cable for Uncle

National Bureau of Standards record station in operation, JANGLE 2.1a.

Exterior of camera installation in C-47, JANGLE 4.1

Structure 3.16B prior to backfill

One of many exposures conducted for JANGLE 6.2

Bailey Bridge, Position #1

JANGLE Uncle emplacement shaft completed

CHAPTER 3. OPERATION TUMBLER-SNAPPER: APRIL 1 - JUNE 5, 1952

PLANNING

Even before the beginning of field operations for BUSTER-JANGLE, in the fall of 1951, the theoretical designers in Los Alamos were thinking about a follow-on series in Nevada in preparation for the future high yield shots in the Pacific. There were a number of design issues that required resolution before making the enormous investment in resources in an overseas operation. The tests which would address these issues became Operation SNAPPER.

Meanwhile, AFSWP was also anticipating what might be accomplished in the way of weapons effects research. In August, 1951, AFSWP advised the Departments of the Army, Navy, and Air Force that the AEC would probably conduct one or more nuclear weapons tests during the spring of 1952. Although the scope of tests had not yet been determined, AFSWP requested that by October 5, 195 (over two weeks before the first shot of Operation BUSTER), the military branches recommend projects for inclusion at the detonations.(Ponton et.al. 1982b:26)

The AEC formally advised the DoD that it intended to conduct a nuclear weapons test series of mostly tower shots in Nevada beginning around the first of May, 1952. During September and October 1951, AFSWP formulated a military effects test program, consisting of about 32 projects, for Operation SNAPPER. The Research and Development Board of DoD approved the program, recommending several modifications; and on January 19, 1952, the Joint Chiefs of Staff approved the revised AFSWP test program. (Ponton et.al. 1982b, 27)

Prior to this SNAPPER approval, data were obtained from BUSTER-JANGLE that indicated significant discrepancies between the predicted and actual overpressure resulting from airbursts. On December 14, 1951, AFSWP recommended to the JCS that a series of nuclear tests be conducted, primarily to measure the overpressure caused by airbursts. On January 10 the JCS approved the recommendation and requested that the AEC assume responsibility for fielding the additional nuclear events. First referred to as the QUICKIE Operation these events were renamed TUMBLER and scheduled to be conducted before May 1, the beginning date for Operation SNAPPER. The two operations were soon combined into one operation called TUMBLER-SNAPPER. (Ponton 1982b: 27)

The ultimate authority and responsibility for U.S. nuclear tests at Nevada, as well as in the Pacific, rested with the President of the U. S. He approved each operation and each test. Presidential approval for nuclear tests was issued by the Nation Security Council, see Figure 2-3.1 for the Presidential approval for TUMBLER-SNAPPER, dated March 28, 1952. The Presidential authority was delegated from the President to the AEC Operations Office during an operation.

NEVADA TEST ORGANIZATION FOR OPERATION TUMBLER-SNAPPER
Carroll L. Tyler was again appointed Test Manager for TUMBLER-SNAPPER by Brig. Gen Kenneth E. Fields, director of the AEC/DMA. As for BUSTER-JANGLE, he had an Advisory Panel under the chairmanship of Alvin C. Graves. The panel included Duncan Curry, T. N. White, Lt. Col. J.B. Hartgering, Thomas Shipman, Maj. N.M. Lulejinan, and C.A. Spohn, see Figure 2-3.2 and Appendix F. Tyler had three deputies: Al Graves, Deputy for Scientific Programs; Col. D. Tate, Deputy for Military Participation and Support; and Seth Woodruff, who was responsible for site management.(Ponton 1982b, 45)

In the fall of 1951, after BUSTER-JANGLE, the commanding general of SWC, General John S. Mills, requested that the responsibility he had been assigned in May 1951* for all military participation on atomic tests at the NTS be reassigned. [*Footnote: See section titled PREPARATION FOR JANGLE in Part II-Chapter 2] On January 18, 1952, the Chiefs of Staff of the three services assigned the following responsibilities to AFSWP: technical supervision of DoD experiments on continental military weapons effects tests, the coordination of military participation (like Desert Rock) with such experiments, and assistance to the AEC. To fulfill this new mission, Test Command, AFSWP, was activated on January 29, 1952, only about 2 months before TUMBLER-SNAPPER. Personnel who had been assigned to the SWC Joint Test Group were transferred to Test Command, AFSWP, with headquarters at Kirtland AFB.(Ponton et.al 1982c: 41-43; AFSWP1954: Vol.5:4.5.1) Col. D. Tate, who was Tyler's Deputy for Military Participation and Support, was from AFSWP. (Lt. Col. Earl W. Kesling from SWC had held this liaison position in NTSO during BUSTER-JANGLE).

As Test Director, Graves had two deputies: Jack Clark for Weapons Development, and Herbert Scoville for Military Effects Tests.(Ponton 1982b: 48) Bill Ogle was responsible for the Weapons Development Program under Clark, and Lt. Col. G. B. Page was responsible for the Military Effects Program under Scoville.

On March 1, 1952, the responsibility for rad-safe at the Nevada Proving Grounds was transferred from the Los Alamos Health Division to AFSWP Test Command under the direction of an Air Force officer, Philip S. Gwynn. AFSWP Rad-Safe had four departments: on-site operations, off-site operations, Indian Springs operations, and logistics and supply. Specifically, the DoD assumed responsibility for the safety and health of all military personnel involved in the Desert Rock exercises and in the weapons effects tests.(Hacker 1994: 73-75) Col. Kenner F. Hertford was appointed director of the Office of Test Operations at SFOO on February 1, 1952 to assist Tyler with the expanding work load. (SFOO Correspondence January 31, 1952)

REECo
In the spring of 1982, in volume 14 of REECo's corporate publication, the "REECorder", a description is made of the time period between RANGER and TUMBLER-SNAPPER. "After the (RANGER) test (series) was conducted all of the personnel who had been brought out here were returned to New Mexico. We did

not have any involvement at the Test Site from about March 1951 until January 1952. During that period of time the Atomic Energy Commission contracted with Haddock Engineering, who performed the construction work, built the Control Point buildings 1 and 2, and did the required construction work for the BUSTER-JANGLE series in areas 7, 9, and 10."

On January 7, 1952 REECo returned to the test site to prepare for operation TUMBLER-SNAPPER to be fielded in the spring of 1952. The new REECo contract was essentially the same as their contract in 1951, see Part II-Chapter 1; but McKee and Brown-Olds was now sub to REECo. The job was mainly to provide supervision for craftsmen hired locally, since the RANGER-era security clearance requirements were relaxed. In January 1952, Harold Cunningham was among those "temporarily" assigned to REECo at the test site, from Brown and Olds at Los Alamos. Frank Roger, the Deputy Manager of the Reynolds Nevada Project, was also among those assigned to the test site in 1952.

During the winter and spring of 1952 site maintenance was provided by the Nevada Company, which was a subsidiary of Haddock Engineering. REECo, with a work force of about 800, had responsibility for construction work directly related to nuclear testing. In June, following the TUMBLER-SNAPPER test series, there was a reduction in force and the supervisory people returned to their parent companies. The craftsmen had been hired for just the duration of the operation.

By spring 1952, the base camp was commonly known as Mercury, after one of the original code names for the site. In 1982 several of the early REECo employees reminisced about the old days (ibid., 3): "I'd like to talk a little about Mercury as it existed in 1951 and '52 which really shows how times have changed with the requirement, on the part of the laboratories, to have better housing and the requirement on the part of the DOE, to meet safety standards. The housing in Mercury, the majority of it, consisted of four- and eight-man hutments. These hutments were constructed out of single-walled masonite plywood. The air conditioning was that you opened up a couple of pieces of plywood in the summertime to get the breeze through them. About every 24 rooms, there was a restroom and shower facility. The heat was an oil stove with a 55-gallon oil barrel sitting outside."... "The thickness of that Masonite was only a quarter-inch thick; in the summer you would prop it up so the wind could blow through the screens, but when you let it down in the wintertime it didn't provide much insulation." ... "A hutment was our first office. Motor-pool had them too."... "Yeah, our first administration office was in one of those. That's all they let us have."..."We had some very ingenious craftsmen at Mercury. Bill Benner was the carpenter foreman for the shops at Mercury. Archie Mellot, whose son worked for us now (1982), converted their eight-man hutment into a two-man hutment so that you could not tell from the outside that anything had been done. Inside they had a shower and carpet on the floor, it was really very, very, plush." ... "The salamander stove in the center (of the room) would run low on oil and would explode now and then. The top would come off and soot would fly all over the room."

A small team of Reynolds people remained at the site from late spring of 1952 until December 1952, when REECo received another AEC contract for site maintenance and construction. At that time REECo started a manpower ramp up in preparation for Operation UPSHOT-KNOTHOLE.

CONSTRUCTION STATUS

On April 19, 1952, during TUMBLER-SNAPPER, the AEC Test Information Office in Las Vegas distributed a press release entitled "General Facts on Nevada Proving Ground Construction." The following items are taken from that 8 page document:

> All authorized permanent construction at the NPG is now largely complete. Little remains to be done on camp Mercury, the Control Point buildings and facilities are now nearing completion, including considerable modification and additions resulting from experience gained during last fall's test program. Considerable semi- permanent and expendable construction is well under way in the Technical Areas, including Frenchman and Yucca Flats.
>
> Permanent and related construction at the NPG since February 1951 will total approximately $8.6 million. This includes construction at Mercury and various specialized structures such as those in the Control Point area. Additional construction takes place in the Technical Areas, including the firing areas.
>
> Camp Mercury provides living quarters for test operations and maintenance personnel, a more temporary hutment camp for construction workers, utilities, administrative offices warehouses, and other supply facilities.
>
> The AEC utilizes private industry for maintenance, operation, and construction at the NPG. Personnel of the AEC's Las Vegas Field Office administer all activity at the Proving Ground, but the performance is by contractors.
>
> Construction work at the Proving Ground is rigorously divided into early stages which may be completed without security classification, and into later stages - usually beginning near the equipment stage - where classification is essential. When the classified stages are reached, all concerned workers involved must be cleared by AEC Security according to AEC standards before being permitted access to the classified work. All personnel of those contractor organizations which are more permanently a part of the Proving Ground establishment (administration, operation and maintenance contractors) are required to be cleared by the AEC Security whether or not it is contemplated that they will have access to classified data.

The peak of construction employment at the Proving Ground was reached in October 1951. Approximately 1500 workers are now employed there, and this number will drop fairly steadily to an anticipated figure of 500 by mid-summer.

Wage rates in all crafts are paid in accordance with current union agreements and are in conformance with Department of Labor requirements as specified in the Davis-Bacon Act. What constitutes overtime and overtime rates of pay for the Las Vegas area are established by negotiation between general and specialty contractors and unions of the area. Employment on cost-plus-fixed-fee contracts at the Proving Ground is now averaging about 54 hours a week.

The Nevada Company of Las Vegas has taken over maintenance and operation of the Proving Ground. Its services to the Proving Ground include: transportation, radiation instruments, utility repair, vehicle and building maintenance, guard duty, utility and sewage operation, and some minor construction work. Included is operation of the mess halls and billeting in Camp Mercury.

The Nevada Company is a partnership organized under Nevada laws for the sole purpose of supplying these services at the Proving Ground. The services had been supplied previously by Haddock Engineers, Limited, and Associates. Included in the new company's (i.e. REECo's) management are executives who had experience providing like services under the Haddock contract.

For the performance of these services, the payment of wage rates are normal for the service performed and with consideration for the Camp's isolated location and periodic operation. Charges for food and quarters are paid by the residents.

OTHER CONTRACTORS AT THE SITE

Many other contractors made substantial contributions at the site in specific areas. The S. R. McKinney and Sons, of Las Vegas, drilled water wells. The Newbery Electrical Company, of Los Angeles, did some electrical contracting. Holmes & Narver (H&N), of Los Angeles, which was a major contractor on the Pacific operations, did construction design and engineering. In addition, some architect-engineering management was handled by the Sukas Nasin Co. of New York.

Additional work on the infrastructure included the following task areas and contractors:

Roads: Foster and McHarg Co., Riverside, CA. and Dodge Construction Co., Fallon, NV.

Concrete Aggregate: Wells Cargo, Inc. and J. M. Murphy Construction Co. of Las Vegas, NV.
Power Plant Building: Claremont Construction Co., Claremont, CA.
Electrical Distribution System: Newbery Electrical Co., Los Angeles, CA.
Facilities in Forward Area: Lembke-Clough and King, Albuquerque, NM.
Electrical Distribution and Signal System: Victor L. Bongberg.
Initial Construction of Control Point Area Facilities: McNeil Contracting Co.
Construction of Facilities, Camp Mercury, First Phase: Lembke-Clough and King, Albuquerque, NM.
Water Supply Lines from Wells to Camp Mercury, and Facilities: Pipeline Construction Co., Riverside, CA.
Erection of Towers in Forward Area: Vinell Co., Inc., Alhambra, CA.
Tower Foundations and Structures, Forward Area: Lembke-Clough and King, Albuquerque, NM.

LIVING CONDITIONS DURING TUMBLER-SNAPPER

On February 11, 1952, field manager Seth Woodruff issued Bulletin No. 5, announcing that effective March 1 the United States Post Office Department would establish a post office at Mercury. Arthur L. Ortiz was appointed postmaster. The Mercury post office was a fourth class office and had the official geographic name of "Mercury, Nye County, Nevada". (AECTIO April 15, 1952) Incidentally, prior to the establishment of the Mercury post office, the address for the site had been P.O. Box 2088, Las Vegas, Nevada.

The dormitory accommodations were not, apparently, sufficiently appealing to attract everyone from the CP area, or forward area, to the base camp to sleep. As a result: "the Test Director ... requested that insofar as possible offices in the Control Point Building not be used for sleeping quarters. Accordingly, all organizations having office and work space allocated to them will refrain from installing beds in such space."

There were, however, "Two dormitories ... available (approximate capacities are 30 beds for male personnel and 8 for female personnel) in the Basement of the Control Point Building. This space is available for those personnel who must of necessity sleep during operational periods in the Control Point."(Clark March 21. 1952 Memo)

During Operation TUMBLER-SNAPPER there were approximately 150 people from Pacific Telephone at the NPG. (Pacific Telephone Magazine July-August 1952: 14) "Cloak and dagger silence curtained Pacific Telephone's work in the Nevada desert from the time the AEC first asked for service into the remote Las Vegas Bombing and Gunnery range in January 1951." The AEC wanted a branch exchange at Camp Mercury and another dial PBX at the Control Point. The working conditions were difficult. "And in the Nevada desert, folks can't catch a bus home every night." (ibid., 15) Western Electric installers worked shoulder to shoulder with Pacific Telephone crews. (ibid., 16)

A report from the Los Alamos J-1 Group described some of their various Personnel and Administrative activities during Operation TUMBLER-SNAPPER. (J-Division 1952; and Kelly, Armand W. 1955) The report provides some idea of the conditions experienced by the people in the field during the operation. It identifies "Significant Problems" and identifies "Suggestions for Future Operations". The following are a few extracts that provide a glimpse into the environment of the time:

> One of the problems involved was the failure of test personnel to process through the personnel office upon arrival and departure. As a result, J-1 Section was unable to give accurate reports to LASL of arriving and departing personnel. Payroll changes (on and off the 54-hr week) often were delayed, pending determinations from the individual or his group as to the exact dates of arrival and departure.
>
> In the early stages of planning, it became apparent that a sizable number of the Test Director's staff (including Military Effects personnel) would have to be housed in hutments. The reactions of participating groups to hutment housing were generally unfavorable; and, as the operation got under way, participating groups indicated that they would much prefer to house eight men in each room of the barracks rather than assign any of their personnel to hutments. This procedure was permitted, and with the completion of four new barracks (one was assigned to the Test Director's staff) it was possible to house all Test Director personnel in barracks. The hutments originally assigned to the Test Director were turned back to the AEC. (LASL Group J-1 1952: 10)
>
> The housing of eight men per room proved to be too crowded for adequate comfort during the extended period of time that most test personnel were required to stay at the NPG. The participating groups preferred this, however, to splitting their groups between hutments and barracks.
>
> Delays and postponements in the procurement of office equipment made it impossible to have it placed in the offices prior to the arrival of personnel. As a result, the equipping of the offices was hurried, confusing, and irritating to the users. Items that gave particular trouble were typewriters of the proper carriage size, file safes, and a mimeograph machine.
>
> With the completion of the new cafeteria, the messing facilities for test personnel were quite adequate and satisfactory. In general, the food was acceptable, and very few complaints were received regarding the food and service. (LASL Group J-1 1952: 9-13)

SOME CHARACTERISTICS OF THE WORKING TEAMS
During the 1950s nuclear weapons testing in both the Pacific and Nevada took place on a campaign (or operation) basis. Teams of people from a multitude of organizations were assembled to design and build the hardware and software that would be deployed in the field for one or more tests. Some of these people might devote a year or more in preparation for one test or for a whole operation. Others might be involved only occasionally.

These teams were composed of individuals with diverse backgrounds: scientists, engineers, technicians, clerks, office support personnel, etc. A team from Los Alamos (and later Livermore) Laboratory might be focused on experiments related to a nuclear device being designed, or analyzed, by their laboratory. Or, a team might consist of persons conducting a type of measurement on a number of nuclear devices on a number of operations. Sandia, EG&G, NRL and other collaborators were often integral parts of the weapon laboratory teams. For the DoD's weapons effects measurements, teams would be formed within AFSWP, Sandia and other DoD contractor companies.

As the time for the campaign neared, most of the people in the teams would "go to the field" to finish building and testing equipment. They were usually supported at the site by technicians and by crafts people. The non-technical personnel were usually hired for the operation by a test site contractor. However, like the laboratory people, many of these individuals returned for successive operations throughout the years. In fact, the test site support personnel often became full-fledged members of the laboratory teams.

There are several characteristics that defined the multi-disciplined teams. By far the most important was that they really were teams in the sense that there was a very strong commitment to collaboration and cooperation in order to succeed. A second was that the device diagnostics and effects measurements involved the development and employment of cutting edge technology. This was done in an environment of strong moral and financial support. The Nation was solidly behind nuclear weapons research and development, and those engaged in nuclear testing shared a sense of duty and commitment.

An environment where people enjoy the Nation's support and feel that what they are doing is important, is just the sort of atmosphere that is conducive to the development of dedicated teams. Those who participated in nuclear testing during the 1950s felt a pride in their work and in their accomplishments. Some even referred to feeling patriotic.

The cutting-edge technologies that were used provided not only a technically sophisticated program, but also an intellectually challenging and satisfying experience that contributed to attracting and retaining the best and the brightest at all levels. People who work at the cutting edge of technology often feel themselves to be a part of an intellectually isolated community. The nuclear testing teams were multi-disciplinary, using a variety of technologies, many of which were cutting-edge. It was the multi-disciplinary approach and opportunities to expand one's own knowledge and experiences that made nuclear weapons testing so interesting to many individuals.

The feeling that the program was important and was accomplished to the best of their abilities engendered an esprit de corps in the test divisions that continued throughout the entire nuclear testing era. Perhaps it was strongest in the early years, but there is no doubt that it played a major role in the many successes for almost half a century.

Los Alamos' J-Division Was A Team

J-Division leadership during TUMBLER-SNAPPER, in the spring of 1952 was as follows:

Alvin C. (Al) Graves, Division Leader
Stanley (Stan) Burriss, Alternate Division leader
John C. (Jack) Clark, Associate Division Leader
William E. (Bill) Ogle, Associate Division leader
Roderick W. (Rod) Spence, Assistant Division Leader

J-Division consisted of the following groups and group leaders:

J-1 Personnel and Administration, Armand W. Kelly
J-2 Physical and Personnel Security, William R. Adair
J-3 Plans and Operations, Duncan Curry, Jr.
J-5 Test Data and Information, Bergen R. (Gerry) Suydam
J-6 Site Facilities, Engineering and Construction, Robert H. Campbell
J-7 Equipment, Engineering and Specifications, Theodore Blechar
J-8 Assembly, Walter E. Treibel
J-9 Robert D. Krohn
J-10 Blast Measurements, Edward Zadina, followed by Francis B. Porzel
J-11 Radiochemistry, Roderick W. Spence
J-12 Neutron measurements, Clyde L. Cowan, Jr.
J-13 Gamma Ray Measurements, Newell W. Smith
J-14 Thermal Radiation, Leslie B. Seely
J-15 Photography, Gaelen L. Felt
J-16 Special Problems, Bob, E. Watt
J-17 Special Components, John C. Potts

The POGO Office and Staff Team

William Ogle had a loosely knit technical advisory group, called the POGO group, which included the leaders of the test-related scientific teams at Los Alamos. The Los Alamos test management personnel, mostly the group leaders, were referred to as the "POGO Staff," and the office as the "POGO Office." Dorothy (Dotty) Whitcomb*, Bill Ogle's secretary, was responsible for the naming this august assembly. [*Footnote: Dorothy Whitcomb became Mrs. Herbert Grier, and retained a close association with the Nevada testing community for over fifty years.]

Some suggested that POGO stood for “People Out Guessing Ogle”, but Robert Campbell has said that POGO really just refers to the cartoon character that identified the enemy as us*. [*Footnote: R. H. Campbell, Private communication, June 26, 2003 JCH].

The POGO publications were the perfect vehicles for Dotty to exercise her considerable cartoon and poetic talents*. [*Footnote: TUMBLER-SNAPPER Pogo Handbook, Also called the Poor man’s SNAPPER Handbook, JOL-7908, Winter 1952]

The POGO group continued on and off throughout Ogle’s tenure in J-Division. (W. E. Ogle retired from Los Alamos in October 1972) Among other things the POGO office was responsible for the interlocks that prevented a firing signal from detonating the device if key experiments were not operational, i.e. were in a “no-go” condition. To describe this Dorothy Whitcomb composed the following ditty:

There once was a creature named Pogo
Who tied himself into the NO-GO
And found, by the way,
To his rue and dismay
That testing was surely a slow-go

And so when he went out to Snapper
(Attired in a garment so dapper)
He put his foot down
And would GO into town
But to the site? - NO-GO
(He’s happier)

TUMBLER-SNAPPER TESTS

The focus of TUMBLER was weapons effects, with the principal interest being the exploration of the relationship between the height-of-burst to the overpressure at the ground. The tests were: Able, Baker and Charlie. SNAPPER consisted of Dog, Easy, Fox, George, and How, which were tests of weapons, or weapons concepts, considered for inclusion in the defense arsenal. As it turned out, Charlie and Dog were relevant to both the development and the effects programs.(Ponton 1982b: 28)

The 8 tests conducted on TUMBLER-SNAPPER are listed in Table 2-3.1, and their locations at the Nevada Proving Grounds are shown in Figure 2-3.3.

Table 2-3.1 Nuclear Tests Conducted During TUMBLER-SNAPPER.

TEST	DATE-1952	TYPE	LOCATION	YIELD (kt)
TUMBLER: DoD/LASL, Weapons Effects				
Able	April 1	Airdrop 793 ft	5	1
Baker	April 15	Airdrop 1109 ft	T-7, Sta. 3	1
Charlie	April 22	Airdrop 3447 ft	T-7, Sta. 3	31
SNAPPER: LASL, Weapons Related				
Dog	May 1	Airdrop 1040 ft	T-7, Sta.3	19
Easy	May 7	Tower 300 ft	T-1	12
Fox	May 25	Tower 300 ft	T-4	11
George	June 1	Tower 300 ft	T-3	15
How	June 5	Tower 300 ft	T-2	14

(DOE 2000, 4-5; Lewis, 1997, 19)

T-1, T-2, etc. indicate tower target area 1 or 2 etc. All of these areas were in the vicinity of the B-J Y on the Mercury Highway north of the Control Point.

The TUMBLER tests Able and Baker, both at 1 kt, provided the DoD with data for a comparison of overpressure at the ground for two different burst heights. Baker and Charlie results provided a comparison which approximated cube root scaling, i.e., (3447 ft/1109 ft) ~ $(31/1)^{1/3}$. Other test results provided additional data for other comparisons.

The AEC was under pressure to open the test site in order to give government and media personnel an opportunity to view a shot. Consequently, the test director, Alvin Graves, suggested to the AEC that shot Charlie be an open shot. State Civil Defense Directors, several governors, 60 civil defense observers, and nearly 300 members of the press, radio, television, and motion pictures were invited to attend the April 22, 1952 test. The Federal Civil Defense Administration (FCDA) played a major role in justifying and administering this first "open shot." (Hacker 1994: 77)

News clippings (Bechtel archives, Las Vegas A-83-0025 Box 9, Folder 10) reported that the first open shot resulted in the first "live" telecast of an atomic bomb explosion. AEC Chairman Gordon Dean advertised this event in a briefing to newsmen invited to the shot as "the most powerful tested in the continental United States." (William L. Laurence writing on April 21 for the April 22, 1952 New York Times) It was a 31 kt airdrop, detonated at 3500 ft above the terrain, and apparently failed to impress all of the media observers viewing the telecast. (William L. Laurence, April 22, at Yucca Flat for the NYT) Hal Boyle, an Associated Press staff writer, reported in the April 24, 1952, Binghamton Press (New York) that "There was a mixed reaction to the A-bomb's video debut." John Crosby, writing for the April 27 New York Herald Tribune noted that, "Even when the picture arrived - which was a sometime thing - it was the dullest possible picture." Newspaper reporters are hard to impress, particularly when they feel that an event has been overhyped.

It wasn't all negative though. William Laurence, who had seen several nuclear explosions, noted after viewing the shot from News Nob in Nevada, that he was impressed by the sight and by the military potential.
(New York Times, April 23, 1952).

Other shots also made the papers. Fox was reported by the May 26, 1952 New York Times "... as the most spectacular yet seen by Las Vegas residents." George, fired on June 1, 1952 was reported in the June 2 Albuquerque Journal as being felt and observed 400 miles away in Southern California. How, a 14 kt shot fired on June 5, was reported in the June 5th Santa Fe New Mexican as having a flash that "was reported seen in Kalispell, Mont., about 735 miles northeast of (Las Vegas)."

An International News Service (INS) story noted that "Rep. Daniel J. Flood of Pennsylvania was an official observer at the recent atom bomb (George) blast in Nevada in which tanks were used for the first time. As acting chairman of the House Army Appropriations subcommittee he was permitted to advance into "no man's land" with the troops." Congressman Flood was impressed. "American soldiery, brilliantly supported by American science, is achieving such progress in offensive and defensive military tactics that the free world can face up any threat of Communist aggression with resolute confidence." Representative Flood continued: "I know whereof I speak for I have just returned from 'Ground Zero.'"
(International News Service (INS) 6/9/1952, Washington, DC)

At 0430 PDT over the "loud speaker system came the music, 'Oh What a Beautiful Morning'". The last ten seconds were slowly ticked off and the bomb detonated at 0500PMT. (6/10/1952 Albuquerque Journal)

Returning to a discussion of the shots from a lab perspective, Dog, or SNAPPER 1, was, among other things, a physics experiment related to the thermonuclear weapons program. It was expected that the data from Dog would contribute to the experiments to be performed on the CASTLE series that was planned for the winter and spring of 1954 in the Pacific. Dog was an airdrop in area 7, fuzed to detonate at a relatively low altitude. The low altitude also made Dog useful to the military for their height-of-burst research and facilitated their ground-based neutron measurements.

Shot Easy was designed to explore a new weapons concept. It was one of a pair. If the first one was successful, the second one would not be required. As it turned out, the second shot of this pair was not required. Easy was fired on a 300-ft tower in area 1 and yielded 12 kt.

The radioactive cloud from Easy encountered very high upper level winds, which carried the bomb debris to the northeast and exposed the Lincoln mine, about 45 miles from ground zero, to contamination. At the time of cloud passage, the maximum radioactivity level at the mine was 0.8 r/hour. A lesser amount was detected at Ely, Nevada; and a trace was observed at Salt Lake City. The Easy shot contaminated the Fox and George firing sites, which resulted in a postponement of those two events. (Hacker 1994, 79)

The experience with Easy convinced the Test Information Director, Richard G. Elliot, that the nuclear test management needed improved communication with the public regarding fallout hazards. (Hacker 1994: 80) This issue was reinforced and would be specifically addressed during the summer of 1953.

SNAPPER-Fox was a particularly exciting shot. It misfired at five in the morning of May 20, 1952, and the deputy test director, Jack Clark, who was also the firing party commander, climbed the 300-foot tower to disarm the device. He had done so once before, on BUSTER-Able, in October 1951, but that device was on only a 100 foot tower. There were a total of six men on what was called the firing team - The Los Alamos test director, Al Graves; Clark; Carroll Tyler; and their colleagues Gaelen Felt of Los Alamos, and Herb Grier and Barney O'Keefe of EG&G.

Barney O'Keefe and John Wieneke (LASL), who were firing circuit experts, accompanied Clark in the climb and the disarming mission. They approached the tower in Clark's automobile with the sun-visors down to shield their eyes in the event that the bomb detonated as they approached. When they were about a mile from the tower they raised the sun visors, because they were no longer concerned about being blinded by the flash of a nuclear explosion that they would not survive in any case. Since the tower elevator had been removed after the firing party armed the device, the three were forced to climb the ladder to the 15 foot square cab. Clark performed the actual disarming operations with Wieneke monitoring his actions while O'Keefe reported by telephone to Graves a description of the steps in the procedure. (Cahn 1952: 17-19)

Diagnostic scientists discovered, a few hours later, that an interlock associated with a crucial experiment had prevented the detonation from taking place. The purpose of the interlocks was to prevent firing the device if the most important experiments were not operating properly, and the interlocks were doing what they were designed to do. The experimental problems were corrected, and Fox was fired on May 25, 1952, with a yield of 11 kt. The Fox radioactive cloud went to the northeast, but with considerably less contamination than that from the Easy test.

Fox and George both explored several weapons physics issues that had been under consideration for some time. They were both tower shots. Fox went at 11 kt in area 4 and George at 15 kt in area 3.

The last shot, How, was a tower shot in area 2 and went at 14 kt. How was a test of a new, more advanced, design in preparation for further exploration during future test series. Shots George and How resulted in very little off-site contamination.

The off-site blast damage claims for TUMBLER-SNAPPER were much less than for either RANGER or BUSTER-JANGLE. This was attributed to the meteorological conditions that prevailed at the time. (Hacker 1994: 81)

DIAGNOSTICS WEAPONS DEVELOPMENT EXPERIMENTS

"Many of the Experiments planned for SNAPPER are included as feasibility, technique, and theory tests for IVY* measurements. And a few - because we ALWAYS measure neutrons." (Ogle 1985:13) [*Footnote: IVY, the upcoming operation, fall 1952, on Enewetak.]

Ogle 1952 and Appendix G are the two references for this section regarding the diagnostics experiments conducted during TUMBLER-SNAPPER, see Table 2-3.2.

Table 2-3.2 Weapon Diagnostic Programs on TUMBLER-SNAPPER.

PROGRAM #	PROGRAM	DIRECTOR*	LASL CONTACT
10	Alpha	D. C. Cook (NRL) E. H. Krause (NRL)	N. Smith J. Malik
11	Transit Time	D. C. Cook (NRL) C. B. McCampbell (Sandia) V. Josephson (W-5)	N. Smith N. Smith
12	Technical Photography	G. L. Felt (J-15) H. E. Grier (EG&G) B Brixner (GMX-9)	 G. Felt G. Felt
13	Radiochemistry	Rodney Spence (J-11)	--
14	Initiators	V. Josephson (W-5)	N. Smith
15	Gamma-Rays	J. S. Malik (J-13) E. Storm (H-6)	 J. Malik
16	Electromagnetic Effects	C. L. Cowan (J-12)	
17	Neutrons	C. L. Cowan (J-12)	
18	Thermal Radiation	H. S. Stewart (NRL)	L. Seely
19	Blast Measurement	H. E. Leander (Sandia) F. B. Porzel (J-10)	F. Porzel
20	Timing and Firing	H. E. Grier (EG&G)	A Embry

(*The divisions at Los Alamos that contributed to TUMBLER-SNAPPER diagnostic programs are those named: J, GMX, W and H. The division name followed by a number, refers to a specific group within the division. NRL is the Naval Research Laboratory)

Alpha

Measurements of alpha for TUMBLER-SNAPPER were under the direction of D. C. Cook and E. H. Krause of the Naval Research Laboratory. Alpha was measured for all shots except for the DoD shot Able. Two experimental arrangements were used. The first was a phosphor-photocell detector used by W. Hall, of the NRL, on GREENHOUSE in the Pacific. "It consisted of a large coaxial phototube RCA type C-7154, immersed in a fluorescent solution of terphenyl in toluene, which was housed in an 8-inch-diameter aluminum can." (Ogle, Tumbler-Snapper, 1952: 21)

The second detector was similar, except that an RCA 5819 photomultiplier was used in place of the photocell. Both types of detectors were fed to oscilloscopes that employed Rossi sweeps, a technique described above in Part I-Chapter 1.

Transit Time
[*See Part II-Chapter 1 section titled TEST DIAGNOSTICS, Transit Time]. C. B. McCampbell, of Sandia, measured the transit time for all TUMBLER-SNAPPER shots, except for Easy and How. Unfortunately McCampbell's equipment failed for shots Charlie and Dog and no data were recovered. D. C. Cook measured the transit time for the tower shots Easy and How. V. Josephson's group (W-5) at Los Alamos also measured the transit time of shots Fox and George. (ibid., 38)

Yield
EG&G measured the fireball diameters as functions of time for all shots using motion picture cameras and high speed still cameras.

EG&G used the following relationship to derive the yield:

$$W = 1.294 \times 10^{-8} \times \rho \times (Dt^{-0.4})^5$$

Yield (W) in kilotons, equals, a constant (1.294×10^{-8}), times the ambient air density (ρ), in grams per liter, times $Dt^{-0.4}$ to the fifth power. D is the fireball diameter in meters at t msec. The term $Dt^{-0.4}$ is actually a variable in the early stages of the growth and becomes more or less constant in the region of light-minimum time. This near constant value is used to find the yield. (ibid., 41)

This formulation of the yield from fireball growth gives good agreement with other yield measurements for most shots except for very low yield detonations. F. B. Porzel at Los Alamos developed an alternative formulation that gave somewhat better agreement for selected shots. (ibid., 42)

The yield was also deduced by NRL from measurements of the thermal flux from each device using recording thermopiles*. The main thrust of this research was to explore the relationship of the thermal yield to the radiochemical yield. The conclusion was that the thermal yield was proportional to the radiochemical yield to the 0.87 power. (ibid., 43) [*Footnote: A set of thermocouples arranged to measure heat.]

The bhangmeter yield, which was deduced from the time to the minimum of the light intensity, was measured on all shots. The time is approximately proportional to the yield to the 1/3 power. The time is approximately 14 msec for a device with a yield of 20 kt. Bergen Suydam, at Los Alamos, developed a calibration curve based upon past data. It is interesting to note that the GREENHOUSE data indicated that the time to minimum is color dependent. Consequently the calibration data were taken with photocells having the same spectral response as the photocells in the bhangmeter apparatus. (ibid., 46)

The measurements of yield by radiochemistry were, of course, performed on every shot and were generally considered the standard. Rodney Spence, the radiochemistry group leader, concluded that the uncertainty in the absolute yield by this technique may be as large as 10 percent. (ibid., 57)

Yields deduced from measurements of the rate of fireball growth were considered the next most accurate. The bhangmeter measurements were thought to be less accurate; perhaps good to 20 percent. Other techniques were in the developmental stage, with unknown accuracies.

Film-Badge Measurements, Total Gamma Dose
Ellery Storm of the Los Alamos group H-6 used film badges to measure the total gamma dose as a function of distance from 1500 to 4000 yards from surface ground zero on all shots except Able, the 1 kt airdrop. The purpose of these experiments was to measure the gamma dose as a function of distance and to test scaling laws for a range of yields. The fraction of the device total energy released as gamma rays is relatively small and is dependent upon the details of the design. (ibid., 48)

Neutron Measurements
The radiochemistry group, J-11, measured the internal neutron flux using threshold detectors placed inside the device. The detectors were specific isotopes with known energy thresholds for reactions involving one neutron incident resulting in two neutrons out. These processes are referred to as (n, 2n) reactions. The resulting isotopes are radioactive, and the total amounts produced in the detonation can be deduced. Neutron fluxes above the various thresholds were determined and shed light on the neutron spectra at the location of the detectors.

The neutron measurements group, J-12, used threshold detectors placed along radial lines extending out from ground zero. After the shot, the induced radioactivity was measured in the laboratory with scintillators and photomultiplier tubes.

ADDITIONAL MEASUREMENTS

Additional measurements were made on specific shots to explore the dependence of yield on the details of the firing hardware, and also to develop new diagnostic techniques. By and large the experiments were successful, and produced the desired data.

Fireball photographs of tower shots show spikes that radiate out along the tower guy wires. John Malik started exploring the physics behind this phenomenon during GREENHOUSE and continued on TUMBLER-SNAPPER. His experiments were facetiously known as the "Malik rope tricks." Malik suspected that the explanation involved thermal radiation absorption by the cables. He confirmed this by observing what occurred to cables that were wrapped in reflective aluminum foil, or that were painted white, and those left with the natural flat dark surface. The spikes occurred only on those cables left with the natural dark surface, which absorbed much more radiation than those wrapped in foil or painted white. (Shelton 1988: 5-25)

Investigation of Fluoresence Under High Gamma Fluxes
This was a series of experiments performed by the Naval Research Laboratory group under the direction of C.V. Strain. The purpose of this project was to investigate the possibility of converting gamma rays into light signals by the use of fluorescers*. [*Footnote: Fluorescers is the technical term used for materials that exhibit fluorescence. Fluorescers absorb radiation of one wave length then, nearly instantly, reradiate at a different wave length. This conversion process starts and stops almost instantly when the input radiation starts and stops.] Fluorescers were placed near the bomb to absorb high energy gamma rays. The light produced by the flourescers was "piped" (by using mirrors) to the recording station. The initial plans were to conduct this on only the Easy shot; but since the data acquired on Easy was not sufficient, it was also conducted on Fox and George.

Hot-Spot Experiment
The purpose of the Hot-spot experiment was 2 fold: 1) to try out photographic techniques and 2) to obtain useful information on the propagation of radiation. This project was conducted only on the How shot. The device was fired atop a 300 foot tower in a massive steel box. The bomb was placed at one end of the box and pairs of 35 foot-long steel pipes were fastened to the box at five distances from the bomb center. The time of the appearance of light at the bottom of each of the 10 pipes was recorded on 2 Bowen cameras placed in a block house 500 yards from the base of the tower.

The ends of the 35 foot pipes were not observed directly. Instead, the collimated beams were directed to 10 mirror piers on the ground near the tower base, and the Bowens looked at the mirrors. The blockhouse also contained 2 framing cameras which looked directly into the tower cam and recorded the progress of light down the outside of the How box. A picture of the experimental set up around the tower cab is shown in Felt 1952:24.

Light Absorption Around Bomb
By determining the amount of light absorbed around the bomb, photographic techniques for yield determination could be made more accurate. The first test run of this project was conducted on Easy by H. Stewart of Naval Research Laboratory. LASL's J-15 conducted this project on Fox, George, and How.

Airblast Measurements
LASL and its contractors conducted measurements of airblast. Two categories of surface loss for the blast wave were considered for making the measurements: Thermal effects and Mechanical effects.

Sound Velocity
Measurements were made by J-10 LASL. They indicated a marked increase in sound velocity prior to shock arrival that was consistent with the temperature measurements.

Dust Loading
Dust loading was measured with beta densitometers by J-10. Measurements showed significant dust loading before and after shock arrival.

Mass Motion Photography
Mass motion photography was conducted by J-10 with EG&G. They photographed motion of smoke puffs in air. Pre-shock dust and smoke was observed on all shots. On Dog, the thermal shock was observed arising from the ground.

DoD PARTICIPATION ON OPERATION TUMBLER-SNAPPER

Approximately 10,600 military and civilian personnel participated at TUMBLER-SNAPPER for the DoD. They took part in the following field activities, which are discussed below:

- Exercise Desert Rock IV - troop activities and support
- Nuclear Weapon Effects Projects
- Participation in AEC Weapons Development Programs

(Ponton 1982b, 36)

EXERCISE DESERT ROCK IV

As for BUSTER-JANGLE, the Sixth Army Commanding General was the Desert Rock IV Exercise Supervisor. He was "-- responsible for Army, Navy, Marine Corps, and Air Force personnel and for providing administrative and logistical support to the exercise troops". His offices were at the Sixth U.S. Army headquarters, located at the Presidio in San Francisco. The Commanding General's deputy was the Exercise Supervisor at the site and "... was designated Exercise Director and Commander of Camp Desert Rock." He was the person who was responsible for liaison with the AEC. This was accomplished through the Test Managers Deputy for Military Participation and Support, Col. D. Tate*, see Figure 2-3.2. The AEC Test Manager reviewed and approved all program activities associated with the nuclear tests at the site, including Desert Rock activities. (Ponton et.al.1982b, 53-54)
[*Footnote: At this time, Tate was Inspector General of Field Command AFSWP.]

The AEC demanded that the DoD assume "complete overall responsibility for safety of troops" and assure the AEC "that such responsibility is completely understood by DoD." In March, the AEC received a letter from Brig. Gen. Alvin R. Luedecke, AFSWP deputy chief which sought "revised ground rules for troop maneuvers". He sought revision "in the interest of indoctrination of ground troops to an extent which would be of value in readying them for the actual use of atomic weapons". Luedecke was advocating closer distances from a detonation than those with which the AEC felt comfortable. He also cited that: "Troops should also be allowed to 'maneuver in the vicinity of ground zero as soon as practicable after the explosion'". (Hacker 1994: 75)

Los Alamos was concerned about the closer proximity proposed by the DoD. Los Alamos insisted that the DoD acknowledge the laboratory's safety concerns with appropriate documentation. Shields Warren, of the AEC's Division of Biology and Medicine, was more conservative and refused to endorse the DoD's proposal. (ibid., 75-6)

Warren recognized that while the DoD agreed to assume the health and safety responsibilities for the military personnel at the test site, the general public held the AEC accountable for ensuring the professional conduct of the operations. He did not think the DoD's proposal was sound. The DoD, on the other hand, wanted troops at what it considered a realistic tactical range from a nuclear explosion and felt that the AEC's reaction was too conservative. Not surprisingly, Brig. Gen. Kenneth Fields, the AEC/DMA director, supported the DoD's position. Ultimately the AEC commissioners approved the DoD plan, but not without considerable soul searching. The operational details were left to the AFSWP Test Command exercise director and the AEC test manager. (Hacker 1994, 76)

Exercise Desert Rock IV, conducted on TUMBLER-SNAPPER, involved approximately 7,350 DoD military participants engaged in observation activities and tactical maneuvers. About 1,500 additional military personnel were needed to support the exercises. (ibid., 53)

A formal observation program, as described for BUSTER-JANGLE, was conducted on four of the eight TUMBLER-SNAPPER tests: Charlie, Dog, Fox, and George. A few members of the Exercise Director's staff observed Able and Baker. The observers at Easy were support personnel assigned to Camp Desert Rock. There were no observers on the last test, How. Observers arrived at the trench areas shown in Figure 2-3.4 about 90 minutes prior to the scheduled shot time. After the shot, the Desert Rock Control Group escorted the observers on a tour of the equipment display area to examine the effects of the detonation on equipment, fortifications, and shelters. (Ponton 1982b: 66-67)

The troop maneuvers were conducted in a manner similar to that on BUSTER-JANGLE. After the detonation, the maneuver troops left the trench area and began their advance, with radiological survey teams preceding the troops to determine the limits of safe advance.

The TUMBLER-SNAPPER Charlie exercise included four C-46 aircraft with 120 paratroopers of the 504th Regiment, 82nd Airborne Division who parachuted into the test area in support of the military maneuvers of Desert Rock IV. (4/21/52 Santa Fe New Mexican (SFNM) story by Bill Becker describing plans for shot Charlie) They took off from the Yucca Flat strip. (Frank H. Bartholomew writing for the SFNM on 4/22/52) The paratrooper troops were dropped in an area centered about 1.2 miles NW of surface ground zero.

A tank maneuver exercise was held soon after the George shot. (Hacker 1994, 77) As in BUSTER-JANGLE, a study was conducted by HumRRO of the psychological reactions of troops participating in the maneuvers. (Ponton 1982b: 70-72)

DoD NWE PROGRAMS AND PROJECTS

Prior to TUMBLER-SNAPPER, AFSWP and the Coast and Geodetic Survey conducted a study of the geology and topography of Yucca and Frenchman Flats. The resulting data were to be used in determining the effects of geological structure on the propagation of the blast wave. Also, both prior to and after TUMBLER-SNAPPER, Sandia conducted 250-pound TNT tests in order to gather data on the variation of pressure with height-of-burst. The data from tests prior to TUMBLER-SNAPPER were used to predict the pressures that would result from, Able, Baker, Charlie, and Dog.

About 750 DoD personnel participated in the effects measurements. (Ponton 1982b: 73) The eight effects programs with their 42 projects were less extensive than those on BUSTER-JANGLE, but their scope was impressive.

OPERATION	PROGRAM	# PROJECTS
TUMBLER-SNAPPER 8 Tests	1) Blast Measurements	9
	2) Nuclear Measurements and Effects	3
	3) Structures	3
	4) Biomedical	5
	6) Test of Equipment and Operations	5
	7) Long-range Detection	5
	8) Thermal Measurements and Effects	8
	9) Supporting Measurements	4
	Total	**42**

The AFSWP Test Command, Military Effects Group, was responsible for the DoD's effects measurements. This was the only operation on which Test Command would participate as it was replaced by AFSWP Field Command, Weapon Effects Test (WET) on August 1, 1952 (see Chapter 6). WET remained the responsible agent for DoD's weapons effects testing at NTS for the reminding duration of atmospheric testing.

AFSWP's measurements involved military and civilian laboratories, universities, support contractors, and the three armed services. (ibid., 74) Nearly all of the participating organizations had also participated on BUSTER-JANGLE. Some of the projects sponsored by AFSWP during TUMBLER-SNAPPER are briefly described in the following.

Airblast and Ground Motion Measurements

On BUSTER, the DoD did not conduct an airblast program, but Los Alamos and Sandia did, see Part II-Chapter 2. "The BUSTER shots were the first well instrumented nuclear tests using operational heights of burst."(Swift 1955: 15) The BUSTER tests "revealed considerable disparity, both in magnitude and in wave form, between the predicted and observed ground level pressures." (Salmon 1953: 15) "Overpressure results obtained at Operation BUSTER were 1/2 to 1/3 of predicted values."(Bourton 1952:13) The data from GREENHOUSE showed a smaller discrepancy from theory associated with the shots in the Pacific. New understanding and theoretical approaches were needed to describe what became known as “non-ideal” air blast. A substantial effort consisting of both airblast and thermal measurements was mounted by AFSWP to address non-ideal airblast on TUMBLER-SNAPPER.

The lower and irregularly shaped air blast waves were having a dramatic effect on strategic planning and on the presumed effectiveness of the limited number of stockpiled weapons. (Lewis 1997:19; Shelton 1988: 5-25) writes: “The nation was losing stockpile effectiveness faster than the fissile material was being produced.”

For the DoD, the general objective of the 4 airdrop tests was to provide data for an empirical height-of-burst chart (and handbook). TUMBLER Able (1 kt @ 793’) was intended to be a repetition of BUSTER Baker at a different HOB (1 kt @ 1118’). It was conducted in Frenchman Flat over terrain that was chosen to have more reflecting ability than BUSTER Baker’s terrain at Yucca T-7 Station 3. To “obtain a smooth, thermally reflecting, and dust-free surface”, the Frenchman Able area “was carefully dampened and rolled prior to the test”.

BUSTER Baker’s T-7 Station 3 site was used for TUMBLER Baker and Charlie and for SNAPPER Dog. “However, several factors contributed to undesired differences. During BUSTER, “the surface was extremely dry and powdery; a walker would frequently sink ankle-deep into the dust”. In “the spring of 1952, winter rains had increased the moisture content of both surface and subsurface soil to a considerable extent.” Finally, at the main blast line, “much of the sagebrush was removed by blading and the many vehicles packed the soil to a more compact condition. All these factors acted to reduce the difference (if any) between the reflecting ability” of the Frenchman and Yucca sites. (Salmon 1953: 26-7)

TUMBLER Baker (1 kt @ 1109’) represented a moderately high burst over BUSTER terrain. TUMBLER Charlie (31 kt @ 3447’), was a scaled version of TUMBLER Baker, “to test scaling laws over a 30 to 1 yield range”. TUMBLER Dog (19 kt @ 1040’) had a “scaled height of burst of about half that of” TUMBLER Able (1 kt @ 793’) and was similar to BUSTER Easy. (Salmon 1953: 26)

In Area 5, a main blast line was developed for Able. Another blast line, at T-7 Station 3 was developed in Area 7 with greater distances from the IGZ for Baker, Charlie, and Dog. The actual distances were different because the actual bursts were not at the IGZ.

	Burst Center with respect to IGZ		
Shot	**X (ft)**	**Y (ft)**	**Z (ft)**
Able	-122	+67	793
Baker	+108	-126	1109
Charlie	-108	-100	3447
Dog	-126	-164	1040

These differences in detonation location adversely affected a number of projects. (Salmon 1953a:28) Fourteen stations at various distances from the IGZ, were instrumented. Not all projects used all stations on all shots. For Able, an instrument trailer was located in an underground revetment 35’x 11’x 2.5’ high at 5,000’. For

Baker, Charlie, and Dog the trailer revetment was located at 12,050'. (Aronson 1952: 32-39)

Seven extensive airblast and ground motion projects were conducted along the blast line, (Ponton 1982b; Killian 2011). In addition, two B-29s successfully dropped 16 canisters for Project 1.1 with pressure gages closer to their planned locations than had occurred on JANGLE(Haskell 1953: 3-17); and smoke rocket photography measurements were made on project 1.5.(Aronson 1952: 23-7, 90-1)

A thermal line was located 25' from the blast line. Project 8.2 measurements along this line represented significant contributions to the non-ideal airblast issues. Temperature-time was measured on the thermal line at ground level, and on towers 10' and 50' by NRDL. "-results indicate that the rapidly fluctuating temperatures produced by the detonation vary markedly from point to point at the same distance from point zero. Severe pre-shock temperatures occurred above grade level only where the incident thermal radiation was sufficient to produce 'popcorning', i.e. exploding of sand by the absorption of thermal radiation." (Broida 1952: 3, 13-20)

Also, "Total thermal radiation measurement made near the ground indicated that, even before the arrival of the shock wave, serious obscuration is produced by ---- 'popcorning' of sand, and smoke produced by the burning of ground litter. The thermal energy received - (by a drop aircraft in Project 8.3)-"was appreciably greater than that received at equivalent distances along the ground. This increase is primarily due to reflection by the ground." (Broido 1953: 3, 14-21)

"Following TUMBLER-SNAPPER, it became apparent that the anomalous behavior was caused by thermal effects. Radiation from the fireball was indirectly producing a heated layer of air in front of the airblast wave, causing what was to become known as a 'precursed' airblast wave form. This thermal-blast interaction would become the subject of intensive study for years to come." (Lewis 1997:19) In addition to heated air in front of the airblast wave, soil in front of the airblast wave can explode, i.e. "popcorn", due to its absorption of thermal radiation. The presence of dust- and dirt-laden air just above the surface can reduce the peak pressure significantly.

Structures

The Structures program was very modest in comparison to JANGLE. Three impressive revetments were constructed for project 3.1: a G-Type, also referred to as a "Russian Revetment"*; a wall revetment; and a pit revetment. The structures were built primarily with the local soil. Timber shoring was employed for the near vertical walls. The G-type revetment had the soil surfaces stabilized. The wall and pit soil surfaces were not stabilized. Four aircraft were placed behind these 3 revetments (2 in G, 1 behind the wall, and 1 in the pit) for each shot. (Schraut 1953: 42, 47, 52) [*Footnote: Schraut (1953: 42) states:"Initial investigations indicated that the G-Type revetment would be the type to investigate according to information then available from the Directorate of Intelligence, Target Analysis Branch." DoD projects were generally focused on offense - how to destroy enemy targets; and the

Civil Defense projects were of course more focused on defense - how to protect people and assets.]

A total of 28 aircraft were subjected to the Baker, Charlie, and Dog detonations both behind the revetments and in the open. The majority of the 28 test aircraft were "obsolete and did not include any foreign types": 16 F-47s; 2 F-86s; 1 F-90; 7 B-17s; 1 B-29; and 1 B-45. The ranges at which the aircraft were placed spanned the range from no damage to complete damage. Each detonation had a somewhat different arrangement of aircraft type, location, and orientation with respect to GZ; however, the entire program was keyed to the 31 kt Charlie. (Schraut 1953: 3-4, 39-41)

Project 3.3 provided the startling sight of conifer trees in the desert. The Forest Service wanted to predict atomic blast damage to forests and to establish relationships between blast parameters, tree motion, and damage. Four pine trees of about 45' height and 1' diameter, from the Mount Charleston area, were placed in concrete foundations, at each of four stations, located 5,000', 6,000', 7,000', and 8,000' from GZ. Emplacement took place at the 2 stations closest to GZ prior to Baker and at the 2 farthest stations prior to Charlie. Each of the stations also had a lollipop: a 4" aluminum I-beam in a concrete foundation which held a 32" diameter disk weighing 380 lb at a height of 14'. A lollipop was considered an "ideal reproducible tree". The trees were spaced to observe individual trees rather than a group or forest. (Brown 1953: 3-4, 19)

This project focused on field-test methods and measurement techniques. It was in preparation for the next tree test on UPSHOT-KNOTHOLE which had a strand of trees. At each station, one tree and the lollipop were instrumented for strain-time and for maximum strain at heights of 1' and at the base of the crown of the tree. Motion pictures were also made to obtain deflection data. (ibid., 3, 9)

Long-Range Detection

Project 7.1a represented the first coordinated effort to observe and record the illusive electromagnetic (EM) pulses. The rise time of the pulse is very rapid, microseconds; and it "starts with the emission of the prompt gamma rays, before the case is shattered". Stations were set up at: the test site in Yucca Flat; Stanford University; Boulder, CO, Alamogordo, NM; Robins, GA; Sterling, VA; McDill, FL; Ramey, Puerto Rico; Maynard Mass.; Kindley, Bermuda; and Camp King, Germany. Various antenna, receivers and recorders were assembled from standard radio equipment and installed at the stations. The station in Yucca Flat was constructed with special equipment within a truck which could move for the different shots. Recordings were made on shots Charlie through How. (Oleson 1953: 3, 11-16)

DoD PARTICIPATION ON LOS ALAMOS WEAPONS DEVELOPMENT TESTS

While Los Alamos was responsible for the weapons development experiments, the DoD participated in one way or another on all of the weapons development shots. This participation varied for the different events. In some cases, the DoD provided support, such as setting up equipment and retrieving data. In others, AFSWP personnel, some of whom were assigned to Los Alamos, served as technical and scientific staff.

The Air Force, particularly the Air Force Special Weapons Center (AFSWC), played a major operational and support role in the scientific and military test programs. Based at Kirtland AFB in Albuquerque, AFSWC used Indian Springs AFB in Nevada as its principal staging area during the test operation. They provided most of the aircraft and personnel required for aircraft operational control, airdrop delivery, cloud sampling, courier missions, cloud tracking, aerial surveys, and weather reconnaissance.

FEDERAL CIVIL DEFENSE ADMINISTRATION (FCDA)

AFSWP invited the FCDA to participate in TUMBLER-SNAPPER to broaden their knowledge of radiological safety operations and to permit them to become familiar with special civil defense scenarios following atomic explosions. After participation in the tests, there were discussions among FCDA, Health and Welfare, and the AEC about the feasibility of obtaining training for State and local civil defense radiological personnel during UPSHOT-KNOTHOLE, which was the next Nevada operation. (Lamoureux 1953, 9)

MILITARY SIGNIFICANCE

By the spring of 1952 the public was beginning to realize that while nuclear weapons were indeed awesome their effects were more limited than some might have imagined. For example, military correspondent Hanson W. Baldwin wrote an article that appeared in the April 24, 1952 New York Times that dispelled some of the myths surrounding the military effectiveness of nuclear weapons. He pointed out a number of problems associated with the military use of atomic bombs: (1) "The problem of delivery." In 1952 this was by fighter-bomber, and accuracy and escape tactics were conflicting. (2) "The type of weapon." Interestingly the services wanted an earth penetrating weapon, and continued to pursue this objective for decades. (3) "The height of burst. It should be low enough to destroy strong points but high enough to prevent major contamination." (4) "Utilization of shock effect. To capitalize upon the demoralizing effect upon the enemy, attacking troops must be as close as possible to the burst and move in very rapidly to seize the objective." This raises serious tactical deployment concerns. (5) "Indoctrination. Intensive indoctrination to avoid exaggeration or underestimation of atomic capabilities is badly needed by the Army." It was also badly needed by almost everyone else. (6) "Numbers of weapons. It is clear that one, two or three atomic weapons used against a strong, disciplined and well dug-in enemy, will not produce any very decisive results. To penetrate a zone of defenses strongly held in depth - like the Chinese thirty-mile-deep defensive line in Korea we must think in terms of many atomic weapons, plus conventional arms." (7) "Field fortifications. The good earth in the atomic age is more than ever the soldier's best friend. Atomic weapons applied to the battlefield put a premium, as in World War I, upon engineers and carefully designed, heavily constructed dugouts with overhead protection."

The military-sponsored programs to measure the effects of nuclear weapons in tactical and strategic situations almost overwhelmed the weapons development efforts of the AEC, Los Alamos and Sandia. The sheer numbers of military people involved in BUSTER-JANGLE dwarfed the weapons development contingent. The AEC and the labs felt that the NPG was theirs and that the DoD should participate on a noninterference basis. In the spring of 1952 the AEC sent the MLC a letter suggesting that the weapons development program could accommodate effects tests of "extreme urgency." Presumably Operation TUMBLER satisfied this criterion.

EXECUTIVE OFFICE OF THE PRESIDENT
NATIONAL SECURITY COUNCIL
WASHINGTON

March 28, 1952

MEMORANDUM FOR: The Secretary of State
The Secretary of Defense
The Chairman, Atomic Energy Commission

SUBJECT: Request of Presidential Approval for TUMBLER-SNAPPER

REFERENCE: Memo for Secretaries of State and Defense from Executive Secretary, NSC, same subject, dated March 19, 1952

Pursuant to concurrence by the Secretaries of State and Defense as of March 21, 1952, the proposal of the Chairman of the Atomic Energy Commission for the conduct of the TUMBLER-SNAPPER series of atomic tests, as outlined in the enclosure to reference memorandum, has been submitted to the President for consideration.

The President has this date approved the request set forth in the last paragraph of the enclosure to the reference memorandum.

/s/ James S. Lay, Jr.

James S. Lay, Jr.
Executive Secretary

(Ref: AEC March 31, 1952):

Figure 2-3.1 Presidential Approval for TUMBLER-SNAPPER.

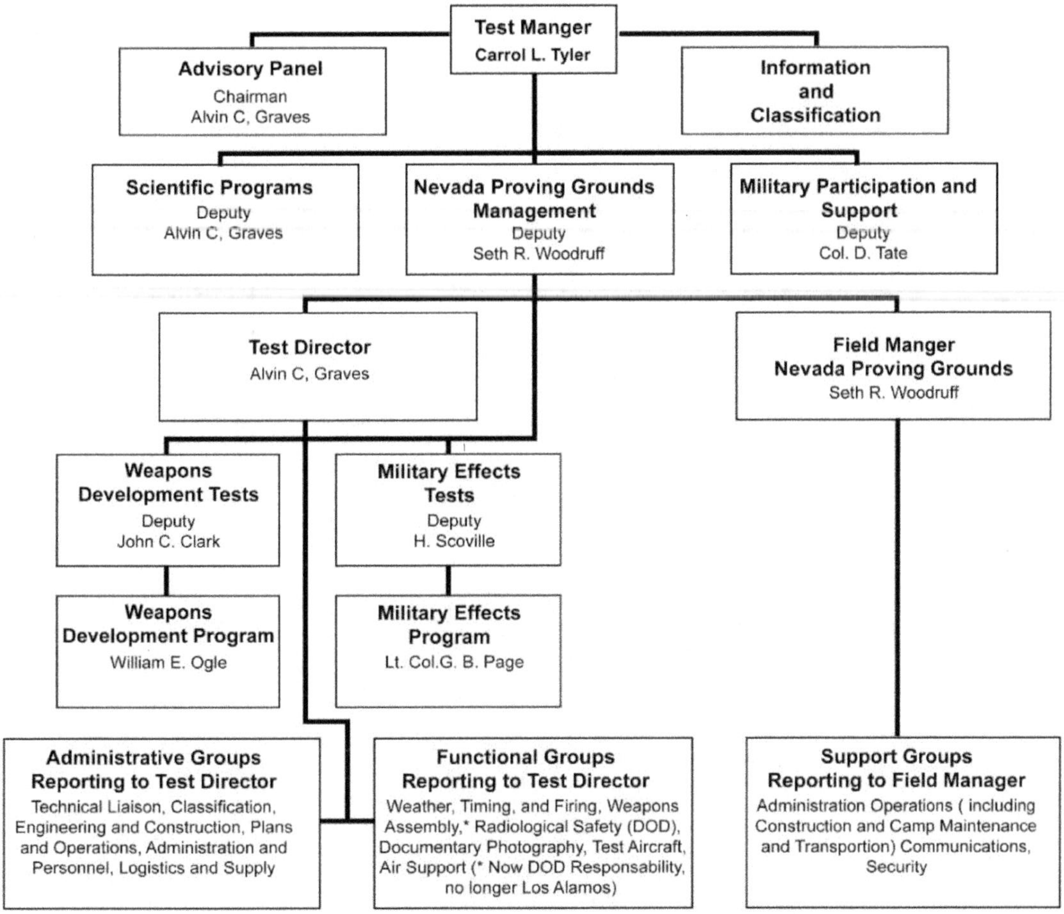

Figure 2-3.2 Nevada Test Organization for Operation TUMBLER-SNAPPER.

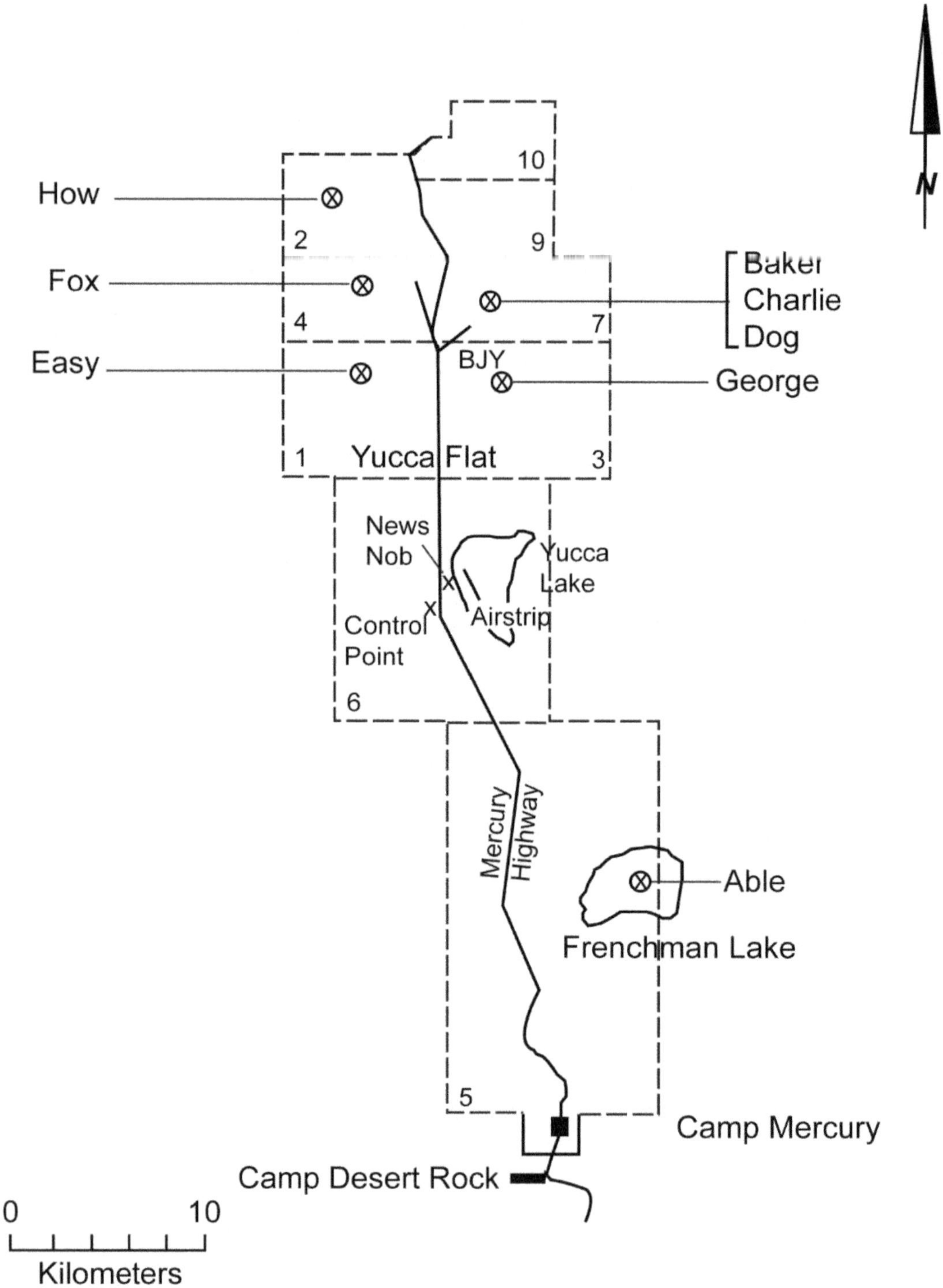

Figure 2-3.3 Location of TUMBLER-SNAPPER Tests.

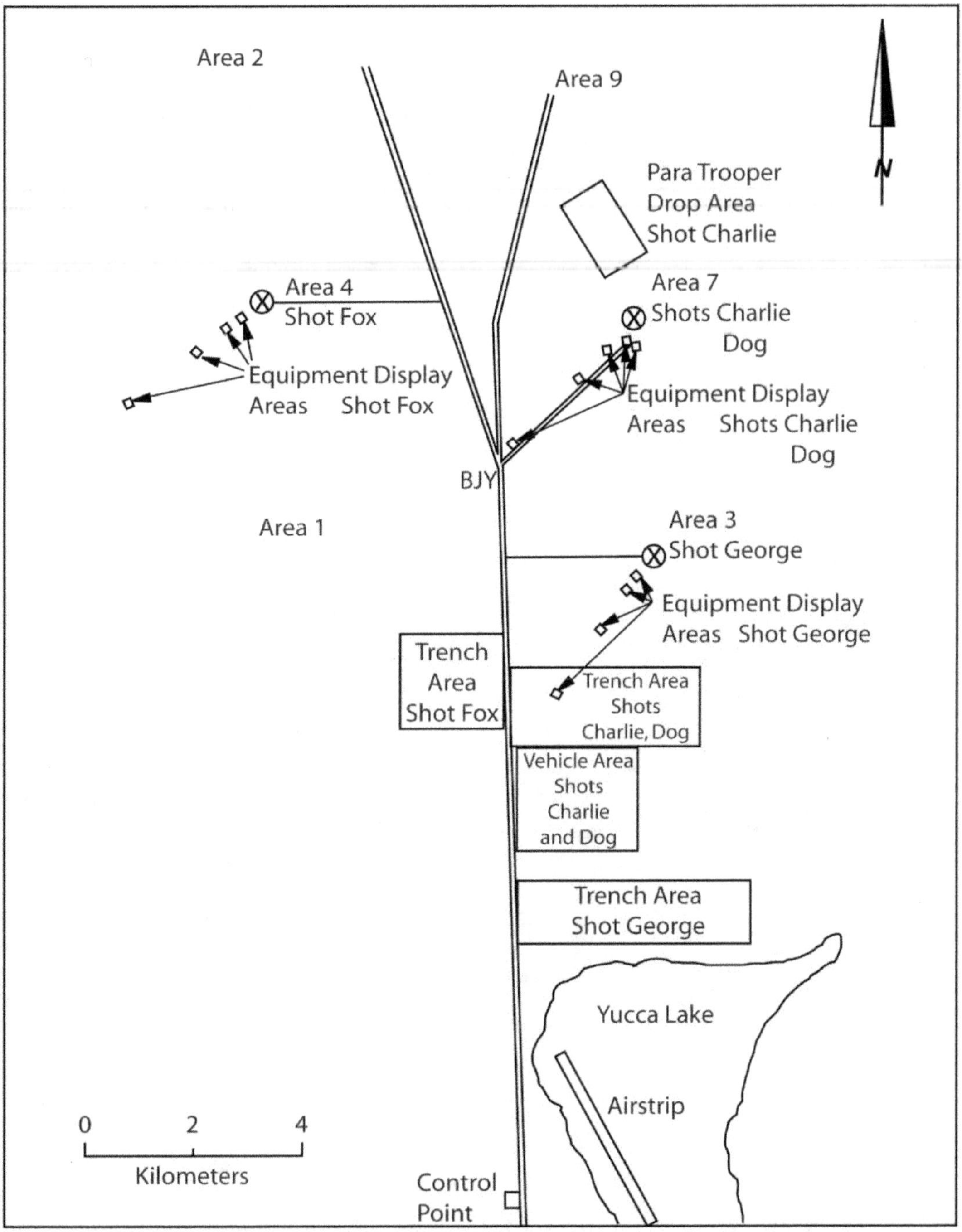

Figure 2-3.4 Desert Rock Exercises IV on TUMBLER-SNAPPER.

Hutment area, looking SE, at time of TUMBLER-SNAPPER

Camp Desert Rock during TUMBLER-SNAPPER

Men in trenches, standing position, pre-shot

Truck with trees at guard station near Mercury

8000 foot station, general view of trees and lollipop

Russian revetment under construction

Positioning F-47 in 16 foot revetment at 4000 feet from GZ

Aircraft being instrumented for blast effects

Looking down shot tower T3 from below cab

Tower T-4, Snapper #2 tower shot

Men climbing instrument tower to emplace instrumentation, TUMBLER-SNAPPER 1.2.

Ground Zero with target circles

Tower for thermal instrumentation with scaffolding, Station 206

CHAPTER 4. A SECOND NUCLEAR WEAPONS LABORATORY IS ESTABLISHED AT LIVERMORE, CALIFORNIA

President Truman signed an Executive Order on January 31, 1950, directing the AEC to initiate a crash program to develop the hydrogen bomb (see Part I-Chapter 7, section titled 1950). Los Alamos did indeed initiate such a program that resulted in a concept conceived by Edward Teller and Stan Ulam to use an atomic bomb to ignite a thermonuclear reaction. However, serious disagreements between Teller and his Los Alamos colleagues, and with Norris Bradbury in particular, eventually resulted in an irreparable rift. As a result, Teller left Los Alamos and began a serious campaign to form a laboratory to compete with Los Alamos across the board, but particularly in thermonuclear weapons research.

With help from E. O. Lawrence and L. Alvarez at the University of California, and with the fortuitous availability of a university facility at Livermore, California, the conditions were right for the establishment of the new weapons laboratory under the auspices of the same institution that operated Los Alamos. Herbert F. York was the first director of the University of California Radiation Laboratory at Livermore. Here is his description of the origin that new laboratory.

A PERSONAL PERSPECTIVE ON THE FORMATION OF THE NUCLEAR WEAPONS LABORATORY AT LIVERMORE, CALIFORNIA. By Herbert F. York*
[*Footnote: From "The Advisors. Oppenheimer, Teller, and the Superbomb", Stanford University Press, Stanford, California, 1976. With permission of the author and the publisher]

Three originally separate strands of events, each a part of the fabric of the superbomb controversy and the events flowing from it, eventually coalesced and led to the creation of a second American nuclear weapons laboratory at Livermore, California, in the summer of 1952. One of these strands was the determination of Ernest Lawrence and Luis Alvarez to involve themselves and the colleagues at Berkeley in some direct, useful, and important way in the American response to the first Soviet A-bomb. Another was the protracted conflict between Edward Teller on the one hand, and Norris Bradbury and his senior staff on the other, about how the Los Alamos laboratory might best go ahead on the H-bomb program. This conflict finally led Edward Teller to conclude that a second laboratory had to be established to do the job adequately. The third factor, was the happenstance that a small group of Berkeley scientists participated in the George experiment at Operation GREENHOUSE, thereby creating at Berkeley a small cadre of young men familiar with the details of thermonuclear weapons design.

On the very day the news of the first Soviet A-bomb became known to the public, Lawrence, Alvarez, Latimer, and others at Berkeley began to ponder the appropriate American response to that event, and to search for ways they themselves might participate in such a response. They discussed the matter among themselves, and they then traveled to other centers of nuclear research to learn the views of other scientists. Among the places they visited was Los Alamos, where they were particularly interested in learning more about Teller's ideas about the subject.

On the basis of these early explorations, they concluded that they should support Teller's proposals for an urgent, high-priority program at Los Alamos to develop a superbomb based on the fusion process, and the Berkeley Group should undertake the design and construction of a reactor which could produce a large excess of neutrons. Teller had explained to them that substantial amounts of tritium - a heavy radioactive form of hydrogen which does not occur naturally - might be needed in the development and manufacture of fusion (hydrogen) bombs, and they knew that the best way to produce tritium was in a reactor specially designed to produce a large excess of neutrons. The GAC, in its famous meeting of October 1949, agreed that the design of such a reactor should be undertaken, but it suggested that the program be carried out by the Argonne National Laboratory, which had very much more relevant experience.

Lawrence and Alvarez were at first disappointed at this turn of events, but they soon responded with an entirely new concept based partly on an idea of Winn Salisbury. It soon became known by its cover name: the Materials Testing Accelerator, or the MTA. The basic idea involved a two step process: first, produce large quantities of free neutrons by brute force; and second, absorb these neutrons in suitable materials to produce any of several desired end products - tritium, plutonium, U-233 (another fissionable material suitable for bombs), or radiological warfare agents.

... Lawrence first asked Robert Serber and me to make theoretical estimates of the production of neutrons in such a device, and then asked me to check them experimentally. I had just received my Ph.D. in physics at Berkeley, and had stayed on at the laboratory as what today would be called a "postdoctoral fellow". My data revealed that suitably large numbers of neutrons would be produced almost no matter what materials were used to construct the primary target for the deuteron beam. ... If tritium were the desired end product, then the secondary target would be constructed so that most of the neutrons were absorbed by lithium-6 atoms. Or, if plutonium were desired, then the secondary target would be constructed so that most of the neutrons were absorbed by U-238 atoms. ... the MTA made it possible to exploit the basic raw material uranium ore much more completely and efficiently than would otherwise be the case.

In the spring of 1950, while I was still refining measurements of the potential neutron yield of the MTA, Louis Alvarez approached Hugh Bradner and me to tell us something about the expanding work on the superbomb at Los Alamos. He had, he said, recently been talking with Edward Teller about the matter, and it appeared that the project could use some assistance from scientific groups at other laboratories. Bradner and I promptly flew to Los Alamos to meet with Teller and others, and we quickly agreed to set up a special group at Berkeley to perform some diagnostic experiments on the first thermonuclear test explosion, then planned for the following spring as the George shot of Operation GREENHOUSE. The group, code-named the "Measurements project," consisted of about forty persons, roughly half of whom were young physicists of the type we would now call "postdocs." In essence, we were to make experimental observations of certain physical phenomena as these unfolded during the first fraction of a microsecond of the thermonuclear explosion.

I recall that several different considerations strongly motivated and inspired me to participate in the hydrogen bomb program. One was my own perception of the growing seriousness of the cold war, much influenced by my very close personal student-teacher relationship with Lawrence. The Sino-Soviet bloc had just been formed; Stalin and Mao both said that it was monolithic and that its goal was world revolution. Another inspiration was the scientific and technological challenge of the experiment itself; it was to be the very first occasion in which a thermonuclear reaction took place on the surface of the earth, and we were to make complex observations extending over a period of less than a millionth of a second. Five years before, I had played a peripheral role in the Manhattan project. I had not participated in the Trinity test (of the first A-bomb) and I had only heard about it a week or so after it occurred. This time I was being invited to participate directly in the heart of the matter. Yet, another strongly favorable consideration was my discovery that Teller, Bethe, Fermi, Von Neumann, Wheeler, Gamow, and other like them were at Los Alamos and involved in this project. They were among the greatest men of contemporary science, they were the legendary yet living heroes of young physicists like myself, and I was greatly attracted by the opportunity of working with them and coming to know them personally. Moreover, I was not cleared to see GAC documents or deliberations, and so I knew nothing about the arguments opposing the superbomb, except for what I learned secondhand from Teller and Lawrence who of course, regarded those arguments as worn and foolish. (I saw the GAC report for the first time in 1974, a quarter of a century later!)

I mention my own motivations because I think I understand them better than the motivations of others. I do so here only to explain, not to justify or to rationalize. I do so because I think I was not untypical, and because my reactions can be expected to have occurred in others in similar situations. Oppenheimer's (11) later reaction to Teller's explanation of the ideas about how to make an H-bomb- "technically so sweet that you could not argue about it" - is a related response in someone vastly more sophisticated than I was. Other reminiscences of similar situations also clearly express the excitement scientists and other humans commonly find in such huge history-making events (12).

Most of the preparatory work of our group was done in Berkeley. However, the pilot setup of our electronics gear required more room than was readily available at Berkeley, so for that we used some space in the former naval infirmary at the Livermore station. The California Research Corporation was already at work on the MTA project at the site, and its working relationship with UCRL made it natural and simple for it to provide us with all the necessary housekeeping functions.

During March and April 1951, most of the members of our special group moved out to Enewetak Atoll, in the Marshall Islands, and we set up our equipment in its final form there, in the shadow of the George device.

On May 8, 1951, at Enewetak Atoll the first thermonuclear test explosion on earth was successfully conducted. The tritium deuterium mixture (13) burned well, and the various diagnostic experiments including that of our Berkeley group - were also successful in recording the various phenomena that accompanied the explosion.

Some of the members of our Berkeley group, after completing the analysis of their data, participated in the general post-experiment discussions and in some of the future planning sessions. No specific plans for further participation resulted from these discussions, however, and so the Berkeley group was disbanded and its members turned to various other projects, mostly pure research in high energy physics.

Edward Teller stayed on at Los Alamos for another six months after the George shot. The next major experiment - the Mike shot of Operation IVY (fall 1952) - was to be based on the Teller-Ulam idea, and he participated directly in the determination of its basic configuration. In November 1951, he left Los Alamos and returned to the University of Chicago. He did so in part because he felt the remaining theoretical work that was still needed to get Mike ready could be done just as well without him - Bethe was already scheduled to be at Los Alamos during the final design period - but mainly because the ancient arguments between him and Bradbury over how to run the laboratory and the hydrogen bomb program continued to worsen. Thus, in the summer of 1951, only some months after he came up with the final, capping suggestion in the series of ideas that led to the invention of the superbomb, he concluded that the establishment of a second, independent laboratory was needed to exploit this new approach in a timely fashion. As Teller put it (14),

> It was an open secret, among scientists and government officials that I did not agree with Norris Bradbury's administration of the thermonuclear program at Los Alamos. Bradbury and I remained friends, but we differed sharply on the most effective ways to produce a hydrogen bomb at the earliest possible date. We even disagreed on the earliest possible date itself, on the timing of our first hydrogen bomb test. The dissension with Bradbury crystallized in my mind the urgent need for more than one nuclear weapons laboratory.
>
> I knew that science thrives on friendly competition, on the fostering of different points of view, and on the exchange of ideas developed in different surroundings. I knew, too, that a single group of scientists working together can easily become fascinated by special aspects of a development - to the neglect of other hopeful approaches. My conviction grew that the safety of our country could not be entrusted to a single nuclear weapons laboratory, even though that laboratory were as excellent as Los Alamos. This

> conviction was hardened by a growing awareness, as our work progressed at Los Alamos and our thermonuclear knowledge increased, that we were pioneering a big new field of weapons development. I began to doubt that one laboratory would be physically capable of handling all the work that had to be done. Weighing all of these ideas and circumstances, I came to the inescapable conclusion that at least two weapons laboratories, working in cooperation but also in the traditional American spirit of competition toward the mutual goal of adequate national defense, were vital to the future of the United States.
>
> I also concluded that I could advocate establishment of a second weapons laboratory most effectively if I were not associated with the existing Los Alamos Laboratory. So, regretfully, I left Los Alamos in November 1951, and returned to the University of Chicago.

Teller soon succeeded in persuading Gordon Dean, who had by then replaced Lilienthal as chairman of the AEC, to consider the matter. At Dean's request the GAC reviewed the idea. Except for Willard J Libby, like Teller a professor at the University of Chicago, and then a new member of the committee, it opposed the idea on the ground that the establishment of a second laboratory would divert talent and resources from Los Alamos and thus slow down the overall program. Very probably, the opposition of the GAC to a second laboratory also importantly involved personal elements. Teller's claim that the competition was a good thing was often expressed in terms which made it clear he felt the Los Alamos leadership was unimaginative, negative, and otherwise inadequate. It was equally clear that Teller felt much the same way about many of the members of the GAC itself, and so it is not surprising that the GAC supported Los Alamos and Bradbury against what they regarded as an unwarranted personal attack.

Teller also sought support of his ideas in the air force. The air force would be the principal user of the hydrogen bomb, and a number of persons at the top of the air force very quickly evinced great personal and institutional interest in the issues being raised. David Griggs, one of the founders of the Rand Corporation and just then chief scientist of the air force; James (Jimmy) H. Doolittle, a much respected retired general and a high level general consultant to the air force leadership, and General Elwood ("Pete") Quesada, the commander of the joint task force that had conducted Operation GREENHOUSE, all became strong partisans of Teller, and helped him make further contact with higher air force officials. As a result, Teller and his ideas were warmly received and strongly endorsed by Thomas Finletter, secretary of the air force, and his special assistant for research and development, William A. M. Burden. They in turn, began to make moves toward establishing a second nuclear weapons laboratory under air force sponsorship, and in 1951 they actually did arrange to sponsor briefly some nuclear calculations Teller was doing at Chicago. For this particular work, they used the facilities of Project Chore, a minor army project that had been going on at the University of Chicago for some years and which could handle classified work. However serious their intentions concerning a full-scale second laboratory may have been, their actions greatly increased the pressure on the AEC either to do something itself or to see its monopoly in the field eliminated. During this period, Teller also was given the opportunity to brief Secretary of State Acheson, Secretary of Defense Robert Lovett, and Deputy Secretary of Defense William C. Foster. The very fact of these briefings, independent of their specific result, put further pressure on the commissioners.

The congressional JCAE again played a crucial role. The chairman, Senator McMahon, was mortally ill at the time, and so the staff director, William Borden, personally conducted on his behalf a campaign for a second laboratory - and against the GAC leadership and its views.

In the meantime, in late 1951, Thomas Murray, the AEC Commissioner most sympathetic to the idea of a second laboratory, got in touch with Ernest Lawrence to discuss the matter with him. Lawrence was already well known to be very sympathetic to the idea of expanding the thermonuclear program. He responded positively, and volunteered to study the matter further himself. Since I had been more deeply involved in the recent thermonuclear program than anyone else at Berkeley, Lawrence in January 1952 asked me what my views were. As a direct result of Lawrence's inquiry, I made a series of extended trips to Los Alamos, Chicago, and Washington, where I discussed the matter with most of the people named above plus a few others, including Army General Kenneth Fields, then the AEC's director of the Division of Military Applications, and his deputy, Navy Captain John T. Hayward. I found the whole affair heady and exciting (I had just turned thirty), and I was readily persuaded to Teller's point of view. I reported to Lawrence that I, too felt it would probably be useful to establish a second laboratory. The idea of doing so at the Livermore site was, for us, a natural one, and we suggested it immediately to AEC authorities.

That specific addition to the general idea changed the nature of the argument. A proposal to establish a second laboratory in existing facilities at Livermore as a branch of the UCRL, as compared to a proposal to simply establish one "somewhere" under an unspecified aegis, clearly meant much less expense and an immediate, if small, cadre of people ready to go to work right away. As a result, and as GAC chairman Oppenheimer later recalled, the GAC and the AEC "... approved the second laboratory as now conceived because there is an existing installation, and it could be done gradually and without harm to Los Alamos (15)." As I recall it, Lawrence and Teller felt at the time that Oppenheimer himself was still really opposed to a second laboratory but that under the new circumstances he had no other choice. Even so, during that year I met with Oppenheimer at Princeton, and discussed the plans for the Livermore laboratory. He received me in a personally friendly fashion, but I cannot recall his being of any particular help.

The precise nature of the plans for the new laboratory, however, primarily reflected Lawrence's ideas about how to go about such things, and deviated considerably from Teller's views of what should be done. In essence, Lawrence firmly believed that if a group of bright young men are simply sent off in the right direction with a reasonable level of support, they will end up in the right place. He did not believe that the goals need to be spelled out in great detail, nor that it was necessary that the leadership consist of persons who were already well known. Teller on the other hand, had become deeply suspicious of the intentions of the AEC leadership, and he therefore wanted something more analogous to the 1943 plans for Los Alamos, that is, a plan for a laboratory that would be led by a large cadre of famous scientists and that would have a well-defined goal.

To complicate matters, during the spring, Lawrence, then suffering from a chronic illness, spent much time away from Berkeley on long rest trips. As a result I was left pretty much on my own to draw up the specific plans for a second laboratory with nothing except the most general guidance from my immediate superior. However, as a result of ten years of close association, I both clearly understood and firmly agreed with Lawrence's approach to "big science," and I generated plans which he always warmly endorsed when he had a chance to review them.

Finally, and in close accordance with Lawrence's (and my) views of the matter, the AEC in June 1952 approved the establishment of a branch of the Berkeley laboratory at Livermore which would assist in the thermonuclear weapons program by conducting diagnostic experiments during weapons tests and other related research, but the question of how soon (or even whether) the Livermore laboratory would actually engage directly in weapons development was left open. The AEC's official planning document (16) described the mission of the Livermore laboratory this way:

> a. Development and experimentation in methods and equipment for securing diagnostic information on behavior of thermonuclear devices and the conduct of such instrumentation programs in support of tests of thermonuclear devices in close collaboration with the Los Alamos Scientific Laboratory.
>
> b. While the work authorized above is the immediate objective of this proposal, the Commission hopes that the group at UCRL (Livermore) *will eventually suggest broader programs of thermonuclear research* to be carried out by UCRL, or elsewhere. (Emphasis added.)

Lawrence felt that that kind of statement of intentions provided an adequate base upon which to build a second weapons laboratory. I would have preferred something more concrete, but I was prepared to accept it as a place to start from. Teller, on the other hand, found the vagueness of the AEC's plans for the Livermore laboratory entirely unsatisfactory. As a result, in early July he told Ernest Lawrence, Gordon Dean, myself, and others that he would have nothing further to do with the plans for establishing a laboratory at Livermore. The Berkeley administration was prepared to go ahead anyway. However, at the insistence of Captain Hayward more than anyone else, intense negotiations were resumed among all concerned. Within days, these led to a firm commitment on the part of Gordon Dean that thermonuclear weapons development would be included in the Livermore program from the outset, and a renewed commitment on the part of Teller to join the laboratory.

The laboratory was launched in September 1952. I became the director, and the Scientific Steering Committee included Teller, Harold Brown, John S. Foster, Jr., Arthur T. Biehl, and a few others. Teller, because of his obvious special status, was given veto authority over the decisions of the committee, but otherwise had no formal authority. Brown was put in charge of the development of thermonuclear weapons at Livermore, and first Biehl and then Foster was put in charge of the development of improved fission weapons. There were some early problems in the administration of the Theoretical Division, but these were resolved on a temporary basis by making Richard Latter, a Rand physicist then temporarily on loan four days a week to Livermore laboratory, the acting head of the division. After about a year, he was replaced by Mark M. Mills, who remained in that position until his death in a helicopter crash at Enewetak in early 1958. These organizational arrangements, although they contained some peculiar elements, worked out very well, and none of the strained relationships that had surrounded Teller at Los Alamos developed at Livermore.

[Author's note: The following reference numbers are as cited in York's text]
(10) The only published account of the MTA project is a "gee whiz" article by Allen P. Armagnac, "The Most FantasticAtom-Smasher," in *Popular Science*, November 1958, p. 108 *et seq*. References to the project can of course be found in Hewlett and Duncan, USAEC *History*, and in Childs, *An American Genius.*

(11) USAEC, *In the Matter of J. Robert Oppenheimer,* The MIT Press, pp.81, 229,251.

(12) For instance, in Teller, *The Legacy of Hiroshima*; Len Giovannitti and Fred Freed, *The Decision to Drop the Bomb*, New York: Coward-McCann, 1965; Astashenkov, *Kurchatov*; and Golovin, I. V. *Kurchatov.*

(13) See also Childs' account of this in Childs, *An American Genius.*

(14) Teller, *The Legacy of Hiroshima*, pp. 54-55.

(15) Hearing transcript, USAEC, *In the Matter of J. Robert Oppenheimer*, The MIT Press,
p. 248.

(16) Director of Military Applications, USAEC, *Thermonuclear Research at the University of California Radiation Laboratory*, AEC 425/20, Washington, June 13, 1952.

(17) Shepley and Blair, *The Hydrogen Bomb*, Excerpts were published in *Life* magazine.

Attachment III is by Raymond Gilbert who was assigned to Livermore when he was in the Air Force. He describes his experiences at Livermore and the test site during the era of Operation TEAPOT.

E. O. (Earnest Orlando) Lawrence.

Gerry Johnson

East Avenue 1950s

Livermore Lab Site 1952

Livermore Lab Site mid 1950s

E. O. Lawrence, Edward Teller, Herb York

Group At PLUMBBOB, outside of Ranier tunnel - Chuck Violet and Willard Libby at left, Gerry Johnson to right of post

Duane Sewell

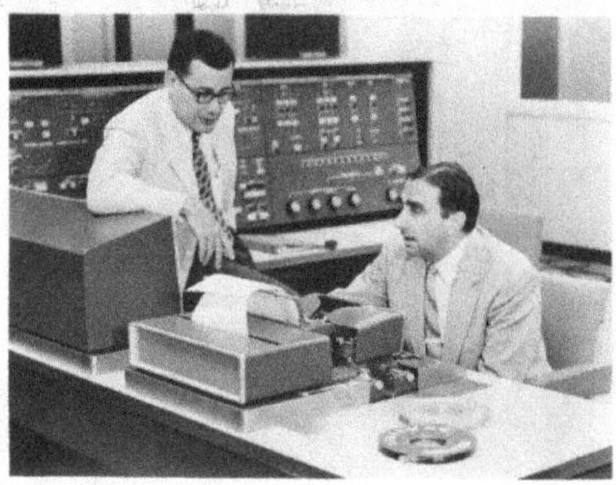

Harold Brown and Edward Teller in computer room

CHAPTER 5. ARMED FORCES SPECIAL WEAPONS PROJECT (AFSWP)/DEFENSE ATOMIC SUPPORT AGENCY (DASA) ORGANIZATION AND LEADERSHIP

1951 REORGANIZATIONS AT HEADQUARTERS

Brig. Gen. H. B. Loper relieved Maj. Gen. K. D. Nichols as commander of AFSWP on January 23, 1951*. Nichols and Loper worked together when they were both members of the MLC in 1949 and the Nichols-Loper report on the estimated damage caused by the disclosures to the Soviets by Klaus Fuchs influenced the President regarding the urgency of initiating a crash program to develop thermonuclear weapons. (DTRA 2002: 80) After Nichols' departure from AFSWP, his successors were not ex officio members of the Military Liaison Committee (MLC) but attended their meetings as observers. (DTRA 2002:99)
[*Footnote: After Nichols' retirement from the Army in September 1953, he became General Manager of the AEC until 1955. At the time of his retirement from the Army, he was their youngest Major General. His early retirement was inspired by AEC Chairman Louis Strauss and President Eisenhower, both of whom felt that Nichols could help the AEC improve their relationship with the military. (Nichols 1987: 297-299)] On July 5, 1952, Loper was promoted to Major General with an effective date of rank of October 12, 1951.

Shortly before his appointment, Loper wrote to the Chiefs of the three Armed Services re-examining the AFSWP mission. His draft, with a few changes, was approved by the Joint Chiefs for AFSWP to provide: specialized training and technical services; coordination with the AEC for storage and surveillance of the nuclear stockpile; coordination with other agencies for planning continental and overseas weapons tests; and for the evaluation of weapon effects from those tests. In addition, AFSWP continued its support role in weapon development, procurement, and assembly.

Loper recognized the increased nuclear activity by the three services. "AFSWP would play a coordinating role under Loper's plan, an interdepartmental rather than a joint agency, '... utilizing established agencies of the Armed Forces to carry out programs. Existing organizational structures will not be duplicated, nor will additional activities be established.'" (DTRA 2002:84)

Operation RANGER was fielded just over a months after President Truman transferred the test site to the AEC. As a result, AFSWP did not have time to develop much in the way of weapons effects programs; and, in addition, they were immersed in planning for Operation WINDSTORM, proposed for the Aleutians. While AFSWC was able to participate fairly extensively in RANGER, AFSWP did not.

In March of 1951, the Headquarters divisions were reorganized into three directorates: Administration, Operations, and Technical Services. No changes occurred in the organization of the divisions themselves.

On November 28, it was announced that the three directorates would report to three Deputy Chiefs of Staff. Also, a new division was added to the Technical Services directorate, which now consisted of four divisions: 1) Weapons Defense Division, 2) Weapons Development Division, 3) Weapons Effects Division, and the new division, 4) Weapons Test Division. (AFSWP vol 4: 3.1.12)

The new Weapons Test Division had as its nucleus the greater part of the former Test Branch of the Weapons Effects Division. The Weapons Test and Weapons Effects Divisions are the ones that were most closely associated with activities at the test site.

1951 AFSWP FIELD COMMAND ESTABLISHED

AFSWP Field Command was established on the first of May and was given the responsibility for Sandia Base, Albuquerque. Prior to this, Sandia Base was administered by Headquarters Sandia Base, under the command of Brigadier General R. M. Montague, USA. On February 12, 1951, Montague was relieved by Brigadier General Leland S. Stranathan, USAF. On May 1, 1951, Stranathan became Commanding General Field Command, AFSWP. (AFSWP vol 4:4.1.2-1.7)

As discussed in Part II-Chapter 2, the responsible agent for military support to the AEC during BUSTER-JANGLE was the Air Force Special Weapons Command (AFSWC) under the leadership of Brigadier General John S. Mills, at Kirtland AFB. In the summer of 1951, Mills established a Joint Test Group lead by Lt. Col. E. S. Kesling to fulfill this responsibility. This group served the Test Manager of BUSTER-JANGLE as military liaison. However, military liaison to the Test Director on BUSTER-JANGLE was provided by Headquarters AFSWP. The Deputy to the Test Director, A.F. Spilhaus, was a consultant to Headquarters AFSWP. The individual responsible for the weapons effects tests, Colonel Max S. George, had been the leader of the Test Branch of the Weapons Effects Division at Headquarters, AFSWP.

After BUSTER-JANGLE, in the fall of 1951, Mills, requested that he be relieved of the responsibility for the Joint Test Group.(AFSWP vol. 5:4.5.1) Subsequently, in mid-January 1952, the Chiefs of Staff of the three services assigned to AFSWP the responsibility for technical supervision of continental military weapons effects tests, the coordination of military participation in such tests, and assistance to the AEC in continental tests. To fulfill this new mission, Test Command, AFSWP, was formed at the end of January, 1952, about two months before TUMBLER-SNAPPER. Personnel who had been assigned to the AFSWC Joint Test Group were transferred to Test Command, AFSWP, with headquarters at Kirtland AFB.

On TUMBLER-SNAPPER, the Deputy to the Test Manager for Military Participation and Support was Col. David A. Tate of Field Command. The Deputy to the Test Director for Military Effects was Herbert Scoville from Headquarters, and the leader of the Military Effects Program was Lt. Col. G. B. Page from the Weapons Test Division of AFSWP Headquarters.

In June 1952, after Operation TUMBLER-SNAPPER, Test Command at Headquarters was transferred to Sandia Base and reported to Field Command, AFSWP under Stranathan. (AFSWP 1954: Vol. 5:4.5.2) During TUMBLER-SNAPPER, Test Command AFSWP consisted of 53 Army, 16 Navy, and 18 Air force personnel. (Ponton et.al.1982c:41) However, TUMBLER-SNAPPER was to be the only operation on which Test Command participated.

DIRECTORATE OF WEAPONS EFFECTS TESTS, FIELD COMMAND (DWET, FC)

On July 7, 1951, Colonel Paul T. Preuss, USAF, reported to Headquarters, Field Command for his new assignment as Special Assistant to Commanding General Stranathan. Preuss was assigned the task of studying the integration of Test Command activities and personnel into Field Command AFSWP at Sandia and into AFSWP at Washington Headquarters. He met with people from the AEC, LASL, AFSWP Headquarters and Field Command. On July 24, 1952, Preuss submitted a report to Stranathan recommending that Test Command be de-activated and that its activities be assumed by a Directorate of Weapons Effects Tests (DWET) reporting to the Commanding General of Field Command. (AFSWP 1954: Vol. 5:4.5.2-4.5.3)

Stranathan and Loper both approved the recommendation. Test Command was disestablished on August 1, 1952; and its responsibilities assumed by Headquarters, Field Command. Simultaneously, the Directorate of Weapons Effects Tests (DWET) was established with Col. Preuss as Director. (AFSWP vol. 5:4.5.3)

The DWET mission was to implement Field Command responsibilities in Continental atomic tests. These responsibilities included:
1) Technical direction of the DoD's weapons effects tests and the DoD's effects measurements on the Lab's shots.
2) Coordination of the military participation and assistance in support of the AEC.
3) Completion of plans, preparations and reports for the technical programs.
4) Coordination of planning for training participation by the Services. (AFSWP 1954: Vol. 5:4.5.3-4.5.4)

In 1953, DWET was composed of:

- An Administrative Division - Comdr. Roger W. Luther,
- A Plans and Operations Division - Col. Leonard F. Dow,
- A Support Division - Lt. Col. E. M. Tolliver,
- The Nevada Proving Grounds Detachment - Comdr. J. J. Lenahan.

E. B. Doll, of the Stanford Research Institute, was named Technical Director of DWET on August 16, 1952. He served in this capacity for continental tests and was a member of the Nevada Test Site Organization through TEAPOT (AFSWP vol. 5:4.5.4-4.5.5). DWET did not change significantly in its core functions during the atmospheric testing era, but there were occasional reorganizations and name changes.

The Support Branch, later called Support Division, assisted DoD participants on all operations until the 1958 test moratorium. In the figures depicting the NTSO organization for the operations UPSHOT-KNOTHOLE through HARDTACK Phase II, it is referred to as Field Command Support or Field Command Support Unit (FCSU), and on HARDTACK Phase II, it was simply called DoD Support.

The Nevada Proving Grounds Detachment was organized in May 1952 as a part of the former Test Command and its activities were inter-related with those of the Support Division. The internal organization was not affected by the transfer to Field Command.

DWET had an extensive portfolio of activities: integration with FC, planning for UPSHOT-KNOTHOLE; and clarifying relationships with the staffs at Headquarters AFSWP, SFOO, and LASL. Pruess announced at an August meeting with personnel of the new directorate that in the future there would be the fullest cooperation with the AEC and LASL. Preuss' earlier conferences that included AEC and LASL had convinced him that there was a need for more cooperation between the various working elements of the AFSWP and the AEC. (AFSWP 1954: Vol. 5:4.5.5)

1953

In January 1953, General Loper, who had served for two years as Chief AFSWP, suffered a heart attack and retired from the Army. Major General Alvin R. Luedecke, who as an Air Force Brigadier General had served as a Deputy Chief in AFSWP, took his place. (DTRA 2002:99)

During AFSWP's early years, the agency relied on other government entities such as LASL, Sandia, the National Bureau of Standards and the Army Chemical Corps for research and development. By the time Luedecke took command in 1953, the Weapons Effects Division was working on preliminary plans for effects measurements to be conducted in collaboration with defense contractors and university laboratories. The use of a wide variety of governmental, industrial and university laboratories ultimately became the permanent operating mode of the agency. (DTRA 2002:100)

AFSWP's operational arm at Field Command held jurisdiction over Sandia Base and all tenant organizations, including the buildings used by the Sandia Laboratory. On-site training was a major responsibility of Field Command, both for weapon assembly teams and for those assigned to test and storage operations. It also had responsibilities for construction, supply, and logistics at Sandia Base, the weapon storage sites, and other DoD nuclear installations. In 1953, it had a staff of 10,250 consisting of 1,550 officers, 7,100 enlisted personnel, and 1,600 civilians. As continental testing increased, whole units of engineering and operations personnel moved from Sandia Base to the Nevada Proving Grounds and began construction and installation of equipment months before an operation. At the same time, Field Command also had direct participation in Pacific tests. (DTRA 2002:104)

In March, 1953, the mission of Field command, and therefore that of DWET, was augmented to include planning responsibilities for Operation CASTLE. In June 1953, additional responsibilities were assigned Field Command by the Chief, AFSWP, in: “tests involving nuclear detonations participated in or conducted by agencies of the United States outside of the Continental United States.” (AFSWP 1954: Vol. 6:4.6.1)

Colonel Pruess served the Test Manager of Operation UPSHOT-KNOTHOLE as Deputy for Military Operations. E. B. Doll, Technical Director of WET served as the leader of the Military Effects group under the Test Director. Doll’s deputy was Col. H. K. Gilbert, also from WET. From the Radiation Branch of Headquarters, Col. Edward Giller and Major Hard also served on Doll’s staff. (AFSWP 1954: Vol. 6: 3.8.1)

Starting with Operation UPSHOT-KNOTHOLE, the DWET assumed the responsibility for the preparation of reports on continental operations. This resulted in the prompt publication of the preliminary test reports of all the DoD projects.

1954 - 1955

On Operation TEAPOT, Colonel Hershell E. Parsons, who had replaced Pruess on October 11, 1954, served as Deputy for Military Operations in support of the AEC Test Manager. Doll again headed the Military Effects Group under the Test Director. There was also a Field Command Support group from WET Field Command at TEAPOT.

On June 20, 1955, Rear Admiral Frank O’Beirne relieved Major General Stranathan at Field Command. The Commanding General, Field Command was now called Commander, Field Command. This assignment also brought about the practice of having the Commander, Field Command of a different service than that of Chief, AFSWP.

1957

On Operation PLUMBBOB, Col. H. E. Parsons from DWET, FC, again served the Test Manager as Deputy for Military Matters. Under the Test Director, the military’s weapons effects work was done by a group now titled Field Command Weapons Test. The Director of this group was K. D. Coleman, from the DWET Test Branch.

Also during PLUMBOB, General Luedecke completed his tour as Chief, AFSWP and retired from the Air Force. He went on to replace Kenneth Nichols as General Manager of the AEC. Luedecke played a leading part in the negotiation of the transfer of most of the nuclear weapon stockpile from the AEC to AFSWP and then to the various Services. At the end of 1952, the nation had stockpiled 841 weapons with a total yield of almost 50 megatons. By the time Luedecke left AFSWP in 1957, the stockpile had grown to 5,543 weapons with a total yield of 17,546 megatons. He was also influential in the formulation of the Atomic Energy Act of 1954. This act removed information primarily related to military applications from the Restricted Data Category, which was reserved for weapon design information and limited to the AEC. A lesser classification became available for weapon effects, and gave the Military Services much easier access to the data they needed. (DTRA 2002:125)

Rear Admiral Edward N. Parker replaced Luedecke as Chief AFSWP on June 10, 1957. He had served from 1952 to 1954 as Deputy Chief of AFSWP under Nichols. The Luedecke years were characterized by "accomplishment and growth for the agency". The challenge for Parker would be to meet the Soviet's new challenge ... "in space". (DTRA 2002:124-6)

A reorganization of the technical divisions at Headquarters was announced in November 1957. The Deputy Chief of Staff for Technical Services was replaced by two newly designated deputies: A Deputy Chief of Staff for Research and Development and a Deputy Chief of Staff for Weapons Effects and Tests.

The Weapons Effects and Weapons Development Divisions were disbanded. The personnel and responsibilities formerly assigned to the Weapons Effects Division and the Weapons Test Division were transferred to the Deputy Chief of Staff for Weapons Effects and Tests who had under his purview five Divisions: Radiation, Blast and Shock, Analysis, Medical, and Weapons Test. The first four of these divisions had been branches of the Weapons Effects Division

1958

The Deputy Test Manager for Military Matters on HARDTACK Phase II was Colonel W. S. Hutchinson, while the leader of the Effects Group was Lt. Col. John W. Kodis (who eventually joined Los Alamos as a staff member). There was also a DoD Support group that reported to the Test Manager and provided support for the military work.

In April 1958, John G. Lewis, a civilian physicist, at AFSWP Headquarters was assigned to the Blast and Shock Division. Lewis continued his association with AFSWP's successors throughout the duration of nuclear testing as a government employee and later as a contractor. John understood both the military and civilian communities of DoD and knew how to integrate work with the AEC Laboratories. Also, his natural negotiation skills and futuristic visions made him a key element in the unity and continuity of the whole nuclear weapons effects community.

FORMATION OF DASA

President Eisenhower proposed a general reorganization of the DoD in early April, 1958. The objective was to unify the Services under the Secretary of Defense who would … “allocate funds among the Services, assign each Service combat roles, select officers for promotion to the most senior rank, centralize all public relations, and, presumably, put an end to inter-service squabbling.” By early August 1958, the Defense Reorganization Act had won Congressional approval. (DTRA 2002:148)

Two weeks after passage of the 1958 Defense Reorganization Act, AFSWP was ordered to conduct a full evaluation of its mission and responsibilities under the new DoD structure. While the draft review was completed in two weeks, the evaluation and coordination process took the rest of 1958. A variety of names for the new AFSWP were considered: Special Weapons Command (SWC), Joint Atomic Support Agency (JASA), and finally, the winner, Defense Atomic Support Agency (DASA). The new mission was approved by the JCS at the end of December and endorsed by Deputy Defense Secretary Donald A. Quarles the following May. As a result, May 6, 1959, is considered the official birth date of DASA and the new Field command. (DTRA 2002: 149; DASA 1959: Part I Cpt 2:1-2, 6)

AFSWP had been an inter-service agency reporting to the JCS, while DASA was established as an agency responsible to the Secretary of Defense through the JCS. The DASA chief, a position rotated among the Services, was selected by the Secretary of Defense upon recommendation by the JCS. (DASA 1959: Part I Cpt 2:1-2)

The new charter gave DASA responsibility for supervising all DoD weapon effects tests, which had formerly been conducted by the individual Services. DASA would also be in control of all the Services nuclear testing budgets, which would be lumped together into a single appropriation.

DASA’s charter had more broadly stated basic functions than AFSWP’s July 1951 charter. However, most of the new functions and responsibilities specified in the DASA charter had, in fact, been given to AFSWP as additional assignments over the years. As for the organization itself: “… there were no apparent differences within the headquarters” between AFSWP and DASA. Admiral Parker agreed to stay on as director until August, 1960. (DTRA 2002:149-160)

DASA Field Command remained at Sandia Base and continued with the same general organization that existed under AFSWP. (DASA 1959: Part III Cpt 16: 1)

The JCS assigned Joint Task Force 7, which had fielded nuclear weapons tests in the Pacific, as a subordinate command of DASA. (DASA 1959: Part I Cpt 2: 2)

Major General Harold C. (Sam) Donnelly, replaced Field Command Chief Admiral Parker on an interim basis, between August 21, 1960 and January 16, 1961*. [*Footnote: Between 1964 and 1968, Donnelly served as the Chief DASA.] In January 1961, Major General Robert H. Booth took over the reins at DASA. He served until his retirement in 1964. Booth’s administrative abilities were challenged by the task of rebuilding a nuclear testing capability that, as Frank Shelton put it succinctly, "... had gone to pot" during the moratorium years. (DTRA 2002:160)

DASA DURING THE MORATORIUM

DASA headquarters remained at the Pentagon until 1967; however, their planning for the future certainly was not stationary. They had 350 service and civilian personnel by the end of December, 1959 and grew through the moratorium and the first 10 months after the resumption of testing, until they had about 570 by the end of December 1962. (DASA 1959: Part I Cpt 4: 1, 2)

By 1960. DASA had undertaken some re-organizations, and the DoD had lost most of their enthusiasm for maintaining a capability to resume nuclear testing in the atmosphere and underwater. Instead, emphasis was placed on underground or outer space tests, with particular attention focused on the development of test instrumentation. In addition, a large fraction of the effort was devoted to theoretical and computational studies and on simulation modeling.

DASA's plans for Vela-Uniform (see Attachment VIII) and Operation JERICO were typical of their underground testing activities. Vela-Uniform was a program focused on the exploration of a seismic detection system that could adequately monitor an international nuclear test ban. It was an outgrowth of the previously planned CONCERTO program, which had been canceled. The first shot scheduled for CONCERTO became the first Vela-Uniform shot, which in 1960 was tentatively planned with a series of 24 underground events. DASA was to carry out the close-in measurements program, the high explosive shots away from the NTS, and furnish support to other DoD agencies participating in the operation. Funding for the work was to be provided by the Advanced Research Projects Agency* (ARPA).

By mid 1961, Vela-Uniform had grown to a planned 35 underground shots; thirteen nuclear and twenty-two with conventional high explosives.

JERICO was planned as a single nuclear detonation in an underground chamber at the NTS. Its purpose was to study the effects of x-rays on various missile components. The fabrication of the test chamber was completed by the end of 1960, and it was estimated that Operation JERICO could be performed within 14 months of the time that authority to proceed was received from Washington. By 1961, this time had shrunk to 9 to12 months.

In June 1961, the JCS requested that DASA review and bring up to date plans for specific programs to be pursued in the event of a resumption of testing. DASA cited two proposed underground shots, Hard Hat and Marshmallow, and suggested tests in other environments, which would be necessary to address Service requirements. With the resumption of nuclear tests by the USSR on September 1, 1961, Hard Hat and Marshmallow were approved with readiness dates of February 15, 1962, and June 15, 1962 respectively.

CHAPTER 6. SANDIA LABORATORY DURING THE FIRST DECADE

PRE-1952

In Part I-Chapter 4, the section titled FORMATION AND EARLY YEARS OF THE SANDIA CORPORATION describes the formation of Sandia and its first president George Landry.

The principal Sandia organizations involved in nuclear testing in the late 1940s and early 1950s were under the leadership of Glenn Fowler and Robert Petersen. Fowler's Field Test Department reported to technical associate director, Robert Henderson. Petersen was the Associate Director of Research and was responsible for Sandia's Applied Physics Lab.

In 1950, AFSWP contracted with Sandia for airblast and weapons effects research. Specifically, Sandia was tasked to model airblast loading on structures using data from high explosives experiments. Petersen's directorate took the lead, with W. Jack Howard as the principal investigator for the team that did the high explosives experiments at Sandia's Coyote Canyon facility south of the main technical area. (Johnson 1997, 43-44; Banister 1994: 2) Jack Howard had a long and distinguished career at Sandia. In the 1960s he was the Assistant to the Secretary of Defense for Atomic Energy (ATSD/AE) and later the Sandia Executive Vice-President.

Everett Cox was also in Peterson's directorate. He was hired to lead the Phenomenology Division and by 1950 was head of the Weapon Effects Department in the Research Directorate. This department had the lion's share of Sandia's involvement at the NTS.

Cox came to Sandia as an authority on long-range blast-wave propagation. The Las Vegas window damage experience during RANGER highlighted the need for Cox's expertise in pressure-measuring devices for the study of air blast refraction and focusing by the upper atmosphere. Cox, with Herb Plagge and Jack Reed, "developed methods for predicting weather-dependent pressures at large ranges for BUSTER-JANGLE and subsequent operations at (the) NTS". Their work used a microbarograph system "for measuring low-amplitude pressure waves at critical locations."

Prior to a shot, microbarographs were set out, and high explosive detonations conducted to assess whether atmospheric conditions could result in damage at long ranges. "This approach effectively avoided significant damage to private property and thus reduced public concern about testing at (the) NTS." (Banister 1994: 14)

Cox's Weapon Effects Department undertook a large program in the spring of 1951 on GREENHOUSE in the Pacific. Experience gained during this operation was quickly applied to Sandia's activities for both the DoD and Los Alamos during BUSTER-JANGLE in the fall of 1951. Among the Sandians who participated were:

Petersen and Cox for planning and management; Harland Lenander and Luke Vortman for pressure gauges and installation; Byron Murphy for accelerometers; D. B. Kelsey and V.V. Meyers for electronic recording; and E. L. Johnson for operational support. (Banister 1994: 3) Attachment II by Byron Murphy describes some of his early experiences at the test site.

The emphasis of this effort was to study blast loading of structures, although free-field measurements were taken as well. It was assumed that by knowing the loading forces, the structural response could be calculated for a wide range of conditions. (Johnson 1997: 43-44)

Data recording on GREENHOUSE was done by people in Cox's department in collaboration with contractors. Following GREENHOUSE, Glenn Fowler's Field Test Directorate (organization 1600 in Sandia parlance) was established, and Harlan Lenander, and the people from Cox's department with recording responsibilities, joined the new directorate. In 1951, the 1600 directorate consisted of two departments: 1610, Test Operations lead by W. T. Moffat and 1620, Instrumentation lead by Glenn Fowler. Fowler was considered "an excellent supervisor who was able to judge well the capabilities of men and give them responsibility in a manner that made them determined to succeed." The "Fowler Rule" was a foundation for test personnel: "No, matter what the rules say, or I say, don't do anything stupid." (Banister 1994:1) The Petersen and Fowler directorates had significant participation at the NTS in 1951, and their personnel continued such participation after subsequent organizational changes during the 1950s. (Banister 1994: viii-ix)

During BUSTER, Sandia was involved in transit time studies, blast and shock measurements, and weapons assembly for Los Alamos's J Division. Walt Tribel and Joe Dawson of Sandia, Jack Clark from Los Alamos, and Barney OKeefe from EG&G comprised the 4-man arming party for BUSTER. The first shot of the operation, BUSTER Able, misfired, and the arming party went to the forward area to climb the 100 foot tower in order to correct the problem. This was accomplished as described in Part II-Chapter 2, section titled BUSTER-JANGLE TESTS.

Sandia made air pressure measurements, including those for terrain effects. In addition, they undertook the tremendous work involved to instrument the Army, Navy, and Air Force structures exposed to JANGLE. Jack Howard was in charge of the crew responsible for recording blast pressure at ground level stations while Harland Lenander and his people supported the DoD structures program. (Furman 1990: 600)

FIELD SUPPORT AND LOGISTICS DIVISION

In November 1951, a new organization, the Material and Field Service Division, was chartered within Field Test to assume Sandia's administrative and logistics support responsibilities for full-scale testing. The new division consisted of four sections which supplied non-technical support services at the site: a self-service stockroom;

receiving and handling services; packaging services; and Field Services.(Seward 1994:H3-4)

The division had about 30 people and remained approximately the same through 1958. During the moratorium, Sandia fielded off-site experiments without the benefit of the NTS contractor infrastructure. This resulted in additional plans and operations, construction, procurement, and budget management responsibilities for the division. (Seward 1994:H3-4)

1952 AND REORGANIZATION Sandia Laboratories director George Landry returned to Western Electric in March, 1952, and was succeeded by Bell Labs Vice-President Donald A. Quarles, who had been supervising the Nike missile electronic-guidance program.

Sandia had a young work force, of about 4,000 (more than double its size of 1,740 when Sandia was formed in November 1949) with an average age of 32. About twenty percent of the total were women and about eighty percent of the whole workforce was from New Mexico. Most of the technical staff were engineers and the largest fraction of those were electrical engineers. One of them found "work proceeding on a six-days-a week schedule under intense, secretive conditions in the face of what seemed ominous Communist threats. There was an immediacy, an urgency, with regard to doing everything possible to be responsive to national policy in growth of the stockpile and variety of weapon types. Cost was of little consequence." (Johnson 1997:52)

Quarles was a well-liked and respected leader. He had an "engaging person-to-person style and interacted with employees on a social as well as professional level ... (and) according to one executive, 'one could sense a change of attitude at the Labs. Like a fresh breeze, the feeling of camaraderie spread through the Laboratory.'" A typical work day for Quarles "began at 6:00 in the morning and ended at 6:30 in the evening. Fortunately his executive secretary, Rosalie Franey Crawford was very durable and provided a stabilizing continuity by serving in the same capacity for seven other Sandia presidents. (Furman 1990:530-1)

"At his first executive staff meeting, Quarles noted that he had learned in Washington that the AEC headquarters had decided Sandia would no longer perform any production of war reserve weapons. The AEC intended to limit Sandia to the production of test prototypes, or to furnish the military with a few custom-made new weapons for use in national emergencies, meaning a direct threat to the" US or its allies in NATO. (Johnson 1997:50) Thus, a shift away from production was in the wind and probably influenced both George Landry's replacement as director of Sandia Laboratories and the changes on the horizon.

A key reason cited for the 1952 reorganization was to place more emphasis on research and development. William MacNair was named vice president for Systems Research (the 5000 organization*). He had been a consultant for Bell

Telephone Laboratories (BTL) and was placed in the Sandia position to strengthen the applied research activities. MacNair reported directly to the President, Quarles. (Furman 1990: 527-9) [*Footnote: Sandia and Los Alamos used different titles within their organizational structures. For example Division Leaders had different levels of responsibility at Sandia and Los Alamos, and later Livermore. This occasionally led to confusion about a person's role. Sandia used numbers, for instance:

5000 (3 zeros)	Vice-Presidency, headed by a Vice-President who reports to the President
5100, 5200, etc. (2 zeros)	Directorate, headed by a Director who reports to a Vice-President
5110, 5120, etc. (1 zero)	Department, headed by a Department Manager who reports to a Director
5111, 5112, etc. (no zeros)	Division, headed by a Supervisor who reports to a Department Manager
5111-1, 5111-2, etc.	Section, headed by Section Supervisor who reports to a Supervisor

Each tier represented about a factor of 5. There were about 5 Divisions in a Department, 5 Departments in a Directorate, and 5 Directorates in a Vice-Presidency. This was based on management theories and studies which indicated that an "effective span of control" consisted of about 5. (Dick Lynch, Conversation, August 30, 2005)]

Los Alamos (as well as Livermore) was organized under a Director. He had a number of Associate Directors. Most Associate Directors were responsible for a number of Divisions. A Division Leader was, of course, responsible for a Division, which consisted of about 150 to 300 people. The Divisions letter designation often reflected the initial of the first Division Leader such as Z-Division for Zacharias or CMB for chemistry and metallurgy - Baker. A division had about 10 groups which were under the direction of Group Leaders. Groups usually consisted of about 20 to 40 or more people. The leaders at each level would usually have alternates, (i.e. deputies) and one or more Associates or Assistants. In some cases, a large group might even have several sections, responsible to "Section Leaders" who managed the work of about 10 people. Livermore used about a factor of 10 as the span of control for their organizations. Management theories and studies were not evoked to arrive at 10. It was rumored at Livermore that Jesus Christ had found that 12 was too many. Such attitudes somewhat reflect the differences in how management was viewed at Livermore and Sandia "in the old days."]

Four Directors reported to MacNair: Director of Research (5100) Stuart Hight; Director of Field Testing (5200) Glenn Fowler; Director of Apparatus Engineering (5300) F. J. Given; and Director of Electronics (5400) L. G. Abraham. Hight, Given, and Abraham were all from Bell Labs. This appears to have been a significant reorganization from the perspective of the nuclear test community.

Sandia was again involved with the arming of Los Alamos' weapons and with making airblast measurements on TUMBLER-SNAPPER.

Shortly after the September 1952 launching of the UCRL at Livermore, Bob Henderson, Richard Bice, and Ralph Wilson of Sandia met with Livermore Director Herb York in order to negotiate Sandia's role in support of the new laboratory. Initially, Sandia provided "Livermore with vital support, especially with full-scale nuclear tests" in the Pacific and Nevada.

In August 1955, "Sandia Vice President* Robert Poole proposed the formation of a laboratory consisting of perhaps 250 employees at Livermore". Approval of the AEC and the Sandia Corporate board came quickly, and "during the fall of 1955, a few Sandians went to Livermore on a temporary assignment to work directly with Livermore on its early nuclear weapons".(Johnson 1997:70-1) The pioneer group of 15 Sandians consisted of: C. E. Barncord, Nora-Bell Byrd, S. Gayle Cain, Clifford O. Erickson, Vernon M. Field, Benjamin F. Fisher, Jr., Wayne A. Grimshaw, Charles A. Gump, William B. Marsh, James McMinn, Robert L. Siglock, Frank J. Thomas, Mary A. VanBrocklin, Orval W. Wallen, and Charlie Winter. (Furman p. 682) In March 1956, Sandia formally established a second nuclear ordnance laboratory with headquarters in an old WAVE's barracks across the street from UCRL. (Furman 1990: 679) [*Footnote: The director of Sandia Laboratories was also the president of Sandia Corporation, similarly the vice president.]

Near the end of 1955, AEC Chairman Lewis Strauss told Mervin Kelly* that "Sandia must provide the ordnance engineering support for Livermore... and it should not be overly conservative in the personnel and facilities assigned to the task." York also urged Sandia to send more than the planned 250 employees. In early 1956, the AEC "directed Sandia Corporation to establish a laboratory at Livermore under the existing contract. Sandia's Jack Howard later attributed the formation of Sandia California to strong personal support from York, who wanted a separate engineering organization." (Johnson 1997: 70-1) [*Footnote: In a sense, Mervin "Joe" Kelly was "the man behind the scene" for Sandia. He was a former student of Robert Millikan at University of Chicago, joined AT&T in 1918, and patented improvements in vacuum tubes and transoceanic telephone service. He became director of research for Bell Labs in 1936 and managed substantial WWII defense work. Kelly took Bell Labs "to first-rank leadership through his insistence that it sponsor fundamental research in addition to empirical, cut-and-try methodology". He employed William Shockley and John Bardeen, who devised the transistor and won the Nobel Prize. Kelly also employed "young researchers in transistor and solid state sciences such as John Hornbeck, Morgan Sparks, and George Dacey, each of whom would, in time, serve as president of Sandia". In 1949, Kelly had recommended the formation of Sandia Corporation to manage Sandia, see Part I-Chapter 5. (Johnson 1997:29)

Kelly was elected to Bell Labs corporate board in 1952 (he later became its chairman), and this marked a turning point in Sandia's history. Although he never served as Sandia's president, he gave the facility personal attention for the remainder of his career. Even later, while serving as president of Bell Labs, he regularly spent one week out of six at Sandia.]

1953
Quarles left Sandia on July 29, 1953, to become Assistant Secretary of the Department of Defense for Research and Development. Later, he served as Secretary of the Air Force, and Deputy Secretary of Defense. Tragically, in 1959, he died of a heart attack on the day President Eisenhower intended to appoint him Secretary of Defense. (Johnson 1997, 49-54)

James McRae was selected by the Board of Directors as Quarles' successor. A native of British Columbia, he received his doctorate from Cal Tech. He joined Bell Labs, served in the Army Signal Corps in WWII, and returned to Bell Labs to serve in upper management. He arrived at Sandia on the first of September.

"Not long after his arrival, McRae scheduled group meetings with all Laboratory personnel. This show of interest and willingness to listen to people and their concerns became a trademark of the Mc Rae administration." (Furman 1990: 638)

SANDIANS AT THE NEVADA TEST SITE

After BUSTER-JANGLE, the NTS was considered a "bit more plush" than the living conditions on the Pacific atolls. This was attributed to the "Nevada climate being less benign" than the Pacific and management wanting to maintain morale. A barracks room held 4 to 6 men. "Many field testers were military veterans, so these conditions caused no shocks." Food was "plain but plentiful." The Steak House opened during PLUMBBOB, in 1957, with a flat cost of $5.00, which was considered quite expensive at that time. (Banister 1994: 2-3)

> The frequency of poker games may occasionally have interfered with worker efficiency. A second factor reducing efficiency was getting up early to watch nuclear tests. Individuals soon developed yield thresholds for observing at NTS. It usually took more than 10 kilotons to get seasoned people out of bed. ... We were usually so busy that finding recreation, besides poker or craps, was not a problem. People tended to hit the sack on Sundays or if shot weather delays developed. The more energetic ones found interesting things to explore such as mines in Nevada. (Banister 1994: 3)

On Nevada operations, Sandia gave "one compassionate leave every three weeks or so. People could return to Albuquerque or, alternatively have spouses come to Las Vegas. ... It might be thought such long absences would cause marital problems, but statistically, field testers seem consistent with other Sandians in this regard. Perhaps field work made them such interesting individuals that it compensated." (Banister 1994: 3)

SUMMARY OF SANDIA'S ARMING AND FIRING (A&F) ACTIVITIES DURING THE 1950s

All of the early nuclear devices, including the weapons employed in the strikes on Japan, required a certain amount of final assembly as a step in the sequence leading up to the detonation. In addition, appropriate fuzes had to be tested and installed in the devices planned for air-drops. Finally the firing sets, that fired the detonators, had to be readied for the test.

Norris Bradbury headed the assembly activities and George Kistiakowsky the arming party on Trinity. The groups and personnel supporting them are given in Appendix B. Z-Division personnel from both Los Alamos and Albuquerque conducted similar activities on CROSSROADS and SANDSTONE. Three groups, lead by Arthur Machen, George Koester, and William McCord were assigned responsibilities for assembly, fuzing, and firing respectively. EG&G was responsible for the timing signals and firing circuits. (Johnson 1994: C-4)

Initially, when Sandia Corporation was formed, the role of the assembly, fuzing, and firing personnel did not change. But, in time, assembly and fuzing responsibilities were phased down, and the firing group's role enlarged.

Edwin Jenkins' group did the electrical inspection of firing components while Joe Heaston's group did the mechanical inspection and assembly in Albuquerque before the test devices were flown to NTS for RANGER. (Johnson 1994: C-4)

By BUSTER-JANGLE, both Jenkins and Heaston had sections in Walt Tribel's Division, which was a part of Glenn Fowler's Field Test organization. They continued to do the electrical and mechanical check-outs on TUMBLER-SNAPPER and UPSHOT-KNOTHOLE. On the latter series Livermore provided their own arming and firing support but asked Sandia to provide such service on future operations. (Johnson 1994: C-4)

Los Alamos and Sandia, and later Livermore, collaborated on Zipper design, the code name for the components that generate neutrons that initiate the fission chain reaction. Early designs used internal initiators, inside the pit, that released bursts of neutrons upon the arrival of the compression shocks. These were limited-life components that required frequent replacement and extensive disassembly of the devices. During the 1950s, both nuclear design labs conceived ideas for generating the neutron fluxes with hardware that was external to the pit. Glenn Fowler somewhat facetiously commented that Zipper referred to their ease of replacement. The external location made it possible to open the side of a weapon, replace the neutron source, and zip it back up. (Johnson 1997:58)

The fissionable material in the earliest implosion weapons was kept separate from the high explosive until the use was imminent. This assured one-point-safety, which was the concept that the probability of a significant nuclear yield in an accident scenario would be vanishingly small. Eventually, however, weapons designs were developed that incorporated inherent one-point-safety. This introduced the era of the so-called sealed pit. These, and other, advances in the sophistication of nuclear designs evolved over time and were the motivational factors for many of the tests in Nevada.

SANDIA SUPPORT FOR THE DOD ON UPSHOT-KNOTHOLE

Sandia supported three projects for the DoD Blast and Shock Measurements Program on UPSHOT-KNOTHOLE.

The first involved pressure-time measurements on shots Annie and Simon, both fired on 300 foot towers. The intent was to record pressures on Annie and to use the Annie data to make predictions for Simon. Measurements from Simon were compared with the predictions. The Ballistics Research Laboratory also participated in this project on both shots, and NOL participated on Simon. (Ponton 1982a: 88)

In the second, Bill Perret and V. L. Gentry measured vertical stress in the ground in a two part project. One part of the project made earth stress measurements at 3 depths and at five ground ranges during Encore and Grable. The second part used arrays of directionally sensitive earth stress and strain gages and accelerometers, which were installed at a distance of about 1200 ft and at a depth of 5 feet for Annie, Encore, and Grable. (Perret 1955:3, 11-13, 32-34)

In the third, Mel Merritt instrumented the windward (toward GZ) and lee sides of a ridge with sixteen pressure gages for Simon…"pressures on the fore-slope were higher - and those on the back-slope lower -- than would have been predicted at the same distance over flat terrain. These effects were caused by a compression wave from the initial upslope of the ridge and a rarefaction (release) wave from the down slope at the crest of the ridge." (Merritt 1954:3, 20, 21)

Sandia's involvement in other future operations would be similar and even more extensive

1955

LASL Test Director Jack Clark asked Sandia's Edwin Jenkins to work directly for Los Alamos during TEAPOT and to provide the A&F support. Jenkins' electrical section which was responsible for fielding the A&F hardware, included Bob Burton, George Duffield, and Ed Holder. They also controlled the A&F interfaces with the EG&G Timing & Control systems and with the nuclear test devices. This A&F organization, with few changes in responsibilities, participated in almost every nuclear test operation since TEAPOT.

Radar fuzes were needed on air-drops, and the High Altitude (HA) shot during TEAPOT, was the last air-drop in Nevada. Thus, the late-time arming activities became mostly electrical in nature, and the name of Jenkins' group changed from "assembly, fuzing, and firing" to "arming and firing", or simply A&F.

"The A&F activities consisted of electrical check-out of the firing system, proving compatibility with the device's detonators, controlling critical interfaces, and following established safety procedures to connect the firing system to detonators that arm the" … device. These basic responsibilities have changed little over the years, but new ones have been added, primarily regarding the design and procurement of firing components and safety.

John Banister and Frank Shelton conducted a number of air blast measurements over desert, asphalt and water surfaces on the MET shot. This work is discussed in Part II-Chapter 10 under Program 1. Luke Vortman was the Sandia director of Program 34 for the FCDA during TEAPOT. Sandia also made measurements in support of several FCDA programs on Apple 1 and 2. (Rollosson 1955: 3-4)

1956

Sandia was reorganized again in 1956, and Fowler was appointed the Vice-President of Research. During this period, Sandia explored various ideas and definitions of basic and applied research that would be appropriate to their roles and missions.

1957

Toward the end of February Fowler presented Sandia management with a justification for "the expansion and formalization of a fundamental research program." Fowler emphasized that "the research organization should actively exploit the results of fundamental research to determine feasibility 'of the new knowledge being applied effectively to the ordnance problems facing the company in the future.'" (Furman 1990: 674)

PROJECT 57 and PLUMBBOB were conducted with the support of Jenkins' A&F section. During PLUMBBOB, a Sandia neutron generator group was also present for the first tests of devices with Zippers.

Bob Burton was one of the people to climb the Diablo tower when that device failed to fire. This incident led to better power monitors and the separation of instrument power, which was used only for test device hardware, from all other power sources. (Johnson 1994: C-6)

During PLUMBBOB, helium-filled balloons were used for the first time to suspend cabs containing the test devices. Arming operations took place in the cabs after the balloons were inflated and hooked to the cab, ready to ascend. A&F signals were sent to the test device through a multi-conductor cable suspended from the cab. (Johnson 1994: C-6)

Sandia also conducted a number of important effects experiments for the AEC during PLUMBBOB. They had their own Test Group during that operation.

1958

James McRae left Sandia in September of 1958 to become vice president of AT&T. Until his death in 1960, he served as a member of the AEC General Advisory Committee. (Johnson 1997: 79)

Julius Molnar of Bell Labs succeeded McRae. After obtaining his Ph.D. from MIT, he worked for the National Defense Research Committee during WWII and joined Bell Labs in 1945 where he worked in electronics and microwave research, becoming vice president for military programs.

"In December 1958, Molnar approved changes in Sandia's name. Sandia Corporation remained the legal name for the organization, but since "laboratory' seemed more descriptive of an increasingly research and development oriented organization, the New Mexico and California sites were referred to as Sandia Laboratory and Livermore Laboratory, respectively." (Johnson 1997:85)

In support of HARDTACK Phase II's underground shots, Sandia initiated a geological research group that included Bill Perret, Jim Shreve, and Byron Murphey. They were charged with exploring ground motion and seismic wave phenomena in support of containment and treaty verification studies as nuclear testing moved underground. (Johnson 1997:85)

The A&F group provided essentially the same support during HARDTACK as they had during PLUMBBOB. Faced with a cutoff in testing the work environment during HARDTACK was frantic. "... as many experiments as possible were squeezed into the schedule:

> Everyone worked extra-long hours. In some one-point safety tests, the A&F hardware was taken to GZ for initial set-up, dry runs were conducted, and the tests were finished in a few hours. Sometimes the A&F team took the hardware to GZ for initial set-up and did not return to the CP until ready to fire.
>
> One tunnel event was done in this way in about 36 hours, including sandbagging the zero room. The A&F group never left the tunnel; they took turns getting a little sleep on sandbags between dry runs. After arming the device, the A&Fers were required to stand by until the sandbagging was finished and the tunnel was evacuated; they returned to the CP, ready for the countdown.
>
> Test hardware became scarce, and the supply of neutron generators was exhausted. Personnel in Albuquerque worked almost around the clock to build neutron generators, using every available part. In fact, in several instances, A&Fers suited up in rad-safe clothing and recovered contaminated neutron generators after a safety test in an effort to reclaim parts. (Johnson 1994: C-8)

George A. Landry, first president of Sandia Corporation

Glenn Fowler

Tom Cook, who would become an Executive Vice-President of the Laboratories, spent his early career in the field of weapons effects.

Luke Vortman led Sandia's Plowshare cratering experiments.

William R. Perret, experimentalist.

Hilt Deselm, Robert W. Henderson, and Walt Treibel.

Left to right, James McRae, Glenn Fowler, Richard Bice, and Everett Cox

Among the Sandians sent to Livermore in 1955-56 were: 1 Charles Barncord, 2 Charles Gump, 3 Cliford Erickson, 4 Benjamin Fisher, 5 Robert Siglock, 6 Gayle Cain, 7 Vernon Field, 8 Charles Winter, 9 Wayne Grimshaw, 10 Frank Thomas, 11 Orval Wallen, 12 James McMinn, 13 Mary Van Brocklin, 14 William Marsh, 15 Nora Byrd.

CHAPTER 7. NUCLEAR WEAPONS 1952-1953

FALL 1952

British Nuclear Test in Australia

At 0924 (local time), on Tuesday October 3, 1952, the British detonated their first nuclear explosive, code named Hurricane, aboard the small River-Class frigate HMS Plym, just off Trimouille Island in the Monte Bello Island group. Trimouille is located at 20° 24'S and 115° 34.2'E, about 85 miles from Onslow, a small coastal town, on the northwest coast of Australia. Being seafaring folk, the British are sensitive to the vulnerabilities of their harbors in general and London in particular. Enormous destruction would ensue, of course, if an atomic weapon concealed in a merchant vessel were detonated while lying at anchor. Sir William Penny, who had been at Trinity, Crossroads, and Sandstone pointed out that little technical data were available on the effects of an explosion in a harbor. The first British test was designed, in part, to explore this possibility. Thus, Hurricane was both a weapons development and a weapons effects test. (Cathcart 1994: 253; Arnold 1987)

While Los Alamos scientists traveled about 200 miles from Los Alamos to Trinity, the British scientists traveled for about 8 weeks by ship, nearly half way around the world, to their first test site. They had a two-ship convoy that consisted of the small aircraft carrier Campania and the Plym. Their route was around Africa, because there were security concerns associated with a transit of the Suez Canal. The nuclear weapon was transported on the Plym, which served as the firing platform when the device was tested. At shot time, the Plym was anchored about a half-mile offshore, in about 40 feet of water. (Cathcart 1994) The device was located about 6 feet below the water line at the time of detonation. (Private communication with Jack Klump* on March 9, 2007). [*Footnote: Jack Klump, of Science Applications Inc., conducted fairly recent numerical calculations on the Hurricane event for the DoD and obtained this information by researching British documents.]

Mike - The World's First Full-Scale Test Of A Thermonuclear Device

On November 1, 1952, the first experimental megaton-range thermonuclear test, code-named Mike, was detonated by Los Alamos at Enewetak during Operation IVY. Mike was a laboratory-sized device and not a deliverable weapon. It went at 10.4 Mt, and was a resounding success. This was the first full-scale shot, but the second thermonuclear event. The first was George on May 8, 1951, during Operation GREENHOUSE. It had a yield of 225 kt. Mike was the largest yield test that the US had conducted up to that time, and George was the second.

Operation DOMINO Tentatively Scheduled For Fall 1953

The labs were taken with the concept of quick test operations, such as RANGER, to address specific issues. Presumably, a quick test series could be done at the NPG between the standard, full-scale, operations.

In the fall of 1952 a test series called DOMINO was planned for the fall of 1953. LASL's Duncan MacDougall said (FWC meeting announcement GMX-801 9 December '52): "...for reasons known to FWC members, the Nevada operation in the fall of 1953 will be restricted to air drops." Graves said that they could do yield measurements. It is likely that J-Division could only do relatively simple experiments in the fall of '53 because they were getting ready to field CASTLE in the Pacific, scheduled for the winter and spring of 1954.

In late January 1953 Bradbury told Fields that DOMINO was "...to be scheduled only if it is required in connection with the urgent exploitation of weapons systems suitable for stockpiling. In concept, the operation will follow that of the RANGER." ... "It is not anticipated that any specific planning on the part of the Los Alamos Scientific laboratory for Operation DOMINO would occur until well after Operation UPSHOT." (DIR - 802. Bradbury to Fields via C. L. Tyler, Manager SFOO)

> It is ... unlikely that the program would exceed five shots and may, of course, be any number less than this - even zero. No tower shots are planned, and all shots would be air dropped.
>
> It is unlikely that the need for a DOMINO program will be decided until approximately July or August, 1953, and thereafter it is probable that a specific program would be submitted only a few weeks in advance of a proposed test date.

Bradbury went on to say that the dates would probably fall between the first of September, for programmatic reasons, and the end of November for weather reasons.

In late April 1952, at the Los Alamos Fission Weapons Committee, Carson Mark suggested that one of the DOMINO shots be the same as the primary for a CASTLE device.

At the 62nd Fission Weapons Committee meeting (May 18, 1953, (SRD)) it was announced that DOMINO for the fall of 1953 had been canceled.

COMMITTEE ON OPERATIONAL FUTURE OF THE NPG

Toward the end of 1952 the AEC established a committee to assess the future of the Nevada Proving Ground. The continental test site had been in operation for two years and there had been three test series, RANGER, BUSTER-JANGLE, and TUMBLER-SNAPPER. AEC Santa Fe Office (SFO) Manager Carroll Tyler felt that it was time to step back and evaluate the safety, the usefulness, and the future role of nuclear testing in the continental United States. Their formal directive was: " ... to review the original purpose of a continental site; to analyze the changes in scope which have developed; to project requirements for the future, including the need for any other continental site; and to arrive at conclusions concerning future utility and use of Nevada Proving Ground." (Ref: LANL Archives: A-99-019 635 Nevada Test Site (5/11/1953-12/29/1955))

The group was called the Committee on Operational Future, NPG*, and it had the endorsement of AEC Division of Military Applications Director, Brig. Gen. Kenneth Fields. [*Footnote: Its members were: SFO Office of Engineering and Construction Director Reuben E. Cole, chaired the new group. The AEC in Washington was represented by Dr. John C. Bugher (Director Of the Division of Biology and Medicine), Morse Salsbury (Director of the Division of Public and Technical Information), Captain J. T. Hayward (Deputy Director of the Division of Military Applications), and Col. V. G. Huston, Captain Harry H. Haight, Col R. F. Campbell, and Donald Mastick (DMA). The AEC New York Office was represented by Merrill Eisenbud. Everett F. Cox represented the Sandia Laboratory. Alvin Graves and John C. Clark represented Los Alamos. James E. Reeves and Given H. Dugger represented SFO. Richard G. Elliott of SFO served as secretary. Darol Froman of Los Alamos and Herbert C. York of Livermore formally reviewed the minutes of the committee meetings.] (Note: 5 of the 12 members were from DMA.)

Darol Froman contributed his views on the name of the continental test site: "I would like to point out that the name, 'Nevada Proving Grounds', is, to my way of thinking, both a misnomer and a confusion. I believe it was the consensus of opinion of those attending the Committee meeting and certainly felt strongly by several others at Los Alamos that the Nevada facility should not be, and was never intended by the AEC to be, a proving ground. ... The original concept as well as the apparent feeling of the Committee is that the facility is a test site similar in concept, although of a completely different magnitude, to R Site at Los Alamos."

Norris Bradbury felt that the AEC was losing sight of the purpose of the proving ground. On January 5, 1953 he wrote to the Committee: "In view of the primary purpose of the entire NPG setup, I am inclined to feel that medical and public relations problems are somewhat overemphasized in the selection of the Committee and that the real reason for the establishment of the proving ground may be overlooked."

"With regard to the 'requirement and reasons' for establishing a continental test site, I do not believe that these have changed at all from the point of view of the LASL. I do believe, however, that the picture has changed enormously from the military point of view, and that the AEC has, in fact, accepted a changed concept. I regard the tendency to use the NPG for the purpose of weapon system tests (the forthcoming gun shot), for civil defense effects tests, for troop indoctrination and maneuvers, and for reportorial press as quite outside the original concept of this site. Indeed this trend, if continued, can force us to abandon this site for no other reason than that the military have taken it over. Even now the use of this site by other agencies is reaching such a level that it may sometime be necessary to recall that this area was actually established at the specific request of the LASL for its own needs."

Dr. Bugher took issue with Bradbury's comments: ..."I would like to have it a matter of record that the oversimplified concept which was entertained originally by the Los Alamos Laboratory concerning the NPG as a backyard quick-testing area was never realistic, and actual operations promptly disproved the soundness of the concept. The costs involved, and the magnitude of the issues concerned, give to such operations a character that involves far more than the details of weapons development. The principle of obtaining the maximum of necessary information from each detonation is unquestionably sound, and this principle, re-expressed implies that effects testing, save under the most unusual circumstances, will be keyed to the development program."

Bugher went on to add ... "I feel that the original concept was approximately correct if one includes in it an understanding that effects, indoctrination, and public display objectives are admissible if they do not interfere appreciably with the primary objectives of weapons research and development."

The committee discussed the interpretation of "interference" with the development program by other activities, such as weapons effects measurements. The DMA representatives said that they always consulted the field organizations regarding interference. The field organizations responded that they interpreted the DMA consultations as simply the question "Can you do this?" DMA apparently disagreed with that characterization of the consultations.

Graves, speaking for Los Alamos, explained that interference often takes the form of extending a test series, which consumes valuable and limited time of the scientific and technical personnel. Almost all of the shots since the RANGER Operation have had a DoD component that taxed the laboratories.

The Los Alamos representatives along with committee chairman Cole pointed out that fiscal constraints limited the housing and support facilities at the site. They suggested that the facilities not be stretched unreasonably, such as putting 8 men in housing designed for 4, by the demands of programs of secondary priority.

The consensus of the committee was that it was important to pursue the secondary programs to the extent possible without unduly interfering with the activities of the primary programs. In short, activities that did not require the NPG should be pursued elsewhere.

A number of other secondary objectives were raised. For example, there was discussion of a possible test under adverse weather conditions. For a variety of reasons, including safety (radiation hazards due to rainout of bomb debris), scheduling, and diagnostic measurement problems, the idea was dropped.

Recognizing that the military was there to stay, the AEC committee explored ways to increase the capacity of the Nevada Proving Ground. In order to define the problem they looked at a number of issues:

> The nature and character of future tests compared to past tests.
> Radiological contamination of the site.
> Stabilization of firing site soil surface.

Need for industrial safety personnel.
Downwind contamination and radiation hazards.

"Dr Eisenbud summarized experience as shown by fallout data outside the 200-mile radius. The highest short-range exposure occurred in Northern Utah following the May 7, 1952, detonation (TUMBLER-SNAPPER Easy, a tower shot with a yield of 12 kt). The total whole body exposure to individuals in this area was equivalent to about one month's exposure to natural background radiation; superimposed was a lung dose due to the inhalation of respirable fission products, this dose being roughly the same order." (Ref. "Report of Committee on Operational Future of the Nevada proving Ground. AEC, May 11, 1953 p. 17) "He found that, when measured against permissible exposure levels and against normal exposure from atmosphere and ground, there was no short range hazard." (ibid., 17)

The report of the Committee, which was released in May 1953, pointed out that there was sensitivity to improving the claims process for the public. "The AEC General Counsel was preparing legislation to give the AEC authority for settling claims similar to that which the Armed Forces have." (ibid., 19)

The Committee concluded that Los Alamos and Livermore would continue to require the continental test site in Nevada.* [*Footnote: Livermore's first shot, Ruth, was fired at the end of March 1953 during UPSHOT-KNOTHOLE.] They also recognized the legitimate needs of the DoD for effects studies and for military training and indoctrination. Finally the Committee anticipated that future use would be as heavy as it had been in the past. In any case, it is clear that the Committee felt that it was essential for the AEC and the laboratories to proceed in a professional and responsible manner with regard to the obligations to the public.

There is a persistent, but unjustified, impression that pressure during the 1950s to get new nuclear weapons into stockpile compromised the government's interest in the welfare of the citizens. It has been asserted that the public's health and safety were not of much importance to the AEC or to the laboratories during this period. (See, for example "Bombs in the Backyard" by A. Costandina Titus. University of Nevada Press, Reno and Las Vegas 1986 and 1989) From even a cursory perusal of relevant documents, however, it is clear that the nuclear testing community was very much concerned with the public health aspects of nuclear weapons tests.

The Committee went on to point out that the NPG did not have the capability in terms of technical facilities or support to accommodate the nuclear test requirements of the early to mid-1950s. It was clear that Los Alamos was feeling overwhelmed at the site by the other users. Livermore and the FCDA both began their participation at NTS during UPSHOT-KNOTHOLE in the spring of 1953. Sandia participated in both the effects measurements and in the development shots. The DoD sponsored effects tests and troop training, including maneuver and indoctrination exercises. The Army proof fired the atomic cannon in the Grable event. One serious complication for both the AEC and the DoD was the large number of un-cleared participants and visitors at the site.

The Committee made a number of recommendations. One was to limit the growth in users at the site. Another recommendation was to do effects measurements on weapons development shots as much as possible. The rationale was that:

- The development shots would be fired in any case and the DoD might as well use the events to make measurements relevant to their needs.
- The number of shots dedicated to effects measurements would be reduced.
- The expenditure of precious uranium or plutonium would be reduced and consequently freed up material that could be used for war reserve weapons.

The drawback for the DoD, of course, was that their participation on the laboratory's development shots was on a noninterference basis, and the fact that the performance of the devices was often, if not usually, uncertain. However, to the extent that the DoD could use development shots, it either cut down on the total number of tests in a series or resulted in additional data for the DoD.

While DoD participation on AEC shots was, in principle, on a noninterference basis, it could not, in fact, totally avoid interfering with the AEC's operations. Just the logistics in Mercury and in the forward areas significantly taxed the AEC's resources.

Another recommendation suggested limiting the yields as follows: Surface or subsurface shots limited to 1 kt. Tower shots to approximately 35 kt, and airdrops to 50 kt until the completion of an assessment of bomb fuzing reliability.

Acting Director of Military Applications, Col. Marcus F. Cooper (USAF), sent a memo to Carroll Tyler on July 7, 1953 with a request to reconvene the Committee to Study the Operational Future of the NPG in the light of developments during and after UPSHOT-KNOTHOLE. A few changes were made. Tyler was appointed chair, Bradbury replaced Graves, and a meteorologist was added. Perhaps most importantly, the DMA representatives were to act in an advisory capacity and not to participate in derivation of the Committee's conclusions and recommendations.

"The report of this Committee, which is to be made to the Director of Military Applications should include recommendations regarding:

1. Establishment of criteria in general terms governing types of tests which may be conducted at Nevada proving Grounds.
2. Review of meteorological criteria governing conditions under which these tests may be conducted, including recommendations as to precautions which are or may be taken to safeguard persons and animal stock in downwind areas as well as advisability of actions taken to decontaminate persons or property outside the Nevada Proving Grounds and LV Bombing and Gunnery Range.
3. Educational or other measures which can be taken to alleviate present existing public concern over dangers popularly conceived to be caused by these test operations.

On September 29, 1953, Tyler submitted to DMA the "Preliminary Report on Continental Tests and Future Utilization of Nevada Proving Grounds." The Committee had met on Sept. 24 and 25 and "... concluded unanimously that:

a. Nuclear weapons development progress requires both a Pacific and a continental proving ground.
b. The value of continental tests is clearly established in papers presented.
c. The continental site should continue to be the Nevada Proving Grounds, which has been used for four successful series and which meets the essential criteria of logistics and safety better than any other continental site known to the Committee.
d. Papers presented, which are to be supplemented by further studies, indicate clearly that operational controls can be strengthened to provide continuing assurance of public safety.
e. A more extensive educational and informational program is required to support continental operations and use of a continental site."

"The Committee recommended unanimously and I endorse to the Commission that:

a. Continental tests be continued at Nevada Proving Grounds; and
b. A more extensive educational and informational program be activated nationally and in the region of Nevada and its adjoining states."

On October 19, 1953, Tyler assigned committee members and advisors several tasks. Among other assignments, Tyler asked Graves and Felt to complete a paper on hot-spot location, surface stabilization, and shot configuration - i.e. balloons, towers or airdrops. Tyler also asked Graves for his views on long vs. short operations, the frequency of shots, the feasibility of phasing in easier shots in place of more difficult shots to meet weather situations, and the effects on personnel of any recommendations. The results of these studies are addressed in later chapters.

NUCLEAR WEAPONS TEST OBJECTIVES

Military Requirements

During the late1940s the military had only the already-obsolete gun-type weapon, Little Boy, and the implosion weapon, or Fat Man, and its closely related descendents as nuclear weapons designs. By the early 1950s there were a number of drivers that greatly influenced the nuclear weapons landscape. The weapons were getting smaller and more efficient. More nuclear material was available than in the 1940s; and thus more weapons could be both tested and built; and there were enormous strides made in new weapons delivery technology.

Nuclear weapons were no longer just Air Force delivered bombs. The Army and Navy wanted their own nuclear capability. It would be only a slight exaggeration to say that the services wanted nuclear systems to match the corresponding conventional capabilities. This meant nuclear artillery shells, bazooka projectiles, missile warheads, depth charges and, of course, all manner of new bombs. In addition, the development in the 1950s of thermonuclear weapons meant that mediocre weapon delivery accuracy could be, at least partly, compensated by higher yields. Also higher yields implied, it was thought at the time, increased military effectiveness.

By the late 1950s it became clear to most strategic planners that yields above about 1 megaton were of little military use, except for strikes against very hard targets. In fact, there were many drawbacks to the largest weapons. Indiscriminant collateral effects and unintended consequences were recognized as undesirable. Also, the physical size made very high yield weapons difficult to handle and limited the number of weapons on any particular platform.

As military analyses evolved and weapons systems became more discriminating it became clear that small, accurate, lower yield weapons were more effective than indiscriminant high yield weapons.

In summary, during the 1950s there was at first a plethora of new ideas and concepts for nuclear weapons and a desire for ever larger yields. This was followed in the late 1950s by a gradual recognition that the idea of a nuclear capability to match each and every conventional capability didn't make military sense. The number of weapons in the stockpile continued to grow during the whole decade, but the total yield of the stockpile started dropping toward the end and continued to decline for the rest of the century.

Nuclear Weapons Development and Effects Tests

Nuclear weapons effects tests needed devices as sources with well-understood yield and outputs. The effects measurements were usually too complicated and expensive to jeopardize with untried devices. Weapons development tests, on the other hand, used experimental devices. By their very nature, the yields and outputs were not known. It often made sense for some simple effects measurements to piggyback on the development shots, but the options were limited during the atmospheric test days and practically disappeared when testing went underground. Also, limited weapons development data were obtained on effects shots; but such data were related mostly to reproducibility. As the DoD settled on a few devices for effects measurements, the design laboratories lost interest in making more than cursory or failure-mode performance measurements on the effects shots.

One-Point Safety

Most nuclear weapons, such as the Fat Man, have high explosives surrounding nuclear material. The explosives are detonated by the action of a number of detonators fired simultaneously. It is very important that a nuclear weapon not go off with a significant nuclear yield in an accident. A weapon is "one-point safe" if the nuclear yield of a weapon detonated at one point, as in an accident, is less than a few pounds equivalent of high explosive. That is to say, in an accident scenario the nuclear yield would be significantly lower than the high explosive yield. There are arbitrary figures to describe the one-point safety criteria that must be met by weapons, and during the 1950s a number of tests were made to examine that feature. Most, but not all, of the weapons tests that had an extremely low yield (or zero yield) were one-point safety shots.

Gun-assembled weapons are inherently not one-point safe. For such weapons mechanical means were used to insure that a nuclear yield would not result from an accident.

Initiation Options

When a nuclear weapon is fired, nuclear material that originally was subcritical is very rapidly assembled into a critical mass. It is necessary then to inject neutrons to start, or initiate, the chain reaction. The hardware that supplies the neutrons is termed the initiator. There are a number of ways to inject neutrons and various trade-offs for the different schemes to accomplish this. The behavior of a particular nuclear weapon design, including the initiator, can only be determined with a nuclear test. Many of the shots in the early to mid-1950s were devoted to testing the efficacy of initiation technology.

Composite Pits

The idea of using Pu239 and U235, rather than just Pu239, in implosion weapons occurred to the weapons designers during WWII (see Part I-Chapter 1 where Oppenhiemer raises this issue with Graves). One reason to use both was that the production of U235 was greater than that of Pu 239. It would be possible, of course, to use just U235 in a weapon. Much of the work during the 1950s was focused on exploring the design parameters to achieve specific yields, weights, and dimensions making the most effective use of U235 and Pu239.

Primaries: Thermonuclear Weapons Triggers

Another line of attack during the mid-1950s focused on the design of fission devices to trigger thermonuclear weapons. Trigger devices are called primaries. Primaries are designed to transfer their energy most effectively to the secondary, and their designs are different than those for a "stand-alone" fission weapon. During the 1950s, much of the testing at the NTS was conducted for the development of primaries.

Full-scale thermonuclear weapons had too high a yield to be safely fired in the atmosphere in Nevada. In the early days they were tested at the Enewetak or Bikini Atolls in the Pacific. When testing went underground in the 1960s it was clear that very high yields could be fired in Nevada. The largest yield underground test at the NTS was Boxcar, fired by Livermore on April 26, 1968, at about 1.3 Mt.

There was not an official maximum yield for Nevada. However, it is doubtful whether 1.3 Mt could have been exceeded, or even matched, after Boxcar. That yield caused an enormous public furor and raised questions about the integrity of near-by Hoover Dam when exposed to the seismic shock of such a large explosion. There was also, of course, concern about possible damage to high-rise buildings in Las Vegas. When the US needed to test a 5 Mt device it was fired as the Cannikin shot in 1971, approximately a mile underground on Amchitka Island in the Alaska Aleutian chain.

PHASES OF DEVELOPMENT FOR NUCLEAR WEAPONS

In 1953 the AEC and the DoD defined six (and then seven) phases to categorize the life cycle of a nuclear weapon. This model served for decades with only minor modification:

Phase 1: Weapon Conception - the DoD/AEC exchange of preliminary information that could lead to a feasibility study of a weapon concept. Los Alamos, Livermore, Sandia and DoD studies were completed independently or cooperatively.

Phase 2: Feasibility - joint AEC, DoD, and contractor investigation of whether a concept could be developed and manufactured. If a weapon were seen as feasible, and desirable, the AEC would issue a Phase 3 authorization for development of the weapon design.

Phase 3: Development - design definition. Engineering design and production planning lead to full-scale mockups for environmental and flight testing.

Phase 4: Pilot Production.

Phase 5: Initial Production.

Phase 6: Quantity Production and Stockpile.

Later, Phase 7 was added: Retirement. (Johnson: 52-3)

Nuclear weapon development tests proposed by Livermore or Los Alamos could support any phase, but generally were related to Phases 1 through 3. Later stockpile confidence testing addressed issues raised in Phase 6. Some AEC related tests were focused on more basic science and did not fall into any particular phase.

CHAPTER 8. OPERATION UPSHOT-KNOTHOLE MARCH 17 - JUNE 4, 1953

PLANNING

In October 1951, the Chief of AFSWP recommended to the Joint Chiefs of Staff that a large military effects test program be conducted in the spring of 1953 at the Nevada Proving Grounds. The effects of nuclear detonations on military equipment as well as on structures and other targets of military significance were of specific interest. In December 1951, the Joint Chiefs of Staff approved the recommendation, subject to a future determination concerning the nature and number of tests. DoD code-named the Operation KNOTHOLE, which was scheduled to begin April 1, 1953. (Ponton 1982a: 31-32)

In early June, 1952, shortly after the conclusion of TUMBLER-SNAPPER, Los Alamos was seriously thinking about the next two test series. It was thought that a continental test series after Operation IVY*, which was scheduled for the fall of 1952 in the Pacific, would be very worthwhile. [*Footnote: IVY, conducted in fall 1952, consisted of two shots at Enewetak. Mike, the world's first high yield thermonuclear explosion, produced 10.4 Mt; and King, the largest fission device yielded 500 kt. Both were Los Alamos devices.] Also, a test operation in Nevada would be desirable prior to CASTLE, the high yield test series planned for Bikini and Enewetak during the winter and spring of 1954. The AEC's 1953 test series in Nevada would be code-named UPSHOT. (Clark 1954: 9)

In addition, the Federal Civil Defense Administration (FCDA) planned to increase their involvement in nuclear effects studies beyond their participation as guests of AFSWP on TUMBLER-SNAPPER. Their program was dubbed Operation DOORSTEP, which included an extensive variety of exposure studies as well as a training course for civil defense personnel. (Hacker 1994, 89, Lamoureux 1953: 9)

By June 1952, the AEC and the DoD had agreed that their spring 1953 programs would be combined into a joint operation code-named UPSHOT-KNOTHOLE. Meanwhile the FCDA had discovered that their budget would not support an independent test program. As a result, they decided to participate, to the extent feasible, on the AEC or DoD tests. (Ponton 1982a:32)

By September 1952, the AEC test program included at least five Los Alamos devices (7 were actually tested) and two for the just-established University of California Radiation Laboratory (UCRL) at Livermore.

Based on recommendations from the armed services, AFSWP formulated plans for the DoD weapons effects test program. After reviews by AFSWP and the armed services, construction for KNOTHOLE began in Frenchman Flat in mid-December for the DoD's extensive effects measurements programs in both airblast and structures. (Ponton 1982a, 31-32)

The DoD plans included a military effects test with a stockpile device as the source and a systems test of a nuclear weapon fired from a 280 mm cannon. Also, the U.S. Sixth Army again planned to participate with Desert Rock V exercises.

NEVADA TEST ORGANIZATION FOR OPERATION UPSHOT-KNOTHOLE

Two main differences of the UPSHOT-KNOTHOLE organization were:

1) Livermore (UCRL), which had been founded less than six months prior, participated for the first time in a nuclear test operation. Since Los Alamos had the test operational experience, it was clear that for UPSHOT-KNOTHOLE, at least, Los Alamos would supply most of the operational leadership. As UCRL personnel quickly "learned the ropes," the two nuclear design labs assumed the roles of equal partners.
2) The Federal Civil Defense Administration (FCDA) participated on UPSHOT-KNOTHOLE with weapons effects programs that were from a civilian perspective. The Test Director formed a third test group in his line, the Civil Effects Tests Group.

Except for these differences, the organization for UPSHOT-KNOTHOLE was essentially the same as that for TUMBLER-SNAPPER.

The Test Manager was again Carroll Tyler. Alvin Graves again chaired Tyler's Advisory Panel, see Appendix F. The other Advisory Panel members were: Thomas Shipman - LASL; Brig. Gen. James P. Cooney - USA; Howard L. Andrews - USPHS; Everett F. Cox - Sandia; Col. Benjamin Holzman - USAF; and Capt. Harry Haight - USN.

With the addition of the FCDA to the test activities, more effort was required to accommodate additional visitors and public exposure. Thus, groups that reported to Tyler were formed for these initiatives, see Figure 2-8.1.

There were three Deputy Test Managers: Deputy for Scientific Operations, Alvin Graves; Deputy for Military Operations, Col. Paul Pruess; and Deputy for Support, Seth Woodruff. As in TUMBLER-SNAPPER, both Woodruff and Graves had two roles: as deputies and as leaders of large organizations with a wide array of functions. Preuss also served in two capacities: as Tyler's military deputy and as senior representative of the Commanding General, Field Command, which placed him as the military commander of all DoD units and personnel on duty within the NPG. (Preusss 1953: 20)

Woodruff was the Support Director and J. B. Sanders his Deputy. A number of branches with different support functions reported to Woodruff. The Field Command Support Unit (FCSU) headed by Lt. Col. E. M Tolliver, USA was one of the branches. The FCSU operated as an agency of the Support director in overall support of the NTSO and also operated under Preuss in the implementation of his responsibilities. (Preuss 1953: 70)

Al Graves was again the Test Director with Jack Clark as his deputy. In this capacity they acted for both nuclear design laboratories. The three groups that conducted the scientific programs reported to Graves: The Weapons Development Test Group with Bill Ogle as Director, for Programs 10 thru18 and 20, which included both the LANL and UCRL diagnostics measurements; The Military Effects Test Group with E.B. Doll as director and Col. H.K. Gilbert as Deputy, for Programs 1 thru 9; and The Civil Effects Group with R.L. Corsbie Director, for Programs 21 thru 29. (Ponton 1982a: 48-51) Graves also had two sections that worked on staff and advisory functions and on support functions.

The test organizations and the major contractors adopted three new operational ground rules for UPSHOT-KNOTHOLE. The first was no detonations on Sundays. The second was that tower shots were to be at least one week apart. And the third was that the interval between a tower shot and an airdrop was to be a minimum of four days, and the time between an effects shot and the cannon shot was to be a minimum of two weeks.(Clark 1954:12)

Clark describes "how things were done" between the Test Director and AFSWP:

> The administrative procedure employed by the DoD and test organization to accomplish the military effects tests is as follows: The military effects test programs originate with AFSWP, which obtains test proposals from the various military agencies. AFSWP integrates and coordinates the proposed program with the predicted characteristics of the device to be tested. The program is then presented to the Research and Development Board (RDB) of the DoD for approval before being forwarded by AFSWP, Washington, to DWET*, Sandia Base (Albuquerque). It then becomes the responsibility of DWET to implement the program and carry out the tests**. The philosophy of the Test director is not to screen the DoD tests*** for their effectiveness or value but rather to coordinate this test** program into the over-all test** operation and it is only from an operational standpoint that an experiment is questioned." (Clark 1954: 20) [*Footnote: Division of Weapons Effects Tests, described in Chapter 5.] [**Footnote: Since the authors use "test" to refer to a nuclear detonation, the word "program" or "project" or "measurement" would be more appropriate here than "test".]

Essentially the same procedures were used by the Test Director for the FCDA activities. Also, the Test Manager used a somewhat similar approach to the various other activities conducted by the military and the FCDA such as: Desert Rock; the FCDA open shot; and other exercises. However, the criterion of paramount importance was that such "add-on" activities not interfere with the test schedule.

PERSONNEL CONDITIONS AT THE NPG DURING UPSHOT-KNOTHOLE

Three days before to the first UPSHOT-KNOTHOLE detonation, the Las Vegas Test Information Office, located at 1235 South Main Street, issued a press release regarding the Nevada Proving Grounds. In it, Camp Mercury is described as providing: office space and living quarters for test organization personnel in both temporary and permanent quarters; utilities, warehouses, mess halls, recreation facilities, motor pool, and other facilities such as administrative offices. The Control Point is described as being a complex of permanent and temporary facilities approximately 20 miles north of Mercury. It is on Yucca Pass, between Frenchman Flat and Yucca Flat, permitting vision into both general areas, see Figure 3-8.2.

The major maintenance, operation and construction contracts that were in force in March 1953 were:
- Reynolds Electric and Engineering Company, of El Paso, Texas, for maintenance, minor construction, and scientific structures support services.
- Federal Services, Inc., for security and other guard services.
- Universal Food Service, Inc., for feeding, housing, and related personnel services.
- Pacific Telephone and Telegraph Co., for communication facilities and service.
- And three construction firms with five contracts covering non-permanent construction.

By UPSHOT-KNOTHOLE, the initial construction associated with the so-called permanentization had been completed. The main construction efforts during the Operation were test related. Facilities were generally considered adequate and smoother administrative procedures helped make life for the test personnel easier.

Preuss (1953: 71) provided the following tally of just the DoD personnel at the NPG during UPSHOT-KNOTHOLE. The numbers cited for each of these groups is probably fairly representative of both the previous and future operations.

The maximum number of AFSWP sponsored personnel stationed at the NPG at any one time was 1334:

	Officers	Enlisted	Civilians	Total
Military Effects Group	179	265	343	787
Field Command Support Unit	24	93	1	118
Units on Mission Basis				
Rad Safety	31	161	17	209
Air Weather	19	51		70
Motor Maint	1	29		30
Others Integrated in Joint AEC-DoD Staff	22	21	2	45
Total Operational	276	620	363	1259
Technical Observers	75			75
TOTAL	351	620	363	1334

Non-AFSWP sponsored military personnel that assisted non-DoD agencies within the NPG under arrangements made other than through AFSWP:

	Officers	Enlisted	Total
Weapons Development Group (AEC)	22	16	38
Civil Effects Group	3	31	34
TOTAL	25	47	72

Associated installations of military personnel, exclusive of observers, troops for maneuvers, and trainees were approximately:

	Officers	Enlisted	Civilians	Total
Artillery Test Unit	46	210	9	265
Camp Desert Rock (Army)	145	1980		2125
Indian Springs AFB	30	410	10	450
TOTAL	221	2600	19	2840

Billeting of AFSWP sponsored DoD personnel was handled by AFSWP's Field Command Support Unit (FCSU). The policy was "to house DoD personnel in compatible groups, maintain units in compact areas, and to minimize movement of personnel. The following space was allocated by the AEC for AFSWP sponsored DoD personnel:

> 5 barracks, capacity 55 each
> 95 hutments, capacity 4 each
> 77 hutments, capacity 9 each
> Total Space - 1277 Beds
>
> Officers and civilians were assigned barracks space insofar as possible. The Hutment Area was divided into two major areas: Officer-Civilian and Enlisted. Space within the two major areas was subdivided into smaller areas so that individual units and projects could be grouped together. The barracks rooms were increased from six beds to eight beds to allow space for technical observers. The above procedure worked to the general satisfaction of all concerned, considering the over-all crowded conditions.
>
> Throughout the operation morale remained exceptionally good considering the circumstances, and disciplinary cases requiring formal legal action were at a minimum ..."
> (Preuss 1953: 72-73)

Protestant and Catholic chaplains conducted Sunday services. They also arranged for the conversion of a temporary hut into a small chapel for private devotions. Private contributions augmented by three hundred dollars from the AEC made possible the acquisition of an electric organ.

Chaplains from Mercury conducted services at Camp Desert Rock and at Indian Springs since these outposts lacked their own religious coverage during most of the operational period. (Preuss 1953: 73)

> Bus recreational trips to Las Vegas were scheduled two to seven nights per week, depending on the test operational schedule, and on Sundays to points of interest including Hoover Dam and Death Valley. In addition, a bus was dispatched to any point within 100 miles at any time for recreational purposes when approximately 20 or more people were interested. Moving pictures were shown each evening with approximately 50 percent of the films (maximum obtainable) being furnished by the Armed Forces Motion Picture Service and the remainder being rented on a profit/risk basis by a group of officers and enlisted men. It is apparent, from experience during the operation, that additional Special Services athletic equipment could have been utilized and that a more comfortable and efficient structure is required for motion pictures."
> (Preuses 1953:74)

AFSWP put 900,000 miles on 206 vehicles and had only one accident with damage exceeding twenty-five dollars. (Preuss 1953:76)

Army Staff Surgeon, Col. Clinton S. Maupin, supervised medical service for the DoD. His responsibilities included: advising NTSO on health and sanitation; conducting investigations of on-site and off-site radioactivity; and advising the Site Radiological Safety Officer. Hospital facilities were available at Nellis AFB, and both the AEC and DoD had ambulances assigned to the site. The Mercury dispensary and ambulance were available 24 hours a day, 7 days a week. Approximately 670 cases were handled, including 28 people who required hospitalization. "A total of 13 accidents were reported ---only three injuries were serious enough for injured to lose time from duty." (ibid., 81)

There was a weekly inspection of "sanitation in the kitchens, cafeteria, recreation hall, and barracks. No outbreak of disease traceable to sanitary defects was observed." (ibid., 79)

UPSHOT-KNOTHOLE TESTS

The eleven tests conducted on Operation UPSHOT-KNOTHOLE are cited in Table 2-8.1, and their locations at the NPG are shown in Figure 2-8.2.

Table 2-8.1 Tests Conducted During UPSHOT-KNOTHOLE

TEST	DATE-1953	SPONSOR	TYPE*	AREA	PURPOSE	YIELD (kt)
1. Annie	17 March	LASL	TOWER 300 ft	3	WEAPONS	16
2. Nancy	24 March	LASL	TOWER 300 ft	4	WEAPONS	24
3. Ruth	31 March	UCRL	TOWER 300 ft	7	WEAPONS	0.2
4. Dixie	6 April	LASL	AIRDROP 6,020 ft*	7	WEAPONS	11
5. Ray**	11 April	UCRL	TOWER 100 ft	4	WEAPONS	0.2
6. Badger**	18 April	LASL	TOWER 300 ft	2	WEAPONS	23
7. Simon	25 April	LASL	TOWER 300 ft	1	WEAPONS	43
8. Encore	8 May	DoD-LASL	AIRDROP 2,423 ft*	5	EFFECTS	27
9. Harry	19 May	LASL	TOWER 300 ft	3	WEAPONS	32
10. Grable	25 May	DoD-LASL	280 mm CANNON 524 ft*	5	EFFECTS & WEAPONS	15
11. Climax	4 June	LASL	AIRDROP 1,334 ft*	7	WEAPONS	61

* indicates height above the terrain

** Initially, Badger was planned to be the 5th shot and Ray the 6th. DoD reports of UPSHOT-KNOTHOLE frequently use the terminology Shot 5 or V5 when referring to Badger and Shot 6 or V6 when referring to Ray. However the 4th shot, Dixie, contaminated the area planned for Badger. As a result Shot V6, Ray, was advanced to 11 April, and became the 5th shot in the operation. Badger, AKA Shot V5, was delayed and became the 6th shot in the operation.

(DOE 2000: 4-5; Lewis 1997: 21)

Jack Clark had an interesting comment related to the pace of testing: "In Operations BUSTER-JANGLE, TUMBLER-SNAPPER, and UPSHOT-KNOTHOLE the trend had been to be more leisurely in the schedule, with the tests for Operation UPSHOT-KNOTHOLE scheduled on a 6 or 7 day interval basis."(Clark 1955: 14) The tests were, of course, considerably simpler in the mid-1950s than they were in later years; but never the less the reference to a test series with 6-7 days between shots as "leisurely" appears quaint when viewed fifty years later.

The general objectives for the 9 weapons development shots were to continue the exploration of initiation physics, to study new ways to improve the efficiency of fission devices, and to explore thermonuclear design details.

The third and fifth shots, Ruth* and Ray*, were Livermore's first and second nuclear tests. [*Footnote: Ruth and Ray were Herbert York's "honorary" aunt and uncle.] They were physics tests of devices that were of particular interest to Edward Teller. The performance of both was disappointing, but the experiments were of adventurous designs exploring the feasibility of concepts that had been discussed for years, but not tested.

The weapons effects measurements fielded on Encore and Grable focused on fundamental blast and radiation studies. Grable was a systems test of a weapon fired from the Army's 280-mm cannon. The detonation point was 524 feet above Frenchman Flat, 11,000 yards north of the artillery piece. Encore and Grable were planned to have the same GZ and to use the same airblast instrumentation.

About three weeks before Encore, an unexpected rainstorm flooded Frenchman Lake and the underground recording bunkers. Fortunately, the equipment was recovered, dried out, and reassembled in time for the shot. (Lewis 1997: 21)

Encore was plagued by a more serious problem than the rainstorm. The Air Force crew dropped the bomb about 840-feet (about 280 yards) south of the intended GZ.* This resulted in a significant loss of data and a subsequent policy of avoiding air drops for important effects experiments. Plans were initiated to use steel towers for military effects tests, but at that time the maximum tower height was only 300 feet. "Important scaled-height-of-burst (SHOB) effects were observed on BUSTER and TUMBLER at ranges of 400 to 1000 $ft/kt^{1/3}$, and these effects would be difficult to study using towers. Later, tethered balloons would replace towers as device support structures." (Lewis 1997: 21) [*Footnote: On Encore, the troops & observers were located 4000 yards SW of the intended GZ. (Preuss 1953: 92) If the actual GZ was 280 yards farther S than planned, this would have placed the troops at an actual distance from GZ of about 3810 yards.]

WEAPONS DEVELOPMENT DIAGNOSTIC EXPERIMENTS

The main experimental objectives for Operation UPSHOT-KNOTHOLE were alpha, yield, and the times of the onset or duration of various nuclear phenomena that occur during detonation. These were the key quantities to be compared with theoretical calculations of device behavior. The associated experiments were tied to interlocks that prevented device detonation in the event the experiments were not operating correctly. The purpose of the interlocks, of course, was to prevent the detonation when it was clear that key data would not be obtained. Other experiments provided additional information on device performance, weapon outputs, or contributed to the development of new diagnostic technology.

The Weapons Development Test Group experiments for both LASL and UCRL were categorized as Programs 10 to 20 as follows:

Program 10 -Gamma-ray measurements: included alpha, transit-time, and the development of new measuring techniques involving gamma rays, as well as some fundamental investigations of the gammas produced by nuclear detonations.

Program 11 -Simultaneity investigations: were measurements of the time sequence of nuclear events involved in the special assemblies that were used in this test series.

Program 12 -Technical Photography: included fireball photography, air-burst-position determination, cloud phenomena, bhangmeters, and the very-high-speed framing camera photography.

Program 13 -Radiochemistry: consisted of the determination of yield, efficiency (fraction of fuel burnt, see Appendix A), compressions, and other nucleonics measurements made possible by the use of rad-chem detectors.

Program 14 -Initiation Measurements: involved the use and interpretation of technology employed in the initiation of a nuclear reaction.

Program 15 -Electrical investigations: included fundamental investigations of electromagnetic signals generated by nuclear detonations.

Program 16 -Unassigned

Program 17 -Neutron Measurements: included fundamental measurements on total neutron flux, the neutron spectrum, and the Tenex* and Phonex* experiments for determining neutron energies. [*Footnote: Tenex is a measurement technique to determine the temperature of the thermonuclear burn region by measuring the broadening of the 14 Mev neutron spectrum caused by the tritium + deuterium (T3 + D2) fusion reaction (see Appendix A, section titled THE FUSION PROCESS). (Ogle 1985: 74). Phonex stands for Photographic Neutron Experiment, where neutrons produce recoil protons in photographic plates. The resulting tracks are measured using a microscope, and yield information related to the incident neutron spectra. (Louis Rosen, private communication)]

Program 18 - Thermal-radiation measurements: covered fundamental investigations on total thermal radiation, thermal power, spectral measurements, and light absorption near the fireball.

Program 19 -Unassigned

Program 20 -Timing and Firing: covered the timing and firing signal requirements associated with the experimental program, such as Go-No-Go devices and special timing signals. (Clark 1954: 20-21)

Programs 10, 12, 13, 15, 17, 18, and 20 were conducted on all tests. Program 11 was conducted on only Nancy and Simon, and Program 14 was only conducted on Annie and Harry. (Clark 1954: 129)

OVERVIEW OF TESTS AND DIAGNOSTICS

Annie, March 17

Annie was an "open shot" and enjoyed considerable publicity because it was the host for the FCDA's Operation DOORSTEP which is described in this chapter's FCDA NWE PROGRAMS AND PROJECTS section. The device diagnostic experiments, alpha, yield and the timing of certain reactions, were successful.

Nancy, March 24

Good data were obtained on all of the principal experiments which were: yield, alpha, time resolved photography (for fireball growth measurements), and time intervals between reactions of the various components of the device. The latter two experiments were particularly complex.

The high-speed photo station for the time resolved photography was located 1500 feet west of GZ. It had blast-proof doors that were to close during the 300 msec that separated the detonation time from the shock arrival time at the station. The blast doors did not work correctly; and while the recording equipment suffered serious damage, the film was recovered and yielded good data.

An interesting photograph, taken about 2 msec after detonation, showed spikes on the lower half of the fireball. (Clark 1954: fig. 5.27, p. 57) Radiation absorption by the tower legs and the three tower guy wires resulted in the observed spikes. The spikes were clearly visible in the high-speed photographs until about 15 msec after detonation. This phenomenon was observed on a number of tower shots.

The time interval experiments employed elaborate arrays of magnetite concrete collimators located on steel towers and on the ground. These collimators provided lines-of-sight from the device components to detectors in Station 300 of Area 4, located 1000 yards east of GZ. This station was also the alpha-recording shelter. At the 15 msec time period a mass clearly moving out of the fireball was identified as one of the magnetite concrete collimators originally located in the tower cab housing the test device.

The cone-shaped wire mesh structure, which is still located near the dirt road leading to the mountaintop* experiment station from the Tippipah Highway, was used to pick up electromagnetic signals from Nancy. [*Footnote: This mountain top station was called the "monastery," and was used on a number of shots for a variety of observations, including photographic and EMP experiments. It had several buildings that were removed in 2005.]

The wind pattern changed after the 9:00 pm D-1 weather briefing to one with a more northerly trajectory than had been predicted. As a result, the Rad-Safe officer located at the Lincoln Mine, about 30 miles northeast of GZ, was instructed by the Deputy Test Director, Jack Clark, to advise the mine superintendent to keep his people indoors for the first few hours after the shot. Also, outdoor activities should be limited for the remainder of the day. This was the first continental test where "nuisance" action was taken by the test organization. (Clark 1954: 61)

Ruth, March 31
This was the first Livermore test. The laboratory was only seven months old when this shot was fired, and they naturally relied heavily upon the Los Alamos test organization. Los Alamos and the AEC ran this test operation, as they had previous Nevada operations, with Livermore participation in the key functions required to field their devices. Livermore people, for example, played important roles in device assembly and arming for their shots. (Clark 1954: 61)

This first test was related to a program that was near and dear to the heart of Edward Teller. The concept was pursued and rejected by Los Alamos in the 1940s as not promising. It was, however, interesting weapons physics and was probably worth some effort. As a first design for the fledgling laboratory it clearly had a high risk of giving a disappointing yield.

Since the calculated yield varied quite a bit during the preparation for the shot, it was decided to go with a 300-foot tower to be on the safe side. (Clark 1954: 61) The detonation took place at 0500 PST in Area 7 of Yucca Flat. Due to its disappointingly low yield, many of the trigger circuits for the prompt diagnostic experiments and effects measurements failed. However, the highest priority experiments, alpha, transit-time, fireball photography, and rad-chem, all gave useful data. (Clark 1954: 62)

About half of the steel tower remained standing after the shot and was a source of great amusement for the Los Alamos scientists and engineers who chided their Livermore colleagues about the new laboratory's prowess in nuclear weapons design. They suggested that Livermore should get either a larger bomb or a smaller tower.

Ray, April 11
Livermore fired the Ray event eleven days after Ruth. It is described here because of its relationship to Ruth. Chronologically it followed the next shot to be described, Dixie, by five days. Ray was closely related to Ruth and was expected to be a low yield shot. Originally this event was scheduled for a 50-foot tower, but because of the uncertainty in the design yield the Test Director decided that it should be fired on a 100-foot tower.

Clark's Weapons Assembly Group did the non-nuclear assembly and transport, and Livermore did the nuclear assembly. (Clark 1954: 69)

Detonation was in Area 4 at 0445 PST of April 11. This time the steel tower was completely destroyed. As with Ruth, the yield was also about 200 tons. Although the yield was lower than expected, all of the key data were obtained.

The radioactive cloud rose to around 13,000 to 14,000 feet and traveled in a westerly direction, passing Death Valley at +3 hours. No significant fallout occurred in populated areas. (Clark 1954: 70)

Dixie, April 6
Los Alamos fired the Dixie shot between Livermore's Ruth and Ray events. This was an airdrop in Area 7. The Los Alamos weapons assembly team moved from the test site to the Sandia Base in Albuquerque for the test.

The Dixie event was fired during daylight, but the target array was both lighted and painted. The lighted array was useful because the drop aircraft arrived over the site while it was still dark. Sandia supplied the operators and equipment for the release-tone signal on the drop aircraft and at the Control Point.

An Air Force B-50-D dropped the bomb at 0730 PST from 29,000 feet above the terrain on a bombing run of 293 degrees. The time of fall was 37.5 seconds with an actual burst height just over 6,000 feet above the terrain. The device detonated about 560 feet east of the intended target with a yield of about 11 kt. The upper level air was saturated, which made cloud sampling difficult at best. (Clark 1954: 65)

Two USAF QF-80 drones were flown on cloud-sampling missions. Four monkeys and 120 mice were aboard the aircraft, and all but two mice survived the missions. One drone just missed the nuclear cloud by flying beneath it. The other drone flew through the lower fringes of the cloud. The rad-chem data from that flight suffered from fractionation* within the cloud. [*Footnote: Fractionation is the separation, and loss, of weapons debris within the cloud. Different elements separate out of the cloud at different rates.]

This turned out to be a common problem on UPSHOT. Rod Spence, the leader of the Los Alamos radiochemistry group "stated that the radiochemical data are subject to considerable uncertainty due to fractionation occurring in the cloud before samples were taken. ... In order to get more reliable radiochemical data it would be necessary to improve the sample collection techniques, probably by having drones or rockets fly through the main body of the cloud at an early time." (FWC 1953: 62nd)

Fortunately the fireball-yield data were good. The yield was particularly important for Dixie because the shot was a comparison with previous events, and the difference in yield between the shots was particularly significant for the theoretical analysis of the device behavior.

Since the shot was fired over a mile above Yucca Flat there was some concern about the effect on automobile drivers on Highway 95. The following announcement was made by radio the night before the detonation:

> In order to protect drivers of automobiles on Highway 95 against any possibility of temporary flash blindness, they will be halted during the hours prior to the detonation at four check points and will be advised of an approximate time at which they should pull to the side of the road, remaining halted until after the detonation.

Fortunately, there were no auto accidents as a result of the flash. (Clark 1954: 67)

Badger, April 18
This shot required rather complex gas handling systems. To accommodate this experimental feature a platform was constructed 13 feet below the cab of the 300-foot tower to house the hardware.

Badger was detonated on April 8, 1953, at 0435 PST with a yield of 23 kt.

The military performed some interesting experiments on flash blindness on five of the UPSHOT-KNOTHOLE events, including Badger. Clark comments:

> In dark-adapted rabbits, unprotected by filters, exposed to the flash of an atomic bomb, burns of the retina were obtained at distances from 2.5 to 28.5 miles with three possible but questionable burns being obtained at 42.5 miles. This does not mean that the flash at these distances would necessarily burn the retina of man. However, retinal burns have sometimes occurred in man on unprotected exposure to the flash at 10 miles or less and retinal burns at greater distances are considered to be entirely possible. (Clark 1954: 74; Ponton 1982a: 108)

The diagnostic experiments performed well. Yield, alpha, and transit time data were obtained on Badger. Also, a new technique for remotely measuring alpha was developed and tested. Unfortunately, some high-speed photographic data were lost due to fallout on station 413 located about 2 miles southeast of the detonation point. (Clark 1954: 74)

The wind situation was not as predicted, and the main part of the nuclear cloud moved slightly south of east, reaching Flagstaff, AZ, five and a half hours after shot time.

Simon, April 25
Simon had an extremely complex suite of experiments, even by the later standards of the 1980s and '90s. There were measurements of various radiations from specific parts of the device as functions of time. Elaborate collimators were designed to permit detectors to view isolated regions without being overwhelmed by background radiation from other parts of the bomb.

One problem that arose prior to the shot involved the development of rapidly closing doors for the photo station (Station 1-380). The doors that were employed on shot Nancy did not work as planned, and the blast wave entered the bunker doing serious damage. It took until the afternoon of the day before the shot to gain confidence in the fast-acting doors for Simon. As it turned out, the doors appear to have closed prematurely with the loss of much of the high-speed photographic data.

The device was detonated on a 300-foot 100-ton (the load it could carry) four-legged steel tower in Area 1 of Yucca Flat at 0430 PST on April 25, 1953. The yield was 43 kt.

As usual for many of the UPSHOT-KNOTHOLE shots there were two 2400-pound TNT charges detonated two hours and one hour prior to zero time for microbarograph calibration.

Good alpha and yield data were obtained.

On-site fallout from Simon, and from Badger, delayed work at Area 3 and resulted in a delay in Harry, which was originally scheduled for May 2.

The radioactive cloud trajectory went just north of Glendale, Nevada, and over Riverside, Bunkerville, and Mesquite, Nevada. Off-site fallout was sufficiently high that the Test Director called for roadblocks at prearranged vehicle monitoring points on Highways 91 and 93. This was the first time that such action was necessary. Vehicle monitoring points were also established at Alamo and North Las Vegas, Nevada, and at St. George, Utah. If a reading of 7 mr/hour or higher was found inside the car the driver was asked to take the vehicle to a previously determined facility for a car wash at AEC expense. Sixteen cars were washed in North Las Vegas and twenty-five cars were washed at St. George, Utah. Apparently the public reaction to this was agreeable. After Simon, however, the Test Manager, the Las Vegas Field Manager, and the Test Director met to develop plans to speed up the emergency procedures for handling fallout should that be necessary in the future. (Clark 1954: 82-83)

The Chairman of the Joint Congressional Committee on Atomic energy invited all senators and representatives to witness one of the last four UPSHOT-KNOTHOLE detonations. As a result, there were fourteen congressmen present for Simon.

Encore, May 8
This was an airdrop DoD effects shot fielded by Los Alamos over Frenchman Flat. Since most of the effects personnel associated with this shot did not have previous nuclear test experience and since this event required very expensive facilities, it was decided to have a complete dry run three days before the full-scale detonation. The dry run, scheduled for May 4, would consist of a nuclear weapon, minus the nuclear components, set to detonate at the actual location and time of day.

Problems were experienced in the dry run. The B-50 strike aircraft was unable to drop the device in three bombing runs. Later the same day there was a signal dry run at the request of the director of the military effects tests. The Test Director and his advisors agreed to dispense with further attempts to drop a non-nuclear dry-run device and scheduled the full-scale drop for May 7.

Weather delayed the shot until May 8, 1953. As usual microbarograph high explosive shots were fired two hours and one hour prior to the nuclear detonation. The off-site blast forecast was for a potential focusing on Las Vegas with no damage at Indian Springs. (Clark 1954: p. 87)

The B-50-D strike aircraft arrived at the proving ground from Albuquerque at 0450 PST and immediately began the arming procedures, which were finished at 0542. The nuclear detonation was very close to the planned time of 0830 PST. The bombing altitude was 19,000 feet above the terrain and the burst height was 2420 feet above the Frenchman Flat terrain. The burst was about 840 feet south of the intended detonation point. (Clark 1954: 87)

The principal weapons development experiments on Encore were yield and alpha. The main focus of this shot, from the Los Alamos perspective, was a comparison of a slightly changed design from one that was shot earlier. The difference in yield was directly related to the design change.

Harry, May 19

Harry, scheduled for May 2 in Area 3, was delayed by contamination from Simon, which was fired in Area 1 on April 25. Consequently Harry was rescheduled for May 16, and weather further delayed the shot until May 19. The slip in Harry forced delays in Grable, until May 24, and Climax until June 1.

Removal of the top layer of soil and depositing it about 100 feet away went a long way toward decontaminating the Harry tower area. This made it possible for the weapons engineers to work at the tower in preparation for the shot.

The NRL scientists had REECo erect a 300-foot tower 400 feet east of the Harry tower to house argon flash lamps for program 18, the thermal radiation program. The concept was that light from the flash lamps would be observed through the region of high intensity radiation surrounding the nuclear explosion. Chemical processes in the atmosphere could then be deduced from the observations.

Fireball yield, alpha, and several time intervals were measured successfully. Good data were also obtained from the argon flash lamp experiment.

As with most of the detonations on Operation UPSHOT-KNOTHOLE there were congressional observers. Thirty-seven arrived on May 16, but after a three-day weather delay only 23 were on hand for the shot. The military had several hundred additional observers.

METEOROLOGY FOR HARRY

Off-site contamination from Harry was a problem, and it is worthwhile to examine the meteorological conditions that prevailed at the time.

Weather briefings on May 15, 16, and 17 each resulted in 24-hour delays. Finally, at the 1300 weather briefing on May 18 the conditions looked promising and an execute order was issued for the morning of May 19. At the 2200 May 18 weather briefing a fallout pattern similar to that on Badger, but shifted slightly northward, was forecast. The cloud was expected to go over Glendale (population 15), Overton (population 600), Logandale (population 350), and Riverside (population 14). "Thus far in the UPSHOT-KNOTHOLE series Glendale, Overton, and Logandale had received no fall-out, but Riverside already had received as much as they should have for this period. The meteorological experts stated that there was a 50 percent likelihood of at least a 10-degree shift in wind direction. Besides the somewhat unfavorable off-site fall-out picture, there was also the problem of cloud sampling with the 6/10 cirrus cover predicted.* [*Footnote: 6/10 cirrus means that 60% of the sky is covered with cirrus clouds.] The Air Operations Officer proposed to send a scout plane aloft at 0230 on D-Day. If the cirrus at that time was well below the estimated cloud height of 42,000 ft, the chances of getting samples were not too remote. With this somewhat dubious outlook the Test Manager decided to proceed with the shot but to watch closely the meteorological observations as they were collected during the remainder of the night." (Clark 1954: 94-95)

Harry was detonated on a 300-foot tower in Area 3 of Yucca Flat at 0505 PDT on May 19, 1953. The yield was 32 kt. The blast was heard as far away as Las Vegas, NV, and Bishop, CA.

FALLOUT FROM HARRY

A shift in the wind pattern sent the radioactive cloud on a more northeasterly trajectory than forecast. As a result, the off-site fallout occurred in the most heavily populated regions east of the proving grounds. The cloud went north of Riverside over the Saint George area of Utah and then moved southeast over Kanab, Utah.

The highest fallout levels on highways occurred on U.S. 93 between Glendale and Alamo, Nevada. As with shot Simon, there were roadblocks on several highways. Approximately 125 vehicles that had picked up radioactive fallout were washed at government expense at Las Vegas, Alamo, and St. George.

The AEC investigated numerous reports of suspected radiation sickness. No illnesses directly attributable to fallout from shot Harry were ever established. However, the fallout situation from Harry and from Simon was sufficiently serious to warrant a review of the criteria under which the last two shots of UPSHOT-KNOTHOLE were fired. (Clark 1954: 99)

Congressman Douglas R. Stringfellow of Utah was disturbed by the anxiety of his constituents regarding the possible danger of the radioactive cloud. After Stringfellow's discussion with AEC Chairman Dean the Commission requested information on the criteria for firing shots in Nevada. Alvin Graves, of Los Alamos, and John Bugher, director of the AEC Division of Biology and Medicine, attended a Commission meeting and were directed to take action to avoid a repetition of the fallout over St. George.

Cloud sampling, for a radiochemical yield determination, was only partially successful because the sampling aircraft had problems locating the nuclear cloud in the general overcast.

Grable, May 25

Grable was a full-scale operational test of the 280 mm, Mk-9, nuclear artillery shell fired from the gun. There were seven objectives for the test:

1. Determine that the atomic shell would function as intended.
2. Obtain a reasonable measure of effectiveness of the atomic shell.
3. Train selected Army personnel in the handling and the firing of the atomic shell.
4. Test the stockpile-to-target sequence.
5. Determine whether the shell would survive normal shipping and handling procedures.
6. Determine if all the details of security plans were necessary and sufficient.
7. Determine whether the ballistic difference(s) between ... (the various types of 280 mm shells)... could be reconciled by computation.

(Clark 1954: 99)

The program for this operational test involved several phases. The last phase was the actual firing of the nuclear projectile at the Nevada Proving Grounds during Operation UPSHOT-KNOTHOLE. The measurements of interest to AFSWP and the DoD were the blast, thermal intensities, gamma and neutron intensities and the overall effects on target structures. Los Alamos was interested in the yield, alpha, and the general confirmation of proper device performance.

The artillery piece was located 10,956 yards south and slightly west of the detonation point, which was 500 feet above Frenchman Flat. The Test Director, who was responsible for the execution of the test, mandated that the gun be fired from the sequence timer at the CP. This would permit both Test Director control and the distribution of timing signals for the diagnostic measurements.
The firing crew for the 280 mm gun was directed to remain in slit trenches or behind a barricade during firing to insure their safety in the event of a conventional explosion of the artillery shell. The Mk-9 device was designed to preclude a nuclear explosion within a few miles of the gun.

Preliminary test firings took place on May 15, 22, and 23. The full-scale test took place at 0830 on May 25, 1953 with a yield of 15 kt. Incidentally, this yield was somewhat higher than calculated. The time-of-flight of the artillery shell was about 18.7 seconds. The actual detonation took place 139 feet south, 86 feet west and 24 higher than the intended target.

Grable was a very popular tourist attraction. Eighty congressmen, the Secretary of Defense, AEC Commissioners, the Secretary of the Army, the Chairman of the Joint Chiefs of Staff, and a number of other observers from the departments of State and Defense attended the shot from a vantage point about 8 miles west of GZ.

The radioactive cloud encountered winds aloft of 100 knots that carried the cloud to the northeast. The maximum off-site fallout that was detected was 7 mr per hour recorded at the Lincoln mine, about 60 miles northeast of the GZ. Minor amounts of fallout were detected in Crystal Springs, Caliente, Panaca, Pioche, and Milford,

Nevada. No fallout was detected at the range of Salt Lake City or Ogden, Utah. (Clark 1954: 105)

Climax, June 4

Differences between the expected and actual yields of a number of the UPSHOT-KNOTHOLE devices led to concern about the behavior of several of the designs slated for test during the CASTLE series planned for the Pacific during the winter and spring of 1954. As a result Los Alamos suggested adding a shot to the present series in order to explore some weapons physics questions.

The proposed shot would be code-named Climax. It would be the largest shot yet in Nevada and would be an airdrop detonated at 1,000 feet or higher. The Commission approved the shot the day before Harry was to be fired on June 2. President Eisenhower had already approved adding the Climax shot on May 27. (Hacker 1994:105)

This was to be a relatively high yield airdrop. The altitude was selected to be low enough to get good alpha data and high enough to reduce fallout to an acceptable level. A burst height of 1350 feet, directly above Yucca Flat alpha station 3-303 in area 7, was selected. The strike aircraft was a B-36 with a planned release altitude of 30,000 feet above sea level.

A less-than-ideal weather pattern and control room technical problems combined to stimulate a Test Manager decision to call for a 48-hour shot delay from the planned June 2nd until June 4th. With regard to the weather the key factors were upper level wind direction, speed, and the probability of down-wind precipitation.

Two drop-aircraft were involved in the test. A B-50 dropped a high explosive charge over the target for microbarograph calibration 35 minutes before the nuclear detonation. The B-36 strike aircraft dropped two 500-lb test bombs over the target prior to the final run. The nuclear burst occurred at 1334 feet above the terrain at 0415 PDT. Good yield data were obtained from both fireball growth and from cloud sampling. The final yield was 61 kt, the highest so far at the NPG.

The high detonation altitude prevented the dust stem from forming and mixing with the radioactive cloud. Consequently, there was very little off-site fallout. (Clark 1954: 117)

DoD PARTICIPATION ON UPSHOT-KNOTHOLE

About 21,000 DoD personnel participated on Operation UPSHOT-KNOTHOLE in the following activities:

- Administration and Support of the Nevada Test Organization
- Exercise Desert Rock V:
  - Camp Desert Rock Troop Support, AKA Permanent Party
  - Damage Effects Evaluation (Conducted by members of Permanent Party)

- Observers, Troop Orientation and Indoctrination
  - Volunteer Officer Observer Program
  - Tactical Troops (maneuvers)
  - Operational Helicopter Tests
- NWE Projects
- Participation In Weapons Development Diagnostics Experiments
- AFSWC Support

(Ponton 1982a: 1, 22-23)

The DoD's involvement with administration and support of the Nevada Test Organization is discussed earlier in this chapter. The other four DoD activities identified above are discussed in the rest of this section.

EXERCISE DESERT ROCK V

Approximately 20,000 of the DoD participants, about 95 percent of the DoD personnel on UPSHOT-KNOTHOLE, took part in Exercise Desert Rock V activities. (Ponton 1982a: 53)

The objectives, organization and conduct of Desert Rock V were similar to those of the previous Desert Rock Exercises I, II, and III on BUSTER-JANGLE and IV on TUMBLER-SNAPPER. (Ponton 1982a:60)

The AEC Operations Manager, Tyler, set "over-all (including radiological) safety criteria at the Nevada Proving Grounds for all participants," but was not responsible for the implementation with regard to the Desert Rock participants. The Exercise Director had that responsibility, see Part II-Chapter 3, section titled DESERT ROCK IV. (Hacker 1994: 95)

Military observers, volunteers, and troops were closer to the ground zeros at shot times on UPSHOT-KNOTHOLE than they were on the previous two operations. " - many of the previous restrictions placed on military participation by AEC were entirely removed. As a result it was possible to conduct realistic close-in military operations with each burst in which military personnel participated." (Bullock 1953: 7)

The commanding General of the Sixth U.S. Army, Lieutenant General Joseph M. Swing, was the Exercise Supervisor for Desert Rock V. Throughout both the planning and operational phases of the exercise, Swing remained at Sixth Army Headquarters at the Presidio of San Francisco. He named army Brig. Gen. W. C. Bullock, the Exercise Director. Bullock was Commander of Camp Desert Rock; and with his on-site presence, supervised the activities of the exercise and support troops. (Bullock 1953:5; Ponton 1982a: 54)

Camp Desert Rock Troop Support, AKA Permanent Party

Bullock states: "Exercise Desert Rock V ... began on 5 January with the arrival of permanent party personnel who were to administer Camp Desert Rock and conduct operations during the exercise." (Bullock 1953:7) The Permanent Party consisted of approximately 2,000 troops on UPSHOT-KNOTHOLE. Their organization and activities during this and subsequent atmospheric operations were fairly similar, and are described here in some detail. (Ponton 1982a: 53)

Operations and Logistics were the two main organizations of Support Troops under Bullock. They were involved mainly with the field exercises and support of troops. Bullock had additional groups for functions associated with: administration; security; comptroller; and staff activities. The latter included public information, liaison, and a visitor's bureau.

Operations included the Instructor Group, the Air Branch, and the Radiological Safety function.

The Instructor Group prepared and conducted the orientation program for incoming troops, observers, and post personnel. The group consisted of four Army officers from four separate continental armies and an Air Force officer, a Naval officer, and an Army medical officer representing AFSWP. (Ponton 1982a: 56-7; Bullock 1953: 11) The following subjects were covered in the orientation briefings: Introduction and Security; Atomic Weapons Family; Characteristics and Effects of an Atomic Explosions; Medical Aspects; Protective Measures and Radiac; Army Delivery Means; Air Force Delivery Means; Navy Delivery Means; Tactical Employment; History of Desert Rock Exercises; and a Seminar with TUMBLER-SNAPPER film. (Bullock 1953: 11-12)

The Air Branch had 5 fixed-wing aircraft and 3 helicopters for: air observation support; air evacuation; courier service; fuel service; and minor aircraft repair. They also supplied aircraft for radiological safety surveys.

Radiological Safety consisted of about 70 members of the Chemical Service Platoon. They:

> planned and conducted the radiological safety procedures used to limit the exposure to exercise troops entering the forward areas. Before each shot, they trained exercise troops in radiological safety procedures. After each shot, they accompanied troops into the forward area; conducted aerial and ground radiological surveys, monitored trenches, equipment displays, and troop maneuver areas; and decontaminated Desert Rock personnel leaving the forward areas. (Ponton 1982a: 56-57)

Logistics was organized into the following sections:
- Engineer - "provided supplies, equipment, and personnel for the construction of trenches and test facilities and the maintenance of" the camp.
- Signal - "established wire and radio communications within the test areas and at the camp". They "also issued and processed film badges".
- Ordnance - "procured, distributed, and maintained weapons and vehicles for the exercise troops and equipment display areas."

- Quartermaster - "provided exercise and support troops with food, clothing, bedding, laundry service, tents, petroleum products, office equipments, and general supplies."
- Transportation consisted of four sections: Office, Commercial Traffic, Supply and Camp Motor Pool.
- Chemical - "provided equipment and supplies in support of radiological safety operations in the forward areas."
- Medical - "provided medical aid men, and ambulances for each observer and troop convoy and established temporary medical aid stations at trench and forward parking areas." They also inspected meat for the camp mess and provided medical care at the camp.
- Dental - "furnished dental advice and care to camp personnel." (Ponton1982a: 57-8)

Although not identified in the organization chart, Bullock also cites the presence of 2 Army bands, which must have been something of a curiosity at the NPG. (Bullock 1953: 6)

"Throughout Exercise Desert Rock V, however, there was a shortage of support troops." Their actual strength never reached more than 77% of the authorized level. "Many of the troops had only 30 days or less of military service remaining upon arrival" at camp. "The constant turnover in personnel resulted in long work hours and sometimes 7 day work weeks." (Ponton 1982a: 58-9)

In addition to the responsibilities just described, teams of officers from the Permanent Party conducted the activities for the Damage Effects Evaluation program.

Damage Effects Evaluation

> The Damage Effects Evaluation Program enabled military personnel to study the effects of nuclear detonations on animals, equipment, and field fortifications. Teams of officers from the Camp Desert Rock Chemical, Engineering, Medical, Ordnance, Quartermaster, and Signal Sections inspected the preshot condition of the display area. The teams then witnessed the shot from the observer trenches. After the shot, the teams returned to the display area to compare their predictions with the actual effects of the detonation. The teams participated in all shots except Ruth, Dixie, Ray, and Climax. (ibid., 77-8)

Observers, Troop Orientation, and Indoctrination

This program --- "was designed to acquaint official observers and troops from the Army, Navy, Marine Corps, Air Force, and other DoD personnel with the effects of nuclear detonations." There were two basic types of DoD military observers: those who were assigned to Camp Desert Rock Troop Support and those who came to the site to just observe one or more shots. There were also a few civilian observers. The program was similar to Observer programs on BUSTER-JANGLE and TUMBLER-SNAPPER and "consisted of lectures, films, preshot and postshot tours of equipment display areas, and observation of nuclear detonations in the forward areas of the NPG". (Ponton 1982a:65-7)

The Army developed safety criteria, in terms of distances from GZ that applied to the placement of both observers and maneuver troops during UPSHOT-KNOTHOLE. The distance safety criteria were established based on a directive dated 5 February 1953 from the Office, Chief of Army Field Forces (OCAFF) which specified protection for observers and troops from thermal ($\leq$ 1 calorie/cm^2 of thermal radiation), blast ($\leq$ 5 lbs/in^2 of overpressure), and radiation ($\leq$ 6 roentgens during UPSHOT-KNOTHOLE, with $\leq$ 3 roentgens of prompt radiation).

Table 2-8.2 Safe Distances From GZ Determined By US Army

Maximum Predicted Yield* (kt)	Distance From GZ (m)					
	Tower Shots		Aircraft-Delivered		280mm Canon	
	Troops in open	Troops in trenches	Troops in open	Troops in trenches	Troops in open	Troops in trenches
1 - 5	4,115	3,200	6,860	5,950	5,030	4,120
5 - 10	5,950	3,200	8,690	5,950	6,860	4,120
10 -15	7,315	3,200	10,060	5,950	8,230	4,120
15 - 20	8,230	3,200	10,980	5,950	9,150	4,120
20 - 25	9,150	3,200	11,890	5,950	10,060	4,120
25 - 30	10,060	3,390	12,810	6,130	10,980	4,300
30 - 35	10,490	3,475	13,720	6,220	11,890	4,390
35 - 40	11,430	3,660	14,180	6,410	12,350	4,580

(Ponton 1982a: 145-8) [*Footnote: *Maximum predicted yield* is generally not reported in the unclassified literature. During the design of a nuclear device, a great many calculations are conducted which provide information about how the yield changes due to changes in different input parameters. For instance, input information about how a particular material in the device behaves under the extreme pressures and temperatures that occur during detonation may not be well known. Therefore, a set of calculations (known as *parametric studies)* would be conducted to examine how the yield changes due to an estimated range of input values that describe the material's behavior. The estimated range used is often taken as what experienced physicists consider to be a *credible* range. Thus, the *maximum predicted yield* is often referred to as the *maximum credible yield* or simply *max cred.*]

The distance criteria resulted in the placement of observers and troops closer to GZ than they had been on the previous two operations. Using Tables 2-8.1 and 2-8.2, Table 2-8.3 was developed to compare the Army's safe distance trench criteria with the actual trench location.

Table 2-8.3 Comparison of Army's Safe Distance Criteria with Actual Trench Location.

TEST	DATE-1953 (from Table 2-8.1)	TYPE (from Table 2-8.1)	YIELD (from Table 2-8.1) (kt)	ARMY'S SAFE DISTANCE TRENCH CRITERIA (from Table 2-8.2) (m)	ACTUAL TRENCH LOCATION (from Table 2-8.4) (m)
Annie	3/17	TOWER	16	3200	3200
Nancy	3/24	TOWER	24	3200	3660
Ruth	3/31	TOWER	0.2	-	-
Dixie	4/6	AIRDROP	11	-	-
Ray	4/11	TOWER	0.2	-	-
Badger	4/18	TOWER	23	3200	3660
Simon	4/25	TOWER	43	3660+*	3660
Encore	5/8	AIRDROP	27	6130	3660/3480**
Harry	5/19	TOWER	32	3475	8960
Grable	5/25	280mm CANNON	15	4120	4570
Climax	6/4	AIRDROP	61	-	-

[*Footnote: Table 2-8.2 gives 3660 m for 35-40 kt Tower Shots. One would expect a slightly larger distance for 43 kt.] [**Footnote: Trenches on Encore were placed 3660 m from the planned GZ. The actual GZ was (840 ft) 256 m S of the planned, so the actual distance from GZ was about 3480 m.]

The actual locations of the trenches in Table 2-8.3 were obtained from Peruss 1953: 92, in a table similar to Table 2-8.4 below. Table 2-8.4 provides an overview of both the numbers of Observers and Maneuver Troops and their locations during the detonations of UPSHOT-KNOTHOLE. Except for Encore, (and possibly Simon), the actual trench locations were either at the Army's safe distance or more conservative, i.e. greater.

Table 2-8.4 Numbers and Locations of Observers and Troops for UPSHOT-KNOTHOLE Detonations. Numbers in parentheses indicate "repeat" observers or troops.

SHOT	LOCATION: DIRECTION /DISTANCE FROM GZ (m)	ARMY Obser. Troop	NAVY Obser. Troop	AIR FORCE Obser. Troop	MARINE Obser. Troop	CIVILIAN Obser. Troop	TOTAL Observer Troop
Annie	SW/3200	303 1181	162 -	11 -	0 -	00 -	535 1181
Nancy	SW/3660	312 2349 (75)	90 -	80 -	17 -	16 (5) -	515 (5) 2349 (75)
Ruth	NO OBSERVER OR TROOP PARTICIPATION						
Dixie	News Nob	60 (5) -			75 -		135 (5) -
Ray	News Nob	33 (15) -	1 -	4 -	25 (20) -		63 (35) -
Badger	SW/3660	266 (60) -	122 -	101 -	122 2084 (15)	10 -	611 (60) 2084 (15)
Simon	S/3660	205 2450 (65)	14 -	340 -	1 -	24(2) -	584 (2) 2450 (50)
Encore	SW/3480*	99 526(50)	14 -	255 -	2 -	7 -	377 526 (50)
Harry	SW/8960	315 2149 (60)	92 -	439 -	12 -	21(2)	879 (2) 2149 (60)
Grable	E/4570	446 2670 (85)	70 -	13 -	29 -	31(2) -	589 (2) 2670 (85)
Climax	NO OBSERVER OR TROOP PARTICIPATION						
TOTAL	Observers Troops	2039 (80) 11325 (355)	555 -	1273 -	282 (20) 2084 (15)	139 (11) -	4288(111) 13409(350) **17697(461)**

[*Footnote: See Table 2-8.2. Trenches on Encore were placed 3660 m from the planned GZ. The actual GZ was (840 feet) 256m S of the planned, so the distance from the trenches to the actual GZ was about 3480 m.]

Volunteer Officer Observers Program

A Volunteer Officer Observer Program was conducted for the first time on UPSHOT-KNOTHOLE. "This program was designed to measure the ability of trained staff officers to calculate safe distances from nuclear detonations and to allow them to experience a nuclear detonation from the distance calculated." (Ponton 1982a, 66) "Officers deemed well trained enough to understand fully the risks … quantitatively as well as qualitatively … "served as volunteers. (Hacker 1994: 96)

The OCAFF again specified exposure values for the Volunteer Officers Program. They specified the same values of thermal and blast exposures as were used for the other observers and troops: thermal ($\leq$ 1 calorie/cm^2 of thermal radiation) and blast ($\leq$ 5 lbs/in^2 of overpressure). However, for the Volunteer Observer Program, the OCAFF raised the permissible exposure to radiation ($\leq$ 10.0 per test, with $\leq$ 5.0 roentgens of prompt radiation, and $\leq$ a total of 25.0 roentgen during the entire exercise). "The Exercise Director authorized these officer volunteers to position themselves closer to the Nancy, Badger, and Simon ground zeros than the distance established for all other exercise troops and to receive a single dose of gamma radiation not to exceed 10.0 roentgens." (Ponton 1982a, 70-71) The officers chose their distance from GZ by calculating the effects of the nuclear detonation according to data in the 1952 manual, "Capabilities of Atomic Weapons*." (Ponton 1982a, 70-71) [*Footnote: Department of the Army, *Capabilities of Atomic Weapons*. Joint Army, Navy, Air Force and AFSWP. Washington, DC:GPO. TM-23-200. Also published as OPNAV Instruction 003400.1 and AFOAT 385.2. Revised edition: 1 October 1952. 162 pages.]

"At shot Nancy, four Army, four Navy, and one Air Force officer volunteers were positioned in trenches 2,290 meters (1.4 miles) from the Nancy ground zero. At shot Badger, six Army and six Marine Corps officers occupied trenches 1,830 meters (1.14 mi) from the Badger ground zero. At Shot Simon, seven Army officers and one Navy officer were located in two trenches 1,830 meters (1.14 miles) from ground zero. These three tests on which the Volunteer Officers Program was conducted were all on 300' towers with yields of 24, 23, and 43 kt respectively. (ibid., 70-71)

Appendix J provides a summary of where troops were located on the different shots during the operations BUSTER thru HARDTACK II. The closest scaled distances were experienced by the Volunteer Officers on UPSHOT-KNOTHOLE.

Tactical Troop Maneuvers
The tactical troop maneuvers are outlined in Table 2-8.4. They consisted of a total of 13,409 troops of which 350 had already participated on one or more tests. These maneuvers were similar to those conducted on Operations BUSTER-JANGLE and TUMBLER-SNAPPER.

The troops entered the forward area by truck or bus convoy a few hours prior to shot time, often with observer participants in the Orientation and Indoctrination Program. The observers and troops "then occupied trenches, from which they witnessed the detonation". (ibid., 76)

There were the two trench areas where the Maneuver Troops and Volunteer Officer Observers were located for the Simon detonation. (Massie 1982: 20) Maneuvers associated with the other tests were similar to Simon in terms of moving from the trenches toward an objective closer to GZ. (Bullock 1953)

After the shot, the troops moved closer to ground zero. They "filed out of the trenches and attacked an objective in accordance with the exercise plans. These troops were accompanied by radiological safety monitors and were preceded by radiological survey teams who determined the limits of safe advance. After reaching their objective, or approaching as close as radiation safety standards would permit, the maneuver troops went to the display area," ... through which they had a guided tour". (Ponton 1982a: 76)

As indicated in Table 2-8.4, (Peruss 1953: 92) a total of 17,697 troops and observers participated in Desert Rock V. Bullock (1953: 9) breaks this number down as follows: Army 13,364; Navy & USMC 2,921; Air Force 1,273; Civilian (all services) 139.

The last troop exercise of the operation, Grable, is also described by Bullock:

> Exercise DESERT ROCK V reached its climax with the detonation of a Mark 9 atomic shell delivered by a 280 mm artillery gun on 25 May. Two BCT's composed of troops from all the continental armies, attacked towards objectives beyond ground zero after the detonation. The Secretary of the Army, two members of Congress, the Chief of Staff of the U.S. Army, the Chief of Army Field Forces, the Commanding General of Sixth Army and 787 additional military and civilian personnel observed the detonation from positions in the troop entrenchment area.

Operational Helicopter Tests

The Marine Corps conducted operational helicopter tests for the first time on UPSHOT-KNOTHOLE. "The tests were designed to investigate factors that would determine the extent to which a helicopter and crew could be used to launch a tactical assault on a predetermined objective following a nuclear detonation." Approximately 40 personnel took part in this program, which was performed on all events except Ruth, Grable, and Climax. (ibid., 77)

These tests involved 3 or 4 H-19 helicopters positioned in a variety of configurations. Some were on the ground, 12-18 km from GZ; some were hovering in the Yucca Lake area; and some were flying at 400' at 8.5 to 20 km from GZ. (Ponton 1982a:77)

> After passage of the blast wave, some of the helicopters flew toward GZ. Near GZ, one helicopter performed a radiological survey of the area, while a second hovered nearby in case of emergency. At other shots, two helicopters landed at an area near GZ to measure and plot the radiation intensities. The helicopters usually returned to Yucca Lake Airstrip, where they were monitored for radiological contamination. After they were cleared, the helicopters returned to Camp Desert Rock." (Ponton 1982a:77)

DoD NWE PROGRAMS AND PROJECTS

As described in Part II-Chapter 6, on August 1, 1952, Chief AFSWP, Major General Herbert B. Loper, disestablished Test Command (which had conducted TUMBLER-SNAPPER) and established within Field Command the Directorate of Weapons Effects Tests (DWET). DWET was located at Sandia Base with Col. Paul T. Preuss its Director. On August 16, 1952, the services of E. B. Doll of Stanford Research

Institute were secured, and he served Preuss as Technical Director. When DWET moved to the NPG for UPSHOT-KNOTHOLE, Doll was Director of the Military Effects Tests Group (AKA, MEG), within the Nevada Test Organization, see Figure 2-8.1. (Preuss 1953:26) Doll's deputy was Col. H. K. Gilbert, also from WET. From the Radiation Branch of Headquarters, Col. Edward Giller and Major Hard also served on Doll's staff. (AFSWP 1954: Vol. 6: 3.8.1) Attachment IV by Edward Giller describes some of his, often humorous, experiences at the test site.

Also reporting to Doll were 9 Program Directors, one for each of the 9 programs conducted by AFSWP during UPSHOT-KNOTHOLE.

OPERATION	PROGRAM	# PROJECTS
UPSHOT-KNOTHOLE	1) Blast and Shock Measurements	12
	2) Nuclear Measurements and Effects	4
	3) Structures, Material, and Equipment	28
11 Tests	4) Biomedical	5
	5) Aircraft Structures	3
	6) Test of Service Equipment and Operations	11
	7) Long-range Detection	4
	8) Thermal Measurements and Effects	14
	9) Technical Photography	3
	Total	**82**

Preuss states that the program directors ... "were delegated wide authority. The basic experimental unit was the project. The entire organization was devoted to supporting the project officers, since they obtained the basic experimental data desired from a test series. The Test Director's support units and staff sections were set up to provide the project officers with the help they needed." (Preuss 1953: 26)

About 1,200 personnel participated on DoD projects during UPSHOT-KNOTHOLE. The DoD placed its emphasis on Encore and Grable. Nearly every project was conducted on at least these two tests, both of which had the same Intended Ground Zero (IGZ). Both were air drops; and their IGZs were significantly different from their actual GZs. For many projects, this difference resulted in: exposure to a nuclear environment either too severe or not severe enough to provide the data required and in the loss of data. Program 3, Structures, with its 28 programs was particularly affected because of its expensive preshot construction and gage placement which were of course set with regard to the IGZ. The total of 82 projects conducted represents the largest number of projects to be conducted during a Nevada operation during atmospheric testing.

Background

TUMBLER confirmed that at relatively low scaled heights of burst, the form of the pressure wave departs considerably from that expected by then available theoretical considerations and small-scale HE tests. The unexpected results "were shown to be associated with the thermal radiation acting jointly on the earth's surface and on the clouds of surface dust, which combined to produce a thermal layer." This layer was shown *qualitatively* "capable of affecting the blast wave by: lowering the peak pressure, increasing the rise time of the shock, and increasing the velocity of propagation of the initial air pressure wave." The increased velocity of the initial air pressure wave "sometimes resulted in an anticipatory pressure wave called a "precursor". TUMBLER also "indicated a marked deviation from the theory of -- (Mach) reflection" at the surface.(Swift 1955: 15) More data was certainly needed to further enhance understanding of the multiple complex phenomena that occurred and to further develop the theoretical models. A successful major effort was conducted on UPSHOT-KNOTHOLE to obtain such data on airblast and ground shock as well as how these effects affect a wide variety of structures.

The post shot UPSHOT-KNOTHOLE reports regarding airblast and ground shock reflect a greater degree of sophistication and understanding of the complexities of nuclear airblast phenomena. The airblast in free-air, (i.e. air that has not been influenced by ground interactions), was (and still is) understood the best. It is a lot simpler than airblast which interacts with the ground. The ground reflected shock waves can influence the character of the free-air shock when they interact. The point at which the reflected shock catches up with the free air shock front is called the *triple point.*

During UPSHOT-KNOTHOLE, similar or perhaps some of the same technical activities were performed by more than one organization. This was common practice by AFSWP and its descendants. The value of having more than one organization conduct similar, or the same, work had a number of advantages: competition, corroboration, different viewpoints, the development of a community, not having to be dependent on just one organization or person, etc. These same advantages were also found in the relationships between the Los Alamos and Livermore laboratories.

Blast and Shock Measurements

A prodigious number of airblast and ground shock measurements were made along 3 blast lines in projects: 1.1a, 1.2, 1.1b, 1.5, 8.12a, and 8.12b.

- The Main blast line acquired data for the HOB-pressure-distance curves being developed by the DoD
- The Smoke line examined the effect that a layer of smoke which was generated just prior to a shot would have on the blast wave
- The Mine Field line yielded data for the large minefield clearance project.

Structures, Material and Equipment

The 28 projects of Program 3 were more than any other DoD program conducted at NTS during atmospheric testing. Twenty two of the projects were conducted on both shots Encore and Grable. Five of these projects made measurements on the other 23 projects. Again, the number of measurements made was prodigious and instrumentation development was also conducted. All of these projects are briefly described in Killian 2011. Their results represented a significant advance in knowledge about structures in nuclear environments. Four of these projects are described next.

Railroad Stock
Project 3.6 was the only one ever conducted that exposed railroad cars. Sixteen items of standard of railroad rolling stock were exposed. The items were placed on small sections of constructed rail bed with ties and exposed between 1520 and 6600 feet from the actual GZ on Grable. Pressure and acceleration were measured, and motion pictures were taken. (Sevin 1955)

Lots More Conifers
Project 3.19 provided an even more startling sight than TUMBLER-SNAPPER 3.3. A tree stand covering approximately 1-1/4 acres was exposed to Encore and Grable. The trees were obtained from Lees Canyon on the Charleston Ranger District of the Nevada National Forest at an elevation of 8500'. The focus was offensive (on targets), so the trees were selected to be ponderosa pine trees of diameter, height, and arrangement typical of a "small managed western European woodlot." European woodlots are characterized by their uniform spacing and the absence of undergrowth and dead limbs. (Between Encore and Grable, debris was cleared from the stand.) The 145 trees were placed in concrete foundations and were planted in a uniform staggered row pattern at 20' intervals which covered an area 160' wide by 320' long, parallel to the blast line with the center of the strand about 6400' from IGZ. The trees averaged 51' in height and 15" in diameter at their base. (Sauer 1954: 3, 21-23)

To span environments from substantial damage to no damage, two duplicate lines of trees were emplaced. The two lines ran radial from GZ, between 5,000' and 8000', with trees planted at 500' intervals. Two pine trees were placed at 1500' from GZ to determine over-damage and two broad-leaf trees were placed at 5500' and 7500'. A pair of pendulums was placed at 5000' and 8000' adjacent to the tree line. These pendulums represented an idealized tree system similar to the "lollipop" used in TUMBLER-SNAPPER, but they had substantially longer periods. (Sauer 1954: 23-28)

Prior to Encore, the natural period of each tree was obtained. Static and dynamic pressure was measured within the stand. Tree stem deflection was measured near the center of pressure of each tree crown for every tree, except the 2 at 1500'. Strain and acceleration and maximum strain were measured on 3 trees. Wind velocity and direction were also measured. Photography consisted of: 3 cameras 240' to the side of the stand to record tree motion; 2 cameras located 610' to the side were for the time-displacement of water vapor and smoke formed by the thermal pulse; and extensive still photography of pre-and post-shot activities and conditions was taken. This was the last tree project at the site. (Sauer 1954: 28-34)

Field Medical Facilities
Two types of field medical facilities and their associated equipment were assembled and exposed during Encore as Project 3.27. Two types of medical facilities were used: Unit A, a composite battalion aid station and regimental collecting station; and Unit B, a composite division clearing station, mobile army surgical hospital, and evacuation hospital. Both had 2 configurations: a standard aboveground (*)

configuration and a dug-in (**) configuration. The objective was to assess the degree of protection to the installation and personnel that was provided by a dug-in position.
At 4,163', 4 units: Unit A*, Unit A**, Unit B*, and Unit B**.
At 9,000', 4 units: Unit A*, Unit A**, Unit B*, and Unit B**.
At 15,000', 2 units: Unit B* and Unit B**.
Unit A was a standard squad tent. Underground construction for each Unit A** was an evacuation 34'4" x 16' x 4'6" deep. Unit B was divided into 4 separate tents: a surgical tent; an X-ray, dental, eye, maxillofacial, and ear, nose and throat tent; a pharmacy and laboratory tent; and a ward tent. Underground construction for each one of the 4 tents of Unit B** was 18' x 52' x 6' deep.

Both A and B units contained all representative items of equipment authorized for those units. They were furnished with equipment to be fully functional. Some of the equipment was operational at the time of the blast. Electricity was obtained from a dug-in generator at each location. (Chapman 1954: 3-66)

The "Motels"
The "Motels", which can be visited today at the site, were four long, low, narrow structures that were constructed for this project (which was fielded by the FCDA). But, this was not a building test. It was a test of partitions and curtain walls under a roof. The motels were 303' 10" in length, 11' 2" high and 16' deep, and open at both ends of the depth (the long length was represented by only the roof, no side walls).The length of each structure was divided into 18 cells. Two motels were located about 6,625' from GZ, and two were located at about 4,400'. Curtain walls with and without openings, of different sizes and shapes were made of various materials (brick, cinder block, reinforced concrete, corrugated steel etc.) and construction techniques. Interior partitions of cinder block, wood, steel, and plaster were exposed. Motion pictures were the main diagnostic. (Ponton 1982a: 104-5; Taylor 1956: 5, 19, 28)

Aircraft Structures
During BUSTER and TUMBLER-SNAPPER, aircraft structures were tested by exposing aircraft parked on the ground at various locations with different orientations wrt GZ. By UPSHOT-KNOTHOLE, the capability to predict NWEs in the atmosphere (though not near the ground) had advanced considerably and was beginning to be used to plan flight patterns for weapon delivery. Also, the advent of thermonuclear weapons raised the issue of how to deliver these higher yield weapons from manned aircraft and maintain the safety of the crew. Purkey (1955:15 states): "Capabilities of present operational aircraft, as now known, will not permit the delivery of weapons above certain sizes; the limit yield is generally based upon allowable thermal or blast damage to the delivery aircraft ---." Capabilities are also dependent on the delivery technique(s) used; but these techniques are bounded by the capabilities of the aircraft in a nuclear environment.

Three projects addressed the issue of NWEs on the delivery aircraft while in flight. Each project used a different aircraft. Manned and drone aircraft were used, and these were the first NTS tests of manned and drone flights during delivery. Subsequent NTS operations would conduct similar tests. The following description of Project 5.1's instrumentation and flight operation is also characteristic of projects 5.2 and 5.3 as well as similar projects on later operations.

Two standard blue naval model AD aircraft were converted to drone aircraft and instrumented for the measurement of NWE and aircraft response on: Annie, Nancy, Simon, Encore, and Harry. One of these craft was flown manned on Annie and the other on Harry. The instrumentation was extensive and included measurements for: burst time; thermal radiation; maximum aircraft skin temperatures; overpressure; acceleration of aircraft; wing bending, shear, and torsion; horizontal stabilizer bending and shear; aircraft altitude, velocity, and pitch; and gamma radiation. Many of these instruments were installed at more than one location on the craft (port, starboard, inside, etc.). For instance there were 19 plates for temperature and 7 gamma boxes. The drone flights were under control of: a ground pilot at take off; the pilot of the "mother" control plane in flight; and a pilot at the radar plotting board at the CP during detonation. After detonation, the mother control plane regained control. Manned aircraft were also used to accompany the drone at some distance and to make observations.

There was an "unplanned" occurrence on Simon. At zero time the drone aircraft were "to be in level flight attitude, tail toward the blast ... simulating an escape position of the craft following delivery of an atomic weapon." (Rogin 1954: 21) One drone was positioned for near maximum weapon effects, and "the higher thermal radiation severely weakened all the blue painted skin on the underside of the wing". Both the port and starboard outboard wing panels were torn off at the time of shock arrival as a result of the weakened skin and combined overpressure and gust effects." The panels were recovered after the test and had not incurred any significant additional damage due to the free fall and subsequent ground impact. "A considerable amount of valuable information on thermal damage to aircraft in flight was obtained from these panels--."(Rogin 1954: 3) One result was that blue (a color that absorbs radiation more readily than white or a reflective metal) was no longer in vogue.

SAC And TAC Operations

The skies were busy during UPSHOT-KNOTHOLE. Strategic Air Command (SAC) participated on all tests except for Ruth and Ray to test and further develop the current capability for IBDA. They established standing operation procedures and training requirements, a training program was established, and SAC crews were familiarized and indoctrinated in atomic operations. Project 6.3 was NOT a "shoestring" operation.

Shot	# of Aircraft/Type of Aircraft	# of Aircrew Persons	# of Technical Persons
Annie	12 /B-29	132	20
Nancy	12/B36	204	20
Dixie	12/B-47	33	20
Ray	12/B50	132	20

Simon	6/B47 6/B-50	18 88	20 20
Harry	12/B50	110	20
Encore	12/B36 8/F-84	204 8	20
Grable	12/B36 8/F84	204 8	20
Climax	7/B 36	110	20

The report indicates the flight paths and formations used. (Keeling 1955: 3,15-9)

Project 6.11 of UPSHOT-KNOTHOLE was the first opportunity for Tactical Air Command (TAC) to participate in an atomic test program and to indoctrinate TAC fighter bomber and tactical reconnaissance pilots in the delivery and effects of atomic weapons. On Nancy, 33 pilots were positioned about 10 miles from GZ to learn about flash effects. During Dixie, seven T-33 aircraft carrying 14 pilots simulated a delivery maneuver. On Encore, a similar number of aircraft and pilots simulated a dive bombing delivery maneuver. About 2 hours after Encore, three RF-80 aircraft made 2 photographic runs over GZ for IBDA purposes. The pilots' reactions to these maneuvers is given in the project report.(Rawlings 1953:3)

Long-Range Detection
These projects tapped four technological fields:

- radio chemistry (project 7.5) (Ponton 1982a: 118-9)
- seismology (project 7.4) (Crocker 1955: 3, 13, 16)
- sound (project 7.3) (Olmsted 1954: 3,21,27), and
- electromagnetic signals from the detonation (project 7.1) (Ponton 1982a:118).

These were the last AFSWP projects for long-range detection during atmospheric testing; however, Los Alamos and Livermore actively pursued similar or related studies throughout the nuclear testing era.

Thermal Measurements and Effects
Project 8.2 was initiated to deduce the intensity of light from a nuclear explosion by attempting to measure the mechanical pressure of the photons on a highly reflective surface. (Bohn 1954: 3, 11)

Two projects, 8.4-1 and 8.4-2, were associated with the creation of fog and carbon smoke screens respectively on Grable. The "smoke line" was instrumented with pressure gages to assess the protection provided by the smoke screens. High winds caused the cancellation of these projects on Encore, but they were rescheduled for Grable.(Engquist 1954a: 3, 6; Engquistb 1954: 3,12-3)

Previous projects that exposed fabrics were aimed at determining the degree of fabric destruction or burn. While fabric destruction is important, a more important question to the soldier was: How much heat was transferred to the backing? This question was the focus for Project 8.6. (Feldman 1955: 3, 15-22)

Technical Photography
Extensive photographic coverage, both still and motion pictures, was conducted for all AFSWP projects before, during, and after tests. For Project 9.1, during Encore there were 193 cameras and during Grable 94.(EG&G 1953: 3, 11-2) Dust presented the greatest difficulty for photography. To overcome this difficulty, Project 9.6 stabilized the soil over 40 selected areas by laying about 700,000 yd^2 (over 145 acres) of 2" thick sand-cement (equivalent to the volume of a cube 100 ft on a side). (Duval 1953: 3, 15-6, 20-3; Shockley 1954: 3, 28-9)

DoD PARTICIPATION ON WEAPONS DEVELOPMENT DIAGNOSTICS TESTS

LASL and UCRL conducted 8 programs with 36 projects during UPSHOT-KNOTHOLE, see this chapter's section titled WEAPONS DEVELOPMENT DIAGNOSTIC EXPERIMENTS. DoD participation was in two of these programs: Program 13 Radiochemistry and Program 18 Thermal Radiation Measurements. The Radiochemistry efforts were conducted by AFSWC and are discussed in the next section.

The Naval Research Laboratory of Washington D. C. conducted all six projects in Program 18:
18.1 Total Thermal and Air Attenuation
18.2 Power versus Time
18.3 Spectroscopy
18.4 Light Absorption
18.5 Case Surface Brightness
18.6 Surface Brightness Investigations
Ponton states: "Of these six projects, detailed documentation has been located only for Project 18.3, Spectroscopy. ... Two spectrometers were located in Building 400, a permanent building near the (CP). ... Three other spectrometers were located in a reinforced mobile structure usually positioned about 3 kilometers from GZ. The spectrometers at Building 400 were manned during the detonations. Spectrometers in the mobile structure were operated by remote control during the detonation." (Ponton 1982a: 130)

AFSWC Support
AFSWC air and ground participants in UPSHOT-KNOTHOLE numbered over 400 at Indian Springs AFB and about 2,000 at Kirtland AFB. AFSWC provided the aircraft and personnel required for the airdrop, cloud sampling for subsequent radiochemistry analyses, courier missions, cloud tracking, aerial surveys, and other air support. These missions often included training of personnel and the development of new techniques for sampling. Emergency air evacuation was also provided.

As in previous operations, the Air Force Air Weather Service provided weather forecasting support. Sixty-eight people staffed this support unit, including those assigned to collect meteorological data out to approximately 200 miles. Weather support personnel were housed in the C.P. and were responsible for the weather briefings to the Test Manager's Advisory Panel.

Radiological Safety

The military was responsible for the UPSHOT-KNOTHOLE rad-safe organization. The Rad-Safe operating unit was responsible for rad-safe activities within the test site and for a distance of approximately 200 miles. The Army Chemical Corp Training Center's 9778th Technical Support Unit provided most of the rad-safe monitoring support for UPSHOT-KNOTHOLE. Los Alamos and the U.S. Public Health Service provided radiation experts to assist in the management of the rad-safe operation. (Hacker 1994: 100)

The New York AEC Operations Office had the rad-safe fall-out measurement responsibilities beyond 200 miles, through the AEC/DMA. The New York Office was also responsible for recommendations to the Test Manager for any action that might be appropriate in light of results of the fall-out measurements.

DOORSTEP - FEDERAL CIVIL DEFENSE ADMINISTRATION (FCDA) PARTICIPATION ON UPSHOT-KNOTHOLE

UPSHOT-KNOTHOLE was the first operation with major participation by the FCDA. (Ponton 1982a:131) Interestingly, the "Memorandum of Understanding between FCDA and AEC for Participation in Full-Scale Nuclear Tests" was issued on November 3, 1954, about a year and a half after UPSHOT-KNOTHOLE. (AEC 707/10) As was typical at the site during the 1950s, work went forward and got done on the basis of informal agreements and on "what worked" among the participants. The bureaucracy could just catch up, as best they could. The more formal documented agreements among government or non-government entities often came (sometimes considerably) later. Like agreements for FCDA participation, the formal agreements usually mirrored the previously made informal arrangements.

The FCDA had three types of programs. The first focused on an observer program for representatives from the government and the media. The second involved field exercises for civil defense personnel and the third involved studies of nuclear weapons effects on civilian structures and materials. (AEC 1954b:2)

FCDA Observer Program - Open Shot

The FCDA played a principal role in the justification and conduct of the first "open shot" on TUMBLER-SNAPPER. They did the same on UPSHOT-KNOTHOLE for the second open shot, which the Test Director suggested be conducted on Annie.

The FCDA also pursued 29 of their 36 technical projects on Annie, and their efforts on that shot were referred to as "Operation Doorstep". The name was thought to be appropriate, since the purpose of the program was to show the people of America what might be expected if an atomic burst took place over the doorsteps of our major cities. (FCDA 1953b: 3)

On March 17 (St. Patrick's Day), 1953, more than 600 observers, including civil defense, AEC and DoD personnel, state and local government officials, and representatives of the Nation's information media witnessed the first test of Operation UPSHOT-KNOTHOLE, Annie, detonated on a 300 foot tower. It had a yield of 16kt. The observers were located at News Knob, about a mile and a half NE of the Control Point, near the edge of Yucca Lake. The shot went smoothly, and no particular problems were encountered. (Ponton 1982a: 38)

Field Training Exercises

Field training under actual nuclear explosion conditions was conducted for civil defense leaders. The training included a course of lectures and field exercises between April 22 and May 5, 1953. Field exercises included both on-site monitoring of radioactivity levels and off-site monitoring of fall out. (Lamoureux 1953: 3, 14)

FCDA NWE PROGRAMS AND PROJECTS

The test requirements of the FCDA were sufficiently different from those of either the AEC or the DoD that it made sense to establish a Civil Effects Test Group, under the Test Director to coordinate the FCDA activities with the other NPG users, see Figure 2-8.1. The FCDA measurements were divided into 8 programs referred to as Program 21 through 29 (Program 25 was unassigned) with a total of 36 projects.

UPSHOT-KNOTHOLE	PROGRAM	# Projects
11 Tests	21) Effects Studies (on Shelters)	3
	22) Radiological Defense and Radiation Effects	4
	23) Biomedical Experiments	17
	24) Structures	3
	26) Civilian Vehicle Tests	2
	27) Fall-out Studies in Near Areas	2
	28) Radiation-telemetering Systems	1
	29) Dosimetry and Radiation Measurements	4
	Total	**36**

As for the DoD projects, only some of the FCDA projects conducted on an operation are described herein. Except for about nine projects, for which reports could not be located, the Civil Defense projects are all briefly described in Killian 2011.

During UPSHOT-KNOTHOLE, the three FCDA projects with perhaps the most human interest (and perhaps the most publicity) were those on Annie that involved: two typical American homes; home-type shelters; and civilian vehicles.

Two Typical American Homes

For Project 21.2, two typical center-hall, two-story frame houses, without utilities (plumbing, heating, or wiring), were constructed at 3500 and 7500 feet from the Annie ground zero. (Byrnes 1953a: 3; Goodwin 1953: 2) "These houses are one of the most common types of American home. At present East Coast prices they would cost about $20,000 each, complete with utilities." (Goodwin 1953: 4) "Exposure of the houses was for public demonstration purposes and to study the gamma-radiation scatter and the effects of thermal radiation and blast on each house. (Byrnes 1953a:3) The houses were sparsely furnished with Government surplus furniture. (Goodwin 1953: 4) Department-store mannequins, provided at no cost by the L. A. Darling Company of Bronson, Michigan, were placed in the rooms and the basement shelters. Each basement included one lean-to type shelter and one corner room shelter. The lean-to shelter used about $40 worth of materials, and the corner room shelter used about $95 worth. About 100 film badges were placed in each basement and the basement shelters to determine the penetration of gamma radiation. (FCDA 1953b 5)

Test Site lore has it that unidentified individuals rearranged a few of the mannequins into compromising positions after the FCDA had placed them in what they considered likely locations in the houses. Actually, what the mannequins were supposed to be doing, regardless of position, probably made little difference to the measurements.

On March 24, 1953, a week after DOORSTEP, the following press release was issued by AEC-FCDA-DoD:

> Approximately forty of the fifty clothed manikins used in the civil effects atomic test on March 17 have been recovered, decontaminated, and readied for public exhibit.
> ... inspection revealed that three un-protected manikins had suffered a wide variety of injuries, mostly fatal, while those in basement home-type shelters emerged uninjured. The unprotected manikins presented mute evidence of the injuries that humans would have suffered under similar circumstances and clearly indicated the necessity for adequate personal protection against atomic attack.
> ... The J.C. Penny Company, in cooperation with I.R. Crandall, Director of Civil Defense for Clark County, Nevada, will make the initial display of these manikins in its store at 6th and Fremont streets in downtown Las Vegas within the next two weeks. At the conclusion of this display, the manikins will be returned to the L. A. Darling Company for a nationwide tour of department stores in principal cities. This exhibition tour has been planned by the Darling Company in cooperation with the FCDA to point up the need for civil defense preparedness.

(AEC, FCDA, DoD, 1953)

Home-Type Shelters
During the fall 1951 BUSTER operation, home shelters, many of which were constructed of wood, were tested in Project 9.1a. The results showed weaknesses in the entrances. New designs were developed with sturdier building materials for exposure on UPSHOT-KNOTHOLE. (FCDA 1953b: 9) In a joint project with the AEC, eight earth-covered home-type shelters were constructed and exposed at ranges from GZ of: 1230 feet (1 shelter with expected over pressure of about 45 psi), 1450 feet (1 at 30 psi), 1800 feet (5 at 20 psi), and 3500' (1 at 2 psi). Different designs and construction materials were used for each shelter, and all had 3 feet of earth covering except for the shelter at 1,450 feet, which had 4 ½ feet. Mannequins were placed in shelters at each range. (FCDA 1953b: 9; Byrnes 1953b: 3-4, 9-10) These underground shelters on Project 21.1 were indeed small. The areas were only about four by four feet.

Civilian Vehicles
Operation HOT-ROD, which exposed 5 sedans, was conducted during RANGER; and its results provided the basis for the civilian-vehicle project on UPSHOT-KNOTHOLE. FCDA had been under great pressure from the public to provide definitive information on protection of persons in vehicles. " - some people apparently misinterpreted the (press) release on Hot Rod, and came to the dangerous conclusion that an automobile is a sort of rolling foxhole for the atomic age." (Goodwin 1953: 8)

UPSHOT-KNOTHOLE was the first operation in which American industry participated as a "doner", rather than as a contractor. Without the cooperation of business and industry, the FCDA's programs would have been far more limited in scope.

The FCDA counted on collaboration with industry to conduct projects 26.1 and 26.2. The National Automobile Dealers Association (NADA) agreed to a joint Government-NADA project and donated some cars as did the Automobile Manufacturers Association (AMA). Both the NADA and the AMA brought the cars to Las Vegas at their own expense and Standard Oil of California provided the gasoline. Civil Defense volunteers drove the cars from Las Vegas to Mercury, and the American Association of Motor Vehicle Administrators smoothed the way for the procurement of inter-state licenses. The Society of Automotive Engineers (SAE) provided technical consultants and set up a 10 member committee of automotive engineers who checked the cars and assisted FCDA personnel in all logistics. After the shot, the SAE assessed the damage to each car. The AEC instrumented the cars for radiation and thermal measurements, and some of the cars were equipped with instrumented dummies supplied by the Darling Company. (ibid., 8-10)

The vehicles were oriented in several directions with respect to ground zero: front, rear, side and angled to the blast. Some gas tanks were full and some empty. Some windows were open and some closed, and some brakes were on and some off. (ibid., 10) A variety of typical passenger cars were tested. All major makes were represented, and ranged in model years from 1936 to 1953. In addition, three mail trucks belonging to the Post Office Department were exposed at 3 ranges. (FCDA 1953b:9)

In addition to Annie, cars were also exposed on Ruth, Dixie, Encore, and Grable. After the tests and post-shot evaluations, the FCDA returned the cars to their owners. (ibid., 3) It would be interesting to know their subsequent history.

Deputy Test Director, Jack Clark, commented in a 1954 report of UPSHOT-KNOTHOLE that on Dixie the FCDA fielded fifty-seven automobiles in several arrays simulating urban parking scenarios. The high altitude airburst Dixie caused much more thermal damage than would have occurred from a tower shot at the same yield and distance. (Clark 1954: 67)

As the civilian-vehicle program demonstrated, the arrangements that the FCDA made with industry could be involved and complicated. The associations became even more so during TEAPOT and PLUMBBOB. The separation of civilian sector interests from government interests, the coordination of activities at the site, the access to information (both ready access to civil defense information and restricted access to weapons information), and the questions associated with financial obligations were significant management issues that required well thought out procedures and execution. (Reeves 1954) Fortunately, the diverse interests were somehow balanced and the involvement of industry at the site was successful.

Other Projects

Four types of instruments for determining the location of a nuclear explosion were designed for project 21.3. The instruments used were infrared detectors, tested on Annie and Encore. The immediate knowledge of the location of a nuclear detonation would be of primary importance to those who plan to relieve a stricken area. (Goodwin 1953: 11) However, such detectors would prove to be unreliable.

New types of radiation monitoring equipment were tested in project 22.2. For example, the feasibility of estimating the yield from gamma-radiation-dose measurements in the milliroentgen region was examined using photographic films. (Greene 1953: 3-4)

Forty-two different drug preparations, which were considered essential in a post-attack environment, were exposed on Project 22.4 in their original glass containers in heavy wooden boxes. Six of these boxes, shielded by up to 2 feet of soil, were exposed on Simon, three at 1200 feet and three at 1800 feet from GZ. Eleven boxes were exposed on Harry by setting them into the ground with the lids flush with the surface. The distances from GZ ranged from 1200 to 7500 feet. Three remaining boxes were used as control samples. "Insulin and vitamin B_{12} were reduced in potency by about 10 and 50 percent", but no others showed any deterioration. (Laug 1953:3, 11, 12, 15, 16, 18)

The effect of neutron exposures was examined on Project 23.17 utilizing 24 dogs and 2760 mice in AEC communal shelters. The shelter for Annie was 1500 feet from the GZ. The same communal shelter that was used on the BUSTER Project 9.1b was used for animal exposures on Ruth, Dixie and Climax. This shelter was 300 feet, 6042 feet, and 2000 feet from the three GZs respectively. Three shelters about 2350 feet from the GZ on Harry were also used. (Bond 1953: 3, 11, 12, 18, 23-25)

Mice were exposed on Project 23.2 to neutron or gamma radiation; and their hearts, blood and spleens examined for bacterial infections. Previous studies had implicated bacterial invasion as one of the important causes of death after radiation exposure. (Silverman 1953: 3, 9)

Both plants and animals were exposed to fast neutrons in projects 23.4, 23.14, and 23.16, to determine the frequencies of different kinds of mutations. The specimens were placed inside lead hemispheres, which were placed at several stations with neutron counts that that varied by a factor of 35,000. (Plough 1954: 3, 11, 14)

Dogs that had been exposed within the communal air-raid shelters during Annie and Harry were studied pathologically and clinically for blast injuries. Also, in project 23.15, two anthropometric dummies were test objects in the shelters for displacement studies utilizing high-speed photography. (Roberts 1953: 3)

Gold, sulfur, and manganese detectors were used to measure neutrons in project 23.17. (Tochilin 1953: 3) In 24.2, gamma measurements were made with film dosimeters. (Deal 1953: 3) For 24.3, measurements versus time of air blast phenomena were made near and within the two communal shelters. (Ruhl 1953: 3)

Surveys were made of the areas adjacent to the Site and up to 30 miles from the GZs. Samples of native soils, plants, and animals were taken after 5 tests "along existing trails and roads which crossed the various fallout patterns at distances greater than 10 miles from" GZ for the evaluation of the possible hazards. A test of a radiation telemetering system was also made in projects 27.1; 27.2; and 28.1. (Rainey 1954: 3; Lindberg 1954:3; and Johnson 1953: 11) The accuracy and practicality of various gamma-ray dosimetry measurement techniques were evaluated in projects 29.1 and 29.4.(Taplin 1953: 3 and Dahl 1954: 3)

UPSHOT-KNOTHOLE SUMMARY

In his summary report (Clark 1954: Chapter 6) Clark mentioned the inadequate housing conditions at the base camp at Mercury. Eight-man rooms in hutments and permanent barracks were still the norm.

With the exception of the inadequate housing, the other facilities were mostly satisfactory. This was the first operation where the necessary infrastructure was in place before the start of the test series. Also, there existed a cadre of people experienced in continental testing.

About 9 months separated the start of UPSHOT-KNOTHOLE (3/17/53) from the end of TUMBLER-SNAPPER (6/5/52). IVY, the two shot Enewetak operation, was in the fall of 1952. The year between the two Nevada series provided much-needed time for planning and construction, and resulted in lower construction costs.

UPSHOT-KNOTHOLE, with eleven events, was the largest series to date. There were a number of firsts. The gun shot, Grable, was the first operational systems test. FCDA participation was, for the first time, extensive; and they added many people to the already over-crowded Mercury.

From a technical standpoint this was also the most complex series to date. The three major programs, Weapons Development, Military Effects and Civil Effects were incorporated in the initial planning for the operation. Also, the inclusion of drone aircraft in a continental operation was new. On-site and off-site safety was the primary consideration with regard to the use of drones. After careful consideration, and the development of a detailed operations plan, the Test Manager gave his permission to include them in the test series.

This was Livermore's first test series. According to Clark, cooperation between the Livermore group and the test organization was "very satisfactory." Livermore was successful in obtaining data, and the new crew gained valuable field experience for their participation in the upcoming operations in Nevada and in the Pacific.

The diagnostic scientists and engineers from Los Alamos, Livermore, Sandia, EG&G and the Naval Research Laboratory made a number of very significant advances in the development of experimental techniques to explore the details of nuclear weapons behavior. These included techniques to make remote measurements of alpha and the electromagnetic spectra, and new methods to measure spatial and time resolved features of the explosion. Also, new diagnostic techniques were developed for use on the high yield shots planned for Operation CASTLE. In addition, the radiochemists developed new methods to diagnose shots using a variety of tracer elements built into the devices.

The theoretical designers were particularly pleased to make major breakthroughs in the efficient use of nuclear materials and in the new understanding of the implosion processes. They also gained valuable insights regarding the applications of new methods to initiate fission reactions during nuclear material assembly. Livermore's tests of Ruth and Ray, while disappointing in the yields, provided valuable information on an unexplored weapons concept. Finally, the pre-CASTLE experiments were successful and provided much-needed data for the up-coming Pacific test operation. (Clark 1954: 134)

Taken as a whole the Test Director and Deputy Test Director felt that this operation was the smoothest yet. "There were no major handicaps; the weather for the first 7 shots caused no delays. Camp Mercury living, although considerably overcrowded, was better than had previously been experienced. There were no major transportation problems. Power and telephone communications functioned very well" and "administrative relations within all factions of the test organization were exceptionally good". (Clark 1954: 132)

Off-site fallout was a significant public relations issue associated with Operation UPSHOT-KNOTHOLE. The Stewart Ranch, located north of the Nevada Proving Ground, filed damage claims for cattle that had been found dead and for horses with burns on their backs. The AEC, after investigation, compensated the rancher for the horses, but rejected the claim for the cattle. There were mysterious sheep deaths in Utah that were attributed by the local ranchers to fallout radiation. The sheep had been grazing approximately 50 miles northeast of the proving ground during the test operation, but were later moved to southern Utah.
(Hacker 1994: 106)

The AEC established an investigative group including state and local veterinary and animal husbandry experts to assess the situation. Pinpointing the causes of the sheep deaths was complicated by a number of factors, but it appeared that fallout radiation was at least a contributory factor. It was thought at the time, however, that radiation was not "a primary nor a major contributing cause of death."
(Hacker 1994: 115)

"The off-site fall-out has been thoroughly investigated since the conclusion of the tests, and considerable thought is being given to further minimizing bomb-debris fall-out in future tests without limiting the yield of the devices tested to lower than those of UPSHOT-KNOTHOLE. Methods of improving the weather forecasting are under consideration and will be incorporated in future tests." (Clark 1954: 132)

Operation UPSHOT-KNOTHOLE was taking place at the same time as the Committee on the Operational Future of the Nevada Proving Ground was drafting their conclusions on the continuing role of the test site. Off-site fallout was clearly an important issue. The Los Alamos Test Division Leader, Alvin Graves, expressed the frustration felt by many when he called for ... "an authoritative statement that would balance the value of testing against the degree of risk." (Hacker 1994: 117) Continental nuclear testing was an integral part of nuclear weapons R&D that was vital to the United States and the West during the Cold War. It was also inherently risky. The considerations that went into the shot execution criteria and the evolution of those considerations and criteria over time are traced by Barton C. Hacker in his authoritative study *Elements of Controversy* for the period between 1947 and 1974.(University of California Press, Berkeley, Los Angeles and London.1994)

Fallout was not an issue confined to southern Nevada and Utah. It was international in scope, and was addressed by representatives of both Britain and Canada at the Tripartite Conference on Radiation Hazards in Harriman, New York, in March 1953. (Hacker 1994: 117)

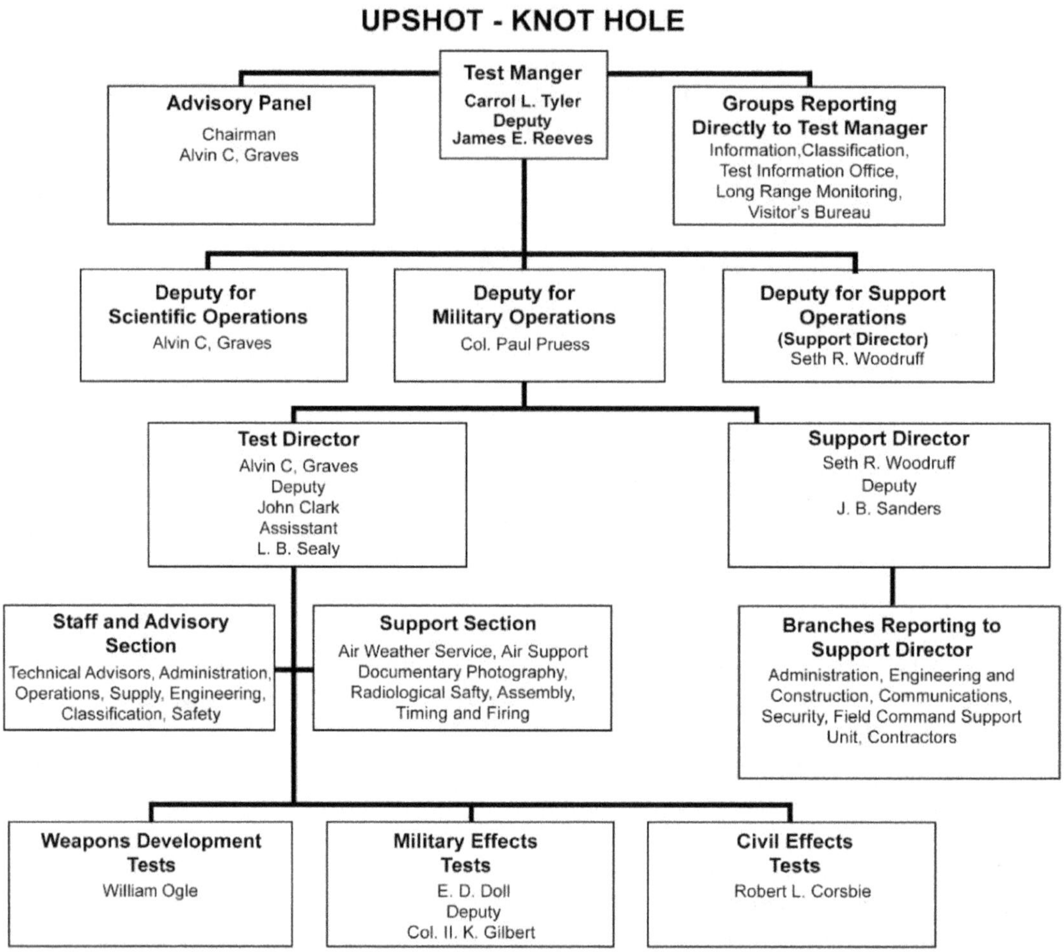

Figure 2-8.1 Nevada Test Organization for Operation UPSHOT-KNOTHOLE.

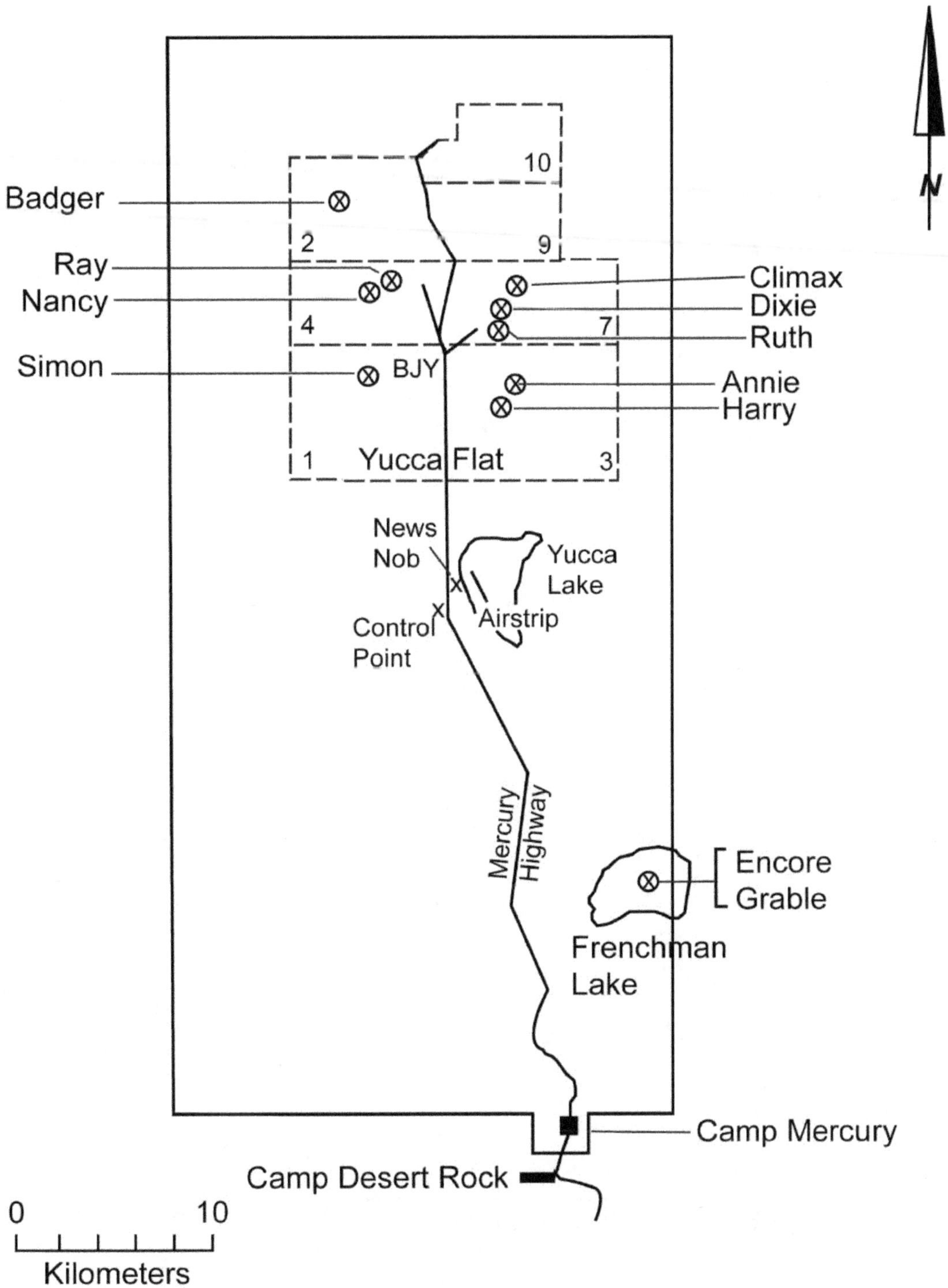

Figure 2-8.2 Location of UPSHOT-KNOTHOLE Tests.

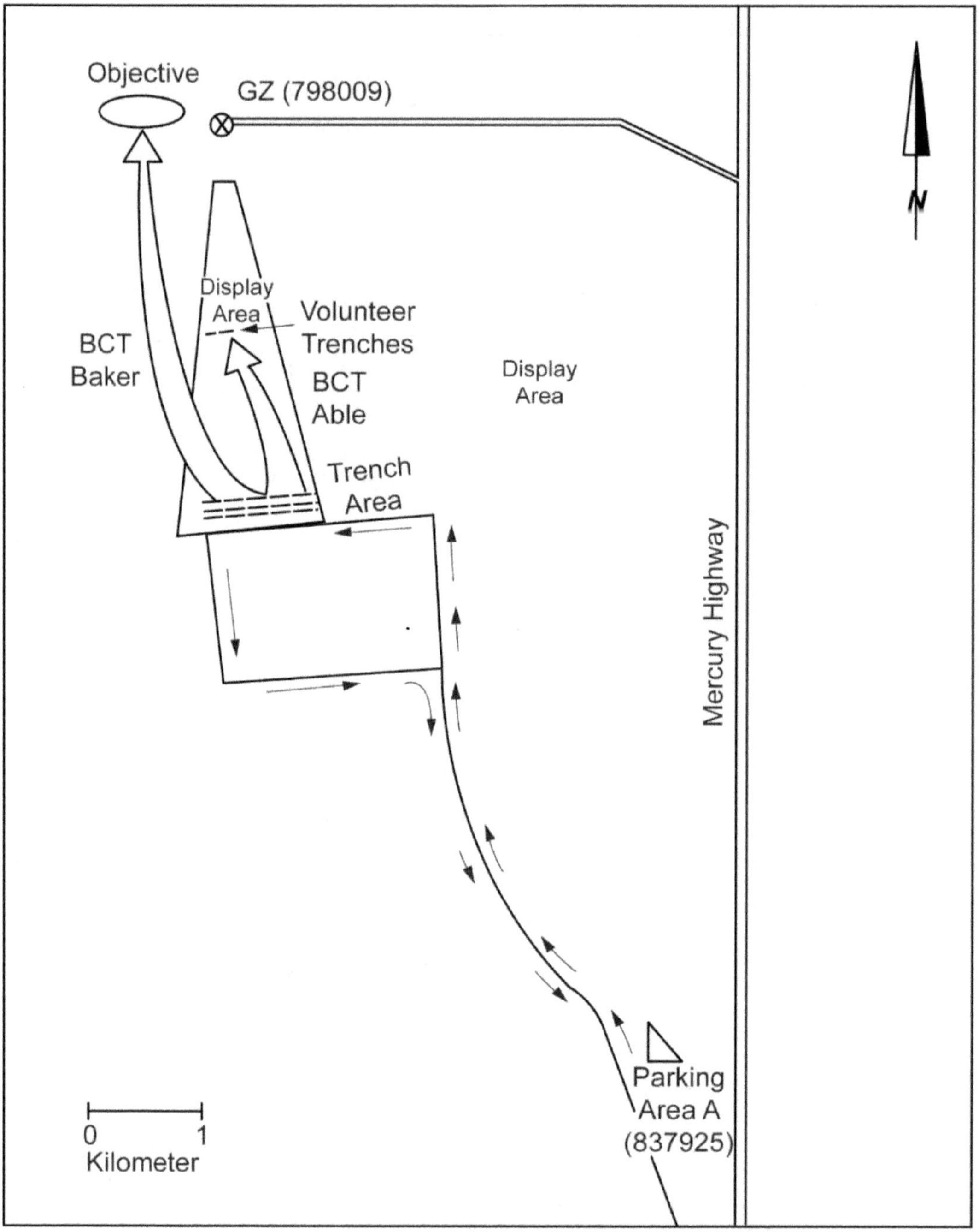

Figure 2-8.3 Desert Rock Exercises V on UPSHOT-KNOTHOLE.

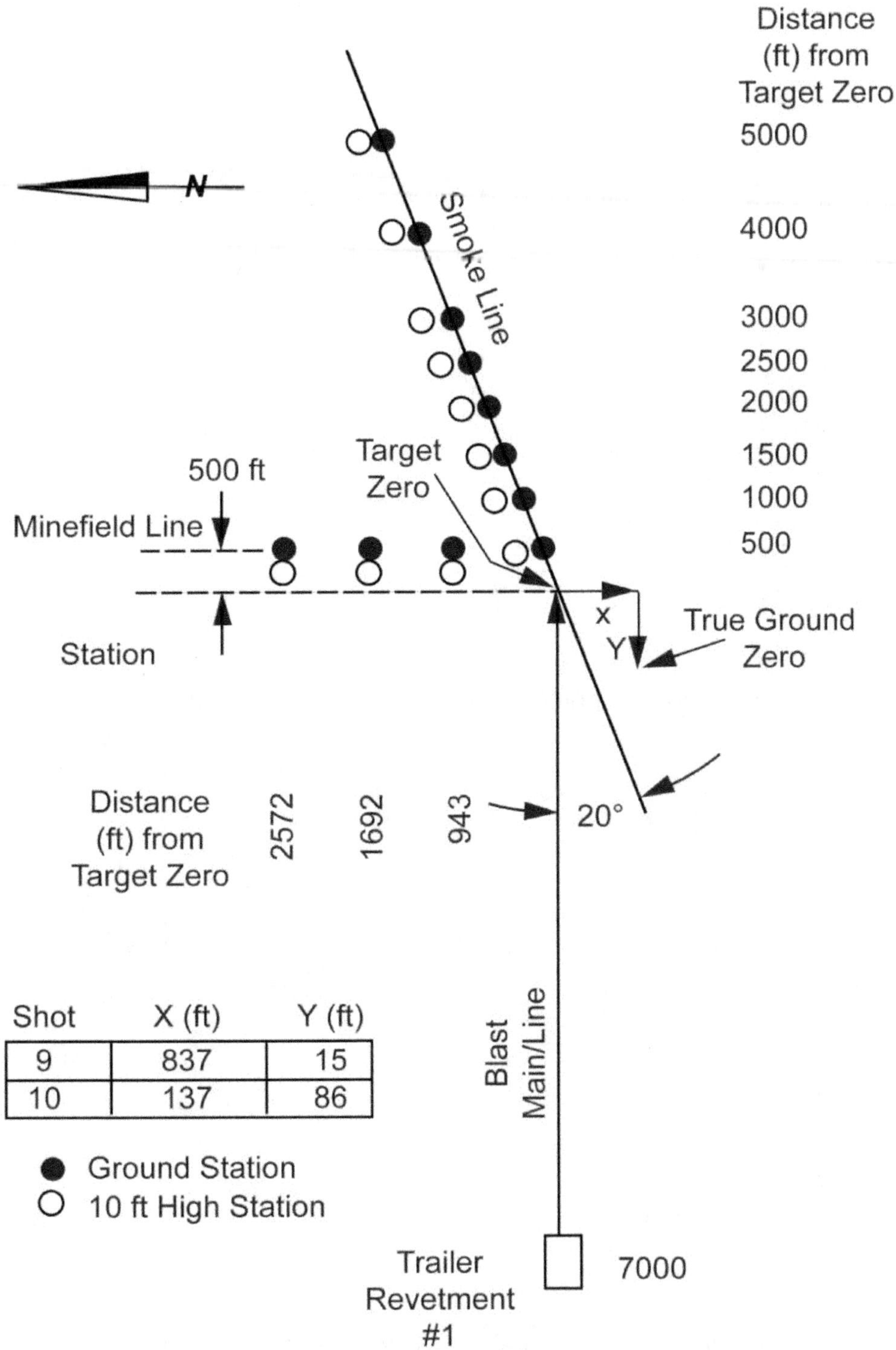

Shot	X (ft)	Y (ft)
9	837	15
10	137	86

Figure 2-8.4 Smoke Line and Minefield Stations for Instrumentation on Encore and Grable.

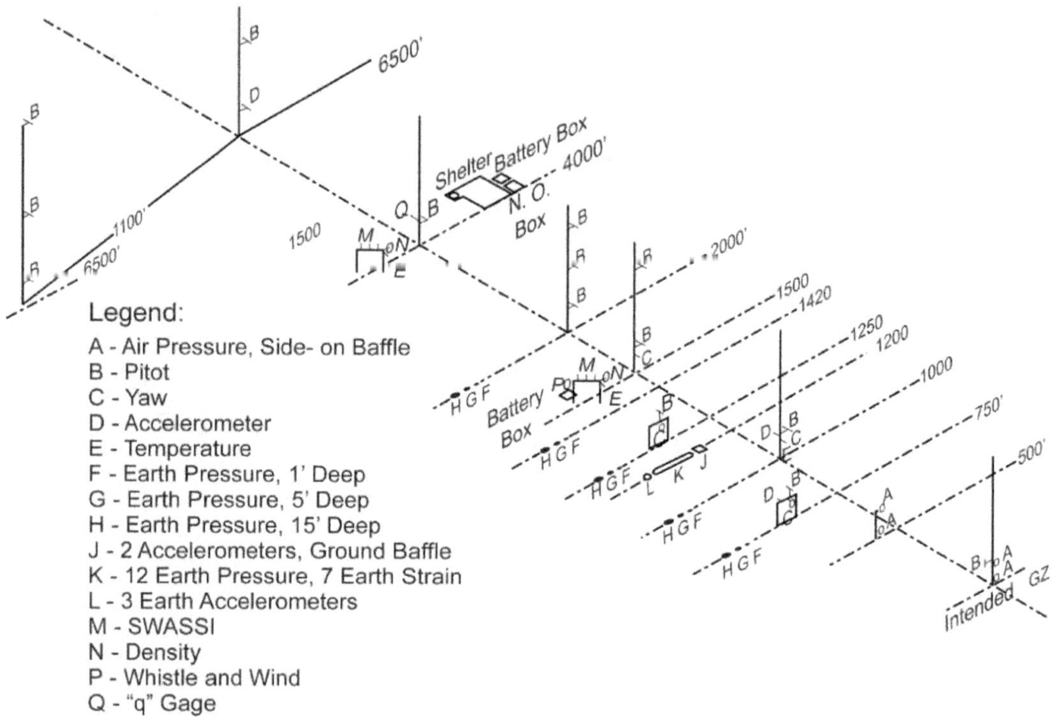

Figure 2-8.5 Sandia Airblast Instrumentation on Encore and Grable.

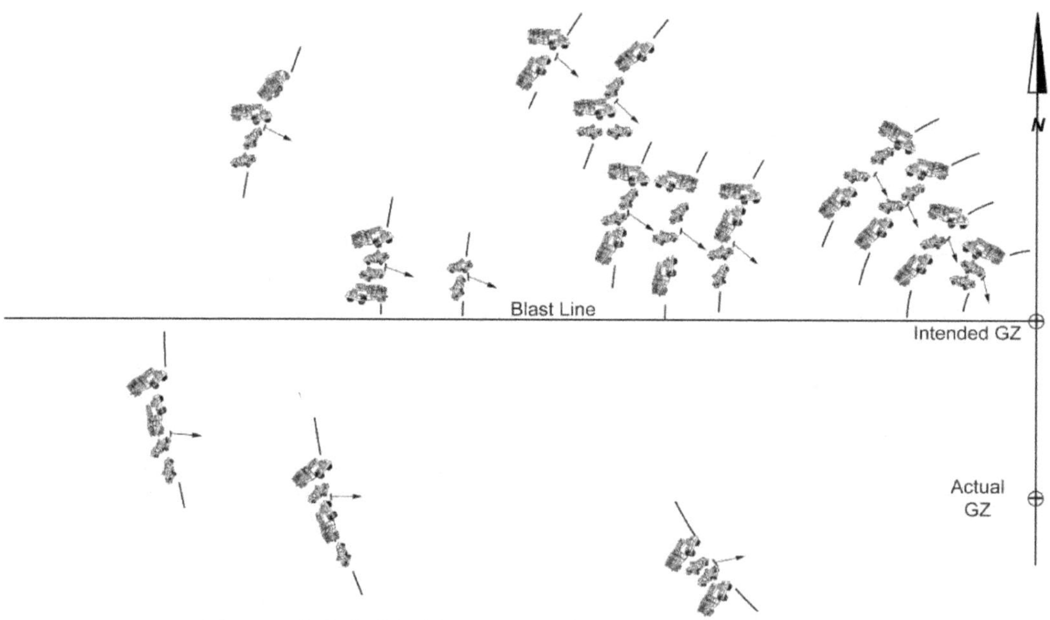

Figure 2-8.6 Field Layout of Vehicles for Encore.

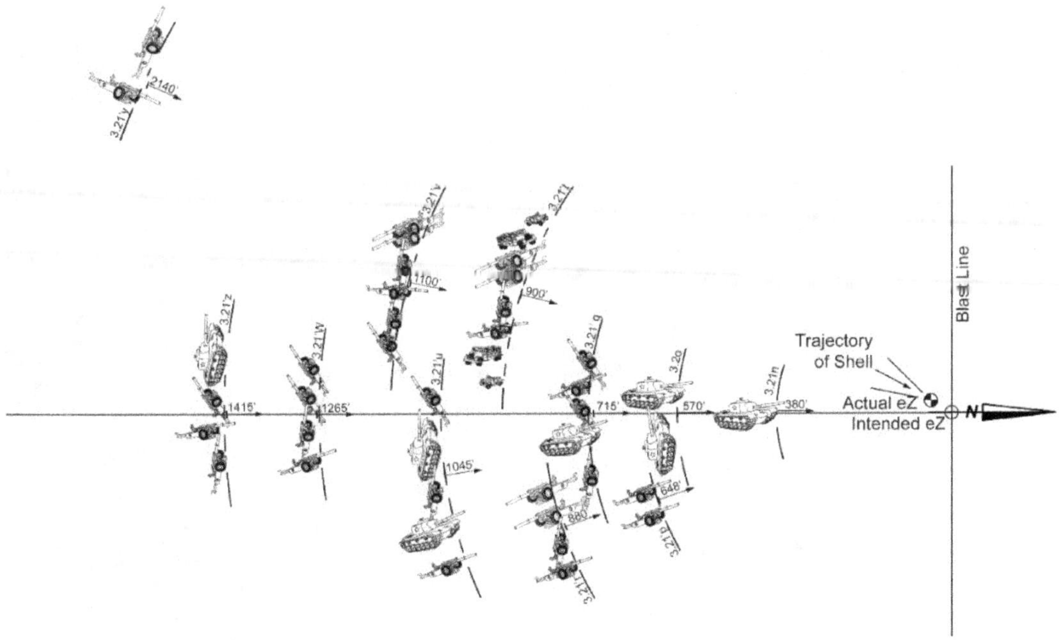

Figure 2-8.7 Field Layout of Ordnance Materials for Grable.

Major General Alvin R. Luedecke

Ed Giller and Harold Agnew (at a later era).

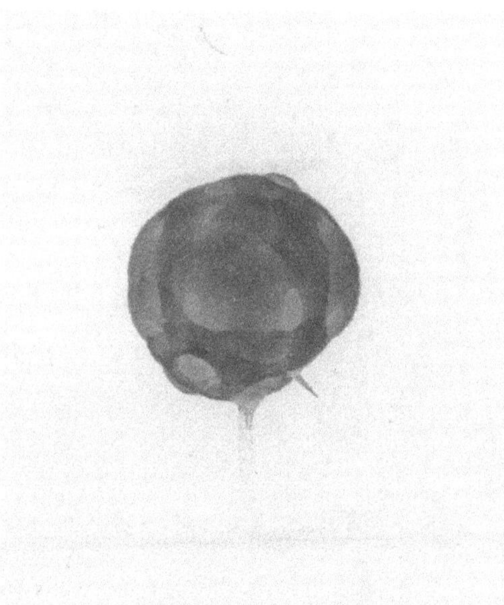

Badger, showing spikes due to guy wires on tower

Trench T-1 used by volunteers, approximately 2000 m from GZ, 7 May 1953

Tree strand, looking diagonally across from tree 18, project 3.19, 5 May 1953

90 mm AA gun, M1A1, at 1500 feet from GZ, project 3.21, 5 May 1953.

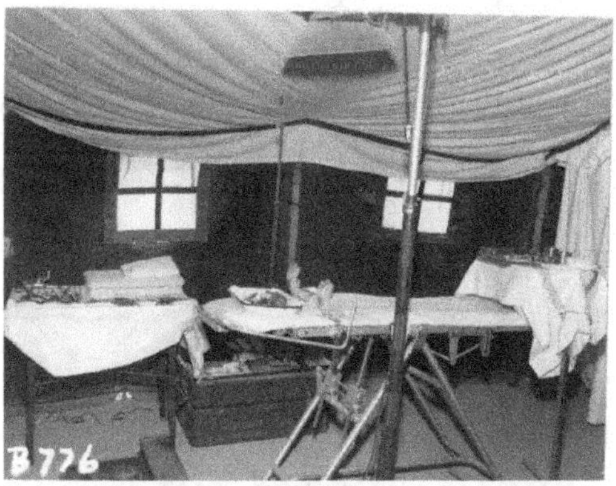

Operating room field tent as set up for test, project 3.27, 25 April 1953

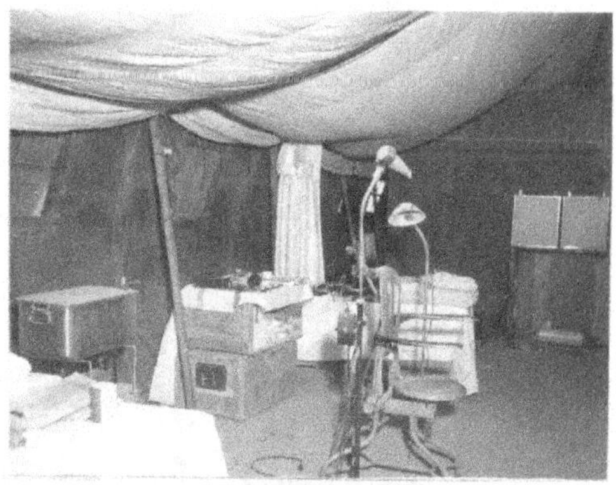

Dental hospital field tent as set up for test, project 3.27, 27 April 1953

Revetment station #228, for data recording, project 8.12a, 6 May 1953

Lt Col Prickett USAF giving a briefing at News Nob.

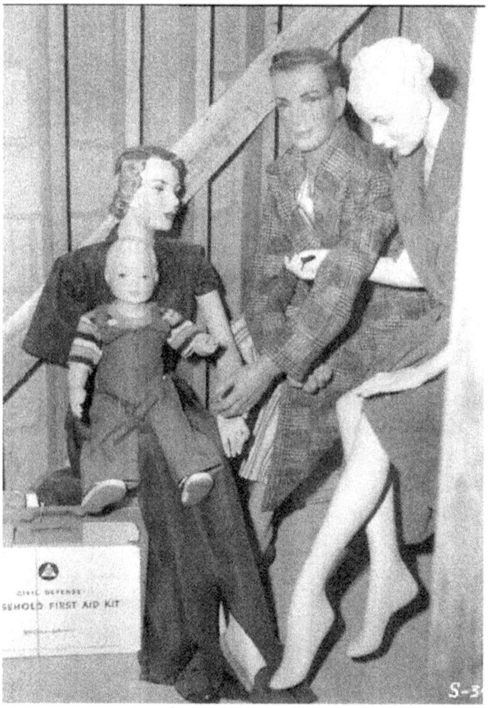

Mannequins in basement shelter on Operation DOORSTEP. A693

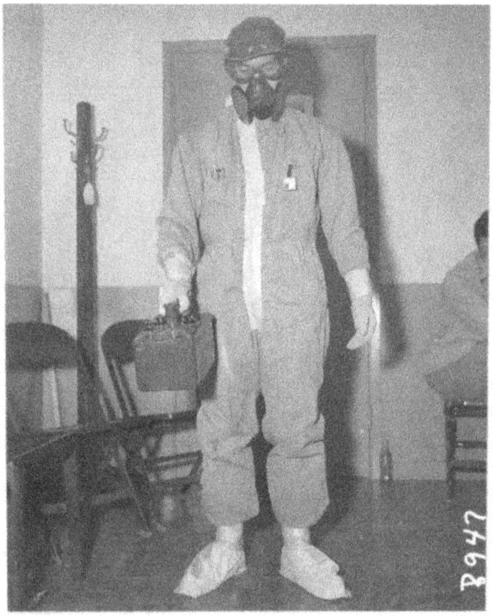

Lt Raymond S. Landen Jr. in Rad-Safe uniform. B947

Mannequins in a house on Operation DOORSTEP. Fenher pg 94

House and car exposed on Operation DOORSTEP. Fehner pg 94

House located 3,500 ft from GZ on Operation DOORSTEP. Fehner pg 96

Troops in trench with fallout raining down, during Simon. 100 Suns

The 280 mm cannon that fired the first and last nuclear projectile at the NTS, May 25, 1953, Grable. The vertical lines to the right of mushroom cloud were generated by flares in order to make airblast measurements. Guide pg 6.

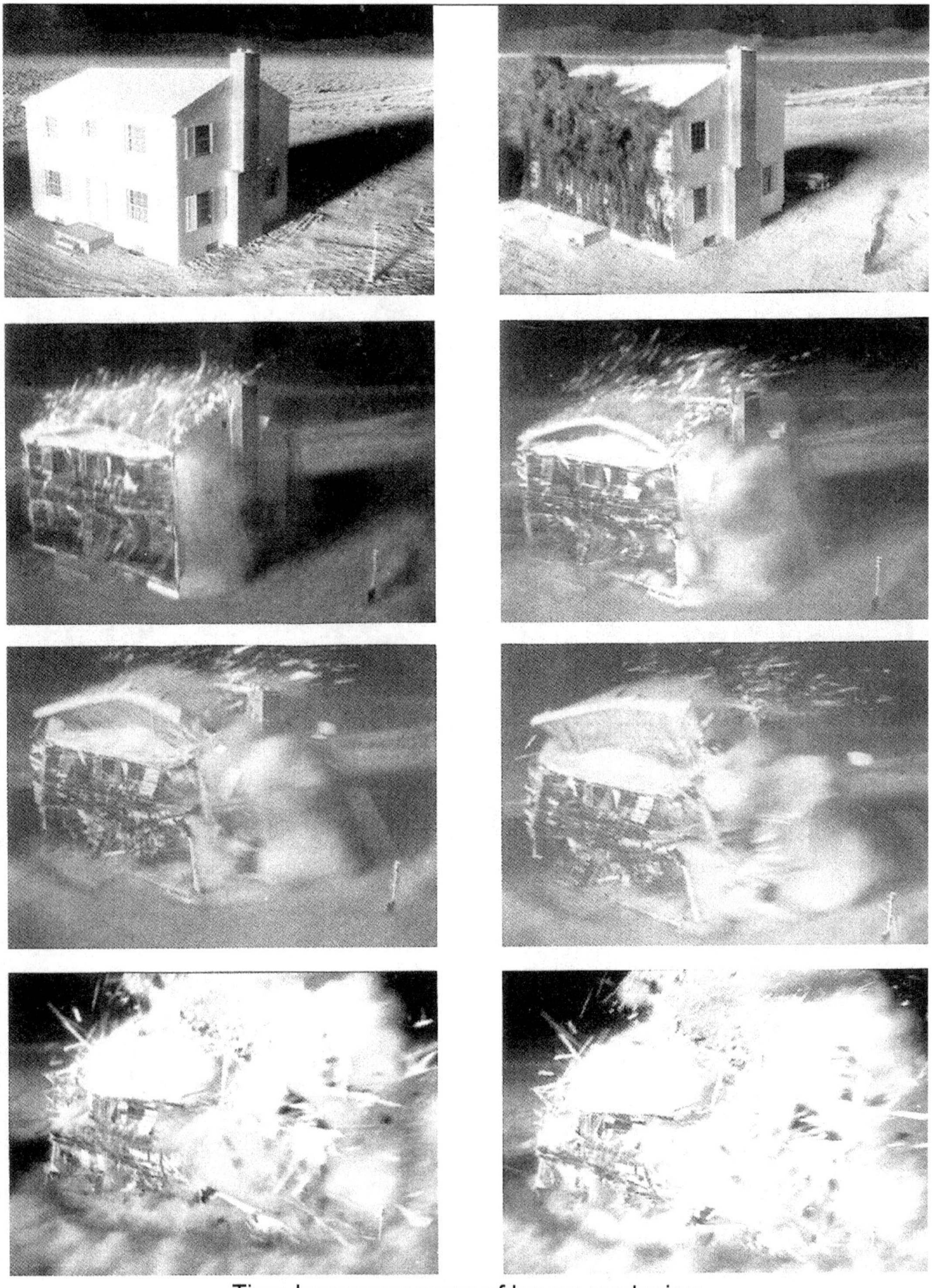

Time lapse sequence of house explosion

CHAPTER 9. 1953-1954

SECOND COMMITTEE ON OPERATIONAL FUTURE OF THE NPG - SUMMER 1953

The fallout issue was the driving reason for reconvening a revised committee on the future of the test site.

On July 7 Acting Director of the DMA, Colonel Marcus F. Cooper, USAF, requested that Carroll Tyler reconvene the Committee on the Future of the NPG "... in light of developments during and after UPSHOT-KNOTHOLE." The composition of the new Committee was changed from that of the first committee. It would report to the Director DMA and would include Tyler as chair, Bradbury instead of Graves, and a meteorologist. The DMA representatives this time around would serve in an advisory capacity.* [*Footnote: Committee membership: Carroll Tyler, Chairman, Manager of U.S. AEC, Albuquerque; Captain Howard L. Andrews, U.S. Public Health Service, Washington, DC; Dr. Norris E. Bradbury, Director, LASL; Dr. Darol K. Froman, LASL (Alternate); Dr. John C. Bugher, Director, Division of Biology and Medicine, U.S. AEC Washington; Dr. Gordon M. Dunning, Division of Biology and Medicine, AEC (Alternate); Dr. Everett F. Cox, Manager of the Weapons Effects Department, Sandia; Richard G. Elliott, Secretary, Director, Office of Information, AEC/SFO, Albuquerque; Dr. Alvin C. Graves, J-Division Leader, LASL; Col. Ben Holzman, Chief of Staff for R&D, AFSWC, Kirtland Base, Albuquerque; Dr. Lester Machta, U.S. Weather Bureau, Washington; Morse Salisbury, Director, Division of Information Services, AEC, Washington. Advisors were: Dr. John C. Clark, J-Division, LASL; Given H. Dugger, Director, Safety & Fire Protection Division, SFO, AEC, Albuquerque; Merrill Eisenbud, AEC/NYOO; William Harris, AEC NYOO (Alternate); Captain William Guthrie, AEC/DMA; Captain Harry H. Haight, AEC/DMA; Col. Paul T. Preuss, DWET, Field Command, Sandia Base, Albuquerque; James E. Reeves, director, Office of Test Operations, AEC/SFO Albuquerque; Dr. Thomas N. White, Jr., Health Division, LASL; Seth R. Woodruff, Jr., Manager, Las Vegas Field Office, AEC/NV, Las Vegas; Dr. Herbert C. York, Radiation Laboratory, U of C. Berkeley.]

The Committee was directed* to make recommendations regarding:

- Criteria governing the types of tests that would be conducted at the Nevada Proving Ground.
- Meteorological criteria for test execution. Include recommendations for actions that may be taken to safeguard persons and animal stock in downwind areas. Also, address the advisability of decontamination of persons or property outside the Nevada Proving Ground and the Las Vegas Bombing and Gunnery Range.
- Educational and communication measures that could be taken to alleviate public concern over the dangers perceived to be caused by the nuclear test operations.

[*Footnote: Author's comment - the direction probably "came from" Cooper but relied significantly on input from the new Chairman, Tyler.]

When the committee met in August, Tyler posed many questions that were raised by the broad charter and suggested assigning groups to research and report on the issues.

Los Alamos sent a report, dated August 28, 1953, to Tyler's committee. Darol Froman wrote a cover letter dated September 2, (TAD-1419) in which he said that the report was prepared by Los Alamos with input from Livermore. The report made the following points:

> The development of atomic weapons of all types involves a composite effort including four major activities, namely, primary experimental research, theoretical investigations and calculations, component development experimentation and full-scale nuclear detonations.

Main purposes (of nuclear detonations):

> To assure the adequacy of a weapon, or warhead, before it enters the national stockpile.
>
> To provide a firm basis for undertaking the extensive engineering and fabrication effort which must be expended to carry a 'breadboard' model to the version satisfactory for stockpile purposes.
>
> To demonstrate the adequacy (or inadequacy and limitations) of current theoretical approaches in order that promising avenues of development may be more fully explored or given lower priority of attention.
>
> To explore phenomena which can vitally affect the efficiency and performance of an atomic weapon but which are not susceptible to prior theoretical analysis of sufficient certainty.
>
> To provide a basis of choice among existing theoretical methods of weapon improvement in order to concentrate the attack along lines of greatest practical significance.
>
> To determine the validity of entirely new and untried principles proposed for application to the production of explosive atomic energy at improved efficiency.
>
> To provide entirely new information pertinent and valuable to weapon development arising simply as a by-product of scientific observation of full-scale detonations.
>
> To gain time in very urgent development programs by substitution of full-scale tests for a portion of a possible but lengthy calculational and experimental program in the laboratory.
>
> To provide, as a by-product, basic scientific information which becomes a part of the stockpile of such knowledge more normally obtained in the laboratory."

Director, Weapons Effects Tests, Colonel Paul Preuss prepared a DoD study of the evaluation of the NPG that was distributed by the committee's secretary, Richard Elliott on August 28. This report described the DoD's role in support of the AEC in nuclear testing: "The coordination of military assistance to the AEC in the conduct of continental tests is a responsibility of the Commanding General, Field Command,

AFSWP, while for overseas tests this responsibility is assigned to the commander of a joint task force." It then goes on to outline the many advantages of the NPG over the PPG such as: options for experimental layout, climate, cost, convenience, etc.

Military training and indoctrination, along with the weapons effects programs, were the principal reasons for DoD participation in the nuclear testing operations. By September, 1953, 37,500 officers and men, mostly from the Army and Marine Corps had participated in the Desert Rock exercises that could only have been accomplished at a continental test site.

The DoD report used the service's meteorology support to the AEC as an example of the cost savings associated with Nevada testing compared to the meteorological support required at the Pacific Proving Grounds. "Weather data required for continental tests is provided by existing U.S. Weather Bureau and the U.S. Air Force observation and reporting nets augmented by six outlying weather stations manned by a total of approximately 30 military personnel. The weather central at the NPG is manned by 40 military personnel."

Due to the isolation of the Pacific Proving Grounds and the scarcity of observation stations in the Pacific, the following facilities were identified as being required to provide the necessary weather information:

- A weather central on Enewetak manned by 40 men.
- A B-29 weather reconnaissance squadron of 12 aircraft manned by approximately 400 personnel.
- Four weather stations on outlying islands, approximately 300 to 800 miles from Enewetak atoll, which employ a total of approximately 80 men and require their establishment from the Hawaiian Islands, 2400 miles distant. This involves the use of LSTs, which initially move approximately 2,800 tons of equipment and supplies, and then subsequent continuous re-supply by Navy PBM aircraft.

The above indicates that 520 personnel and 12 four engine aircraft are required in the Pacific. In Nevada, only 70 personnel and no aircraft are required.

The DoD also outlined the enormous effort in terms of ships, aircraft and manpower that was required to provide military security to the Pacific Proving Grounds. There was no similar requirement for operations in Nevada.

Criteria For Future Continental Tests - Graves and Felt

Alvin Graves and Gaelan Felt drafted a paper on the *Criteria for Future Continental Tests*, in which they examined the relationships between fall-out and variations in yield and wind structure (wind speed and wind shear as functions of height, stability of the wind structures, and lapse rate), the surface conditions under the shot, and the height of burst, see Appendix K. They developed several conclusions and recommendations:

> More data are needed: Surface observations of fall-out; wind observations at 1000 ft intervals should be made at least every 4 hours during the 24 hours before a shot. Shots with yields between 15 and 25 kt should be fired on

> 500 ft towers. Data will assist in subsequent formulation of a policy on burst height as a function of predicted yield. The present investigation indicates that no appreciable reduction in hazard will result from changing the burst height of 50 kt devices from 300 ft to 500 ft.

The last sentence implies that the recommendation for firing shots of 15 to 25 kt on 500-foot towers was a very conservative approach until additional data were available, analyzed and appropriate recommendations formulated. Graves and Felt had attempted to quantify the hazards, but recognized that more data and analysis were required. (Los Alamos archives: A-99-019-635 279-4 Committee to Study NPG 11 Sept. 1953, and Hacker 1994: 119) Clearly the leaders at the labs were sensitive to the public health concerns of nuclear testing.

The criteria set forth in the Graves-Felt paper was approved by the Commission on June 30, 1954 and formed the basis for AEC 141/25, see Appendix K.(Felt 1954) The criteria from AEC141/25 are given in Appendix K as is a formalism for addressing potentially hazardous shots.

Second Committee On Operational Future Of The NPG Meets Again
The Committee met again on September 24 and 25. Tyler sent a preliminary report on September 29 to Brig. Gen. Kenneth E. Fields, who succeeded Col. Cooper at the AEC/DMA. The conclusion was that continental testing should continue in Nevada, but that the AEC must do a better job of informing the public with regard to the rationale for nuclear testing. Also, better public information was recommended regarding the hazards of radioactive fallout and the steps that the AEC was taking to mitigate exposure to the public. (Preliminary Report on Continental Tests and Future Utilization of Nevada Proving Grounds)

Actually, there was little new in the report. Radiological safety had always been a driving consideration in on-continent testing, and the AEC and the laboratories did not detonate a device if the predictions of off-site fallout exceeded, or even closely approached, the accepted standards of the day. In practice they leaned toward the very conservative side. That is one of the mysteries associated with the sheep deaths. While there is little doubt that sheep were exposed to fallout the levels were thought to be so low that it was not possible, in the 1950s, to use radiation alone to explain the observed deaths. In any case, explaining the cost-benefit trade-offs to the public was difficult at best. By the 1960s, testing had gone underground. Eventually, fallout was virtually eliminated by successful underground containment.

Tyler sent a tasking memo to the Committee members and advisors in mid-October, 1953. Among other assignments Tyler asked Graves and Felt to complete papers on radioactive hot-spot locations, surface stabilization in the vicinity of ground zeros, and the options and trade-offs associated with airdrops, towers and balloons for device placement. He also asked Graves to give his views on the issues related to long vs. short test series, the frequency of shots, the phasing in safe shots in place of more difficult shots to meet weather situations, and the effects of various operational options on personnel. Edward Cox, manager of the Sandia weapons effects department, was asked to prepare a paper on the likelihood of fuze failures on airdropped devices.

Cox also prepared a report for the Committee dated November 23, 1953, that addressed blast pressures in Las Vegas resulting from nuclear testing. (Los Alamos National Laboratory Archives File 334. Draft copy of Report dated February 1, 1954)

SOVIET AND U.S. TESTS OF THERMONUCLEAR DEVICES

On August 12, 1953, the Soviet Union conducted its first thermonuclear explosion, Joe 4, at their Semipalatinsk test site. It was fired on a 90 foot tower, and had a yield of 400 kt. (Mikhailov 1999:16) Andrei Sakarov (Sakharov 1990: 182) indicated that this design had little potential for extrapolation to much higher yields. He referred to this as a "second-idea" device. His "third idea" probably corresponded, at least approximately, to the concept developed by Teller and Ulam in the United States.

1954 - CHANGES AT THE SANTA FE AND ALBUQUERQUE AEC OFFICES

Carroll Tyler resigned from the AEC on February 15, 1954. He had retired from the Navy at the rank of captain in the submarine service in 1947. That summer, at the recommendation of Vannevar Bush, Tyler joined the AEC's new Office of Santa Fe Operations as manager.

From 1955 to 1958, Tyler worked for Skidmore, Owings and Merrill engaged in supervising construction of the U.S. Air Force Academy near Colorado Springs. In 1958, he again became involved with the test site as Manager of REECo, supplying support operations. He retired from REECo in 1962. (Press release: AEC June 7, 1968)

On February 16, 1954, Donald J. Leehey became Manager of the AEC Santa Fe Office (SFO). He retired from the Army with the rank of Colonel in 1946, after being responsible for the construction of military facilities in Washington, Oregon, and Idaho, and after serving in Europe. He worked in New York for construction companies prior to his appointment with the AEC. Leehey served as manager of the Santa Fe Office until September 30, 1955. (Press release: AEC July 18, 1955)

Operation TEAPOT was the only operation conducted during Leehey's tenure as Manager of the Santa Fe Operations Office (SFOO) of the AEC. While Tyler served as Test Manager for all of the previous four test series in Nevada, Leehey and subsequent SFOO managers did not assume the Test Manager position. Rather, they delegated the Test Manager's responsibilities to James E. Reeves for TEAPOT, PLUMBBOB and HARDTACK PHASE II.

A veteran civil engineer, Reeves joined the Corps of Engineers in 1930 and served at Oak Ridge during part of WWII. He joined the AEC at Albuquerque in March

1952 as Deputy Director in the Test Organization. Then, in July Reeves became Acting Director. On January 9, 1953, when Tyler was still Manager of SFOO, a press release was issued that described the appointment of Reeves as Director of the Office of Test Operations of the AEC's SFOO in Albuquerque. Reeves served in this position until 1957. (AEC Press Release January 9, 1953 and AEC Press Release June 14, 1957)

In 1957, Reeves was promoted by Kenner Hertford (who became Manager of the Santa Fe Operations Office in 1955) to the position of Assistant Manager for Test Operations in the AEC's ALO. (The Santa Fe Operations Office (SFOO) had by now become the Albuquerque Office (ALO)) In this capacity, Reeves acted for the ALO Manager in the direction of continental and off-continental test programs at the AEC's test sites. (Press release: AEC June 14, 1957)

OPERATION CASTLE IN THE PACIFIC

The CASTLE test series consisted of six shots, with five in the megaton range. This was an immensely important operation that significantly influenced the course of U.S. strategic nuclear weapons development throughout the Cold War.

On February 28, 1954, which was about six months after the Soviet Union's Joe 4 test, the US began Operation CASTLE with the firing of an advanced thermonuclear device, Bravo, at Bikini. (DOE 2000: 4) Bravo had a yield of 15 megatons, about twice the expected yield, and was the largest yield test the US ever conducted. The Bravo device used for the first time a new thermonuclear fuel, which eliminated the need for cryogenic equipment. The (n, 2n) and tritium production cross-sections of the new fuel were not well known and turned out to be larger than anticipated. As a result there were more neutrons and tritium than predicted and the yield went high. (DTRA 2002: 117)

Unfortunately there was a serious fallout exposure from the unexpectedly high yield of Bravo that extended well beyond the keep-out area set up for Bravo. A Japanese fishing boat, Lucky Dragon, received fallout; and members of the crew were hospitalized on their return to Japan. This tragic event received considerable publicity; and along with the Soviet Joe-4 test, resulted in international attention being focused on the implications of a nuclear arms race. The public health debate over the significance of the radioactive fallout continued for many years and lead to renewed efforts to negotiate restrictions on nuclear testing.

THE NEVADA PROVING GROUND BECOMES THE NEVADA TEST SITE AGAIN

Los Alamos's Bradbury and Froman had repeatedly told AEC officials that they did not think that "Proving Ground" left the proper connotation for the actual purpose of the AEC's Nevada site. They eventually prevailed and on December 31, 1954, SFOO Manager Leehey announced that: "the name of the Commission's test facilities in Nevada will be changed to the 'Nevada Test Site' with the abbreviation NTS."

The history of the names for the nuclear testing area in Nevada can be summarized as follows:

- On July 8, 1951, the Nevada site was officially named the Nevada Test Site (NTS) by the AEC. Before that time it had no official name, but Nevada Test Site was one of the names frequently encountered and is used here for the period of RANGER. NTS was the name during BUSTER-JANGLE.
- In early 1952, the name Nevada Test Site (NTS) was changed to the Nevada Proving Grounds (NPG). NPG was the name used during TUMBLER SNAPPER and UPSHOT-KNOTHOLE.
- The name Nevada Test Site (NTS) was adopted on the last day of 1954. This name stuck throughout the remainder of the nuclear testing period and is still used today.

CHAPTER 10. OPERATION TEAPOT: FEBRUARY 18 - MAY 15, 1955

PLANNING

In early October 1953, AFSWP recommended that a high-altitude test be conducted to provide information related to air defense. Subsequently, AFSWP requested the services to submit proposals for Operation TEAPOT, which had been announced by the AEC for the autumn of 1954. (DASA 1960a:15)

At the January 21, 1954 meeting of Los Alamos' Fission Weapons Committee (FWC), Chairman Duncan MacDougall announced that: "There is a faint possibility that Livermore may be able and willing to carry out a brief pre-TEAPOT series in November 1954. If this materializes, it would be desirable, in the interests of shortening the TEAPOT series, if LASL also made a couple of its simpler experiments at that time. The Director has therefore asked us to consider whether we could be ready to make two simple shots (airdrops, yield and alpha measurements only) in November 1954, instead of in February 1955." Meanwhile, Livermore was planning to fire their shots on towers.

Furthermore it was suggested that Livermore would provide the test director for the pre-TEAPOT series while LASL would provide similar services for Operations TEAPOT and REDWING.

"In consideration of 'manageable size' for TEAPOT, Graves pointed out that it is his belief that the Commission will tend to believe that approximately eight shots is a limit for a Continental operation, independent of how easy or difficult they are to make. Graves said that the present firm DoD requirement for TEAPOT is for a high altitude shot at about 2 kt." Also, DoD would like surface and underground shots (JANGLE-fashion) at 10 kt., but this is presently not very likely to happen. Civil Defense will probably not sponsor any particular shot, but simply distribute houses, etc. around shots otherwise scheduled." (FWC March 1, 1954)

A preliminary discussion of TEAPOT between the SFOO and the weapons development laboratories was held on February 14, 1954 at Enewetak, just before the first shot of Operation CASTLE.(Clark 1955: 15) The ground rules for Pre-TEAPOT and TEAPOT included the understanding that as far as possible the shots would be airdrops. LASL had been planning on piggybacking on the Livermore pre-TEAPOT series; but even if Livermore were to back out, Bradbury said that LASL should plan on doing a couple of airdrops. (FWC March 1, 1954)

By early spring of 1954 there appeared two TEAPOTS: TEAPOT I in the fall of 1954 and TEAPOT II in the spring of 1955. Operation DIXIE was added for the fall of 1955 and JULEP proposed for the fall of 1956 in Nevada, with Operation REDWING planned for the spring of 1956 in the Pacific. By the fall of 1954 the schedule evolved to show Operation TEAPOT for the winter-spring 1955. (Reeves 1955: 1-8)

The purpose of this litany is to demonstrate the state of flux of the test program over the several years that it took to plan and execute an operation. It was driven primarily by the requirements of the nuclear weapons development program and the needs of the military establishment for weapons effects data. Secondarily it was driven by the Civil Defense concerns and by budgetary considerations.

By May 19, 1954, pre-TEAPOT had disappeared. At the 76th FWC meeting an announcement was made that LASL and DMA had decided to cancel their participation in the pre-TEAPOT operation. TEAPOT was to begin in February 1955 and include a total of 5 or 6 LASL shots. Livermore planned on 2 shots. Also, a post-TEAPOT series in Nevada was planned for the fall of 1955. At this time Los Alamos planned to field their shots as airdrops.

Also on May 19, the JCS approved an AFSWP program for TEAPOT. The approval was subject to review and availability of funds. The program included three tests: 1) a high altitude shot of about 1 kt at 40,000 ft, 2) a surface shot of at least 10 kt, and 3) an underground shot of approximately 10 kt at a depth of 135 feet. The AEC, however, balked at the suggestion for 10 kt surface and underground detonations, stating that the two shots were not compatible with the safety criteria that had been established for future operations at the NTS. (DASA 1960a: 16)

There were enough problems with fallout from tower shots, and a number of ideas of ways to reduce the resultant radioactive debris were explored. One was to put a shallow lake at ground zero to reduce the dust entrainment. On July 16, 1954, Jack Clark sent a memo to Ralph Johnson, SFO, pointing out that putting a large pool of water directly under the detonation point was both difficult and costly and that a test of the concept didn't seem feasible for TEAPOT. He did, however, point out that asphalt stabilization at ground zero might serve to reduce off-site fallout. TEAPOT experiments could perhaps furnish data on the relationships between yield and height of burst, surface material, airborne activity, and off-site fallout.

In the time period prior to TEAPOT, DoD was deeply involved in issues that had arisen from the advent of thermonuclear weapons. Two issues: 1) How to deliver the higher yield weapons, i.e. what flight patterns would be safe for the pilot and crew; and 2) How to destroy an incoming plane, or missile, carrying a weapon, i.e., what yield needed to be placed where with respect to the incoming vehicle. These two issues both require knowledge of the level of nuclear effects, in particular airblast and thermal radiation that can be withstood by a delivery craft in flight. The reason that the DoD wanted the 10 kt surface shot was to test drone aircraft in an airblast field where there was only a single pulse. An airblast field with multiple pulses occurring from the main shock in air and the pulses reflected from the surface was a complicated situation. In the mid-1950s, a multipulse-airblast field interacting with an aircraft was not a situation amenable to accurate analyses. A single pulse was complicated enough. In fact there was considerable uncertainty about where and when the reflected shock(s) catch up with the initial free air shock. These single pulse issues are further discussed in the section titled DoD NWE

PROGRAMS AND PROJECTS later in this chapter. Here, it is sufficient to say that, AFSWP prepared a new program plan.

The new plan: retained HA, the High-Altitude shot; reduced the yield of the underground shot to about 1 kt and called it Effects Sub Surface, ESS; and, with caveats (regarding the acquisition of single pulse data on early TEAPOT shots) agreed to use a 400' tower for testing NWE on drones. This shot was named MET, Military Effects Test. These 3 test names generally use all capital letters, and they are acronyms. (Lewis 1997: 24-5)

Although the majority of the military effects measurements would be conducted during HA, ESS, and MET, a number of important effects experiments could be carried out only on AEC development tests. Therefore the new AFSWP test program also made maximum use of the AEC shots. The JCS and the Secretary of Defense approved the revised DoD test program in August 1954. (DASA 1960a: 16)

The August 9 minutes of Los Alamos' 80th FWC meeting reflected the DoD interest in the test of an air defense weapon of a few kt fired at about 40,000 feet. LASL suggested using a weapons development shot for this, which would reduce the number of tests by one. However, Los Alamos wanted both yield and alpha measured if the shots were to be combined. In addition, Los Alamos wanted to fire no later than about a half hour prior to sunrise in order to have optimum conditions for the observation of Teller light,* while the DoD wanted to fire in daylight for the best conditions to make blast measurements (such as smoke trails). [*Footnote: Edward Teller suggested that light would be produced by gamma rays passing through the atmosphere. This is the so-called Teller Light. He further pointed out that this light could be employed to measure alpha. However, Teller light is only visible when a shot is conducted in the dark.]

LASL proposed to the DoD that two of the lab's devices, which would be tested during TEAPOT, be utilized for the HA and MET shots. The predicted yields of these devices were well within the limits desired. The DoD agreed to the proposal with the reservation that stockpile weapons could be substituted if last-minute calculations indicated that the performance of the experimental devices would not meet military specifications.

On August 18, 1954, the AEC approved the 1955 nuclear test series in Nevada. (AEC 1954: Meeting # 1020 AEC707/5) The operation would be conducted under the criteria approved by the Commission on June 30, 1954 and set forth in AEC141/25, see Appendix K. As described in the last chapter, Gaelen Felt and Al Graves prepared a paper that formed the basis for AEC141/25. A formalism of addressing potentially hazardous shots is also given in that paper and briefly cited in Appendix K. (Felt 1954)

On December 7 Reeves and Graves sent a memorandum through Leehey to (AEC/DMA) Brig. Gen. Kenneth E. Fields describing the specific proposed test program, planned for the spring of 1955, for Commission approval. Reeves and

Graves presented the shot proposals sponsored by Livermore and the DoD as well as for Los Alamos.

The memorandum proposed six shots for LASL, three for UCRL, two for the DoD, and one sponsored by both LASL and the DoD for a total of twelve shots. As it turned out there were a few surprises and two shots were added to the schedule, bringing the total to 14. Except for MET, which was fired in Frenchman Flat, all were detonated in the vicinity of the B-J Y (the major fork in the Mercury Highway about 9 miles north of the Control Point); in Areas 2, 3, 4, 7, 9, and 10, see Figure 2-10.2.

PUBLIC RELATIONS AND INFORMATION

Public relations and the release of information to the public were exceedingly important activities from the beginning of nuclear testing. During TEAPOT, PLUMBBOB, and HARDTACK II, these activities became even more extensive and better coordinated among the NTS user groups. Appendix L briefly describes the mission of public relations and specific activities during TEAPOT, PLUMBBOB, and HARDTACK II.

NEVADA TEST SITE ORGANIZATION (NTSO) FOR OPERATION TEAPOT

Figure 2-10.1 shows James Reeves' organization for Operation TEAPOT. (Reeves 1955, 28-9) W. W. Allaire was Reeves' Special Assistant. Al Graves, who had been Test Director for all of the operations on which Tyler was Test Manager, moved into the position of Scientific Advisor to Reeves and continued as Chairman of the Advisory Panel for Reeves. See Appendix F for other members. Reeves had only one deputy, Col. H.E. Parsons who was Deputy for Military Operations. Two individuals, the Test Director and the Support Director, plus the Test Operations Staff reported directly to Reeves.

Jack Clark was the Test Director with Gaelen Felt as Deputy. Clark had four scientific test groups: Los Alamos Weapons Development lead by Bill Ogle; UCRL Weapons Development lead by Duane Sewell and Vern Denton; Military Effects lead by E. B. Doll; and Civil Effects lead by R. L. Corsbie. These leaders of the scientific groups also served as Technical Advisors to Clark and Felt. In addition to the test groups and Technical Advisors, Clark had two groups that performed various staff and support functions. Seth R. Woodruff, Jr., was the Support Director with Joe B. Sanders as deputy. (Reeves 1955, 30).

Technical Staff

The weather, fallout and blast prediction groups reported to the Test Manager, but their primary responsibilities were in support of the Advisory Panel. The blast prediction group included about seven Sandians, and often as many additional people from other laboratories. Sandia developed a computer, named RAYPAC (Ray Path Analog Computer) for TEAPOT that predicted blast pressures for various locations prior to shot time. The predictions turned out to be reasonably accurate when no off-site blast focus was forecast. However, when a focus was expected the blast predictions were often way off. It turned out that blast patterns are extremely sensitive to weather details.

Blast pressure measuring instruments were located at CP-1; Indian Springs; Las Vegas; Boulder City; Caliente; Lund; and Tonopah, Nevada; St. George, Utah; and Bishop and Inyokern, California for all TEAPOT shots. Portable instruments were available to be installed at additional locations for specific shots. (ibid., 70-71)

AEC - Los Alamos Responsibilities

Los Alamos's J-Division had the major responsibilities for a number of the activities that are associated with conducting an operation. This was the case since RANGER when Los Alamos was essentially the only organization testing. As the load of these activities increased the Manager of the Santa Fe Operations Office established an Office of Test Operations, largely to relieve Los Alamos of as many operational and administrative test-related responsibilities as feasible. Thus several support units that had reported to the Test Director were transferred from the Test Director to the Support Director's or to the Test Manager's staff. "These were: (a) responsibility for arranging for the Advisory Panel, (b) the meteorological unit, (c) off-site rad-safe, and (d) microbarographic unit. In addition, the security control procedures for all photographic activities within the proving ground, which had until then been the responsibility of the Test Director, were transferred to the Test Manager's staff. Likewise, coordination of all radio frequencies had been the responsibility of the Test Director's Plans and Operations unit. This function was transferred to the Support Director's communications section." (Clark 1955: 12-13)

TEST OBJECTIVES FOR OPERATION TEAPOT

By the winter and spring of 1955 the design of primary weapons in the United States was quite sophisticated compared to a decade earlier. The state of knowledge of implosion hydrodynamics, was rapidly expanding and new concepts were being explored and tested. The rapid advances in computers were instrumental in this expansion of knowledge. Boosting was an accepted technology and was frequently employed to enhance the yield of the new designs. New materials were going into fission weapons, and the resulting designs required full-scale tests to verify the calculated predictions. Also, a number of the TEAPOT experiments were expected to support Operation REDWING, the very extensive Pacific test series planned for the spring and summer of 1956 at Bikini and Enewetak.

TEAPOT TESTS

Table 2-10.1 gives the 14 tests conducted during Operation TEAPOT, and Figure 2-10.2 shows the location of those tests at the NTS.

Table 2-10.1 Operation TEAPOT Tests.

TEST	SPONSOR	DATE 1955	TYPE	AREA	PURPOSE	YIELD kt
Wasp	LANL	02/18	Airdrop 762'	7	Weapons	1
Moth	LANL	02/22	Tower 300'	3	Weapons	2
Tesla	UCRL	03/01	Tower 300'	9	Weapons	7
Turk	UCRL	03/07	Tower 500'	2	Weapons	43
Hornet	LANL	03/12	Tower 300'	3	Weapons	4
Bee	LANL	03/22	Tower 500'	7	Weapons	8
ESS	DoD/LANL	03/23	Crater -67'	10	Effects	1
Apple 1	LANL	03/29	Tower 500'	4	Weapons	14
Wasp Prime	LANL	03/29	Airdrop 737'	7	Weapons	3
HA	DoD/LANL	04/06	Airdrop 36,620' MSL*	5	Effects & Weapons	3
Post	UCRL	04/09	Tower 300'	9	Weapons	2
MET	DoD/LANL	04/15	Tower 400'	5	Effects & Weapons	22
Apple 2	LANL	05/05	Tower 500'	1	Weapons	29
Zucchini	LANL	05/15	Tower 500'	7	Weapons	28

(DOE 2000, 4-7; Lewis 1997, 26)
[*Footnote: MSL, Mean Sea Level]

For the first time, 500' towers were used in this operation. The 36,620-foot HA shot was the highest nuclear explosion to date and the highest ever at NTS. Considerably higher tests were conducted at Enewetak Atoll and Johnson Island during HARDTACK Phase I. The 67-foot depth of ESS was the deepest buried test to date.

ESS

No yield measurements were made on ESS, but the DOE lists the yield of ESS as 1 kt. (DOE/NV 2000:2-3, 6-7) The resulting crater had the following average dimensions: Volume, 2.62×10^6 ft^3; Diameter 284.8 ft; and Depth 89.6 ft. (Schuster 2001: 29, 30)

Col. H. E Parsons, USAF, the Acting Director of Weapons Effects Tests, in a memo to the Manager of the AEC Santa Fe Operations Office, outlined the issues associated with the ESS fallout considerations. He estimated that the maximum cloud height would be less than 4,500 feet above the ground. The low cloud height implied that the significant fallout would occur within 2.5 hours after the shot. It was recommended that the shot be fired when the average wind speed was less than 20 mph. (Parsons 1954; Reeves 1954)

Hadr (High altitude dry run)

On March 25, an airdrop of a non-nuclear device was made over Area 1 with a burst height of 38,000 feet above mean sea level. The detonation occurred at 0900 PST.

The purpose of this non-nuclear shot was to make comparisons with HA. (Reeves 1955: 83)

Apple 1

Apple 1, a weapons development test, failed to give satisfactory performance. Because of the importance of this device to future development tests, a second device designated Apple 2 was added to the schedule as Shot 13. (DASA 1960a: 20)

PREPARATION AND DETONATION OF A DEVICE: 1955-1958

This section describes the procedures that were undertaken to detonate the nuclear devices during TEAPOT.

Transportation of Nuclear Devices

Nuclear devices that would be detonated on towers or balloons were transported to the Test Site by a Carco Air Service C-47 (DC-3) airplane with a Carco Twin Bonanza as escort. The flights avoided populated areas to the extent possible. The complete SOP is described in a document *Procedure for Transportation of Nuclear Devices from Los Alamos to NTS by Air*, and issued by the Santa Fe Office Manager on January 24, 1955. The intention was to land at the Yucca Flat airstrip. But, rains and a rough runway surface resulted in a move of operations to Indian Springs as an alternative until the Yucca strip was again made operational. (Clark 1955: 73)

Device Assembly

The Assembly and Arming Unit was staffed with personnel from Los Alamos, Livermore, Sandia, and the DoD. It was under the overall leadership of E. L. Jenkins of Sandia. Livermore people assembled the Livermore devices, which were all tower shots. Los Alamos people assembled the Los Alamos devices for the LASL tower shots. Sandia people, under the direction of Harvey North, assembled the airdropped devices in the ballistic cases. The Army Corps of Engineers assembled the shallowly buried ESS device as a training exercise.

Signal Dry Runs

Signal dry runs started about four days before the scheduled shot date and continued up to about eighteen to twenty-four hours prior to shot time, when recording station buttoning-up began. (Reeves 1955: 96)

Final Shot Preparation

Two days before the shot the Test Director issued an operation plan that addressed a general description of the test, the execution responsibilities for rad-safety, personnel control, vehicle traffic control, communication and electric power management, and a schedule of specific steps in firing the device.

Emplacement
Normally a device was not emplaced in the tower, or aircraft in the case of an airdrop, until the day before the planned shot and the Test Manager's advisory panel had decided that the weather looked favorable for the test. For airdrops the final arming was completed after the last practice bomb run, or about four minutes before release. On a few special occasions the test devices were mounted in the towers several days prior to the shots, but this was avoided whenever feasible. (Clark 1955: 95-96)

Neutron Sources
A source of neutrons is used to initiate a chain reaction when a critical configuration of the fissionable material is achieved.

Fuzing
A tower event had a sequence timer that fired the shot, which was activated from the control room. Experiments that were considered vital to the success of the test were interlocked in order to prevent the firing signal from reaching the device if crucial diagnostic experiments were not functioning properly.

"At zero time the fire cam switch closed, furnishing voltage to the fire lines. If the circuit was complete through the interlocks, the fire relay in the zero rack was actuated and its contacts furnished voltage ... The X-unit capacitor bank was discharged into the detonators, and the device exploded." (Clark 1955: 82)

For airdrops, the sequence timer that was used for the tower shots was replaced with a fuze system. At a preselected time after weapon release, to ensure a safe separation of the device from the aircraft, the weapon was "armed." This was followed at another preselected time by a fire signal. The times were selected to fire the device at the correct altitude above the terrain. A baro switch was set to produce a fire signal at a safe height to preclude a ground contact explosion in the event of a failure of the primary fuzing circuit for the high altitude shots.

A radar fuze was used as a backup for the lower altitude Wasp and Wasp Prime events. Also, a regular (war reserve) nuclear weapon TX-12 ballistic case was used for the two Wasps drops. The piezo electric switches, which produced a contact burst in the event of a radar fuze failure in a war reserve weapon, were disconnected. (Clark 1955: Cpt. 5)

LASL'S DIAGNOSTIC EXPERIMENTS
Eight programs of diagnostic experiments were conducted by Los Alamos during TEAPOT: Hydrodynamic Yield (e.g. Fireball analysis); Radiochemistry; External Neutron Measurements; Gamma Ray Measurements; Initiator Diagnostic Measurements; Photo physics (e.g. Bhangmeter, Temperature and Opacity measurements); Reaction History; Thermal Radiation and Spectroscopy. (Clark 1954: 4-5)

The DoD participated in the radiochemistry program by conducting cloud sampling. They also participated in the Thermal Radiation and Spectroscopy program with five projects: High Temperature Measurements, High Altitude Measurements, Time Interval Measurements, Spectroscopy, and Disturbed Air Element.
(Clark 1955: 38-42)

On TEAPOT, a large effort was devoted to the development of diagnostic techniques. This was not at all unusual. Perhaps half of the experiments were designed to test new diagnostic concepts. (Clark 1955: 37)

UCRL'S DIAGNOSTIC EXPERIMENTS

Livermore's 3 tests involved 5 diagnostics experimental programs: Radiochemistry; Reaction History (and transit time); Photo physics (including fireball and bhangmeter measurements); External Neutron Measurements (and neutron spectra), and Technical Photography. These five programs were similar to those conducted by LASL. (Clark 1954: 5-6)

ON-SITE RAD-SAFE

The on-site rad-safe team was under the leadership of Lt. Col. T. Collison and was staffed with rad-safe trained military personnel from the various services.(Clark 1955: 20)

DoD PARTICIPATION ON OPERATION TEAPOT

A total of about 11,000 DoD personnel participated in Operation TEAPOT. About 8,000 of these took part in Desert Rock VI. About 2,000, of the eleven thousand were administrative and support personnel while the remaining 1,000 participated in effects measurements and operational training projects. (Ponton et. al 1981b, 48)

Desert Rock VI consisted of three programs: 1) Observers, 2) Troop Maneuvers, and 3) Technical Service. The Observers and Troop Maneuvers were fairly similar to those on past operations. However, new projects for Desert Rock Exercises that were of a more technical nature were conducted in the Technical Service programs. Desert Rock VI was a fairly extensive effort with many facets and is described in Appendix M.

DoD NWE PROGRAMS AND PROJECTS

Approximately 1,500 military and civilian participants conducted the DoD NWE projects during TEAPOT. The major project activities were on 3 of the tests: ESS, HA, and MET.

There were 7 programs with 57 projects conducted by the DoD during TEAPOT.

OPERATION	PROGRAM	# PROJECTS
TEAPOT 14 Tests	1) Blast Pressure Measurements	13
	2) Nuclear Radiation Effects	11
	3) Effects on Equipment and Structures	10
	5) Aircraft Structures	5
	6) Electromagnetic Effects and Tests of Service Equipment	7
	8) Thermal Radiation Effects	8
	9) Supporting Measurements	3
	Total	**57**

Some of the major NWE efforts conducted during TEAPOT are described in the following.

Drone Aircraft For Investigation Of Lethal Effects

From the DoD perspective, "The most important and critical single project of Operation Teapot was Project 5.1, which utilized three drone aircraft to investigate the lethal effects of blast on aircraft structure in flight". This project, conducted by Wright Air Development Center (WADC) was part of the intensive study then underway regarding the use of atomic warheads for continental air-defense. It was necessary to determine the nuclear yield necessary to destroy an enemy aircraft or missile. In the planning for this project, it was decided to use the F-80 drone because "it was the only available proven jet drone." (Purkey 1958a:12) MIT conducted an analysis to determine the conditions for destruction and near destruction of the F-80. A single-peak, ideal shock wave was desired without reflections or other spurious signals that would lead to complex effects on loading and resonance. A single peak shock would also be representative of an antiaircraft detonation at altitude.

A surface burst could eliminate reflections, but the yield would need to be about 10 kt in order to provide a lethal range large enough to accommodate the accuracy of drone positioning. The AEC would not approve a surface burst of this size because of the fallout issues such as discussed earlier in preparation for JANGLE. "Operational problems* eliminated the possibility of using a high air burst for the conduct of this project." [*Footnote: The problems are not stated, but were possibly associated with the accuracy of both the detonation point and the drone positioning.] The remaining possibility was for a relatively low tower shot, if the expected reflected wave could be eliminated or reduced. (DASA 1960a: 83)

Results from UPSHOT-KNOTHOLE ... "indicated that the reflected wave might be greatly accelerated during its passage back through the region heated by the fireball." The acceleration could have the effect of causing the reflected wave to merge* with the initial outgoing wave in the region directly over the burst resulting in a single shock. The height of the tower would be the lowest permissible for the yield in consideration of off-site-fallout restrictions. The DoD wanted a yield of at least 20 kt, in order to meet the operational limitations imposed by position errors. A relatively simple developmental nuclear device having a reasonably reliable predicted yield on a 400' tower was ultimately chosen for this shot, MET. (ibid., 83) [*Footnote: On a tower shot, after the initial downward directed portion of the (essentially spherical) shock wave reflects from the ground surface, it travels upward. In its upward journey, it travels through a region of heated air. Shock waves travel faster in heated air than in cool air. Therefore, a reflected shock can "catch up to" the initial upward directed portion of the shock wave.]

Two projects that measured free-air pressures were instrumental in validating the predictions made for MET regarding where the reflected and direct shocks merge into a single shock. The first, Project 1.1, involved instrumented canister drops over the shot point to measure free-air overpressure as a function of time and distance directly over the shot point on both Turk and Apple 1.(Haskell 1955: 3) The second, Project 1.2, used a "rocket-smoke-grid" and "direct shock" photography. Sixteen smoke rockets that formed a grid* directly above the burst were used on Turk and 20 on Apple 1 and MET. [*Footnote: A grid of smoke lines was produced by the rockets fired at 60° to 70° on two sides of GZ.]

Only Turk data were utilized since the yield of Apple 1 was lower than anticipated. Although the canisters were not ideally positioned, some data were obtained, "which suggested that, even if the reflected shock did not truly merge with the incident shock directly above the burst point, it would be of sufficiently low amplitude that the drone project could be conducted satisfactorily on" ... MET.
(Moulton 1955: 15-6, 19-20, 23-7)

Purkey (1958a: 42) makes an interesting point about the importance of this project. Since the conditions of MET "were tailored to the requirements of Project 5.1, the authority to recommend delay or postponement of the shot was given to the Project Officer." He came very close to using this seldom granted authority.

Four QF-80 aircraft were modified as drones and extensively instrumented. Their planned positions at detonation time were based upon damage criteria: drone number one should encounter light damage, but remain capable of continued flight; the second drone should see severe damage with possible failure; drone three should be so severely damaged that it would not remain capable of normal flight; and the fourth drone would be a spare.

Each drone had two director aircraft. One director maintained control while the second served as a spare. Each drone had its own flight pattern, and when the drones approached GZ, control was transferred to the CP. The director aircraft regained control after the drone passed GZ. Each drone was assigned two chase aircraft that would shoot it down, if necessary. In addition, the chase aircraft provided photographic coverage of the drones. A dry run of the telemetry system and instrumentation was successfully conducted on Bee by drone number three.

Purkey (1958a: 45) described what happened on MET: "Drone 1 made a successful takeoff but went out of control shortly thereafter and crashed." Drones 2 and 3 took off on schedule at 2 minute intervals. By this time, the spare drone was readied and started its takeoff run. "Just before the flying speed was attained the director flamed out on the runway." "Drone control was transferred to the director for the original Drone 1 -". The spare drone "became airborne successfully, but not before it had veered off the runway, jumped a ditch, and traveled about 300 feet across the desert". After the exciting takeoff, "all drones were under complete control and entered the flight pattern satisfactorily."

Free-air-shock photography showed "a relatively clear reflected shock merging with the incident shock approximately 2,600 feet above burst zero". "The coalescence of the incident and reflected waves extended over a 1,000' radius about the vertical through air zero." At this time, the lowest drone was at 3,800' above burst zero; SUCCESS!

Three additional but less programmatically important, as well as less operationally "exciting," Aircraft Structures projects were conducted on TEAPOT.

Investigations Within the Fireball
Intercontinental ballistic missiles were nearing reality by 1955, and some consideration was given to potential defense mechanisms. One project by WADC was designed to provide preliminary information regarding the thermal lethality of a nuclear explosion on an incoming missile. It had 2 parts: 1) a lethality study (exposures inside the fireball) to determine the thermal lethality of a nuclear fireball on basic missile structures such as spheres and cylinders; and 2) a thermal-shock study (exposures beyond the fire ball) to determine the thermal shock resistance of various ceramic materials exposed to thermal flux intensities that were similar to those experienced by an Atlas missile on reentry. One of the most important results of this project was the demonstration that specimens exposed within the fireball could be so exposed and recovered. This important and extensive project is described in Killian 2011. Another even more extensive fireball exposure project was conducted on PLUMBBOB and is discussed in the Part II-Chapter 12.

MET - A *Tour de Force* On Non-Ideal Airblast
Results from shot Harry on UPSHOT-KNOTHOLE, which was a low burst conducted over a desert surface, showed that in the precursor portion of the airblast:
- Overpressures were substantially below those predicted by a simple theory (so-called Ideal theory), but damage to drag-sensitive targets was much greater than expected for the measured values of overpressure.
- Dynamic pressures were equal to or greater than ideal and much greater than would be calculated from the measured overpressure.
- Dust behind the shock front was extremely pronounced. (DASA 1960a: 25)

A primary emphasis of TEAPOT was to make measurements in those regions where the relationship between overpressure and dynamic pressure remained questionable.

Consequently, the DoD decided to measure various phenomena over different surfaces that would be more or less prone to develop a precursor. Three different surfaces were selected:
Desert Soil - a dusty- precursor-forming surface
Asphalt - a non-dusty-precursor-forming surface, and
Water - a non-dusty-non-precursor-forming surface (i.e. close to an "ideal" surface) (DASA 1960a: 25)

Based on the UPSHOT-KNOTHOLE Annie and Grable shots, the most significant precursor region for TEAPOT MET was estimated to extend to about 3,000'. Three blast lines were fielded on MET; one for each surface, are shown in Figure 2-10.3b. A circular asphalt surface from a prior operation is also shown as Figure 1-10.3a. This area was used for Bee which served as a preliminary trial for much of MET's instrumentation. The water surface had a depth of only 3 inches at shot time. This was due largely to the scarcity of water at the NTS which was compounded by the shot delays prior to MET with more people at the site for longer times than anticipated. (DASA 1960a: 26-8, 31)

MEASUREMENTS ON MET'S 3 SURFACES

NEL made measurements of the velocity of sound from +20 msec until shock arrival at heights of 1.5, 3, and 6 feet. Gages were placed at the 2000-foot range over water, at the 1000 and 2000-foot ranges over asphalt, and at the 1000 and 2000-foot ranges over desert soil (project 1.5). (McLoughlin 1955: 3, 9-11, 23)

Following a trial run on Bee, SRI measured static overpressure and dynamic pressure versus time at different distances on MET. Ten instrumentation stations spanned the length of the water and asphalt surfaces while 17 stations were fielded on the desert line out to 4,500 feet. Gages were located at ground level, and on towers at 3, 10, and 40 feet for this project,1.10. (Saachs 1957: 5, 17-20)

Sandia measured the particle velocity, air density, pressure, and yaw* vs time three feet above the surface. [*Yaw and pitch are respectively, the horizontal and vertical angles of the shock afterflow.] Similar measurements were made on Turk at one location and on MET at distances of 2,000 and 2,500 feet along each of the 3 blast lines. In addition, extensive pitch measurements were made at several heights up to 40 feet, project 1.11. (Bannister 1958: 5, 11, 14-22, 26-8)

The NOL fielded three-component force gages for the measurement of transient aerodynamic drag loads on 3 and 10 inch diameter spheres over the 3 surfaces, project 1.12. (Kornhauser 1958: 30-4) They also determined the atmospheric density by measuring the attenuation of an electron beam exposed to air and dust in project 1.13.(Gordon 1957: 12)

BRL Project 1.14b, designed to measure the pressure as a function of time, was conducted on 12 events. One conclusion was that ... "Precursor formation over asphalt was more pronounced than over the desert or water surfaces."(Bryant 1955: Abstract, Ch. 2)

NRDL fielded projects focused on the thermal environment. Project 8.4b involved measurements of the thermal radiation over the three surfaces and a number of other materials. Interestingly, the results indicate that there are significant differences in thermal properties between tower and air bursts. "The air bursts have higher thermal yields, higher peak irradiances, higher peak temperatures, and different pulse shapes than tower bursts." As might be expected, the thermal properties of an air burst vary with altitude. "The higher the altitude, the shorter the time scale, the larger the fireball, and the lower the thermal energy. The peak temperature is little changed". (Hillendahl 1959: 5, 8-9)

In addition, air temperatures were measured, on project 8.4e, prior to airblast arrival at several heights up to 10 feet at the 2000' stations. Rather surprisingly, the maximum air temperatures above ambient were found to be more than 1,500°C. The results indicated a definite trend of lower temperature with an increase of height, and there were indications that the air was not uniformly heated. (Inn 1957: 3, 14-5)

Other Measurements of Free-Air Pressure

Two other airblast projects were conducted on HA. Project 1.9 used smoke puffs as a means of observing particle position as a function of time at several locations on HA. A parachute-retarded canister produced the puffs by simultaneously firing twelve M-15 grenades. While the canister functioned well, the drop was inaccurate. (Reed 1956: Abstract, 7, 9, 13) The second project, 1.3, used 3 microbarographs and 3 milli-barographs by Sandia and 10 BRL Very Low Pressure gages to measure P-max (maximum pressure) of the free air and the wave reflected from the ground on HA. (Reed 1955: 11, 15-21)

The ESS Crater

John Lewis writes:

> Sponsored by the Office, Chief of Engineers, OCE, I had previously prepared a study showing the merits of the ADM (Atomic Demolition Munitions). By 1954 ADM development was in progress and our studies showed that rapid deployment of the ADM as a buried burst could be of considerable advantage to U.S. forces in western Europe if they were about to be overrun by Warsaw Pact ground forces.
>
> The Corps of Engineers was studying rapid burial of ADMs using truck-mounted augers to drill emplacement holes, and had a prototype auger under development that could auger a 1 foot-diameter hole in soil (not rock) to about 90 feet. I suggested that the ESS shot of TEAPOT be modified to 1.2 KT (this would allow direct comparison with JANGLE "S" and "U" events of the same yield) and be positioned in a 1-foot-diameter hole with a 67-foot depth-of-burial.

The emplacement of the ESS device was prepared by a Corps of Engineers combat battalion using engineer field equipment. They operated from their own encampment at Camp Desert Rock just outside the entrance to the NTS at Mercury, Nevada. (Lewis 1997:25)

SRI detonated 6 spherical 265 lb TNT charges at the JANGLE Uncle site prior to TEAPOT. Four of the detonations were below ground and served as trial runs of instrumentation and emplacement techniques as well as providing data for crater size. (Lewis 1958: 15-6)

The primary objective of ESS was to determine the effect of charge depth on the crater characteristics. In addition, the DoD was interested in the relationship between nuclear and TNT explosions for cratering efficiency. Both objectives required accurate measurements of the crater. (DASA 1960a: 152)

Project 1.7 had a variety of techniques for making ground motion measurements. There were:

- 28 ground motion gages emplaced between 200 and 600 feet from GZ. Most were at a depth of 10 feet.
- 4 surface airblast pressure gages;
- 2 horizontal stress gages;
- 46 dynamic displacement measurements at 300 feet from GZ; and
- 40 monuments to obtain total displacement at ranges from 180 to 500 feet.(Sachs: 1958: 13-14, 40-4)

The crater formed by the ESS detonation on 3/23/55 was measured by project 1.6 as were the pre- and post-shot locations of sand columns*.(Schuster 2001: 29-30) [*Footnote: Before the detonation, slanted columns of colored sand were placed in the ground along a line running through surface zero. The pre-shot location of these columns was carefully surveyed. In October 1955, residual contamination from the detonation had decreased to an acceptable level, and excavation of the ESS crater began. The columns of colored sand (which were usually located beyond the region where material was ejected to produce the crater) were uncovered, and their positions were surveyed and compared to their pre-shot positions. The displacements that occurred along the columns were thus determined. (Ponton: 86) Characteristics of the crater (such as lip height and slope of the crater walls at different azimuths) could be related to the final displacements. This sand column technique was to be used on many subsequent cratering events.] Figure 2-10.4 shows the features of a crater along with descriptions of these features. The ESS crater had the following average true crater dimensions: Volume 2.62×10^6 ft^3; Diameter 284.8 ft; and Depth 89.6 ft. (Schuester 2001:29-30).

FCDA PARTICIPATION ON TEAPOT

Industrial collaboration with the FCDA on nuclear tests began during UPSHOT-KNOTHOLE. If industry needed information not otherwise available and if there were a civil defense value to be derived from it, FCDA considered sponsoring the proposal; but, industry paid its own way. (Reeves 1980:256) In planning for TEAPOT, requests for participation were made by industry to FCDA and by FCDA to industry. During TEAPOT, over 150 industry associations, institutes, and companies participated in the FCDA field exercises and technical test projects. The participation ranged from sponsorship by the industry of the full cost of projects, including construction and provision of project personnel, to the supplying of equipment and materials. Over 100 industry personnel participated in FCDA projects at the site during TEAPOT, most of them during Apple 2. (Petersen et.al. 1955: 23-26)

FCDA's participation on TEAPOT again consisted of three parts: <u>(1) Open Shot</u>; <u>(2) Field Exercises;</u> and <u>(3) Technical Projects</u>. (Reeves 1955:101) Although technical projects were conducted by the FCDA on most of the TEAPOT tests, the extensive combined activities of Open Shot, Field Exercises, and Technical Projects that were conducted by the FCDA during Apple 2 were known as Operation CUE*. The technical projects on CUE were much more extensive than those on DOORSTEP. Although PLUMBBOB would have more FCDA projects, the TEAPOT projects spanned a much wider spectrum of civilian life and were of considerable human interest. Therefore a significant number of CUE projects are discussed in this section. [*Footnote: In the Observers Handbook for the open shot, Val Peterson, Administrator of the FCDA, stated that --- "the time has come for a renewed effort, for a restudy of local civil defense needs in terms of new information, and for a greater effort to show the people of America how they can best prepare for individual and family protection. This test program can be our "Cue" for such a renewed effort". (Peterson 1955: 3)]

Open Shot

The Open Shot and Operation CUE were originally scheduled for Zucchini, the last test of TEAPOT. However, because of the changes in scheduling of devices, it was rescheduled for Apple 2. The Joint Office of Test Information - Open Shot, opened to the public at the Las Vegas High School auditorium on April 15th for registration by general observers and media participants. Thirteen hundred persons registered. The program began on Saturday April 23rd with briefings by members of the NTSO and senior representatives of the participating agencies. On Sunday the 24th, a pre-shot tour of the FCDA test area was held as well as another orientation briefing and box lunches. On the 25th, briefings were again held in Las Vegas High School. VIPs arrived and were briefed as was a group from the media. (Reeves 1955: 104-105)

The morning briefing on the 25th indicated that the weather would be unacceptable for firing on the 26th, and the shot was postponed for 24 hours. This was the first of an eight-day postponement. At 9:00 PM, a Test Organization news bulletin was posted in selected hotels and motels in Las Vegas concerning the probability of the next day's shot and the departure time.

Three of the postponements took place at the early morning weather briefing just prior to shot time while the observers and media representatives were already waiting at the NTS Observer Area. The observers were fed at the Observer Area by Civil Defense workers participating in the Field Exercises. Apple 2 was eventually fired at 05:10 PDT on the morning of May 5. Five hundred hearty observers and media representatives were still on hand for the "Cinco de Mayo" event. Some who could not stay for the shot returned to the site to join the D+1 tour of the FCDA test area. (Reeves 1955: 93-97, 105)

On May 6, the observers departed Las Vegas for a third tour. They: arrived at the Firing Area; participated in inspection of the post-shot conditions and were briefed; received preliminary report material; were fed in the Firing Area; returned to Las Vegas; and participated in a question and answer session.
(AEC Secretary 1954a: 20-22)

Field Training Exercises

The FCDA had the perspective that:

> Persons who have engaged in atomic test operations know that familiarity with nuclear devices breeds assurance as well as considerable respect. Administration believes it essential to provide a nucleus of key civil defense personnel to the States and cities who have gained the assurance that can only be developed by experience with an actual atomic explosion. The civil defense field exercise is a method by which a start can be made on the important job of attaining this objective.

(Peterson et. al. 1955: 68)

The Field Exercises are well described by the following press release issued April 22, 1955 by the Joint Office of Test Information, High School Auditorium, Las Vegas, Nevada. This press release, which exhibits the tone of media reporting in the 1950s, does not take into account the 8 day delay in Operation CUE and how the Field Exercise personnel spent those days. However, anyone who has experienced a test delay or even a snowed-in airport can readily imagine how those days were spent.

> 350 CD VOLUNTEERS
> READY FOR EXERCISE
>
> --- More than 350 Civil Defense volunteers from 42 states and Hawaii today have moved into the Atomic Energy Commission's Nevada Test Site here to participate in the "Operation Cue" CD field exercise planned in conjunction with the pre-dawn explosion of an atomic device on Tuesday (April 26).
>
> "Pick and shovel" workers in the Nation's CD force, the field exercise participants have been drawn from the great cities and small hamlets of the

country. In many cases, they are highly skilled volunteers who have trained for disaster duty for the past four years. ---

In sharp contrast to the glitter and comfort of Las Vegas 65 miles from here, male field exercise participants are quartered in simple hutments and the 55 women volunteers in a communal dormitory in this rocky, barren camp.

The CD workers began Friday a rigorous five-day schedule of events with a general briefing by Maj. Gen. Clyde L. Dougherty, Civil Defense director of Detroit Mich., who arrived here two weeks ago to lead the field exercise.

On Saturday, the CD group took a half-day tour of the firing area and made an inspection of the houses, utility and industrial installations and other equipment to be subjected to the atomic blast. They visited "Position Able", the position of the wind-swept desert from which they will observe Tuesday's explosion* -------

Nine CD services - sanitation, engineering, police, rescue, mass feeding, communications, warden, fire and health - will participate in the exercise.

On Shot day, plans call for field exercise participants to move into the test area at 2:00 AM.

After the shot, General Dougherty will lead a small reconnaissance team of CD participants into the front area as soon as AEC-directed radiological monitors say it is safe. For other exercise participants, there will be breakfast served in the area where some 1,000 official observers - including many high government officials - will watch the detonation.

The most important phase of the field exercise, however, will come on Shot-plus-one, when all CD services will be in action amid the rubble of the test line. Police will guard the installations, rescue teams will remove radiologically safe mannequins from the blasted structures and the mass feeding teams will prepare luncheon from more than 1,500 persons in one of the largest demonstrations of its kind to be staged in the nation.

Communication facilities manned by CD personnel will be in operation throughout the firing area. Medical teams will provide first aid, and sanitation teams will perform the same services they would in a disaster. (JOTI-55-T-15)

[*Footnote: During the detonation, participants were actually divided among three positions:

1. The Observer Area: Here were members of the Mass Feeding Group (who would serve coffee), a team from the Communications Group, and two teams from the Police Group.

2. Position Able: The majority of participants were here, 37,000 feet (~ 7.0 miles) from GZ. After the shot, they withdrew to the Observer Area. The Mass Feeding Group set up to prepare breakfast, and other groups performed functions as assigned.
3. Position Baker: This was a trench with a sandbag parapet with 21 participants, of whom 6 were women, and nine media representatives. It was located 10,500 feet (~ 2 miles) from GZ. It had a telephone and two radio links for communication with other elements of the Field Exercise group and the Test Manager. The participants' jeeps were parked to the rear of the trench position which were used after the shot, when this group also withdrew to the Observer Area. (Reeves 1955: 101)]

Two additional training exercises were conducted during TEAPOT: On April 13, 1955, a group of 24 nuclear technology professionals with clearances arrived at NTS for a 3 week exercise that extended into 4 weeks. This Project 38.2 exercise was more detailed and "hands-on" than the UPSHOT-KNOTHOLE 22.1 exercise had been. (Goeke 1955)

For Project 38.5, on April 21 forty-nine people with state and local radiological defense responsibilities departed CA in a "controlled convoy", arrived at Overton NV (about 55 miles NE of Las Vegas) where they checked into accommodations. The 49 men formed into 3 groups: Monitor, Control-center, and Mobile-laboratory. The next days, they spent getting ready for CUE: conducting monitoring and communications exercises, monitoring a contaminated area south of Alamo (about 85 miles north of Las Vegas), and collecting and analyzing soil, plant, and air samples from contaminated areas. On May 2, they returned to their home bases. (Tolan 1957)

FCDA NWE PROGRAMS AND PROJECTS
The FCDA's major technical effort was on the Operation CUE Apple 2 event, where all but 5 of its 47 projects were conducted.

OPERATION	PROGRAM	# PROJECTS
TEAPOT 14 Tests	30) Evaluation and Documentation of Radiological Contamination	3
	31) Response of Residential, Commercial, and Industrial Buildings, and Materials to Nuclear Effects	5
	32) Exposure of Foods and Food Stuffs to Nuclear Explosions	6
	33) Biological and Medical Investigations	3
	34) Shelters for Civil Populations	4
	35) Utilities, Services, and Associated Equipment	5
	36) Mobile Housing and Emergency Vehicles	2
	37) Fallout Studies	4

	38) Civil Defense Radiological Effects Studies	5
	39) Program Instrumentation and Photography	10
	Total	**47**

It is hoped that the hardy reader or casual browser will obtain a glimpse of: the magnitude of the CUE undertaking, the significant construction work; and the diversity of ideas and methods that comprised the FCDA field work. Perhaps the reader can also imagine the bustle of activities that occurred at the site in the final days of preparation for CUE, see Figure 2-10.5.

Houses

Project 31.1 consisted of five pairs of houses, ten in all. These houses became a focus of public attention, as well as a venue for other projects.

TYPE OF PAIR OF HOUSES	DISTANCES FROM GZ (feet)	ESTIMATED OVER-PRESSURES (psi)
TWO-STORY BRICK & CINDER BLOCK HOUSES Two-story, basement, center-hall wall-bearing with 8" masonry walls. 2 basement shelters each. Similar to the frame houses on Annie in 1953.	4,700 10,500	5 1.7
ONE-STORY FRAME RAMBLER HOUSES Wood frame rambler type, built on a poured-in-place concrete slab at grade. Bathroom was designed as an aboveground shelter with 8" thick reinforced concrete walls and ceiling.	4,700 10,500	5 1.7
ONE-STORY PRECAST CONCRETE HOUSES Made of 6" thick pre-cast lightweight expanded shale-aggregate reinforced concrete wall and partition panels.	4,700 10,500	5 1.7
ONE-STORY CONCRETE BLOCK HOUSES Built of reinforced lightweight aggregate concrete blocks.	4,700 10,500	5 1.7
REDESIGNED TWO-STORY WOOD FRAME HOUSES Similar in size and layout to the houses tested in 1953. Based on 1953 findings they were redesigned to strengthen the structure so far as possible within an increase of approximately 10 percent in building cost. Three types of basement shelters each.	5,500 7,800	4 2.5

(Randall 1961:13-17)

For comparison, housing in areas of the country subject to hurricane winds of up to 120 miles/hour are designed to resist overpressures of approximately 0.25 psi.

Diagnostics were conducted by project 39.2 which consisted of visual inspection, and extensive still and motion picture photography. Also, about 1500 dosimeters were placed in the houses and basements. Pressure and total thermal energy were dynamically measured. (Rollosson 1955: 3-4, 11-3)

Shelters

Projects 34.1 and 34.3 consisted of seven types of shelters that were constructed and tested on Apple 2, and 2 of these types were tested on Apple 1.

SHELTER TYPE	SHOT	DISTANCE FROM GZ (ft)	INSTRUMENTATION
BASEMENT LEAN-TO			
Brick house	Apple 2	4,700	1 Pressure
Brick house	"	10,500	
Frame house	"	5,500	
Frame house	"	7,800	
BASEMENT CORNER ROOM			
Brick house	Apple 2	4,700	
Brick house	"	10,500	
Frame house	"	5,500	1 Pressure
Frame house	"	7,800	
BASEMENT REIN-FORCED CONCRETE			
Frame house	Apple 2	5,500	1 Pressure
Frame house	"	7,800	
BATHROOM REIN-FORCED CONCRETE			
Rambler house	Apple 2	4,700	1 Pressure
Rambler house	"	10,500	
UTILITY			
Masonry	Apple 2	2,250	
"	"	2,750	
"	"	3,750	
Reinforced concrete-	"	2,250	1 Pressure
(poured in place)	"	2,750	1 Pressure
"	"	3,750	1 Pressure
Reinforced concrete-	"	2,250	
(precast)	"	2,750	
"	"	3,750	
BASEMENT EXIT			
Closed	Apple 1	1,360	
Partly Open	"	1,350	
Open	"	1,350	
Closed	Apple 2	1,270	
Open	"	1,270	
Closed	"	1,470	

Open	"	1,470	
GROUP Structural Biomedical Structural Biomedical	 Apple 1 " Apple 2 "	 1,050 1,050 1,050 1,050	 A B C B
BLAST LINE See TEAPOT Project 39.2	 Apple 1 " Apple 2 " " " " " " " "	 1,050 1,350 1,050 1,270 1,470 2,250 2,750 3,750 4,700 10,500 15,000	 1 Pressure 1 Pressure 1 Pressure 1 Pressure 1 Pressure 1 Pressure 1 Pressure 1 Pressure 3 Pressure 3 Pressure 1 Pressure

A = 3 pressure & 1 noise
B = 12 pressure, 1 noise, 2 temperature, & 1 dynamic pressure
C = 3 pressure, 1 noise, & 1 acceleration

The first 4 of the shelter types given in the above table were placed inside one of the houses described in Project 31.1 above. The 3 different Utility shelters were made of three different materials and tested at 3 locations. They were aboveground shelters that could also be used as storage shelters. They might be connected to a house through the basement walls.

None of the Basement Exit type of shelters were attached to a house basement, but they were constructed underground at basement level. The doorway that would have led from a basement to the Basement Exit shelter was constructed as a concrete wall. The Basement Exit shelters were all constructed the same, but they were tested with their exit doors in different configurations: open, closed, partly open.

The 4 Group shelters were designed to accommodate 50 persons, at about 5.75 ft^2 per person (less than 2'x3' per person!). They were constructed the same except for an interior partition. One of each pair of the Group shelters was modified by a reinforced-concrete partition dividing the shelter into two chambers, each 12'x12'x8' high. (Vortman 1956a: 13-29)

Warehouses and Chemical Plant Control Room
Two each of three metal industrial building types, suitable as small warehouses or shops, were erected for Project 31.2. Standard construction was used; they were not designed to resist atomic blast. Each type of building was placed at 6,800' and at 15,000'.

ARMCO buildings - frameless steel buildings, 24'x 36'

Behlen buildings - frameless steel buildings, 28' x 32'.
Butler buildings - rigid steel-frame, 24' x 48'.

Also a part of 31.2 was a chemical plant control room building. It was constructed of gypsum with a steel-reinforcing web was designed to protect delicate chemical and electronic controls in processing plants. Union Carbide and Carbon Corp. tested this structure at 5,500'. Gamma radiation film badges were placed in the buildings. Postshot displacements were determined. (Peterson et. al. 1955: 31-2)

Static vs. Dynamic Loads
Design criteria for buildings were conventionally determined by doing calculations and tests with static loading. An issue which would plague the effects community for many years is: If a structure was designed to withstand certain static loads, what dynamic loads can it withstand? Several common types of structural beams, designed with different load bearing capacities, were exposed to the dynamic loading by the blast wave in Project 31.4. (Clark 1955: 9-10; Peterson 1955: 33-4)

Non-Ideal Shock on Structure
Before TEAPOT, "no successful experimental evaluation had been made of the effect of a non-ideal shock front on the blast loading of a structure". To do this, a structure 6'x 36'x 6'high was constructed for Project 34.2 and subjected to the blast wave from Turk at a distance of 1850' from GZ. "In addition to the 42 gauges on the structure, there were 6 gauges ----to measure the pressure wave incident upon the structure." (Vortman 1956b: 3-4, 11)

Mobile Homes and Emergency Vehicles
Seventeen different companies provided mobile homes for Projects 36.1 and 36.2: nine were located at 10,500', seven at 15,000', and one at 16,000' on CUE. A variety of designs and sizes were tested that ranged from: 17', no separate bedroom at 2700 lbs to 42' 10" with 2 bedrooms at 8600 lbs. They were oriented in various directions with respect to GZ; and at the time of detonation, they had variations of windows open and closed and doors open and closed.

Eleven emergency vehicles were also exposed: 1 at 1470', 2 at 4700', 2 at 10,500', and 6 at 15,000'. They were oriented either facing or broadside to GZ and were fully equipped for their missions:

American Gas Association
- Gas service truck
- Heavy-duty gas repair truck

American Telephone & Telegraph. Co.
- Two installers' service trucks

Fire Apparatus Manufacturers Assoc.
- Aerial-ladder truck
- Rescue Service truck

Edison Electric Institute
- Earth boring machine
- Heavy-duty line truck
- Light-duty service truck

Seagrave-Hirsch, Inc.
- Fire department pumper

Willys Motor Co.
- Jeep fire engine

The main diagnostics of these 2 projects were personal inspection and photography. (Shaw 1957: 25, 31, 33-8, 42, 46, 52-3) Civil defense officials had become "accustomed to thinking of damage in terms of four zones:
A damage: Building almost completely destroyed.
B damage: most buildings damaged beyond repair.
C damage moderately damaged buildings that must be vacated during repairs.
D damage: partially damaged buildings that need not be vacated during repairs."
Post shot, the mobile homes and emergency vehicles were evaluated in terms of these damage zones. (Shaw 1957:9)

Foods
Six categories of foods were selected for exposure in the 6 extensive projects, 32.1, 32.2, 32,2a, 32.3, 32.4, and 32.5, that addressed the effects of nuclear irradiations, blast and thermal, and fallout on food and its packaging. The selection of foods was based on the largest volume and most frequent use in the American diet. The food categories were: bulk staples, heat processed food in cans and glass, commercially packaged beverages, meat and meat products, semi-perishable foods and food packaging, and frozen foods. Over 100 different foods with a gross weight of about 15 tons were exposed during CUE. There were 908 cases of canned products and 3800 individual uncased units. (Laug 1956: 3, 9, 10-1; Peterson 1955: 40-1) Also, 3 projects in Program 38 addressed the effect6s of radioactive fallout on food stuffs and packaging.

Materials Found Near and In Homes
Thermal ignition and the response of materials found near in homes were investigated in Project 31.5 by extensive exposure tests of six large classes of objects. These classes ranged from wood products (like stakes and poles) to windows and window coverings, fabrics, plastics, Oxyacetylene units and flammable-liquid drums. (Laughlin 1957: 5, 6; Peterson 1955: 35-8)

Heavy-Duty Machine Tools
In 1954, the AEC was seeking recommendations for stockpiling "special materials, special processing equipment, ---and special machine tools --- which may be required for reconstruction or replacement of facilities damaged or lost through enemy action". Representative examples of large and weighty heavy-duty machine tools were exposed on Project 34.4: mills, lathes, pressure vessels, steam ovens, and the weightiest, a hydraulic press of 49,000 lbs. They were exposed on foundations in and near the brick house and in some of the Project 31.2 buildings. Portable generators were used to provide pre-shot evaluations of the machinery, and similar tests were conducted post-shot. Pressure measurements that were made along the Blast Line were used. (Sparks 1956: 11-3, 20, 32)

Missiles
When a building is struck by a blast wave, missiles (e.g., pieces of building material) can be ejected as the building comes apart, or as objects within the building are accelerated. The ballistic properties of low-velocity missiles, which are produced inside various types of houses, were studied on Project 33.4. Also, some attention was given to missile production: 1) outside, but in the vicinity of houses and 2) in small home type shelters. Missile traps were used inside, outside, in front of, in back of and beside the structures. (Bowen 1956: 3, 13, 15-6, 27-8; Peterson 1955: 423)

Electrical Equipment
Electrical equipment, representative of that in urban residential and commercial areas, was tested by the Edison Electric Institute on Project 35.1. Duplicate electric-power installations consisting of transmission, substation, and distribution equipment were constructed at 4,700' and 10,500'. The transmission line (installed on steel towers) and the substation equipment were representative of equipment for large industrial plants. Power lines ran from poles to some houses, and parts of the systems operated during the detonation. "The damage was confined to the transmission and distribution circuits at the 4,700-ft area and --- the equipment could have been easily and quickly repaired. In the same area, typical homes were completely destroyed."(Wood 1965: 4, 9; Peterson 1955: 46-8)

Civilian Communications Equipment
CUE was the first test of civilian communication equipment, which is "generally designed for lower cost and less rigorous service" than is military equipment. Civilian equipment was placed in the 7 houses for Project 35.2. Also exposed were 150 foot AM (audio modulation) broadcast guyed antenna towers, the communication antenna towers, and the sirens (which are not electrical). "Within and near these houses, communications equipment products were placed in situations that approximated, as closely as possible, the placement of such equipment in commercial buildings and in homes." As possible, identical products were exposed at two distances, where severe and light damage were expected. The equipment "was generally more resistant to nuclear explosion damage than the structures in and near which the products were exposed". (Williamson 1955: 3-4, 14-5, 17-8, 45)

Liquefied Petroleum Gas (LPG)
In Project 35.4a, the LPG Association exposed three different types of LPG containers and systems that were typical of those found at home and at storage, industrial, or utility plants. These systems were: a bulk storage plant, 8 domestic installations, and eight 500 gallon capacity systems. (Tucker 1956: 3, 13, 16-27)

Natural Gas
In Project 35.4b, the American Gas Association exposed a variety of installations for piped natural gas including: underground installations (steel and cast iron piping, regulators, meters, gauge boxes, oil seals); piping from underground installations to connect 3 houses; and piping within the houses. The underground installations

were made “in accordance with current gas industry practice”. Two emergency vehicles, a heavy repair truck, and a customer service pickup truck, were located at 15,000’. The trucks were completely equipped with tools, supplies, and two-way radios. (Cornfield 1965: 12-5, 30, 36, 44)

Records, Materials, and Storage
Records materials and records storage equipment were exposed on Project 35.5 in the houses, behind a wall, and in the free field at locations between 500’ and 10,500’ on Apple 2. The storage equipment exposed was extensive and consisted of nearly everything one might have found or imagined in a well supplied office of the 1950s. Evaluation of damage was post shot inspection and photography. (Nat’l Records Mgt. Council 1956: 5, 14-6)

Biomedical Animal Exposures
To study the effects of blast pressure on animals in Project 33.1, 4 shelters were used in Apple 1 and 2 shelters in CUE. A total of 277 animals were exposed, including: 66 dogs, 52 rabbits, 52 guinea pigs, 63 rats, and 44 mice. An interesting but perhaps not surprising result was: “The geometry and design of the several structures markedly influenced the magnitude and character of the internal, compared with the external, pressure-time phenomena.”
(White 1956: 3, 15, 24, 43, 48)

Project 33.2 used rats in an interesting approach for studying the effect of noise from a detonation on the learning process.(Hirsch 1956: 11-2, 17, 21-2, 27-8)

PROJECT SUNSHINE AND FALLOUT STUDIES
Around July, 1953, radio strontium became the subject of special study, driven by AEC Commissioner Willard Frank Libby*, which came to be known as Project Sunshine.[*Footnote: Libby was a physical chemist whose career encompassed work at various universities, industry, the Manhattan Project, and government. He received the Nobel Prize in Chemistry in 1960 and was a key figure in leading research on fallout.] Perhaps due to his influence, the FCDA, the DoD and Los Alamos increased their studies and measurements of fallout.

FCDA Fallout Projects
On TEAPOT, UCLA’s School of Medicine became even more active in this arena with the 3 biological projects on TEAPOT. For Project 37.1, a study was undertaken of the factors that influence the biological fate and persistence of fallout in areas adjacent to the NTS. The factors included: plants, animals, characteristics of the fallout (e.g., particle size, type of material), and the fractionation of fallout material as it may vary with distance from GZ. (Lindberg 1959: 5, 15)

During Operation CASTLE, “it was found that skin burns on humans resulted from fall-out material that was deposited on the skin and that all or nearly all of the injury resulted from the beta components of the radiation.” For Project 37.2a, special film-packet holders were built in order to measure the probable beta radiation dose from fall-out that arrives at the germinal layer of human skin (which lies directly beneath the outer layer of skin). They consisted of filters, such as Mylar and aluminum that attenuated the fallout about the same as does the outer layer of human skin. These film packets were placed about 10 to 160 miles from GZ on Turk, Apple 1, Met, and Apple 2. (Dickey 1957: 3-4, 9-12)

Chemical dosimetry techniques were evaluated for Project 39.6. Experiments were conducted on neutron induced residual gamma radiation and on the feasibility of using 2 types of chemical dosimeter systems for the measurement of both fast neutron and gamma radiation. Approximately 20,000 dosimeters of the two types were used. (Taplin 1955: 3-4)

In Project 37.2, an extensive radiometric survey provided greatly increased detail, accuracy, and distance information of fallout patterns. Isodose rate and time-of-arrival contour maps were obtained for seven tower and four balloon shots. The predominant particle size of fallout was also determined on several arcs for each fallout pattern. (Baurmash 1958: 5-6, 18, 35, 82)

Fallout surveys and instrumentation development, which often were done together, were vigorously pursued by 9 projects. Three additional projects also provided photographic coverage for fallout projects. Telemetering techniques for fallout information was again tested, as were air zero locators*. [*Footnote: air zero locators attempted (i.e., they were never successfully developed) to determine the location of the detonation.]

Off-Site Fallout

After the 1 kt Wasp shot on February 18, 1955, the Arizona State Civil Defense Director reported to Test Manager Jim Reeves that people in Parker, Arizona, were upset about fallout detected by the Parker Chief of Police. People in Blythe, California, and Yuma, Arizona also reported fallout following the Wasp event. Wasp was an airdrop that was detonated at an altitude of 760 feet above the terrain. This was a very high altitude for such a low yield detonation and as a result the fallout was, in fact, negligible. Never the less the test organization immediately responded with a team dispatched to the communities to investigate and to explain to state officials in California and Arizona the precautions taken to reduce fallout to a minimum.

After the 14 kt Apple 1 tower shot on March 29 four residents in the Alamo, Nevada, area reported "uneasiness, burning and stinging sensations." The resident's symptoms apparently went away after a brief period, following discussions with representatives from the test organization. Fortunately these were the only two shots to generate much off-site fallout interest. (Reeves 1955: 43)

As was customary, the Civil Aeronautics Administration (CAA) had close liaison with the Joint Test Organization and diverted aircraft away from the vicinity of the test site near shot times and away from radioactive clouds that followed detonations. All test-related aircraft had code names (one was "Sawmill") to indicate to the CAA and NTS air traffic controllers that they were cleared to enter the test area.

An interesting fallout observation was made by FCDA scientists on TEAPOT. They noted that a large fraction of the fallout from the Moth event could be removed from a sample of earth material by a magnet. The explanation for this is that for shots on high towers, where the fireball did not touch the ground, the mass of the tower was a significant contributor to the fallout. In fact, the total fallout was forecast to be approximately proportional to the mass of metal consumed; and Moth was used to derive a first estimate of the proportionality factor. (Reeves 1955: 59-60)

The amount of fallout is proportional to the mass of solid material engulfed within the fireball. It was initially assumed that the fireball, which expands spherically in air, retained its spherical shape as it expanded in the ground. However, sound and shock waves (like the front of the fireball) travel faster in the ground than in the air. A new model, called the "squashed fireball theory", used this feature of ground shocks and predicted a greater volume in the ground engulfed by the fireball than was predicted by the simple air and ground spherical expansion. The contribution of the mass of the tower was added to the mass of the ground encompassed by the fireball to complete the improved fallout prediction. (Reeves 1955: 64-7)

PERSONNEL CONDITIONS AT THE NTS DURING TEAPOT

Construction

During UPSHOT-KNOTHOLE, it was realized that the facilities at Mercury should be expanded to provide more adequate living quarters and working space for the large numbers of test personnel who would be participating in future operations,. To provide these facilities, contracts were awarded for the construction of:

- Dormitories, including six for men and one for women
- Assembly Building
- Air Weather Building
- Addition to Buildings 120, 121, and 122 for office space
- Modification to steam generating plant
- Water well 5C in Frenchman Flat
- Additional water storage
- Connect warehouses no. 3 and 4
- Additions to the motor maintenance building

(Reeves 1955:117)

A wide variety of specialized test facilities and structures were also constructed for the experiments and measurements conducted during the TEAPOT tests. The 300 and 500 foot towers were acquired in two phases: fabrication and erection. The erection phase took between 75 and 95 days for each tower. There was also specialized construction for: quick closing blast doors, blast links and vacuum lines, tower reinforcing to withstand nearby adjacent blast, and reinforced concrete buildings projected above ground surface. Construction requirements for a full-scale operation normally presented problems associated with meeting tight schedules. The contractors were delayed to some extent due to unusually severe winter weather conditions for Nevada in December and January. Nevertheless, no postponements or delays were occasioned during TEAPOT due to construction. (Reeves 1955: 120-2)

Security
While the AEC was responsible for overall security at the NTS the DoD handled security matters for the military. The badge office issued 32,500 badges during Operation TEAPOT, not counting the open shot observers or the large numbers of Desert Rock troops who entered the NTS by convoys on shot mornings. (Reeves 1955: 125)

Safety
The Support Director, who was also the Manager of the Las Vegas Field Office, had the overall health and safety responsibility at the NTS. Each participating organization had the health and safety responsibilities for their own employees and was responsible for the actions of their employees. The operational standards were spelled out in AEC manual 550. (Reeves 1955: 126)

The safety statistics are as follows:

Man-hours	2,500,000
Reported Lost Time Injuries	3
Days Lost	42
Motor Vehicles Mileage	500,000*
Accidents	19*

(Reeves 1955: 126-7)
[*Footnote: This does not include DoD mileage cited below by the Field Command Support Unit as 847,000 miles.]

Telecommunications
The AEC was responsible for most of the telecommunications support at the site. The telephone service consisted of a 200-line automatic exchange at Mercury and an additional 200-line automatic exchange at the Control point. There were 20 long-distance circuits connecting Mercury with the Las Vegas exchange of the Southern Nevada Telephone Company.

There was a secure Teletype Comcenter in building 102 at Mercury staffed by the AEC.

There were 230 VHF-FM mobile units, 44 base stations, and 33 handie-talkie units on nine networks for on- and off-site radio communication. In addition there was a VHF-AM air-to-ground system for communications with air-born assets within about 200 miles of the site.

Off-Site Rad-Safe operations used a HF-AM radio network between Mercury and eleven base stations located at Glendale, Alamo, Caliente, Ely, Eureka, Lincoln

Mine, Tonopah, St. George, Cedar City and Beaver. In addition they had twenty mobile HF-AM units. (Reeves 1955: 128-31)

Air Transportation

Carco Air Service, based in Albuquerque, provided regular passenger and airfreight service between Las Vegas and Mercury and special flights to and from other locations. (ibid., 134)

The Nevada Wing of the Civil Air Patrol provided air support and communications assistance to the FCDA. Forty-two airmen participated with twenty-one airplanes and three helicopters. (ibid., 103)

Field Command Support Unit

The Field Command Support Unit (FCSU) was organized at Mercury by AFSWP with the mission to provide support functions in three categories to the military at the site:

1) Pay of personnel, issuance of travel orders, military discipline, and accounting for and control of DoD property and funds.

2) Control and maintenance of DoD vehicles, local procurement, provisions for recreational and medical services, and the accomplishment of Government documents necessary to move passengers and freight by commercial carriers.

3) Assistance in specific areas as requested by the AEC. This included provision of certain transportation and clothing for the Visitors' Bureau, conduct of religious services, and (perhaps of most importance to everyone at the site) furnishing of motion picture service.

The organization of the FCSU at Mercury had:
Logistics Division - provided for supply and procurement and transportation.
Support Division - provided for special services, reproduction, work orders, personnel, billeting and finances, and
Branches - provided the functions of Chaplains, Provost Marshal, medical and safety services.

A total of 134 people performed these FCSU functions during TEAPOT. FCSU maintained 146 generators and 334 vehicles that operated over 847,000 miles during TEAPOT. It also provided 975 medical treatments. (Reeves 1955: 113-4)

Mercury

REECo supplied the housing, feeding, and related personnel services. Mercury had twenty-one single-story dorms for men and three for women. Nineteen of the men's dorms could accommodate six people per room and two dorms could accommodate two people per room. Hutments provided additional men's housing. The women's dorms accommodated two per room. Trailers leased from the Bureau of Reclamation provided additional women's housing. The total capacity was just short of 2900. (Reeves 1955: 136,138)

Since the test organizations preferred to house their people together, the dormitory and hutment allocations were negotiated prior to the beginning of the test series. To aid in planning, forecasts were made of population versus date for personnel associated with the Test Manager's staff (i.e. the AEC) and the Test Director's staff (i.e. the labs). The peak camp population during TEAPOT for the Test Managers staff plus DoD Field Command Support Unit, FCDA*, and NTS contractors was about 1100. The Test Director's personnel, which also numbered around 1100, included the combination of: LASL Weapons Development, UCRL Weapons Development, Military Effects Group, Civil Effects Test Group. These figures do not include the Desert Rock population. (Reeves 1955: 136, 141-2) [*Footnote: This FCDA group was more involved with visitors such as civil defense groups from different cities who participated in exercises and with public information aspects than was the Civil Effects Test Group. The CETG was part of the Test Director's organization and conducted the CD effects projects.]

Mess Halls #1 and # 2 with a seating capacity of approximately 300 each were operated on a cafeteria basis prior to March 1, 1955. Meals were still served at a cost of one dollar. On March 1, 1955, Mess Hall # 2 was converted from a cafeteria-type service to ala carte and short order service. This innovation met with success and provided a different type of food and service for test participants at times other than the normal operating hours of the cafeteria.

A snack bar was operated at the Control Point, serving what they described as light food. On days preceding a shot, this facility remained open on a twenty-four hour basis. During the construction phase of the operation, a mobile food trailer with hot lunches for test area workers was operated in the forward area on an "as required basis" during extra shift hours. Box lunches were made available to people working in the forward area. Hot coffee, rolls and doughnuts were available on shot mornings at the observer areas.

REECo maintained limited facilities to ease the hardships of the primitive test site existence. These included: a barber shop; Western Union Telegraph; check cashing service; Laundromat; laundry and dry cleaning collection station; recreation hall; a public service gas station; official bus shuttle service between Mercury and Las Vegas; and a travel reservation office. Recreation hall furniture included ping-pong and pool tables. Outdoor recreational facilities included horseshoe courts, volleyball and basketball courts, and a softball diamond. (Reeves 1955: 136,139)

During off-hours the AEC briefing room, seating approximately 330, was used as a theater that showed motion pictures provided by AFSWP's FCSU.

A licensed physician supervised a first-aid dispensary, with a combined civilian and military medical staff. Ambulance service was available twenty-four hours a day.

Fire and police protection facilities were maintained through an operating contractor. There were two fire stations, one at Mercury and one at the Control Point area.

Eleven men were regularly employed and were augmented by eighteen volunteer firemen.

Police officers, with an office in Mercury, were duly authorized Nye County law enforcement officers. During operational periods, officers were on duty twenty-four hours a day. Speeding in the forward area was the greatest problem confronting the police. (Reeves 1955:139-40)

The Support Director had the responsibility for assigning and dispatching vehicles from the AEC motor pool. Responsibilities included: procurement of vehicles and equipment, repair and maintenance, and related functions pertaining to transportation requirements and vehicle operation. The various test organizations submitted their vehicle requirements prior to the test series. To the extent possible, government-owned vehicles were employed and augmented when necessary by lease from commercial sources. During construction activities, there was a particularly heavy demand. There were 837 motor vehicles of all types under the control of the AEC motor pool: 286 sedans; 333 pickups; 37 panel trucks; 17 power wagons; 60 jeeps, station wagons, and carryalls; 2 ambulances, 4 fire trucks, 21 buses, and 75 other vehicles for a variety of special needs. During the operation, these vehicles traveled over ½ million miles. (Reeves 1955: 134-5)

Carco Air Service, the Albuquerque-based charter company that flew between Albuquerque and Los Alamos, based two single-engine aircraft at the Municipal Airport in Las Vegas. These furnished air transportation to Mercury and to other points as special missions or requirements developed during the operation. A regular schedule was also established for flights between Las Vegas and Mercury. This service was utilized for the transportation of critical freight, passengers on official business, and instrumentation and collection of data at remote stations. Several flights were also conducted for occasional investigations and terrain surveys and for special travel required by the Test Organization.
(Reeves 1955: 134)

The thirteen major maintenance, operation and construction contracts in force during TEAPOT included:
- REECo, maintenance of utilities and other base camp operations, minor construction, and field support of scientific participants.
- Silas Mason, architect and engineering services.
- Olympic Commissary, feeding, housing, and related personnel services.
- Federal Services, Inc., security and other guard services.
- Pacific Telephone and Telegraph Co., communication facilities and service as required.
- Three contracts are for construction at Mercury of dormitories, modification to steam plant and warehouses.
(USAEC-DoD 1955: 1-2)

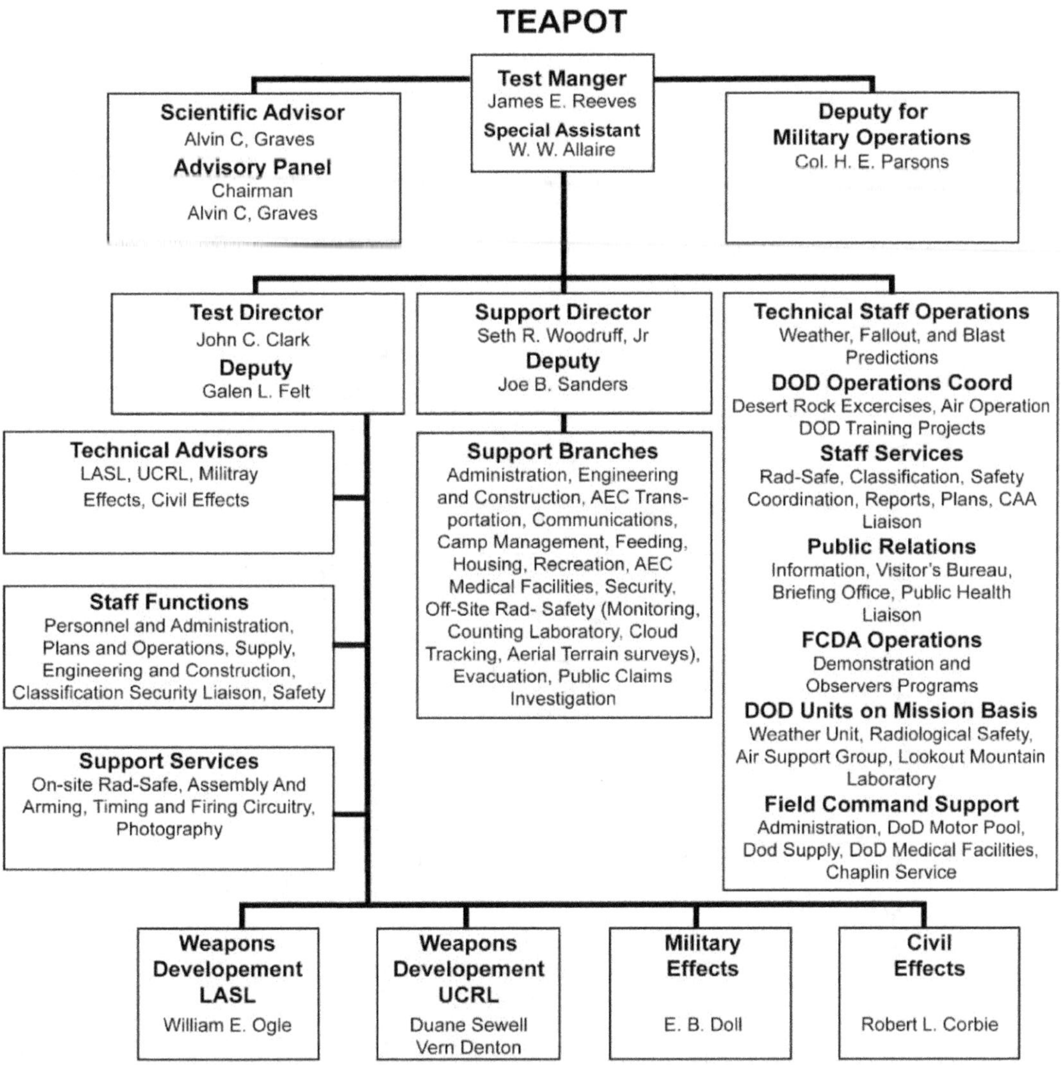

Figure 2-10.1 NTSO for Operation TEAPOT.

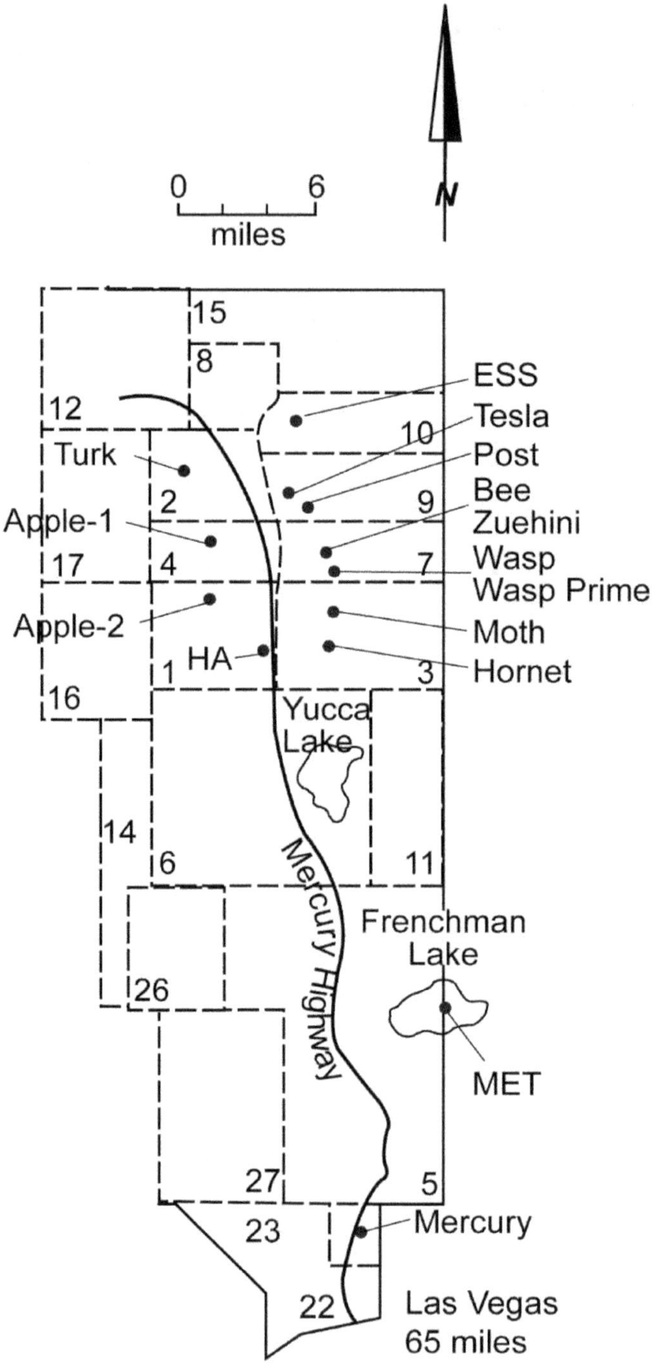

Figure 2-10.2 Location of TEAPOT Tests.

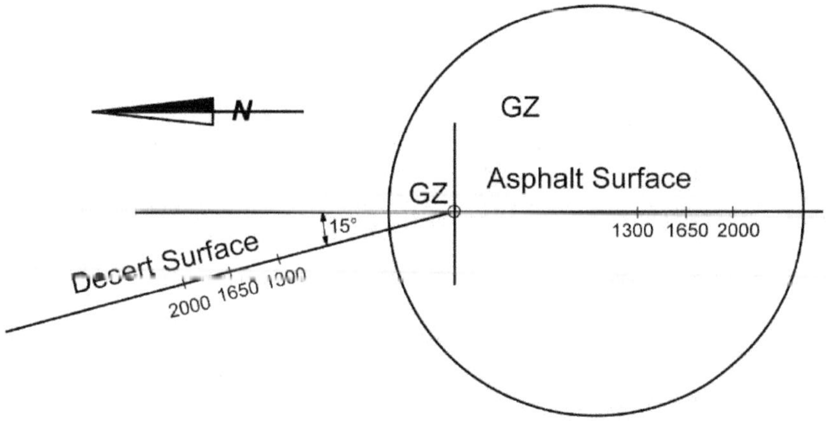

Figure 2-10.3a Pre-existing Asphalt Surface Used for Bee

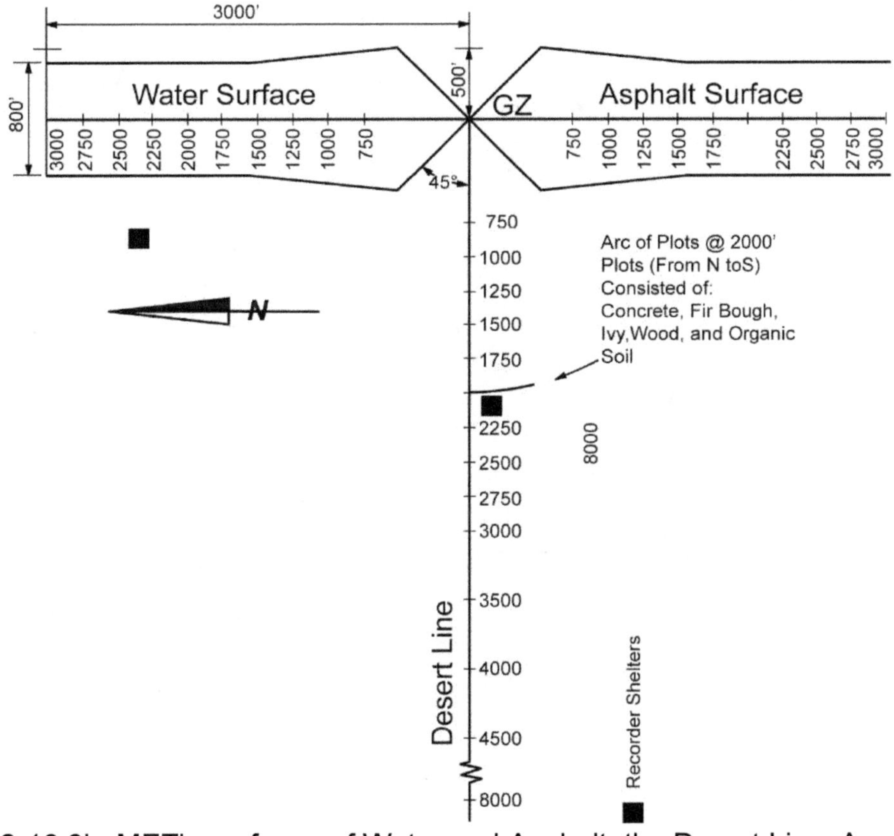

Figure 2-10.3b MET's surfaces of Water and Asphalt, the Desert Line, Arc of Plots at 2000', and Recorder Shelters.

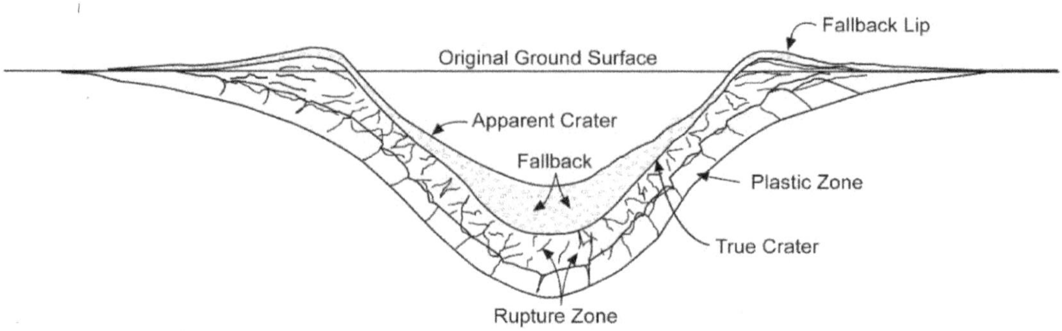

Figure 2-10.4 Crater terminology

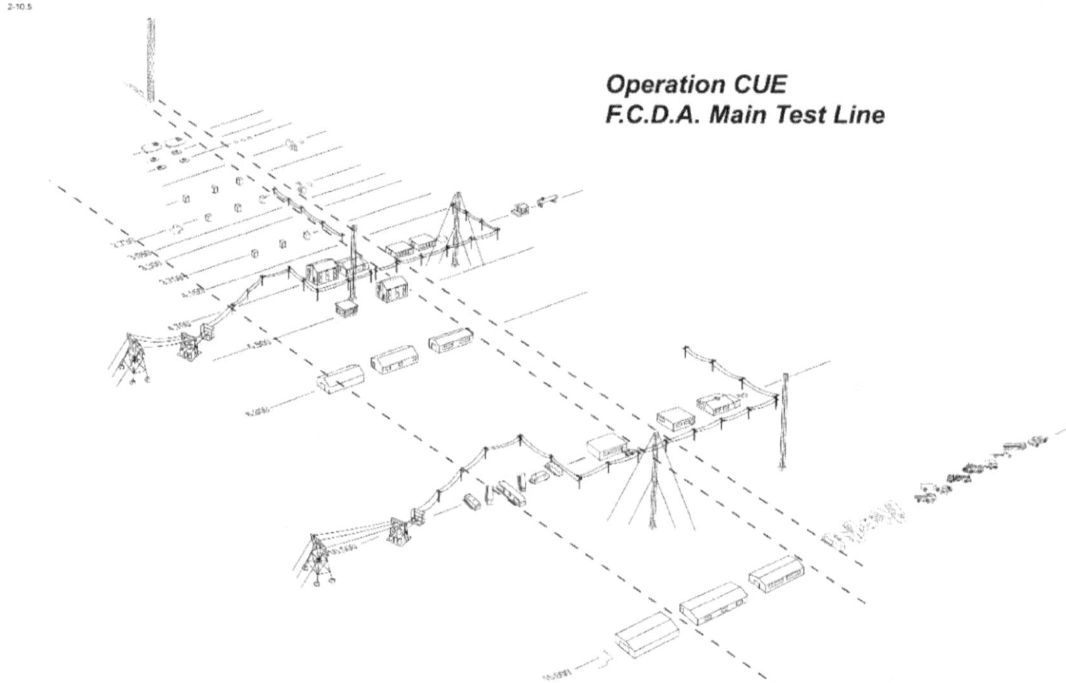

Figure 2-10.5 Operation CUE, FCDA Main Test Line, not all projects are shown.

PLK 231-3 - Indian Springs AFB during TEAPOT, from 7500 ft.

PLK397-10 - Control Point during TEAPOT.

PLK 2-6 - 18Jan55 - North corner of Hut area, camp Mercury in snow

Livermore personnel actively used recreation areas. Because of the Leg and footwear, this photo was probably from a Pacific operation; but the authors understand that the rec room at NTS during TEAPOT would have been similar during celebrations of success.

PLK 243-7 - Electronic equipment in forward area, Project 3.10.1

PLK 8-4 - 19Jan55 “Bird Cage” going down ESS shaft

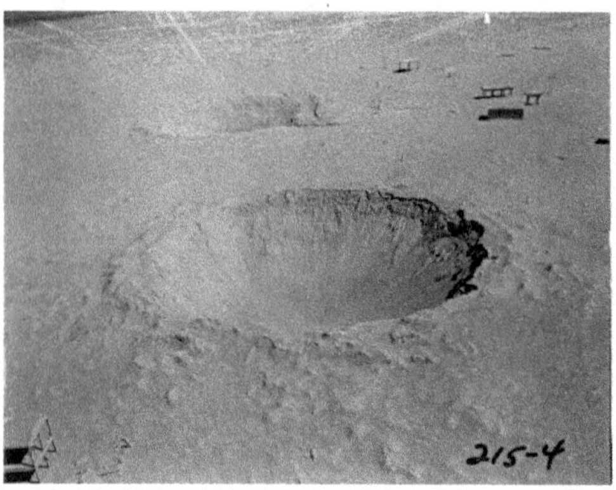

PLK 215-4 - ESS crater with JANGLE crater in background

PLK 113-2 - Project 8.3 Progression of Smoke Screen from top of Hornet tower.

117-3 - 24Feb55 21/2 ton Reo truck 1200 ft from Shot #2 GZ

PLK 45-4 - 8Feb55 ESS weapon saddle with men attaching crane hooks.

PLK 46-3 8Feb55 - ESS weapon being lowered in shaft.

PLK 313-1 - Flare array, generated at t=0, for photography of airblast Project 9

PLK 127-5 - Partially flooded water section near MET tower and tower array for project 5.5 in background.

12-1 - Project 1.7.1 26Jan55 Cables laying in trench

Purple Disc - Mannequins preshot, Operation CUE.

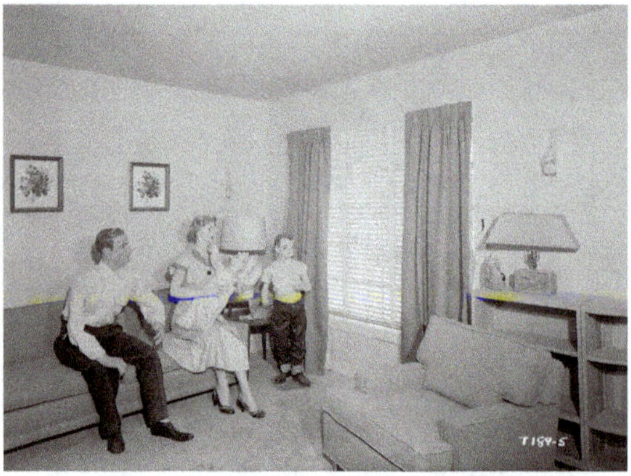

Purple Disc - Mannequins preshot, Operation CUE.

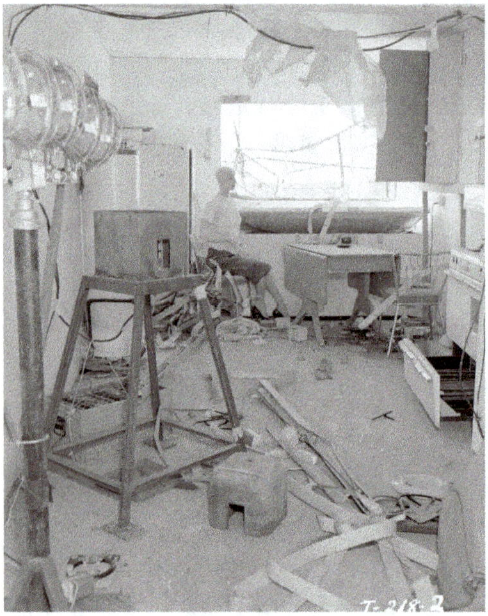

Purple Disc - Mannequins postshot, Operation CUE

Civil Defense Trailers, Operation CUE.

From left: Duane Sewell, Herb York, Walter Gibbons, Bill Ogle, and Jim Reeves

CHAPTER 11. 1955-1957

GROOM LAKE - AREA 51

Groom Lake, or Area 51, was an Air Force facility devoted to the operational testing of hardware and is of interest in the context of the NTS for two reasons. The first is that as close neighbors to the test site they had to be considered when assessing the possible effects of radioactive fallout following a nuclear detonation. The second reason is that the AEC acted as cover for the top-secret facility. The Air Force tried to leave the impression that this was just another facility within the AEC's nuclear test site. Area 51 also went by the code name of Watertown. Many documents dating from the late 1950s and 1960s refer to resources going to or coming from Watertown.

Groom Lake, adjacent to the northeast corner of the Nevada Test Site, is a dry lakebed very much like the Frenchman or Yucca dry lakes. It was used during World War II as an outlying training strip in support of Nellis Air Base. The Lockheed Skunk Works resurrected the site in the spring of 1955 as an airstrip and test area for their top-secret high-flying U-2 surveillance aircraft. The Groom Lake work force reached about 1,800 by 1965.

In 1958 Public Land Order 1662 was enacted. This resulted in the withdrawal of 60 square miles for use "...by the Atomic Energy Commission in connection with the Nevada Test site." This included an expansion of the already restricted airspace above Groom Lake. During the 1960s, the 5000 foot runway was lengthened to 8500 feet; and the restricted airspace over Groom Lake was expanded to 22 by 20 nautical miles.

1955

Planning For Future Operations

On March 23, 1955, the Director of Military Applications notified the Chairman of the MLC that Operation DIXIE, which had been planned for the fall of 1955, would be made a part of Operation REDWING, in the Pacific, during the spring and summer of 1956.

About three weeks before Zucchini, the last shot of Operation TEAPOT, Norris Bradbury suggested to SFO Manager D. J. Leehey, that the next Nevada test series follow REDWING (May - July 1956 in the Pacific) by about a year. The idea was to have a test series with 6 to 8 shots generally similar to those on TEAPOT. (Memo. April 25, 1955, Bradbury to D. J. Leehey, Manager SFO)

Bradbury's April 25th memo to Leehey also suggested the "DoD explore with us and carry out actual tests of the use of balloons to suspend atomic test devices in place of our tower techniques..." (Memo. April 25, 1955, Bradbury to D. J. Leehey, Manager SFO)

Hertford Becomes Manager of SFOO
On October 1, 1955, Major General Kenner F. Hertford (retired from the Army Corps of Engineers on July 31, 1955) became Manager of the Santa Fe Operations Office. From March 1948 until November 1952, Hertford held assignments in Albuquerque; first as deputy Commander of Sandia Base and then as Chief of Staff and Deputy Commander for the Army at AFSWP.

BALLOON SUSPENSION FOR TEST DEVICES

By the spring of 1955, it was clear that it was desirable to have the test devices higher than could be conveniently achieved with towers. The higher the better as far as fallout was concerned. However, from the perspective of the diagnostic and effects scientists, it was best to have the explosions lower (but not too low, for some experiments) and in fixed positions. Airdrops could, of course, be high enough, but it was difficult to have diagnostic equipment near the devices. Also, the burst locations were too uncertain for experiments that required well-defined lines of sight to the bombs or a specific level of effects. Tethered balloons to hoist the test devices seemed to be the answer. They could be high enough and almost as fixed as the shot cabs at the tops of towers.

Los Alamos and Sandia test people got together in early November to explore the possibilities afforded by balloon suspension. The idea looked plausible, and on November 14, Hertford wrote to Sandia President James McRae asking for help in a feasibility study of balloon technology to position test devices high enough to significantly reduce fallout. Hertford noted that the optimum height of the detonation would be about one and one-half fireball radii above the ground. The requirements translated into the capability to lift 1,000 pounds to 1,500 feet. (K. F. Hertford to James W. McRae, ETP: WWA-989, November 14, 1955)

Hertford wanted the technology in hand for a possible test series called PILGRIM, tentatively scheduled for the fall of 1956, which, was delayed until the spring and summer of 1957 with a name change to Operation PLUMBBOB.

The Sandia study [Feasibility Study of Balloon Suspension of Nuclear Devices, 55-56-52, dated April 4, 1956] concluded that lifting nuclear test devices with balloons appeared feasible and that the study should be continued with a test of the technique in Nevada during the next operation. Both LASL and UCRL agreed that they should each have their own balloon areas; LASL in area 7 and UCRL in Area 9.

The AEC identified Holmes and Narver as the Architectural Engineers for weapons test structures at the NTS for FY1957 and urged Sandia to communicate directly with them. (Memo. Ralph P. Johnson, ALOO, to James W. McRae, ETP: MFS-3094, May 15, 1956)

By mid-August 1956, James Reeves in a memo to the LASL and UCRL test organizations mentioned that the weight specification had gone up to 2,000 pounds;

and that although the altitude had officially remained at 1,500 feet, the height could be extended to 1,750 feet. (James E. Reeves to Alvin C. Graves. Status of Balloon Suspension Test, ETP: GHC-415, August 14, 1956)

There was a dry run of a balloon shot at an altitude of 1,000 feet at the NTS on March 1, 1957 with an extensive suite of typical nuclear weapon test equipment. On March 21, James Reeves sent DMA Director, Brig. Gen. Alfred D. Starbird, a status report.

> 1. It is feasible to operationally use a 67 ft diameter balloon to carry up to 2000 pounds to a height of 1,500 feet.
> 2. Balloon position and height can be reliably controlled from Control point, and monitor of position and cable tensions is practicable and reliable.
> 3. Balloon position can be maintained well within desired limits in winds up to 20 miles per hour. At 500 feet it can be limited to a movement of plus or minus 4 feet at winds up to 25 miles per hour.
> 4. Balloon cable system can withstand winds up to 45 miles per hour without breakage.
> 5. Wind speeds in excess of 45 miles per hour will probably cause failure of up-wind guy cable but remaining cable system can restrain balloon in winds up to 55 miles per hour.
> 6. Manual deflation system plus operational techniques can probably prevent disaster conditions if winds exceed 55 miles per hour. ...
>
> ....
>
> 8. Automatic deflation system will reliably prevent escape of balloon and payload from Proving Ground in the event of failure of all four cables.

Device preparation for firing, including arming, could be accomplished with the device aloft. (TWX Reeves to Starbird, March 21, 1957)

Additional data on wind-induced position accuracies were reported on April 10, via TWX, from Sandia to the various test organizations: "Horizontal movement at 1,500 feet with a payload of 2,000 pounds on a 67 foot diameter balloon: Plus or minus 16 feet in wind variance of 20 to 24 MPH, plus or minus 12 feet in wind variance of 18 to 22 MPH, plus or minus 2 feet in wind variance of 10 to 12 MPH." (TWX from H. G. Laursen, Sandia Corp. Mercury, NV to distribution. April 10, 1957)

Ultimately, for Operation PLUMBBOB, balloons were used to loft payloads of 1800 to 4400 pounds to altitudes of 500 to 1500 feet above the ground. Twelve of the 24 full-scale nuclear shots fired above ground were suspended from balloons.* [*Footnote: The first balloon shot was the Livermore Lassen event; a 0.5 kt device fired on June 5, 1957.] Facilities to handle balloons were built in Yucca Flat areas 7 and 9, and in Frenchman Flat area 5. Each installation consisted of the three guy winches 3,000 feet from ground zero housed in buried concrete bunkers. One of the three bunkers also housed the main winch whose cable went along the surface to ground zero, through a sheave, and then vertically upward to the balloon cab. All winches could be operated from either the Control Point (CP) or from ground zero. The main winch adjusted the altitude while the guy winches controlled the horizontal position. The balloons for PLUMBBOB were manufactured by General Mills, Inc. of Minneapolis, Minnesota, and came in two sizes: 67-foot diameter for the lighter loads and 75-foot diameter for the heavier loads. The guy wires were 5/16 inch or 3/8 inch and the mains were 7/16 inch or ½ inch for the smaller and larger balloons respectively. New cables were used for each shot.

The device cabs were fabricated at the laboratories and mock-ups of the assemblies were made before shipment of the cabs to Nevada. Assembly of the various components, except for the nuclear device, was done at the NTS Sandia compound in Mercury. The cab and a 16-foot by 16-foot plywood shelter were moved to the ground zero area about four days prior to the planned shot. The balloon was inflated at a large concrete pad in the neighborhood of the "B-J Y" the evening before the event. It was hauled, along with a 16,000-pound concrete mooring block, to the test area. A dry run with lead in the cab was made to determine the appropriate main and guy wire settings.

Following the dry run, the cab was removed from the balloon. Approximately three hours prior to shot time it was reattached, and device insertion and arming began. One member of the balloon crew was assigned to GZ in support of the arming party. Two television cameras looking vertically upward from GZ were displayed at the CP to permit the winch operator, observing lights on the bottom of the cab, to adjust the horizontal position.

There were questions about what would happen to a balloon that broke free from the main and guy lines. To address this question, a 67-foot balloon with a 2,000 pound cab was released at the surface. It rose to 5,000 feet above the terrain where it ruptured as it was designed to do. The cab and balloon struck the ground 5,800 feet from the release point at a velocity of 120 miles per hour. (Johnson, G. 1957: 53-57)

An ingenious device was developed to vent the helium in the event of a balloon getting loose. An automobile tire was fixed to the top of the balloon. The tire contained a baro switch that could connect a battery to a heating element on the balloon surface beneath the tire. The heating element would burn through the balloon shroud and release the gas. The mechanism could be activated by simultaneously pressing two buttons on the control console. As a backup, the baro switch would activate the heating element at 2300 feet above the terrain. If all else failed the balloon was designed to rupture above about 5000 feet above the surface. (Johnson 1957:90)

The United States fired a total of 25 balloon shots; 24 at the Nevada Test Site and one in the Pacific (It was a 1.7 kt LASL/DoD shot at an 86,000-foot altitude, code-named Yucca, in the vicinity of Enewetak on April 28, 1958.) The last balloon shot in Nevada was Santa Fe, a LASL event on October 30, 1958. There were plans for balloon shots both in the Pacific and at the NTS following the 1958-1961 moratorium. But the rationale was vague; and the motivation appeared, at least to Livermore, to be undefined political objectives. (TWX John S. Foster Jr. to Brig Gen

A. W. Betts, AEC/DMA February 19, 1962. LANL Archives 353.4 Balloons) The post-moratorium balloon shot proposals were ultimately dropped when President Kennedy rejected the concepts proposed by the AEC. (TWX, March 6, 1952 A. W. Betts USAEC to LASL, Sandia Corp, ALOO, AEC Mercury. LANL Archives 353.2 Balloons)

All of the balloons used so far had been approximately teardrop shaped. During the 1958 to 1961 nuclear testing moratorium H. G. Laursen and his colleagues at Sandia did some development work on aerodynamic (blimp-shaped) balloons for the AEC. These also went by the name of aerocap balloons. The purpose of these new designs was mainly to provide a high-lift capability for nuclear tests in the Pacific when high winds were forecast. As it turned out, they were never used. There were no U.S.* balloon shots after testing resumed in the fall of 1961. (J. H. Wendell, LASL, Memo to Distribution J3-W-29, February 17, 1959) [*Footnote: The French used balloons for many of their nuclear weapons tests in the Pacific in the 1960s and 1970s.]

The following quote from a Sandia proposal indicates that there were tentative plans to resume atmospheric testing at the NTS in September, 1961, with balloon suspension of the nuclear devices.

> Underground testing of nuclear devices imposes severe limitations on the rate at which tests can be accomplished and so far the diagnostic measurements of device performance have been unsatisfactory. To provide an accelerated testing program with meaningful diagnostic data will require a return to atmospheric testing. Above ground atmospheric testing, to reduce local fallout, can be accomplished most readily by using spherical balloons at Nevada Test Site for yields up to 100 kt and later by using the Enewetak Proving Ground for megaton yields. (Proposal for Atmospheric Testing with Balloons. S. P. Schwartz, Sandia Corporation to A. W. Betts USAEC DMA. October 2, 1961)

In early January of 1962 the AEC asked the laboratory directors for proposals for a March 1st balloon shot at the NTS. (TWX. January 5, 1962 AEC to Directors (Archives A-99-019))

RADIATION AND CLEANUP

The test area cleanup following a balloon shot was considerably easier than for tower shots. Decontamination options for tower shots were to remove the top 18 to 24 inches of soil or to bring in clean soil to place on top of the contaminated dirt. The first option was impractical because of the vulnerability of buried signal cables. The second option was adopted, with 8 to 12 inches of soil usually providing adequate shielding.

The balloon shots left essentially no on-site fallout. Rather, the principal contamination came from soil neutron activation under the device. The main contributor was Na-24 with a 14.9-hour half-life. This was produced by neutron capture in Na-23 present in the soil. Radiation levels at H+6 hours could be higher than 100 r per hour. When the radiation level declined to about 100 mr per hour the long-lived activity predominated. (Johnson 1957: 64)

The most sensitive area regarding radiation was the so-called "greenhouse" pad where the balloon cab was outfitted at the launch site. Typically about 50 hours work was required at the pad and radiation levels of 10 mr per hour or less were necessary. (Johnson 1957: 64)

An experiment was designed to reduce the neutron activation in the soil by using a surface layer of colemanite ($Ca^2B^6O^{11}$-$5H^20$) and road mix to capture thermal neutrons moderated by the top layer of soil. A one-foot thick pad was tested on the Hood and Owens shots and resulted in a ten-fold decrease in soil activation. The pad appeared to survive unscathed. Another thicker redesigned pad was tested on Wheeler with disappointing performance. It was concluded that more study was necessary before it could be said that a feasible design was in hand. (Johnson 1957: 65)

FALLOUT

Concern over fallout was further highlighted by the release in June 1956 of two reports on the biological effects of radiation; one from the U.S. National Academy of Sciences and another from the British Medical Research Council. (Hacker 1994: 185) The reports generally agreed in their conclusions and were relatively gentle with the AEC and their conduct of tests in Nevada. The stickiest point from the AEC's and the lab's perspective was the National Academy's suggestion for a more restrictive integrated dose to the off-site population. Alvin Graves, characterized it as a limit of three TEAPOT series in ten years. (Hacker 1994:185) The Academy report also recommended more restrictive standards for on-site workers. Since the AEC and the labs were already quite conservative when it came to radiation exposure, both on-site and off-site, the new lower standards would not pose an insurmountable obstacle. Lower limits would, however, reduce the margin of error and constrain planning. (Hacker p. 186) Ultimately the new off-site limits were labeled as guidelines and would not officially go into effect until after the 1958-1961 test moratorium. One immediate effect of the spotlight on fallout concerns was to increase the urgency of the technology to fire the devices either higher or contained underground.

Astonishingly the Air Force objected strenuously to more restrictive radiation exposure guidelines for sampler aircraft crews or for aircraft decontamination personnel. They viewed them as entirely too costly. Fortunately, Dr. Thomas Shipman, from the Los Alamos Health Division, was successful in defending the more restrictive past practices that were ultimately employed. (Hacker 1994: 187)

The AEC was interested in the global implications of nuclear fallout and in 1953 created a program, called Project Sunshine, to address the broad, worldwide, issues. AEC Commissioner Willard Libby was an influential and strong advocate of work regarding fallout, see Chapter 10, section titled PROJECT SUNSHINE AND FALLOUT STUDIES. Libby suggested that that operation PLUMBBOB might yield data relevant to the global fallout questions. (Hacker p.180-188)

In addition there was a lot of discussion at AEC headquarters about the extent of PLUMBBOB and whether some of the higher yield shots should be moved to the Pacific Proving Grounds. Ultimately, the AEC opted to recommend keeping the full series in Nevada and at the end of December 1956, President Eisenhower approved the AEC proposal for PLUMBBOB. (Hacker 1994:189)

WIGWAM

On May 14, 1955 at 13:00 PDT, a 30 kt nuclear test was conducted in the Pacific, about 400-500 miles SW of San Diego, CA (about 29°N, 126°W). Like Trinity, this was a single shot operation, and the shot and operation were both known as WIGWAM. It was a DoD effects test with a LASL device. It was conducted to "investigate the vulnerability of submarines to deep nuclear weapons, and the feasibility of using depth bombs in combat". The device was suspended from a barge; and at the detonation location, which was at a depth of 2000 ft below the surface, the depth of the ocean is about 16,000 ft. The test was witnesses by 6800 personnel on 30 ships. "A 6 mile tow-line connected the fleet tug, Tawasa, and the shot barge. Suspended from this line at varying distances from the barge were three "Squaws"- sub-scale submarine-like pressure hulls equipped with instruments and cameras." (http:www.radiochemistry.org/history/nuke_tests/wigwam/index.html, 2/12/2007)

YEAR-ROUND OPERATION OF THE TEST SITE

The campaign style of testing, where an organization was assembled to go to a remote test site in order to fire a number of nuclear test devices over a period of months and then return to a home base, started with Trinity and was subsequently followed on CROSSROADS in 1946 and SANDSTONE in 1948 before testing was started in Nevada in 1951. Nevada was not a remote site like Bikini or Enewetak. In fact its major attraction was that it was conveniently close to the laboratories and to AFSWP. It was viewed by Norris Bradbury at Los Alamos and later by Herb York at Livermore as a back yard test site for experiments that could be performed without the enormous task force requirements necessary for operations in the Pacific.

Thus it was not surprising when in September, 1955, General Starbird broached the issue of year round NTS use. Staff from the AEC/DMA had talked to Max Roy (W-Div. Leader and Assistant Director for Production at Los Alamos) and Duane Sewell (Director of Scientific Operations at UCRL)*. [*Footnote: Sewell managed Livermore's nuclear test operations for CASTLE (1954), TEAPOT (1955), and REDWING (1956) and was Scientific Advisor to the Test Manager for HARDTACK Phase II.] Roy and Sewell expressed a "willingness to consider possibilities of continuously operating the Nevada Test Site. They feel that with more freedom in arranging test schedules and with some reduction of effort required of the Laboratories at the Test Site the plan may have considerable merit." (September 28, 1955 Col. Alfred D. Starbird, Dir. DMA (Drafted by Lt. Col. Vance Hudgins, USMC) to Donald Leehey, Mgr. SFO)

The Starbird memo requested a study of the merits to be presented to the Commission for action. It requested views on the following: Benefits (mainly to weapons development); How the NTS would be operated; Estimated cost; Effects on PR; How civil defense and effects tests could be handled; Notice of tests; Notification of industries; Yield limitations, if any; Effects on civil aviation.

The mode of one year in Nevada and one year in the Pacific had many disadvantages. Starbird noted the following:

> "1. Unnecessary delays waiting for the next test series.
> 2. Unwieldy number of tests may be proposed for a series
> 3. Firing low yield shots in the Pacific is uneconomical
> 4. Labs are drained of many key personnel for long periods during test series."

Since Hertford succeeded Leehey at the beginning of October it fell to him to reply to Starbird. In a November 1955 letter to Bradbury, Hertford described the DMA request to make a "detailed investigation" of the merits of continuous operation of the NTS. Hertford quoted the original request: "The increasing number of tests envisioned both of stockpile weapons and new designs for ascertaining safety, operational reliability, and vulnerability, as well as for development purposes, plus the accelerated time scales which the Department of Defense believes necessary, raises the question of advisability of operations of the Nevada Test site on a year-round basis." (November 17, 1955. Letter from K. F. Hertford, Mgr. SFOO to Bradbury)

Starbird also asked others to comment. These included: LASL; UCRL; Sandia; EG&G; Field Command, AFSWP; LVBO (Las Vegas Branch Office of REECo); USWB (U.S. Weather Bureau); USPHS (U.S. Public health Service); DBM (Div. Of Biology and Medicine at the AEC); and NYOO (New York Operations Office of the AEC). NYOO was interested in and consulted on radiation problems. (Hines 1962: 154)

Bradbury responded to Hertford in mid-January, 1956, with the following general conclusions:

> "The status of NTS should always be such as to permit rapid use for appropriate nuclear tests in the event of technical or other emergency, or sudden new ideas. ... The LASL does not believe that a regular, periodic, scheduled use of the NTS on some forecast basis, e.g., one test per month and a test every month, will provide the most economical use of technical manpower.
>
> ...

The number of varying activities which can be foreseen for the NTS suggests that its administrative activities will have to be essentially on a permanent basis in any event."

With regard to specific questions:

"We have not felt that the weapons development program has so far been handicapped by the 'test series' type of operation."

"There are and will be undoubted occasions when the ability to test immediately some new idea or some practical problem will expedite weapon development."

"It is our belief that J-Division operations as we now carry them out would not be feasible under a continuous operation scheme. It is probable that some quite different type of operation involving permanent residence in Nevada would be required, and it is rather likely that it would come under some other contractual operation. In particular, the University of California is not likely to look with favor upon a sizeable group of people permanently stationed in some state other than California or New Mexico."

With respect to disadvantages of the present method of testing:

"We are unaware of any 'unnecessary' delays to any weapon program which have arisen out of the method of testing."

"The number of tests is frequently 'unwieldy' but spreading them out in a continuous operation is not likely to make the test program easier on the actual participants."

"It is probably true that the conduct of an isolated low yield shot in the Pacific is more expensive than the conduct of the same shot in Nevada. However, with only one testing group and with the necessity of doing some shots in Enewetak in any event plus the public relations problems in Nevada, it is not clear that dove-tailing small shots among big ones at Enewetak is really much more costly."

"The Laboratory is certainly drained of key test personnel for considerable periods under the present system. However, it is believed that a true continuous operation system would remedy this only by establishing a separate test group in residence in Nevada." (January 13, 1956 Letter to K. F. Hertford, Mgr. AEC, Santa Fe Operations Office, ABQ, NM. From Bradbury. Ref. November 17, '55 letter on continuous operation of the NTS.)

In early April director of the SFOO Test Division, James E. Reeves, sent a letter to the labs, EG&G, AFSWP and various AEC offices forwarding a request from the AEC/DMA for a feasibility study on the scheduling for continuous operation at the NTS. Reeves also outlined General Starbird's directive of Sept, 28, 1955. Reeves attached a draft of a transmittal memo to DMA with a draft summary report, dated March, 27, 1956, for comment. (April 2, 1956. Letter from James E. Reeves, Dir., Test Division, SFO to distribution (directors, EG&G, AFSWP, SAN, LAAO, Dir DBM. Ref. November 17,'55 letter requesting feasibility study on the scheduling of test operations at the NTS.)

Two concepts were developed by Hertford and Reeves for Starfbird:

"Concept 1 - To test when and as required by the developmental laboratories any and all types of devices regardless of yield, ..." i.e. Complete continuous operation.

"Concept 2 - To continue to combine the more difficult type of shots, either from an instrumental or off-site fallout standpoint, into a series type of operation to be conducted approximately every other year and not oftener than once a year, but also reorganize the test facilities and organizations involved so that the easier or simpler type of shots could be conducted on the 'as come' basis."

Costs: Should do a cost comparison in the future. But, "... it is not believed that cost would be the determining factor."

Organization and Operation - NTS: "To conduct tests on an 'as required' basis plus series test operation, consideration should be given to the establishment of a permanent test organization to the extent of manning certain key positions, such as Test Manager and others considered feasible to provide the appropriate continuity of planning and effort necessary to maintain the highest degree of readiness."

Ultimately the report endorsed concept 2. "Many benefits are to be gained by testing on an 'as required' basis plus the series type test operations,"

"Both scientific laboratories (LASL and UCRL) desire that the status of NTS should always be such as to permit rapid use for appropriate nuclear tests in the event of technical or other emergency, or sudden new ideas." ... Also, concept 2 "would spread the effort devoted to weapons development more evenly over the year, thereby avoiding to a great extent the large peaks in effort which the present program requires."

"EG&G states it could do a technically better and more economical job if tests are conducted on a year-around basis."

Sandia saw no significant problems with year-around operations in Nevada.

Field Command AFSWP felt that ... "tests on an as-needed basis should be a distinct advantage to development programs and should serve to expedite the development of new weapons." ... "year-around operation can be advantageous to military effects interest as well ..."

NTS: "From the standpoint of the NTS Support Contractor Organization and level of effort, the adoption of tests on a continuing basis of 'as ready' plus the periodic test series appears to present several advantages over the alternate year series testing." "The contractor would be in a better position to establish permanent positions with better qualified personnel, ..."

ALOO (also referred to as AOO): "... is conducting various other studies that would have a bearing on the conduct of test operations. These are:

The formulation of a plan with procedures for the Las Vegas Branch, Test Division, AOO, to assume the On-Site Radiological Safety functions."

"Study of methods and procedures whereby Las Vegas Branch, Test Division, AOO, would assume responsibility for all supply functions ..." (except scientific equipment).

With regard to public relations, it was felt that ... "tests at a fairly uniform rate might become commonplace and cause little excitement."

Generally, the Nevada offices supported concept 2.

Those involved with civil effects tests also supported concept 2.

There arose the question of whether the photographic industry actually needed notice of nuclear tests in advance*. The New York Operations Office responded: "The photographic industry's committee on radioactivity considers that notice of several months is the most useful service (the) AEC can give them. It is indicated that the industry can continue to operate under the year-around operation without long advance notice but with greater difficulty and at increased costs." ... The "photographic industry needs daily predictions of fallout areas for several days after each Nevada shot." The "apparent attitude of the industry appears to be they are prepared to accept their losses from an intensified test program, assuming it is required in the National interest."

"Comments from the Civil Aeronautics Administration indicate that, should tests be conducted more frequently, the present process as utilized by the CAA for disseminating suitable advance information and NOTAMS (Notices to Airmen) to the flying public would have to become more routine and simplified." (June 7, 1956, K. F. Hertford to Brig. Gen. Alfred D. Starbird: Transmittal for report on NTS operation in response to Sept. 28, 1955 request.)

In mid April Bradbury told Reeves in a memo that Los Alamos generally concurred with the substance of the report. (April 18, memo Bradbury to Reeves)

COMMITTEE ON THE USE OF NEVADA TEST SITE

Kenner Hertford wrote to Bradbury in late December, 1955, to announce the establishment of "a permanent committee to consider future usage of the Nevada Test Site, (and consider) primarily proposals to conduct activities other than full-scale nuclear tests." Secondarily the new committee would advise SFOO on the physical locations and facilities for new proposed activities at the NTS. This was stimulated by requests from both UCRL and LASL to test rocket motors at the site. "In view of the expanded usage of Nevada Test Site from a backyard test area for use by LASL to multiple usages involving a diversity of activities and several agencies, it is considered most important that a group be established to give broad consideration to future usages of the Nevada Test Site."

The membership was as follows:

Ralph P. Johnson, SFOO, Chairman
James E. Reeves, SFOO, Vice Chairman
Alvin C. Graves, LASL
Raemer E. Schreiber, LASL
G. W. Johnson, UCRL
Hayden Gordon, UCRL
L. E. Hollingsworth, Sandia Corporation
R. A. Bice, Sandia Corporation

W. W. Alllaire, SFOO, was the Executive Secretary. J. C. Clark and R. W. Spence of LASL, T. C. Merkle of UCRL, and E. F. Cox of Sandia were alternate members. (Letter from Ralph. P. Johnson to Schreiber and others, ETP:WWA-1657, January 4, 1956)

The first meeting was held on January 10th at the test site.

Pluto And Rover

"The Chairman opened the session explaining that in view of the increased usage of the Nevada Test site for activities other than the full-scale tests for which it was originally established, Santa Fe Operations considers it appropriate that proposals for additional activities be reviewed carefully to determine compatibility and whether such additional activities will interfere with one another." The committee focused on the nuclear ramjet testing facility proposal from Livermore (code named Pluto) and the proposal for a nuclear rocket facility from Los Alamos (code named Rover). They were considering potentially suitable locations for these two new programs. The two areas that received the most attention were the Cane Springs - Jackass Flats region, adjacent to the southwest part of the test site and the region just west of Yucca Flat. "The Panel generally agreed that the rocket propulsion test facilities to the west of Nevada Test Site would not constitute interference with other foreseeable activities."

Real Estate

"The Committee recommended the AEC attempt to procure the entire strip of land immediately west of Nevada Test site. Even though initial phases of the program may be conducted in the Jackass Flats area, it may be necessary to conduct subsequent phases in the canyons further north." (Minutes of the first Meeting of committee on use of Nevada Test Site. January 10, 1956, W. W. Allaire, SFOO)

Raemer Schreiber recommended in a January 24 letter to Ralph Johnson that the AEC make a vigorous attempt to "... acquire all of the land within the bombing range west of the present NTS area up to the north boundary of the present NTS area." He went on to say that the N (Reactor) and J (Test) Divisions at Los Alamos had agreed to a modus operandi for reactor testing at the NTS, whereby J. C. Clark would be the Test Director and R. W. Spence would chair a joint J and N technical team responsible for the scientific programs. (Schreiber letter to Ralph Johnson and distribution, Los Alamos, N-112, January 24, 1956)

In the Spring of 1955, bids were being sought for the beginning of construction in the Jackass Flats Technical Area 25 for Pluto and Rover nuclear ramjet and nuclear rocket projects.

During the spring and summer of 1956 the AEC explored options for acquiring additional land for three purposes. One was to serve as a buffer area to restrict off-site civilian activities in sectors that might be exposed to fallout from the nuclear tests. The focus was mainly in land to the west, north and east of Yucca Flat. The second interest was in land to the west that would be suitable for test activities associated with the nuclear rocket or ramjet programs. The third interest was in land to the northwest, near Tonopah, Nevada, for Sandia's ordnance testing. Ultimately the AEC was successful in acquiring the land that they needed.

GRAVES SUFFERS HEART ATTACK

On or about January 4, 1956 Los Alamos Test Division (J-Division) leader Alvin Graves suffered a heart attack and was on medical leave for several months. William E. Ogle took his place on the NTS utilization committee. Ogle was alternate division leader and assumed leadership of the test activities until Grave's recovery. (Letter Bradbury to Hertford dated January 5, 1956)

Graves was back at work by October 1956.* [*Footnote: Graves died of a heart attack in July 1965, and was succeeded by Ogle as division leader.]

OPERATION PROJECT 56

The first safety experiments conducted at the NTS were LASL's 4 tests between November 1 1955 and January 18, 1956. The objective of each test was to determine the nuclear yield, if any, of the devices with the high explosive detonated at one point, as in an accident scenario. United States nuclear weapons are certified to be "one-point safe." A nuclear yield limit equal to that of 4 pounds or less of TNT is a fundamental factor in the U.S. definition of "one-point safety."

Early nuclear weapons, such as those employed during WWII, were only one point safe before assembly of the nuclear material and the propellant or high explosive. It was not feasible to assemble a Fat Man-type weapon aboard a strike aircraft; and as a result, it was not considered one-point safe. More modern weapons, such as those developed in the early 1950s, were one-point safe and could safely be stored after being assembled. This had important military implications since the assembled weapons were ready for a rapid response to a strike order. Another interest in one-point safety tests focused on questions of clean-up of the fissionable material following a nuclear accident.

James Reeves, head of the SFOO test division, had been selected test manager for Operation 56 in October. Joe Sanders, Las Vegas Branch Office manager, was the AEC support director. Jack Clark, Los Alamos Test Division, was the test director. (Hacker 1994: 173)

All four tests took place in Area 11, just east of the dry Yucca Lake bed. These shots were all conducted as surface detonations and had the inspiring names of: Project 56 Number 1, Project 56 Number 2, Project 56 Number 3, and Project 56 Number 4. All resulted in little or no yield.

PROPOSAL FOR TESTING UNDERGROUND

Dave Griggs and Edward Teller had conceived the concept of containing the radioactivity from a nuclear detonation and discussed the idea at a 1953 conference in Livermore. The focus of the conference was Plowshare, the program aimed at the peaceful uses of nuclear explosives.

Livermore sent Reeves a letter in late winter 1956 outlining the virtues of underground shots. These were primarily the elimination of radioactive fallout and weather delays. On April 4 Reeves forwarded a copy to Ogle for comments.

Ogle later recalled: "During the period of 1956 - 1958, the concept of doing nuclear weapon testing underground received more and more attention, especially by Edward Teller, as a possible solution to some of the test ban debates." He focused mainly on the fallout problem. (Ogle 1985: 50)

FALLOUT CONCERNS

By early September, 1956, the AEC Division of Biology and Medicine was circulating a draft of their proposed NTS safety criteria for the civilian population in the vicinity of the Site. They proposed that the maximum exposure from fallout be 3.9 roentgens for any one-year and 10 roentgens for any consecutive 10 years. The 3.9 figure for any one year was the previous criteria and DBM recommended that it be retained. The DBM noted that the 10 roentgen was conservative and not particularly restrictive in that during the previous five years of testing in Nevada the highest total accumulated exposure to any community had been about four roentgens (about 15 people living at a motor court received about seven roentgens). (Office Memorandum K. F. Hertford to Norris E. Bradbury and H. F. York, ETT: RHG-804, Sept. 14, 1956)

SECOND MEETING OF THE COMMITTEE ON THE USE OF NTS

The second meeting of the Committee on Use of Nevada Test Site was on October 25, 1956 in Las Vegas and focused on underground testing. Present were James E. Reeves (ALOO, Vice Chairman); G. W. Johnson (UCRL); L. E. Hollingsworth (Sandia Corporation); R. A. Bice (Sandia Corporation); and C. L. Lindquist as secretary. Ogle and Schreiber (LASL) were absent. Alvin Graves was listed as a visitor, presumably because he had not assumed yet all of his Los Alamos J-Division leadership responsibilities following his early January heart attack. Other visitors were: Charles B. Read and Edwin B. Eckel (USGS); M L. Merritt (Sandia Corporation); and Charles E. Violet (UCRL).

The principal topic was "the selection of a geologically and operationally suitable area for underground atomic detonation." The main issues included:

a. "Test site safety.
b. Contamination of ground water.
c. Physical damage to building and structures in proximity of shot.
d. Damage to wells, springs, etc.
e. Possible seismic effect - might act as catalyst or trigger for earthquake.
f. Possible damage from public information viewpoint."

The USGS suggested some likely areas and was instructed to make a survey to address whether or not those locations might be suitable for underground testing. (Minutes of Second Meeting of Committee on Use of Nevada Test Site, October 25, 1956, C. L. Lindquist November 2, 1956)

OPERATION REDWING IN THE PACIFIC (MAY TO JUNE 1956)

Operation REDWING in the spring and summer of 1956 was the next test series in the Pacific after CASTLE in 1954. Seventeen shots were fired in REDWING. This is equal to the total number fired in all of the previous Pacific test series. As usual for operations in the Pacific, a military task force fielded REDWING.

Army Major Gordon L. Jacks commanded the radiological safety unit of Scientific Task group 7.1. After retirement from the Army Gordon Jacks became the Los Alamos Test Division's (J-Division) senior representative permanently located in Nevada.* [*Footnote. Gordon Jacks was a military staff member at LASL from July 1, 1955 to May 28, 1959. After retirement from the U.S. Army Jacks joined the LASL J-Division on March 4, 1966. He retired from LANL on January 6, 1984.]

REDWING had its share of weather delays. Cherokee, the second in the series and the first thermonuclear airdrop (3.8 Mt) was delayed 21 days to satisfy cloud, wind and visibility criteria. Sidney Singer, of Los Alamos, remembers assisting Roger Ray, an army major on assignment to LASL, in the creation of a doggerel to commemorate the event.

> What day is this? I wish I knew.
> I think that it is Minus Two.
> Or maybe it is Minus Four -
> Or - heaven help us -something more.
> One thing we'll know when it is done:
> The day before was Minus One.

(Sidney Singer, Private communication, May 19, 2003)

REDWING had six shots, mostly in the megaton region, at Bikini, and eleven lower yield shots at Enewetak. Only one megaton range test, Apache at 1.85 Mt, was fired at Enewetak. "Joint Task Force 7 prepared to fire at either Bikini or Enewetak as conditions allowed; like TEAPOT, REDWING scheduled ready (dates) rather than firing dates. High yield tests imposed far more stringent limits on weather and wind than did smaller devices. Essentially the task force planned to fire the smaller shots while waiting for the right weather to conduct the high-yield tests."
(Ref. Hacker, Elements of Controversy, p. 179)

Only one REDWING test, Tewa, resulted in troublesome fallout. A last minute wind shift took the cloud from the 5-megaton Bikini shot over Enewetak and exposed over 600 test personnel to radiation levels above the 3.9-roentgen limit. (Hacker p. 179-180)

PUBLIC RATIONALE FOR THE 1957 OPERATION PLUMBBOB

The AEC and the laboratories were forever searching for ways to explain to the public the reasons for continued nuclear weapons testing, particularly in the continental United States. In August 1956 Kenner Hertford proposed to General Starbird a set of themes for Operation PLUMBBOB to be used with the media as the rationale for yet another test series.

> Nevada tests are essential to the protection of the United States and the Free World.
>
> They are in the national interest, being essential to weapons development, military capability and utilization, and Civil Defense.
>
> Each Nevada detonation is fully justified as to national necessity and as to public safety. (A later version said only small yield devices are tested in Nevada.)
>
> Nevada tests are closely controlled to assure on-site and off-site public safety with regard to test effects, and such safety has been achieved.
>
> The U.S. public, in view of the need for continental tests, can accept both the low degree of resulting exposure to radiation fallout and the degree of inconvenience sometimes caused by fallout or other effects.

(August 24, 1956 Hertford to Starbird. Memo: Nevada Test Interim Information Program. Themes for the 1957 test operation)

COMMON TEST ORGANIZATION

Norris Bradbury and Herbert York, the Directors of Los Alamos and Livermore respectively, met on September 7, 1956 to discuss, among other things, a common test organization for the NTS. A week and a half later York sent Bradbury a letter outlining some concepts for a common test organization.

There were three groups involved:

> Planning Board: 3 members each from UCRL and from LASL plus a chairman. (7 total). Alvin Graves to be chair. Vice-chair from UCRL. "This board will be a permanent board and will consider planning and coordination for both the Pacific Proving Ground and the Nevada Test Site. Some of the functions of the board will be to designate test areas for the laboratories, to develop shot schedules, to develop criteria for the safety of operations of either test site, ..."
>
> Advisory Board: "The function of the Advisory Board is to form an opinion and to recommend to the Test Manager whether the weather is acceptable for firing a specified shot subject to the criteria laid down by the Planning Board". (Note: This is what was eventually called the Test Controller's Panel. The Test Controller is what York called the Test Manager, as it is used herein, see NTSO figures. However, the NTSO figures use the term Advisory Panel, not Board.)
>
> Office of the Test Director: "The Test Director's Office is designed essentially to coordinate the activities of all groups and agencies connected with a specific test. It is responsible for the development and execution of the operational plan. ... it is to be understood that the Test Director's function is primarily to resolve conflicts and to make sure that safe practices are followed with respect to the operation of the proving ground. Under the Test Director will be an associate Director from UCRL and one from LASL".
>
> General comment: "The general intent of the organization being proposed is to permit each laboratory to operate as independently as possible within the organizational areas defined by the Planning Board".

(September 18, 1956 Letter from Herb York to Norris Bradbury regarding Proposed Test Organization, or common test organization.)

Bradbury had suggested that the first Test Director of the common test organization, for the next two years, be from UCRL. Plausible candidates included Duane Sewell and Gerry Johnson. At this time Johnson was the responsible individual for nuclear test activities at UCRL. Duane Sewell, who clearly had a great deal of responsibility for Livermore programs, was effectively the deputy director at Livermore, but E. O. Lawrence, who oversaw Livermore from Berkeley, eschewed titles and as a result neither Sewell nor Johnson had official titles.

Bradbury, with input from Ogle and Graves, replied to York: "Our position tends to remain the same; and we would like to see a strong test director with a strong organization, and we tend to hope that UCRL will undertake to furnish that organization." They offered specific comments on York's three groups:
Planning Board: "We generally agree on the philosophy of a Planning Board, but feel that circumstances will require representation of the DoD, FCDA, etc. ... Also: restrict this discussion to NTS ... "
Advisory Group: No comment.
Office of the Test Director: "We would tend to make the Test Director's office much more responsible than indicated in your letter." e.g. 'The Test Director's office is designed to support and coordinate the activities of all groups and agencies connected with NTS bomb tests.'"

Further General Comments: "The general intent of the organization being proposed is to furnish the Test Manager (Test Controller in recent parlance) a Test Director's organization that will coordinate and direct the operations of all users of the NTS, will collect and transmit to the appropriate agency the requirements of all those users, and yet will permit each laboratory to operate as independently as possible with respect to its own observational programs on its own shots." (Letter. Reply to York from Bradbury, September 24, 1956)

THE NTS PLANNING BOARD

The AEC and the other NTS stakeholders generally accepted Los Alamos's suggestions for the composition of the Planning Board and in mid November the Test Manager, with the concurrence of the organizations concerned, appointed the following people to the new board:

Chairman - Alvin C. Graves (LASL)
Vice-Chairman - Duane Sewell (UCRL)
DoD Member - Col. Kenneth D. Coleman
Alt. DoD Member - Cdr. Alfred H Higgs
LASL Member - William E. Ogle
Sandia Member - Richard A. Bice
Alt. Sandia Member - Melvin L. Merritt
Test Director - Gerald W. Johnson
UCRL Member - Duane C. Sewell

(Letter A. C. Graves to James E. Reeves (Director, Test Division ALOO), November 13, 1956, JDO-572)

The first meeting of the Planning Board for Nevada Weapons Tests was actually on October 26, 1956, before the final membership was settled. This was the day after the second meeting of the Committee on Use of Nevada Test Site, which had focused on the feasibility of firing shots underground. Several individuals were members of both groups.

The meeting mainly focused on how to conduct tests in order to minimize the exposure of surrounding populations to radiation. One of the Board's initial tasks, assigned to a sub-committee under the leadership of Dr. A. Vay Shelton, was to prepare essentially the same study of the operating criteria as that commissioned for the Committee on Use of Nevada Test Site. (Minutes of the Planning Board for Nevada weapons Tests enclosed in letter from Alvin C. Graves to James E. Reeves, Director Test Division ALOO, November 13, 1956, JDO-572)

REVIEW OF NTS OPERATING CRITERIA

Back in 1953, the Manager of SFOO established a Committee to study the operational future of the Nevada Proving Ground. Operating criteria for firing nuclear devices in Nevada during Operation TEAPOT (February to May 1955) were developed from the work of that committee. The AEC mandated that the criteria be reviewed before the next Nevada test series (PLUMBOB, spring to fall of 1957).* [*Footnote: There were four safety tests of essentially zero yield fired in late 1955 and early 1956 called Operation 56 that did not count in this context as a "test series."]

Hertford said that "It appears a restudy of these criteria ----- is most appropriate at this time, particularly in view of new modes of testing now under consideration, i.e., use of balloons, 1000-foot towers (which were never used), and contained underground shots; the probability of more or less year-around tests at NTS; and, the requirements for Commission reconsideration before the next test." (Office memorandum entitled "Review of NTS Operating Criteria." K.F. Hertford, Manager Albuquerque Operations Office (ALOO) to Norris E. Bradbury (LASL) and H. F. York (UCRL), June 13, 1956)

On October 31, 1956 Kenner Hertford "...proposed that UCRL and LASL establish a joint group to review the operating criteria prescribed by the Commission for Teapot ..." (Memo to UCRL and LASL from Hertford). This was an appropriate task for the new NTS Planning Board to assign to a subcommittee.

The second Board meeting was held November 13, 1956 in Albuquerque and considered:

Shortage of funding for Operation PILGRIM.
Need for Area 8 and a second balloon area.
Additional facilities required by scientific groups.
Fallout
Duration of PILGRAM
Completion of current construction.
Limitation of yields and test configurations.
The following maximum yield versus height criteria were recommended to the Test Manager:

- Surface and sub-surface (if the shot breaks the surface) 1 kt
- 300 foot tower 25 kt

- 500 foot tower 50 kt
- Air drop (fireball not to touch the ground) 80 kt.

Reeves responded by asking Ogle how LASL would cope with test limitations in Nevada of 15 kt on 500-foot towers and 100 kt on balloons or airdrops. Ogle took this up with his colleagues at Los Alamos on November 23. Those present included: D. Froman, Jane Hall, Carson Mark, Albert Petschek, Duncan MacDougall, Harold Agnew, and William Ogle. At that time they were considering shots for Operation PILGRIM, which became PLUMBBOB, executed during the late spring, summer, and early fall of 1957.

"There was ... some discussion as to whether it would be profitable for this laboratory (LASL) to restrict itself in the proposed manner independent of Washington decisions. This discussion was on the basis that a 35 kt shot on a tower would on the average be delayed between one month and two months because of weather. This apparently leads to a waste of effort on the part of J (test) and T (theoretical design) divisions while waiting and hence delays preparations for the next operation. Such a delay is obviously very hard on the morale of the test personnel associated with the shot. There was no real conclusion on this question."

"Livermore had sent a TWX answering the same question from their point of view, their answer essentially being, 'We cannot put up with it and accomplish anything.'"

Starbird had asked whether the higher yield shots could be taken to the Pacific during 1957 without delaying the 1958 Operation HARDTACK test program. Both Livermore and LASL expressed great reluctance to do anything overseas during 1957. (Ogle notes, JO-275, on November 23 meeting in Bradbury's office)

The third Board meeting was held December 12, 1956 at the Radiation Laboratory in Berkeley. They considered:

Reports from the weather and fallout sub-committees;
Reviewed fallout and balloon heights;
Program 57 (a safety shot);
Establishment of a Fallout Prediction unit for PILGRAM;
Limitations of yields on air bursts;
Aluminum towers;
Planning housing and other services;
Status of construction;
Construction in CP area.
A PILGRAM (PLUMBBOB) schedule was prepared with LASL shots early in the series and UCRL in the later part.

The fourth meeting was held on January 24, 1957 in Las Vegas, NV; and the following subjects were discussed:

A report from AFSWC regarding the air-to-air missile;
Revised shot schedule,
Sandia's one-point tests;

Helicopter platforms;
Fireball chemistry;
The underground shot;
Justification for PLUMBBOB;
The CETG program.

Gerald Johnson from Livermore proposed an underground shot on PLUMBBOB. The purpose was to "establish the feasibility of doing a test and obtaining the required diagnostic measurements and to prove that devices of higher yield could be detonated in that manner. They planned to measure yield from both shock propagation (hydrodynamic yield) and by radiochemistry on earth samples.

LASL mentioned that they were considering the detonation of two one-point shots underground in 18-inch diameter, 500-foot-deepholes. Only alpha and yield would be measured. (Minutes of the fourth meeting of the Planning Board, January 24, 1957 p.5)

The fifth meeting occurred at the NTS prior to the Charleston test to review the safety measures involved. It is discussed in Part II-Chapter 12. (Reeves 1980: 143-6)

1957

In early February Brig. Gen. A. D. Starbird, Dir. DMA, passed along an MLC proposal to test an Honest John missile with a nuclear warhead at the NTS. On February 26, Bradbury responded: "The Los Alamos Scientific Laboratory does not concur that such tests are either useful or relevant to the question of overall missile or warhead performance. We do not believe that such tests should be conducted at NTS where they substantially interfere with the basic purposes of the test site. We further are of the opinion that, if the DoD insists on such tests, they should find an appropriate proving ground of their own and conduct such tests on their own responsibility and cognizance.

If the AEC and DoD decide that such tests must be attempted at NTS if feasible, I suggest you ask the Nevada Planning Board to examine the feasibility of such tests under conditions of adequate assurance of public safety." (Los Alamos archives)

CHAPTER 12. OPERATION PLUMBBOB: MAY 28 - OCTOBER 7, 1957

OPERATION PILGRIM EVOLVES INTO OPERATION PLUMBOB

The proposed spring and summer 1957 test series in Nevada was originally named Operation PILGRIM. In the fall of 1956, at the first meeting of the Planning Board for Nevada Weapons Tests, May 1 was established as the start of the operation and June 1 as the date for an air-to-air missile weapons effects shot that caught the fancy of the DoD.

Much of the focus of the test planning community during the fall of 1956 was on fallout prediction and hazards associated with the proposed PILGRIM shots. Interestingly the Planning Board "agreed to try to limit fallout on people outside of the test site to 1R (or 2500 megacuries) in one year." Livermore's Duane Sewell expressed the opinion that the maximum total fallout from a single shot should be limited to about 600 megacuries in order to reduce the jeopardy to continued use of the site. The issues surrounding the fallout questions clearly were not resolved and the board, just as clearly, intended to continue its analysis. (Malik 1956: 3)

In the middle of January, 1957, PILGRIM became PLUMBBOB. (It was referred to as Pilgrim on January 18 (Memo George Cowan to Graves, J-11-466) and as PLUMBBOB on January 24 (January 24 NTS Planning Board minutes, Graves JDO-654, March 20, 1957))

PLANNING

The review, evaluation, and coordination of project proposals for Operation PILGRAM took place during April, May, and June, 1956. The preparation phase of PILGRAM began the same fall. (DASA 1959: Part II, Chapter 7-5, p22)

The Laboratories were focusing on proof tests of antisubmarine and air-defense warheads as well as on weapons physics tests to explore new primary and two-stage design concepts and components. The main objective was still on ever smaller and lighter weapons. In addition, there would be five safety tests to augment Project 56, the safety test series fired in November 1955 and January 1956.

During the planning phase of PLUMBBOB, it was decided that two of the weapon tests being planned by the AEC, Priscilla and Smoky, could accommodate most of the DoD's main objectives. Therefore, while Priscilla and Smoky are not considered to be DoD tests per se, the DoD participation on both was significant. The only significant objective that could not be addressed by the AEC's planned tests was the detonation of a nuclear weapon during air-to-air missile delivery.

The DoD's main objectives for effects measurements on PLUMBBOB were to:
- Study the effects of terrain on blast and shock phenomena. (Smoky)

- Obtain data on loading and response of structures in pressure regions above 50 psi. (Priscilla)
- Measure neutron induced activity in soils. (Priscilla)
- Study electromagnetic pulse signals from detonations that had relatively little or no shielding. (Priscilla and various weapons tests)
- Obtain nuclear and blast criteria for the F-89D aircraft
- Obtain effects data on military equipment and on large biological specimens. (Priscilla and various other weapons tests)
(DASA 1979: 5)

NEVADA TEST SITE ORGANIZATION (NTSO) FOR OPERATION PLUMBBOB
The AEC, in collaboration with LASL, UCRL, and AFSWP developed the organizational structure for PLUMBBOB indicated in Figure 2-12.1. (Reeves 1980, 17-20) Such collaboration certainly existed on previous operations, but it probably was even more important on PLUMBBOB, which was the longest in duration of the atmospheric test operations conducted at the NTS.

James Reeves again served as Test Manager with William Allaire as his deputy. As on previous operations, Los Alamos's Graves chaired the Advisory Panel which reported to the Test Manager. A list of this large, 23 member, panel is given in Appendix F.

The Planning Board also reported to the Test Manager and consisted of Chairman Graves plus 6 additional members, also listed in Appendix F. It held four meetings prior to PLUMBBOB. The Board played a significant role in the review of radiation safety standards for on-site and off-site populations and for the general configurations of the PLUMBBOB tests. A fifth meeting was held prior to the Charleston event.

The only Deputy on Reeves' staff was Col. H. E. Parsons who served as Deputy for Military Matters. He had held essentially the same position on TEAPOT as Reeves' Deputy for Military Operations. As on TEAPOT, Parsons facilitated integration of the Defense Department and AEC programs. Note that the NTSO did not necessarily adhere to the same titles for what appear to be about the same jobs in previous operations. There are undoubtedly interesting, but undocumented, stories regarding the rationale for new titles such as the change from Deputy for Military Operations to Deputy for Military Matters.

The right hand side of Figure 2-12.1 indicates the groups that reported to the Test Manager in the areas of Administrative Staff, Technical Staff, and Other Functions. In the Test Manager's Technical Staff, the significant changes from TEAPOT were: the addition of the positions USGS (US Geological Survey) Coordinator and USPHS (US Public Health Service) Coordinator. These two positions reflect the beginning of underground testing and increased public concern over nuclear fallout.

In his Test Manager's Report for PLUMBBOB, Reeves states: "The organizational concept included considerations of maximum delegation of authority and responsibility to the Test Director and scientific groups, support organizations and the AEC and DoD support organizations." The positions held in previous operations by Seth R. Woodruff, Jr. (Field Manager NPG, Support Director) do not appear for PLUMBBOB. Instead, there are two units titled AEC Support and DoD Support. The Director of AEC Support was Allaire with Max E. Smith as his Deputy. The Director of DoD Support was Lt. Col. E. Wilson, see Figure 2-12.1.

Although expanded to include more participants, the Test Director's organization does not appear to have significantly changed from previous operations. Gerald W. Johnson was the Test Director; the first one from UCRL. Sandia's Don Schuster served as Johnson's Associate Test Director, and Col. William S. Hutchinson was the Military Assistant to Johnson. Two sets of groups for functional and administrative matters reported to Johnson. Johnson had six Deputy Test Group Directors:

* LASL Test Group - John Clark
* UCRL Test Group - H. B. Keller with C. E. Violet as alternate
* Field Command Weapons Test Group - K. D. Coleman with A. H. Higgs as alternate
* Civil Effects Test Group - Robert L. Corsbie with two assistants: FCDA Assistant - E. R. Sounders and AEC Assistant - L. J. Deal
* Sandia Test Group - James H. Scott with G. W. Rollosson as assistant
* Project 57 Test Group - James D. Shreve

This was the first operation on which Sandia had its own test group. This group would conduct programs directly for the AEC. Sandia staff would also participate in the other test groups as they had done in the past.

There were extensive non-military observer activities for shots conducted during PLUMBBOB which were designated as "Open Shots". Although the AEC and DoD participated in these activities and had their own invitees, the FCDA was perhaps the prime "mover" and participant for these activities which are described later in this chapter.

PLUMBBOB TESTS

Table 2-12.1 shows the 29 tests conducted during Operation PLUMBBOB, and Figure 2-12.2 shows their locations.

Table 2-12.1 Nuclear Tests Conducted During PLUMBBOB.

TEST	SPONSOR	DATE 1957	TYPE	AREA Or HOLE	PURPOSE	YIELD kt
Boltzman	LANL	05/28	Tower 500'	7	Weapons	12
Franklin	LANL	06/02	Tower 300'	3	Weapons	0.140
Lassen	UCRL	06/05	Balloon 500'	9	Weapons	0.0005

Wilson	UCRL	06/18	Balloon 500’	9	Weapons	10
Priscilla	LANL	06/24	Balloon 700’	5	Weapons	37
Coulomb-A	LANL	07/01	Surface	3	Safety	Zero
Hood	UCRL	07/05	Balloon 1500’	9	Weapons	74
Diablo	UCRL	07/15	Tower 500’	2	Weapons	17
John	LANL/DoD	07/19	Missile 20,000’	10	Effects	~2
Kepler	LANL	07/24	Tower 500’	4	Weapons	10
Owens	UCRL	07/25	Balloon 500’	9	Weapons	9.7
Pascal-A	LANL	07/26	Shaft -500’	U3j	Safety	Slight
Stokes	LANL	08/07	Balloon 1500’	7	Weapons	19
Saturn	UCRL	08/10	Tunnel	U12c.02	Safety	Zero
Shasta	UCRL	08/18	Tower 500’	2	Weapons	14
Doppler	LANL	08/23	Balloon 1500’	7	Weapons	11
Pascal-B	LANL	08/27	Shaft -500’	U3d	Safety	Slight
Franklin Prime	LANL	08/30	Balloon 750’	7	Weapons	4.7
Smoky	UCRL	08/31	Tower 700’	2	Weapons	44
Galileo	LANL	09/02	Tower 500’	1	Weapons	11
Wheeler	UCRL	09/06	Balloon 500’	9	Weapons	0.197
Coulomb-B	LANL	09/06	Surface	S3g	Safety	0.300
Laplace	LANL	09/08	Balloon 750’	7	Weapons	1
Fizeau	LANL	09/14	Tower 500’	3	Weapons	11
Newton	LANL	09/16	Balloon 1500’	7	Weapons	12
Rainier	UCRL	09/19	Tunnel - 880’	U12b	Weapons	1.7
Whitney	UCRL	09/23	Tower 500’	2	Weapons	19
Charleston	UCRL	09/28	Balloon 1500’	9	Weapons	12
Morgan	UCRL	10/07	Balloon 500’	9	Weapons	8

(DOE 2000, 8-11; Lewis 1977, 31; Reeves 1980, 48)

Some Highlights of the Tests

The tests conducted during operation PLUMBOB included a number of firsts and a record. The record was Hood that, at 74 kt, was the largest yield ever fired in the atmosphere in the continental United States. It was suspended 1500 feet above the terrain on a tethered balloon.

The firsts were:

Lassen, 6/5/57, was the first shot with the atomic device suspended from a tethered balloon. A total of 13 balloon tests were conducted during PLUMBBOB. Eleven balloon tests were conducted at the NTS on HARDTACK Phase II.

John, 7/19/57, conducted by the Air Force, was a weapon system delivery test where an air-to-air missile was fired from an F-89 aircraft. The missile traveled 4.24 kilometers in 4.5 seconds (about Mach 3) before detonating with a yield of about 2 kt at 20,000 ft. The aircrew of the F-89 received 4 R, but no radioactivity was observed on the ground.

Rainier, 9/3/57, which was conducted by Livermore, was the first U.S. nuclear detonation with a sizable yield* (1.7 kt) that was contained underground. The Rainier test included diagnostic experiments in the zero room (that contained the device) to assess nuclear performance. Samples obtained post shot from Rainier also provided the first radiochemical yield from geologic samples. (DASA 1979: 5) [*Footnote: Saturn, Pascal-A and Pascal-B were underground shots that were detonated before Rainier, but they were safety tests with practically no nuclear yield.]

Of the 29 nuclear tests conducted during PLUMBBOB, 5 were safety tests: Coulomb-A (LASL), Pascal-A (LASL), Saturn (UCRL), Pascal-B (LASL), and Coulomb-B (LASL). The UCRL safety test, Saturn, was the first nuclear test conducted in a tunnel, but it had "zero" nuclear yield. Two LASL safety tests, Coulomb-A and -B, were fired on the surface. The other two safety tests by LASL, Pascal-A and -B were fired in unstemmed 500-foot cased holes. These were the first underground tests in holes of any significant depth. Robert Campbell recalls Pascal-A as the: "Biggest damn Roman candle I ever saw. It was beautiful. Big blue glow in the sky. I was up in the CP office, and that was fired from a little handset, out at the B-J Y." (Carothers 1995: 20)

The reason for firing the LASL safety shots in unstemmed vertical holes was to reduce, but not necessarily to completely eliminate, fallout. While they were successful in substantially reducing fallout there was a lot to gain by stemming the holes, as was done in later years.

UCRL concentrated their underground efforts on tunnels while LASL did the same on vertical holes. This line of work continued on HARDTACK Phase II in 1958. Underground tests became the standard for Nevada shots after the moratorium. Underground tests were the only venue after the Limited Test Ban Treaty outlawed atmospheric shots in 1963.* [*Footnote: The LTBT was signed in Moscow on

August 5, 1963 and ratified by the Senate on September 24, 1963. "It prohibits the parties from carrying out nuclear explosions in the atmosphere, in outer space, under water, or in any medium for which 'such explosions causes radioactive debris to be present outside the territorial limits of the State…'" (Thomson 1999:34-7)]

Several tests were particularly noteworthy. For example:
Priscilla. The DoD and FCDA both had extensive airblast and structures effects measurements on Priscilla that emphasized pressures above 50 psi. All of their structures programs were only on Priscilla. Measurements of neutron induced activity were also obtained.

Diablo failed to fire. On June 28, 1957, after the device had been armed, crews entered the area around the base of the tower to remove equipment that was not essential to the test. One of these pieces of equipment was the winch that operated the tower elevator. Power and cables were disconnected, and the winch removed. Unfortunately, the same circuit that supplied power to the winch also supplied power to the device cab on top of the tower.

The countdown reached zero, the firing signal was sent, and nothing happened. When the problem was discovered, it was decided that the only available option was to send a small team to the top of the 500' tower to disarm the device. Five men were selected: Barney Rubin, Forrest Fairbrother, and Walter Arnold, from Livermore; Edmund Tucker from EG&G; and Robert Burton from Sandia. After driving to the base of the tower, Rubin, Fairbrother, and Burton began the 500 foot climb. Arnold and Tucker stayed at the base as support for the climbers and to report their progress, via phone, to the CP.

The climb to the cab took 35 minutes, the final safety checks and disarming the device took about 5. When it was over, Burton called to Arnold at the base of the tower and asked to have the winch brought back in. They certainly didn't want to climb back down. (Livermore photographic display at NTS)

The Test Director called a meeting to assess the situation, and it was determined that either no or very low AC voltage was available at the zero rack in the tower cab. It was further determined that the AC power was effectively removed at the time the elevator and transformer were removed from the vicinity of the tower. Since it did not seem advisable to just replace the elevator and transformer, the arming party was forced to climb the tower to disarm the device. It turned out that the AC power for the device was connected to the utility power rather than to the instrument power and thus was disconnected when the salvage party removed the utility power.(Johnson 1957: 74) This incident led to better power monitors and the separation of instrument power, which was used only for test device hardware, from all other power sources. (Johnson 1994: C-6)

Saturn, which was fired on August 9, 1957, was designed to explore the feasibility of firing one-point safety tests contained underground. Two three hundred foot tunnels were mined concurrently with the Rainier tunnel. One for Saturn and one for a future shot. The containment was provided by blast doors, which proved satisfactory. Alpha was measured in order to diagnose the device behavior.

Smoky was a 44-kt Livermore shot, detonated August 8 in Area 2 atop a 700-foot tower. The military used Smoky to further its study of the effects of topology, rough and sloping terrain, on airblast and equipment. The skies were busy during Smoky with 84 aircraft, used mostly for Desert Rock troop maneuvers.

Fizeau was an unusual Los Alamos event in that it had a large vertical line-of-sight (LOS) pipe. On October 4, 1956, a meeting that included Sandia and Holmes & Narver was held to discuss the feasibility of fielding a 400 to 500-foot long two-foot diameter vertical pipe. The bottom of the pipe terminated in a concrete instrument shelter approximately 20 feet below ground level. Originally called the "Totem Pole", the name was changed to "Peace Pipe" on January 10, 1957.

This test was fired on a 500 foot steel tower, with the Peace Pipe located in the center. It was supported laterally every 50 feet with ¾ inch round steel rods connected to each tower leg. The device was suspended from the tower cab in order to reduce the amount of steel between the device and the top of the Peace Pipe located 100 ft below the cab floor. The recording shelter consisted of reinforced concrete, 28' square in plan view and 23' 6" deep with the top surface level with the natural grade. The top of the concrete was further protected with 2 feet of rock fill. The Peace Pipe extended to within 7 feet of the bottom of the recording shelter, with one side of the pipe accessible to a room for recording equipment.

Rainier. In November 1956 Livermore's G. T. Pelsor requested a 1,700 foot tunnel with four 90-degree bends terminating in a room located 90 feet from the main drift and 22 feet from the drift at 90 degrees to the main drift, see Figure 2-12.3. The cross-section of the tunnel was about 6 feet wide by 8 feet high in order to accommodate mining equipment. The tunnel designation was U-12b. U-12a was a 600-foot long tunnel near U-12b that the USGS used to conduct high explosive tests for geologic surveys during February 1957. (Reeves 1980: 228)

The reason for the four 90 degree bends in the Rainier drift, which led to the zero room (where the device was placed), was the expectation that ground shock would close off the drift and provide containment. This concept worked for Rainier, but the bends were not successful for later shots.

The French used a similar design for their first tunnel tests in granite at In Ecker, French Algeria, in the early 1960s. This probably didn't work well since hard granitic rock has a tendency to fracture and leak radioactive gases rather than form a seal.

F. B. Porzel directed measurements of the rate of shock growth in the rock. The Armour Research Foundation was the contractor doing the work. This was follow-on to the work that the Armour Foundation did on the underwater Wigwam test in May 1955. From the rate at which the supersonic shock wave travels in the medium, the hydrodynamic energy and total yield can be calculated. The shock growth measurements and yield calculations were successful. (Johnson 1957:63)

Many organizations participated with Livermore in the Rainier test. They include: the U. S. Geological Survey, Broadview Research Corporation, Sandia Corporation, Stanford Research Institute, Engineering Research and Development Laboratories, EG&G, and the U. S. Coast and Geodetic Survey.

This turned out to be an exceedingly successful and important test. Livermore demonstrated that testing underground was indeed feasible. They also showed that ground shock damage was manageable. The PLUMBBOB Test Director's report estimated shots up to a megaton could be fired in Area 12 without undue damage. Eleven years later Livermore fired the Boxcar event, at 1.3 Mt, in shaft U20i, about 15 miles northwest of Area 12.

Rainier was an exceedingly important Nevada shot. It had a significant influence both on the future testing programs as well as on international arms control negotiations.

The Grueling Schedule
PLUMBBOB had been a long haul. April 15 was the first readiness date for Project 57, but weather delayed it until April 24. Many of the LASL and Sandia people, who participated in Project 57 also participated in PLUMBBOB.

The first readiness date for PLUMBBOB Boltzman was May 15, and it was finally detonated on May 28. The last detonation of PLUMBBOB was Morgan on October 7, about 1 week short of 5 months from May 15. Weather was the chief culprit. Ultimately there were about two and a half months of delays attributable to unsuitable weather. Another month and a half was taken up by technical delays. The breakdown is as follows:

Delay due to unfavorable balloon weather	7 days
Delay due to unfavorable fallout weather	64 days
Delay due to unfavorable blast prediction	4 days
Delay due to technical reasons	45 days
Total =	120 days

The fallout weather delays started at the beginning of the operation with Boltzman having an 11 day delay followed by Franklin with a 7 day delay. The longest, Shasta, had an 18 day delay, and Whitney had a 9 day delay. The most significant technical delays were Diablo with 15 days and Franklin Prime with 14 days. (Reeves 1980: 48)

The Advisory Panel, who advised the Test Manager about the feasibility of proceeding with a scheduled event, conducted their meetings on site, at the CP, any time of the day or night as necessary to expedite each test. It was very common to convene a meeting near midnight then start another at 3:00 AM. Due to the duration of Operation PLUMBOB and the concurrent work responsibilities of Panel members in their regular jobs, the full membership certainly did not attended every meeting. (Reeves 1980:147) However, at least one representative from each of the key testing technical disciplines was required to be present.

There were 186 meetings held by the Advisory Panel between May 15 and October 7. It often happened that the delay for a test would extend into the time the next test was scheduled to be detonated. In these cases, both tests would be addressed by the Advisory Panel during their deliberations, as a "dual capability".

On a day of a detonation, Advisory Panel meetings might begin for the next shot as early as the same afternoon if the next test was "ready". Sometimes there would be a number of days before the next test was "ready", and the Advisory Panel would have a bit of a break.

Unexpected delays in the test schedule are very expensive because they result in idle time and then a "hurry up and wait" routine, which can be exhausting for those involved. Interestingly, the record indicates that if there were any doubt about the safety of those on or off site, the shot would be delayed. There is no evidence indicating that the Advisory Panel was ever pressured to proceed with a test against their better judgment.

PROGRAMS CONDUCTED ON PLUMBBOB

Each of the six test groups that reported to the Test Director conducted a series of programs. The following table identifies the programs by number. (Harris 1981b, 45)

Table 2-12.2 Programs conducted during PLUMBBOB.

PROGRAM NUMBERS	SPONSORINGG GENCY	GROUP PERFORMING PROGRAM	PURPOSE
1-9	DoD	FCWT Test Group	Weapons Effects
10-19	AEC	LASL Test Group	Weapons Development
21-26	AEC	UCRL Test Group	Weapons Development
30-39	FCDA	Civil Effects Test Group	Weapons Effects relative to Civilian Defense
41	AEC	Sandia Test Group	Weapons Effects and Weapons Development
50-53	DoD	Desert Rock: Army, Navy, Marine Corps, Air Force	Weapons Effects: Observations, Operations, and Technical Projects
62, 63,64	AEC	Sandia Test Group	Weapons Development
71-74	AEC	Sandia Test Group, PROJECT 57	Weapons Development

The seven sets of programs are briefly described in the order indicated in the above table.

DoD NWE PROGRAMS AND PROJECTS - PROGRAMS 1 - 9

There were 8 programs and 47 DoD projects on PLUMBBOB.

OPERATION	PROGRAM	# PROJECTS
PLUMBBOB	1) Blast and Shock	10
	2) Nuclear Radiation Effects	11
24 Tests	3) Effects on Structures	8
	4) Biomedical	3
	5) Aircraft Structures	5
	6) Electromagnetic Effects and Tests of Service Equipment	5
	8) Thermal Radiation Effects	4
	9) Support Photography	1
	Total	**47**

Some of the major NWE effects projects conducted during PLUMBBOB are described in the following. Although Priscilla was a weapons development test, the DoD used it for more of its NWE projects than any other shot.

Blast and Shock

While the 6 blast and shock projects on PLUMBBOB were extensive, they did not compare to the TEAPOT tour de force. Increased understanding was gained, but there were no major breakthroughs.

Priscilla was the first test that focused on the high-pressure region; up to 1,000 psi overpressure and 650 psi dynamic pressure. This focus of Projects 1.1, 1.3, and 1.4 by BRL, SRI, and Sandia (Bryant 1962, Swift 1960b, and Parret 1960) was stimulated by a need to understand the effects of the newer high-yield thermonuclear weapons and the increased hardening of potential military targets.

Measurements were also made by Project 1.1 in the low-pressure region (0.1 - 1 psi). (Bryant 1962) These low pressure measurements showed a large variation in maximum pressure and indicated that temperature and wind velocity can substantially change a shock wave at pressure below 1 psi.

Project 1.2 by NOL tested prototype HARDTACK I instrumentation suspended by balloons and rocket launched parachutes. (Hanlon 1957: 5,9)

An important result by SRI's Project 1.4 was that ground shock was reduced by about 70% within the first 20 feet depth.(Swift 1960: 3, 5, 13, 30-3)

Program 1.7 used 68 steel cylinders with flexible aluminum diaphragms to measure peak pressures between 50 and 200 psi with emphasis on 100 psi. (Bultmann 1960a: 5-25)

Airblast On Different Terrains
On Smoky, three projects focused on how the shock wave was affected due to rolling terrain, steep slopes, and rough terrain and on how the modified blasts affect vehicles. Five so-called Hill and Dale blast lines were used:
Line 1 - Flat Terrain line was used as the control for comparison with other lines.
Line 2 - Rolling Terrain line over small, gentle hills.
Line 3 and Line 4 - over ridges with approximately symmetrical front and back slopes.
Line 5 - Rough, Mountainous Terrain line.
For Project 1.8a, BRL instrumented all 5 of the Smoky Hill and Dale blast lines and tested a new prototype dynamic gage.(Bryant 1959: 4, 9-18)

SRI placed overpressure and dynamic pressure gages above the surface along Lines 1 (5 stations) and 3 (6 stations), project 1.8.
(Swift 1959: 117, 127-132, 135-137)

Fifty-one military vehicles, two M48 (WWII) tanks and 49 other lighter weight vehicles (jeeps, ¼ ton trucks, and utility vehicles) were exposed on Smoky along Lines 1, 2, and 5. Revetments with vehicles were exposed to explore the protection that they would provide. Information was developed such as the range and placement of the vehicles, the type of terrain; the resulting displacement; and the resulting level of damage. (Bryant 1959: 4-5, 9-23, 42)

Ground Motion Spectra
For many design purposes, it is desirable to specify ground motions in terms of shock spectra. For project 1.9, an instrument was developed by Air Research and Development Command to measure the spectra of the ground motion waves.
(Halsey 1959: 4, 11-15)

Nuclear Radiation Effects
Three projects investigated the neutron-induced radioactivity in three types of American soil, project 2.1.(DASA 1960c: 15, 20-22) Fifteen elements were exposed for project 2.2 that represented "significant elements in the soils of the world."
(DASA 1959: 11, 17-26)

Soil activation was studied in project 2.3 with neutron detectors, buried detectors, film badges, and chemical dosimeters.(DASA 1960b:5, 11, 16-20) All together about 1,500 measurements of the free field neutrons were made using gold, plutonium, neptunium, uranium, sulfur and other detectors at different ground ranges.

A test of communications techniques and equipment was fielded for project 2.7 prior to use on John. The hardware included gamma ray* detectors and transmitters, in underground bunkers near GZ, and receivers and general support equipment at the CP. The attenuation of EM waves was measured after 9 shots. [*Footnote: The attenuation of EM waves after an explosion depends on the number of free electrons present (i.e. on their production rate, removal rate, and conductivity). The production of free electrons occurs when gamma rays interact with air.] (Handscome 1962: 11-12, 16-17, 21)

Effects On Structures

Priscilla was the only shot to have the full complement of structures tests. In addition, two projects provided support to the structures projects: BRL, project 3.7, provided instrumentation to measure air blast and ground shock loading and the response of structures.(Meszaros 1960b: 5, 13, 23) The Waterways Experiment Station obtained data for Project 3.8 on the character and some physical properties of the natural soil to a depth of 200 feet in Frenchman Flat.(Goode 1959: 13, 15, 18-9, 23, 29-32)

Project 3.6 was an extensive project that involved the construction of 10 domed and arched structures by the Air Force and Office of Civil and Defense Mobilization to determine blast loading and structural response. This was a very comprehensive undertaking for both construction and instrumentation. Construction materials consisted of reinforced concrete, aluminum, and structural steel. The structures were located at 1180'(5); 1600'(4); and 2030'(1) on Priscilla. They were instrumented with hundreds of gages by projects 3.7 and 3.8. The ruins of some of the domes can be seen at the site today. (Bultmann 1960b: 5, 15-6)

Glass bottles were tested as materials for ground-shock isolation in project 3.5. Two test and one comparison structure were fielded by the Navy. The glass bottles which were stacked around the outside of the two test structures (at least some of the bottles appear to have contained alcoholic beverages) reduced accelerations to about 28% of their free-field values. (Vaile 1960: 5, 11, 21)

Biomedical

Project 4.1 - In Trimble, MI, about 1 October 1956, the breeding for 1,500 Hampshire-Landrace pigs began. Since pigs have a gestation period of about 4 months, they were about 3 months old on the first of May and weighed about 65 lbs. This was a reasonable weight for one person to handle. At this stage the pigs were shipped to the NTS. Pigs gain weight fast. After another month they weigh about 100 lbs, "a weight that makes handling in the field exceedingly difficult". (McDonnel 1961: 18) Also, larger sized uniforms were required for those pigs that were tested with them.

The animal holding facility was on the south side of Frenchman Flat, between the Mercury highway and the Frenchman Flat access road. This area had: "accessibility to a plentiful water supply, excellent natural drainage, good access roads, and isolation from Camp Mercury."

Normally, late spring at the NTS is a time of moderate temperatures and humidity (< 10%). Unfortunately, 1957 did not have a normal spring. There were days with temperatures below 60° F, rain, fog, and even icing. If a large number of the pigs had caught cold, the program might have been curtailed or eliminated.

Ultimately, pigs were exposed in June to the Franklin (6/06), Wilson (6/18), and Priscilla (6/24) shots. Also, four exposure sub-projects that used 710 of the pigs were conducted. (McDonald 1961: 17-20, 24, 118, 147, 158) Attachment IV by Ed Giller provides additional "personal perspectives" regarding pigs.

Aircraft Structures

Naval tactics involved nuclear weapons delivery against hostile submarines. The HSS-1 aircraft used in project, 5.1, was the first helicopter instrumented to obtain experimental flight results at the NTS. (Walls 1960: 5, 11, 18)

The Navy was also considering the use of blimps for attack delivery of the Mark 90 and Lulu Anti-Submarine Warfare (ASW) weapons. Project 5.2 was focused on the response characteristics of the Model ZSG-3 blimp when subjected to a nuclear detonation in order to establish criteria for safe escape distances. Presumably the results would be generally applicable to all other lighter-than-air types.

Two blimps (K-46 and K-77) were extensively instrumented for participation in 15 shots during Operation PLUMBBOB. K-46 was to be used only for moored tests on the ground and K-77 was to be used only for in-flight exposures.(Gilstad 1960)

In April 1957, K-46 was flown from Lakehurst, NJ to the NTS. Two days after arrival, K-46 was destroyed when it was torn from its mooring mast on Yucca Flat by a violent windstorm. Instrumentation was recovered and used on yet another blimp, K-92 that was in NJ at the time. Toward the end of May, K-77 was flown to Nevada and was moored to a mast at a horizontal range of 18,200 feet from the Franklin GZ. Following the passage of the shock wave, K-77 became detached from the mast due to failure of the mooring cone and could not be re-moored. It was destroyed. Fortunately K-77's instrumentation equipment was not damaged.

K-92 was flown to the NTS two days before the scheduled date for Wilson. The firing date was repeatedly postponed; and 7 days after arrival, K-92 was destroyed when it was torn lose from its mooring mast at Yucca Flat by a violent dust devil. Again, all instrumentation equipment was recovered without damage.

Finally, K-40 was assigned to the project and equipped with limited instrumentation. It was exposed to Stokes as a free balloon at a level attitude approximately 300 feet above ground. Mooring lines were released about 20 seconds before shock arrival to obtain free-body response data. Immediately following shock arrival, the blimp's envelope ruptured forward of the car, and the blimp crashed but did not burn. Apparently the fact that the shock velocity in helium is not the same as the velocity in air came as a surprise to the experimenters. (Ogle 1985:96) The result was that the shock inside the envelope was not balanced by the shock outside. Fortunately, the instrumentation was salvageable, but the data were "not adequate to predict all of the critical-response parameters for blimps". (Gilstad 1960) These events clearly demonstrated that the instruments were rugged! Mercifully, blimps faded from the scene as nuclear weapons delivery platforms.

An Example Of Advances In Nuclear Weapons And Delivery Systems
The A4D-1 was a naval carrier-based light weight single place aircraft. It had an assigned delivery capability for a wide variety of nuclear yields that extended into the megaton range. The special weapons were carried externally on the fuselage centerline bomb rack. Nuclear field tests were sought for verification of the theoretical analyses regarding the A4D-1's response characteristics, in particular its wing design. Two A4D-1s were instrumented and flown for Project 5.4, with the second craft used primarily as a backup.(Walls 1958) Note, PLUMBBOB was approximately 20 years after the weapon drops on Japan which were about like Trinity in yield (15 kt Hiroshima and 21 kt Nagasaki (DOE 1994) and used a large B-29 and its crew for delivery. Note that the A4D-1 was a "light weight single place aircraft" based on an aircraft carrier that could deliver a megaton.

John
The John shot was a full-up test of the Air Force's F-89-1capability for delivery of an air-to-air rocket with a nuclear warhead. An instrumented F-89J accompanied the F-89-1 delivery aircraft. At launch time, the instrumented F-89J and the delivery aircraft banked in opposite directions; and each performed typical escape maneuvers. This F-89J also participated in 13 other shots in various positions relative to the detonation point, project 5.5. (Stalk 1960: 4, 11,19)

Minefields
One hundred and fifteen mines of 15 types from the US and foreign countries* were exposed on Priscilla project 6.1. [*Footnote: The foreign mines were from Britain, Denmark, Italy, USSR, Belgium, Germany, and France.] The mines were deployed in arrays at ranges from 920 to 5320 feet from GZ. Most of the emphasis was placed on determining:
1) Reliability of actuation predictions,
2) Effects of DOB,
3) Extent of sympathetic actuation (the actuation of a mine caused by the explosion of another mine), and
4) Whether special methods were needed for prediction of mine actuation in the precursor region. (Deeds 1960: 5, 13-4, 16-7, 21, 24, 27)

Another minefield project, 6.2, was designed to see whether the magnetic fields from atomic detonations could cause mine fuzes to detonate or alter the fuze's sensitivity. (Haas 1962: 5, 11, 28, 44-6)

Navigation Aids
During the 1950s, navigation relied heavily on the LOng Range Aids To Navigation (LORAN) system which was extensively used before satellites for obtaining one's position. Position was determined from the intervals between signal pulses that were received at 1 location (like a ship) from 2 (or more) widely spaced radio transmitters. NAROL, the inverse of LORAN, used 1 atomic detonation as the transmitter and 2 receiving stations, spaced a goodly distance apart. The basic principal of both Loran and Noral is that the velocity of electromagnetic waves is essentially a constant. The difference in propagation time of the bomb pulse over

two paths is a measure of the difference in length of the 2 paths. These data together with knowledge of the location of the 2 stations can be used to calculate the location of the detonation.

An IBDA system based upon NAROL was under development and was fielded as project 6.4 on all of the shots except Shasta, Whitney and Morgan. Receiver stations were set up at Albuquerque, NM (540 mile range), Vale, OR (480 miles), and Rapid City, SD (830 miles).(Houghton 1958: 5, 11-3, 20)

Investigations Within The Fireball

Project, 8.3b, is a continuation of important work within the fireball that was begun on TEAPOT by Wright Air Development Center (WADC), see Chapter 10. This work was applied to the structural vulnerability of ICBMs. Instrumentation for making measurements within fireballs was tested and information on thermo mechanical effects, including ablation was acquired.

An electrically instrumented plastic sphere with a time-history instrumentation system was placed 100 feet from the Priscilla ground zero above an open hole 8' deep. The sphere was not intact after the shot, but its major components were recovered from 200 to 600 feet from their original position.

Twenty-four specimens were exposed on Smoky. They consisted mostly of spheres and cylinders of different materials. All but one were within the fireball. Two specimens were instrumented cylinders with missile-like nose cones that were attached to a tower and oriented with their noses toward the device. Both had inserts of different materials whose ablation characteristics were of interest. Twenty-one other specimens were supported by a special cable array that connected to the 700 foot Smoky tower like another set of guy wires but with much less angle. This allowed specimen to be located within 150' of the device. All but 2 of the specimen were within 450'.

On Smoky, all but one of the specimens were recovered post shot. The tape transport system from 2 of the 5 tape recorders operated satisfactorily. Three tapes yielded signals. The velocity-distance impact gages yielded apparently reliable velocity versus distance data. The state of the art of recording fireball phenomena was advanced from this test, but there is always the desire to get closer and to understand more. (Cosena 1961: 5, 17-20, 31, 35)

Supporting Measurements

As project 9.1, the Lookout Mountain Laboratory Group from Hollywood, CA, provided support for the technical photography of the DoD programs and projects. They provided documentation of the detonations for historical purposes as well as release to the press through the Joint Office of Test Information (JOTI).

WEAPONS DEVELOPMENT - PROGRAMS 10 - 26

There were four general objectives of the Los Alamos and Livermore nuclear weapons test programs on PLUMBBOB. The first was to test designs of smaller and lighter weapons. The second was to test designs that used the least amount of plutonium and uranium. The third was to explore design options that would enhance the safety of nuclear weapons in an accident scenario. The fourth objective for PLUMBBOB tests was to lay the groundwork for higher yield designs that would be taken to the Pacific for full yield tests. (Johnson 1957:.9)

EG&G Support

EG&G supported the technical sponsoring organizations on all of the PLUMBBOB shots. Additional tasks were in collaboration with the AEC, Sandia, Holmes & Narver, and the Lovelace Foundation. Timing and Firing were perhaps the most visible areas of support. They also provided Bhangmeter measurements for early-time yield estimates.

EG&G also provided technical photography with, for example, data on fireball growth for one of the most important yield measurement techniques. In addition to taking data EG&G also analyzed the results and provided technical interpretations.

Reaction history (alpha) was measured on the Los Alamos shots by EG&G. This was a durable responsibility of EG&G, and they continued in this role through the last test in 1992. In addition, EG&G developed new detector technology.

Perhaps the major challenge that EG&G faced during PLUMBBOB was the short intervals between shots and the large number of nuclear tests. For example there were as many as ten dry runs in a day. This didn't come as a surprise, but never the less it taxed the personnel and the equipment to the limit. It was clear that EG&G would need greater resources to field this many shots on this kind of schedule in the future. (Reeves 1980: 71)

ARMING AND FIRING DRY RUNS

> The Arming Organization was responsible to the Test Director for arming and disarming the nuclear weapons and devices. ... Dry runs were conducted in order to check that equipment operated properly while connected in the same manner as at shot time. There were four different type runs conducted: Regular, Power, Frequency and Hot Run. On a Regular Run, timing signals were available to all experimenters, but participation was voluntary based on a dry-run need. The Power Run was mandatory for all experimenters and established that sufficient power would be available at zero time. The Frequency Run was mandatory and proved that all experiments operated successfully free from RF interference. The Hot Run provided the weapons people a chance to test the gas plumbing and live pit under simulated shot pressures. (Johnson 1957: 71-73)

The interlock process was particularly elaborate on complex shots. The interlocks were in the firing circuit to prevent firing if critical experiments or processes were not operational. For example, the reaction history experiment was a key part of every shot. If for some reason the experimental apparatus was not working properly, the firing sequence was interrupted. Each interlock was tested individually in both the go and no-go configurations.

> Arming operations consisted of making final checks and connections to the device and associated equipment preparatory to firing. All arming activities were carefully performed with the aid of check sheets, which thoroughly enumerated each operation. Progress of the Arming party was reported to S-3 (Planning, under G. P. Stobie of Sandia) at specific check points. Final arm connections were completed as close to zero time as practical and only with the permission from the Test Director or Associate. This allowed experimenters the maximum time to secure stations and vacate the forward area. It also gave the weather panel an opportunity to evaluate data obtained closer to shot time thereby minimizing the chance of disarming due to changing weather conditions. (ibid., 73)

Following the equipment checkout at GZ, the leader of the arming party would contact the Test Director to obtain permission to make the final connections that completed the arming process.

After the final arming activities at ground zero, the Salvage Party entered the area to remove the equipment to be recovered before the shot. On tower shots this included the elevator hoist and power transformer. On balloon shots the Salvage Party removed the greenhouse*, bridge crane, and the power generator. (ibid., 73) [*Footnote: The "greenhouse" was a portable structure used to house the balloon cab during the preparation for the experiment. On later shots, in vertical holes, the greenhouse was the name given to the structure that housed the "rack" that contained the device and experiments while they were being assembled, prior to their being lowered to the bottom of the hole.]

Following the arming operation and the removal of the equipment to be saved at ground zero, the Arming Party and the Salvage Party would return to the CP. On the way to the CP the Arming Party would stop at the timing distribution station and take the final steps necessary to prepare for the shot. (ibid., 73-74)

At the CP, the arming party would make the final equipment checks and preparations for the shot and await permission to fire from the Test Manager.

Disarming, when necessary, was approximately the reverse of the arming process. There were two reasons for disarming: 1) Shot delay due to weather or technical difficulties and 2) A misfire or device related problem. Disarming, of course requires access to the nuclear device. For 3 events: Buster Able, 100' tower; SNAPPER Fox, 300' tower; and PLUMBBOB, Diablo, 500' tower, the arming party returned and climbed the tower to access the device. The elevators were replaced for the return of the arming party from the top of the tower.

FCDA PARTICIPATION ON PLUMBBOB

Open shots and field training exercises were the chief interests for the FCDA on Operation PLUMBBOB. The FCDA did not use a code name analogous to the CUE or DOORSTEP, names of programs fielded on previous series; but they did have an extensive array of programs.

Open Shots

About 1,350 observers were witnesses of a shot during PLUMBBOB. (Reeves 1980: 415) This number would have been larger, perhaps by a factor of two, had there not been the weather delays. Ultimately, there were 47 postponements on the open events; and as a result, many found it impossible to remain long enough to see the shot. (ibid., 261)

Observers came from a variety of professions. There were 334 newsmen, including 21 foreign correspondents. There were 394 from the military, the AEC, and Congress. In addition, the DoD invited all foreign countries with whom the US had formal defense agreements to send observers; and about 130 came. (ibid., 414)

The following AEC press release of May 10, 1957 describes the invitation. The list of countries that were invited (like Pakistan) or not invited (like India and Israel) is noteworthy and indicative of international affairs at the time.

> Observers from the military services and/or the Civil Defense organizations of 47 nations are being invited by the appropriate United States Government Agencies to witness specified nuclear detonations in the PLUMBBOB test series scheduled to begin May 16 at the Nevada test site of the United States Atomic Energy Commission. All tests in the series will be of low-yield fission devices essential to the defense of the Free World. --------
> All of the 47 nations invited to send either military or civil defense observers have also been informed that news media representation from these countries may be present to report at least one of the series open to reporting by United States media. -------------
> The purpose of inviting the attendance of observers and news media reporters from these 47 nations is to familiarize them with United States nuclear weapons testing policies and operations, especially safety procedures.
> The countries invited: "Argentina, Australia, Belgium, Bolivia, Brazil, Canada, Chile, Costa Rica, Cuba, Denmark, Dominican Republic, Ecuador, El Salvador, Federal Republic of Germany, France, Greece, Guatemala, Haiti, Honduras, Iceland, Iran, Iraq, Italy, Japan, Luxembourg, Mexico, The Netherlands, New Zealand, Nicaragua, Norway, Pakistan, Panama, Paraguay, Peru, Philippines, Portugal, Republic of China, Republic of Korea,

Spain, Sweden, Thailand, Turkey, United Kingdom, Uruguay, Venezuela, Viet Nam.

The dates of shots to be observed by news media: May 16, June 25, June 27, July 9, July 18, July 28, August 8, August 19, September 1."

(Office of Test Information 1957)

This press release, as well as others, was issued by the Joint Office of Test Information (JOTI). The FCDA and the DoD supplied public information specialists and secretarial staff to these joint offices, which were operated by the AEC. (Reeves 1980:261)

Approximately 200 CD observers witnessed at least one of the 10 open shots. One group had 31 foreigners from 10 countries and included members of the NATO Civil Defense Committee, the Swedish Civil Defense Director, and designers of the French and German shelters. Observer groups for the other open shots included people with national, state and local civil defense responsibilities. (ibid., 261, 414)

There were other observers as well, from organizations that the AEC or DoD felt would benefit from the program. The list included: the Boy Scouts of America, the U.S. Public Health Service, local law enforcement agencies, and mining and industrial groups. Also, AEC Operations officers invited a number of individuals, who would not normally be involved in the tests, from across the nuclear weapons complex. (Reeves 1980: 308-10, 413-6)

Field Training Exercises

Two groups of combined state and local radiological defense personnel attended unclassified training sessions. The two groups totaling 56 trainees convened from August 8 through September 3. Both groups conducted surveys of fallout fields from earlier PLUMBBOB shots, and one of the groups participated in Program 35 activities, which included decontamination, monitoring, and instrumentation evaluation. Finally, the FCDA produced a training film that demonstrated the techniques used in monitoring fallout levels. (Killian 1958)

FCDA NWE PROGRAMS AND PROJECTS - PROGRAMS 30 - 39

By 1957, industry had cumulatively paid about $2.5 million for information that had been sponsored and obtained by FCDA. When the information that industry wanted also had value to FCDA, FCDA would accept the monies and conduct the work. Over 200 companies had participated, and their contributions ranged from providing materials for exposure to sponsoring major experiments. (Reeves 1980:256) In a manner similar to that used for US industry, France and Germany sponsored U.S.-fielded civil defense shelter projects on PLUMBBOB.

There was no formal FCDA operation like DOORSTEP or CUE on PLUMBBOB, but most of the CD projects were conducted at least on Priscilla. All of the U.S.

structures projects, including shelters, were fielded on Priscilla. The French and German shelters were tested on Smoky.

A number of the Civil Defense projects were originally planned for more shots than the after-shot record indicates. (Corsbie 1957) This was undoubtedly because of the delays that occurred during PLUMBBOB.

OPERATION	PROGRAM	# PROJECTS
PLUMBBOB	30) Shelters for Civilian Population	10
24 Tests	31) Structures, Equipment, Devices, and Components	5
	32) Radiological Countermeasures	4
	33) Biological Assessment of Blast Effects	6
	34) Physical Response to Blast Loadings	5
	35) Radiological Defense Techniques	4
	36) Radiological Defense Operations	5
	37) Radio-Ecological Aspects of Nuclear Fallout	6
	38) Effects of Radioactive Fallout on Foodstuffs	4
	39) Instrumentation and Dosimetry	13
	Total	**62**

Domes

An extensive project of 3 domes and a drawbridge-type door was conducted with the Air Force on Priscilla as DoD Project 3.6 and FCDA Project 30.1.(Neidhardt 1957) These projects involved considerable construction and instrumentation efforts. The domes were located at pressure levels of 20, 35, and 70 psi; and they can be seen today at the site. Armour Research Foundation measured deflection and strain versus time on the domes as Project 30.5a. This required the construction of two new instrumentation bunkers; one near the 70 psi dome, and the other near the 35 and 20 psi domes.(Brittain 1957: 3-4, 8-14) In addition, as Project 30.5, BRL made extensive airblast and ground shock loading measurements on the domes. (Meszaros 1960a:5)

Parking Garage

The largest shelter was a square, reinforced-concrete, dual-purpose underground parking garage 87 feet on a side, with nine interior columns 29 feet on center. It was constructed as Project 30.2 at 1600 feet from the Priscilla ground zero, where the over pressure was predicted to be about 35 psi. The entrance was a vehicular ramp along one side of the structure. The roof slab was 3 feet below grade; the walls one foot thick; and the exposed wall along the ramp was 4.5 feet thick for radiation protection. Diagnostics included blast and radiation instrumentation.(Cohen 1961a: 15, 13, and 24)

Family Shelters

Prior to the proliferation of megaton weapons, family type shelters were developed to protect occupants from peak incident overpressures of about 15 psi and to provide radiation protection equivalent to about 3 feet of earth cover. Project 30.3

tested a second generation of shelters for overpressures of 30 psi or more. Space was provided in the shelters for storing supplies to sustain 6 persons for 7 days.

Three shelters were fielded at the 65, 48, and 30 psi overpressure levels. As a shelter was entered, stairs led to an entrance corridor with two 90° bends at a level about 6.5 feet below the surface. After the second bend was a hallway about 5' long with storage on one side that led to the main shelter area, which was about 10' x 7' x 6.5' high*. [*Footnote: This was a floor area of slightly more than 10 ft^2/person] Five and a third feet of earth covered the eight-inch concrete roof, through which protruded an antenna block, an air-driven exhaust ventilator, and an additional vent pipe. A price for this shelter was not quoted, but it was obviously much more expensive than the family shelters tested during BUSTER. This design was the state-of-the-art in family shelters at the end of atmospheric testing in the 1950s. (FitzSimons 1957: 3-10, 14-8, 25-6)

Permanent deformations and time resolved deflections were measured in the 30 psi shelter. Radiation instrumentation was "comprehensive and included both gamma and neutron measurements" conducted by projects 39.1 and 39.9.

French and German Shelters

United States policy was to furnish unclassified information on nuclear effects to the governments of allied nations. The French and Germans used their own resources to develop shelter designs and test plans for projects that provided data for the FCDA as well as for themselves. Both countries engaged Ammann & Whitney as their US agent at the test site.

Five French (Cohen 1962, Project 30.6) and nine German (Cohen 1962, Project 30.7) shelters of significant size were tested on Smoky. These were substantial projects on the part of the French and Germans scientists and engineers as well as the field support at NTS. The structures tested appear to certainly rival and in some cases even exceed the US shelters being tested. These two projects represented substantial efforts and generated considerable public interest. They are briefly described in Killian 2011, and details are provided by the Cohen reports.

The Mosler Safe

The Mosler Safe Company contracted with Ammann & Whitney to design a vault of rectangular structure 12 feet 7 ½ inches by 8 feet 3.4 inches by 8 feet high. "The walls and roof slab were 18" reinforced concrete lined with a ½ "-thick steel plate. The vault was anchored into a large (23' 1 ¾" x 33' 9") mat foundation from 2 to 6 feet thick to prevent it from overturning. The structure, Project 30.4, was radial to and 1150 feet from the Priscilla GZ. "The 7 ½ ton door was mounted on a steel box frame weighing 14 ½ tons and was placed facing GZ." The vault with the door closed was essentially gastight. The longitudinal center line of the structure was radial to GZ. Sandia provided air pressure and maximum deflection measurements for the vault. It came through the test with flying colors. "Although the exterior hardware was destroyed, the door was easily opened by a person familiar with its operation." The door and frame has been removed, but the remains of the vault can be visited at the NTS today. (Cohen 1961b: 3, 11-2)

Personnel Experience Shots in a Shelter
The shelter for project 32.3 was 25 by 48 feet and had a minimum earth-cover thickness of 3 feet, which provided an average shielding reduction of a factor of about 10,000. It was 5200 feet north of Diablo, 4.75 miles north of Kepler, and 2 miles north of Shasta. During the pre and post shot periods of these tests, personnel inside the made a variety of measurements to evaluate shelter performance. They also tested several methods for determining the radiological environment outside the shelter. (Strope 1959: 5, 13-4, 37)

Prior to Diablo, three plots between about 200 and 1000 yards from the shelter were identified as reclamation areas. These three plots were each about 500 ft on a side, and were worked after the shot with heavy earth moving equipment to evaluate decontamination techniques. The work began after the radiation had subsided to a safe level, which was assessed by monitors who were in the shelter at detonation time. (ibid., 17)

Sixteen people occupied the shelter during Diablo. Fallout arrived at the shelter about 6 min after the burst, and higher-than-expected radiation readings forced work outside to be postponed for two days. Fourteen people left the shelter about eight hours after shot time, leaving two people to continue with the data collection.

The post-detonation schedule called for two 2-man monitor teams to leave the shelter, man jeeps, and survey two of the reclamation areas. When the radiation subsided to an acceptable level the grading and scraping began and monitoring continued. The work continued for about 5 hours, after which the shelter was closed and personnel returned to the CP Rad-Safe area.

Unfortunately, the yield of Kepler was less than anticipated and the fallout was negligible. As a result, no useful data were collected. Five people occupied the shelter during Shasta and fallout arrived about 10 minutes after the detonation. The post-shot work outside the shelter started about an hour after the event.

Flying Dummies
Anthropomorphic dummies and equivalent spheres (idealized models having an acceleration coefficient about equal to that of the dummy) were fielded as Project 33.3 on Priscilla and Smoky to determine the velocity-time and distance-time histories resulting from blast winds. Photo triangulation was used for recording the movement. (Taborelli 1959: 5, 17-9)

Burro and Monkey Exposures
Burros (project 39.6a) and monkeys (project 39.6) were exposed to the same environments during Wilson and Fitzeau. These projects were designed in order to obtain comparisons between reactions to radiation exposures between large and small animals. (Kuhn 1958: 1-6; Pickering 1958: 5, 11-4; Pickering 1959: 5, 12-5)

SANDIA TEST GROUP - WEAPONS DEVELOPMENT PROGRAM, PROJECT 41
Sandia conducted three AEC sponsored projects during Operation PLUMBBOB that were concerned with early time phenomenology.

Project 41.1 Fireball Studies had three main objectives:
- To explore material properties in the vicinity of a fireball,
- To explore weapon component properties in the vicinity of a fireball, and
- To advance the basic understanding of fireball physics.

Project 41.1 activities were conducted were Boltzman, Shasta, and Fizeau. (Harris 1981: 174)

Project 41.2 Weapons Vulnerability had two main objectives:
- To increase the general knowledge of the vulnerability of nuclear weapons to nuclear bursts, and
- To demonstrate the ability of Sandia to build a nuclear device hardened against nuclear effects and as similar as possible to an operational nuclear weapon.

The tests on which project 41.2 were conducted were Shasta and Fizeau. For Fizeau, hardened nuclear devices were located atop television towers within several hundred feet of the Fizeau device (which was on a 500 foot tower). The recording station was located on Red Rock Butte. (ibid., 174-5)

Project 41.3, Neutrons Versus Altitude, was designed to determine the effect of ground terrain on the free-field neutron flux during the Wilson event. Wilson was detonated from a tethered balloon at a height of 500 feet. This project was conducted by attaching foil neutron detectors on the ground and on the vertical mooring cables of polyethylene balloons that were located at:

900 feet from GZ at an altitude of	700 feet,
1,800	1,200 feet,
2,700	1,500,
3,600	1,500 feet. (ibid., 175)

DESERT ROCK, EXERCISES VII AND VIII - PROGRAMS 50 - 53
The DoD conducted Desert Rock Exercises, Desert Rock VII* and VIII*, on Operation PLUMBBOB. Although not known at the time, these were the last Desert Rock Exercises. [*Footnote: The difference between Exercises titled VII and VIII appears to be only the change in fiscal year (FY) which occurred on July 1, 1957. Archived references prior to July 1, 1957 refer to Desert Rock VII; and those after July 1, 1957, refer to Desert Rock VIII.]

Desert Rock Exercises VII and VIII were extensive operations, considerably more complex than those that occurred earlier in the decade. The Army had Program 50, with 9 projects; the Navy had Program 51 with 3 projects; the Marine Corps had Program 52 with 3 projects; the Air Force had Program 53 with 10 projects; and the Royal Canadian Army and Air Force conducted 3 programs, but they did not have program numbers. There were: Troop Maneuvers on Hood, Smoky, and Galileo;

Troop-Observer indoctrination Projects; Training Projects; Technical Service Projects; and Operational Training Projects. These are all described in Appendix N.

A Military Critique of Desert Rock VII and VIII

Both DoD and AEC personnel worked very hard to make the NTS operations run smoothly. Good communication and understanding was much more difficult to achieve with the Desert Rock exercises than with AFSWP's effects activities. Desert Rock involved thousands of men who were not involved with the technical success of the operation. Also, almost none of the Desert Rock personnel had clearances and were therefore stationed at Camp Desert Rock, rather than at Mercury. The nearly 5 month duration of Operation PLUMBBOB due to weather and technical delays undoubtedly added to the frustration experienced by Desert Rock personnel. The frustration of Major General John Binns was expressed in his report regarding the PLUMBBOB exercises:

> Due to limited state of training of participating troops and the many restrictions necessarily imposed by the Atomic Energy Commission, unrealistic and artificial requirements which preclude valid conclusions were imposed on the tactical maneuver at the Nevada Test Site.
>
> In view of the experience gained at Exercise DESERT ROCK VII and VIII, it is imperative that the Army make every effort to obtain a maneuver site and low-yield tactical weapons to be utilized by an Army Commander to train troops and test atomic tactical doctrine without the inhibiting restrictions imposed by technical testing and instrumentation therefore. The goal would be to train and test units of at least the size of Battle Group, supported by small yield atomic weapons. (Binns 1957)

SANDIA PROGRAMS 62, 63, 64

Many of the Sandia projects in these programs focused on nuclear test technology. For example, balloon suspension systems and cab designs were extensively tested on PLUMBBOB. Also, Sandia was responsible for zipper design; and they made measurements of the neutron outputs of these components, which provided the neutrons to initiate the nuclear chain reaction in a weapon. In addition, they worked on the development of high speed telemetry that was used to transmit the diagnostic data to the recording stations.

PROJECT 57 TEST GROUP - WEAPONS DEVELOPMENT PROGRAMS 71-74

A high explosive test, code-named Project 57, was conducted on April 24, 1957 in order to study the hazards from accidents that involved nuclear weapons with plutonium. Although the explosion occurred prior to the nuclear tests on PLUMBBOB, it is considered to be a part of that operation because some activities associated with Project 57 occurred during the nuclear testing activities.

G. Johnson, the Test Director of PLUMBBOB, assigned James D. Shreve as Deputy Test Group Director for Project 57. Thus, the Project 57 Test Group operated in a manner similar to the operations of the other Test Groups. (Shreve 1958, 20; Harris 1981, 41)

Project 57 was started when the Albuquerque Operations Office asked Sandia to assume responsibility for arranging the experimental program and selecting a site. The location was in the Groom Lake region in what was known as Area 13, not within the NTS but just over the border within the Nellis Bombing and Gunnery Range. This site was selected because it had a low plutonium background, was reasonably flat for about 50 square miles, and could be used for an extended period of time on a non-interference basis with both weapons testing and the Area-51 programs. (Shreve 1958, 19)

The four objectives of Project 57 were to:

- Estimate immediate and long-term distribution of plutonium and gain an understanding of how this distribution comes about.
- Conduct a biomedical evaluation of plutonium-laden environments.
- Investigate relevant methods of decontamination.
- Evaluate alpha-field-survey instruments and monitoring procedures.

These four objectives had program numbers 71-74 on PLUMBBOB and were headed by individuals who reported to Shreve. (ibid., 20)

Program 71 - Particulate Physics: More than 4,000 sticky-pan collectors were distributed over an area of about 43 square miles. Air samplers, balloon-borne precipitators, soil samples, and photographic methods were used. M. Cowan of Sandia lead this programmatic effort.

Program 72 - Biomedical Field Study of Plutonium Inhalation: This program studied environmental short-term and chronic rates of exposure by using two groups of animals: one exposed to the radioactive cloud and the other placed in the contaminated zone. J. N. Stannard of the University of Rochester directed this program.

Program 73 - Plutonium Decontamination: This program investigated techniques for plutonium removal from large land surface areas, from concrete and asphalt pads, and from materials used in construction such as concrete, stucco, brick, aluminum, and steel. Decontamination techniques were: washing, vacuuming and steam cleaning, plowing, leaching with water, fixation and subsequent removal. E. A. Pinson of AFSWC directed this program.

Program 74 - Surface Alpha Monitoring: This program was conducted on the soil and brush in the area as well as on concrete slabs that had been placed in Area 13 adjacent to the sticky pan collectors. R. E. Butler of Sandia directed this program. (Harris 1981, 177-8)

As of 2004, the area of the Project 57 test had not been decontaminated. (Martha DeMarr, Private communication April 8, 2004)

MUSTER BADGES AND AREA SWEEPING

Security force personnel secured the forward area, just beyond the Control Point, the evening before a planned shot. Individuals entering the area were issued muster badges. The security force then started the sweep of the forward area issuing muster badges as they went along to those already in the area. Those in the forward area were requested to leave before the arming party started arming the device in the early morning hours of shot day. As individuals left the forward area they would turn in their muster badges to the security force at the forward area check point; guard station 300 for operations in Yucca Flat. The Test Manager's permission was required for a group to remain in the forward area after the arming party started the arming procedure. It was required that all muster badges be returned to the security force before initiation of the firing sequence on shot morning.

Usually two hours prior to the scheduled shot time the area between Camp Mercury (Guard Station 200 at the crest of the divide between Mercury and Frenchman Flat) and the CP was closed, as was the gate to the CP. This was usually after the arming party returned to the CP. At that time the muster inventory should have been complete and the final security sweep initiated from north to south to confirm that the area forward of Station 300 was clear of people. When the security force confirmed that the forward area was clear the Test Manager's Panel continued with the assessment of the weather conditions in preparation for the final permission to fire. The Test Manager usually gave the Test Director permission to fire the device about a half hour prior to shot time.

ENGINEERING AND CONSTRUCTION SUPPORT

Engineering and construction support came principally from Reynolds Electrical and Engineering Company (REECo) and Holmes and Narver. Holmes and Narver's function was to supply required design, establish construction schedules, conduct field inspections, and administer contracts of the lump sum contractors. Reynolds's function was to operate and maintain the camp and other permanent test site facilities; perform construction that could not practically be done by lump sum contractors; and to provide field support required by various using agencies. As an added task, Reynolds established a Rad-Safe organization to provide monitoring services and film badge analysis.

AIRCRAFT PARTICIPATION

Both the DoD effects measurements and their Desert Rock programs required participation by many aircraft. In addition, aircraft were used for cloud tracking and sampling and other special program activities. There were no aircraft used on 3 of the 5 safety shots, Pascal A, Saturn, and Pascal B. Both Coulomb A and B used 7 aircraft. The numbers of aircraft ranged from 14 on Laplace to 84 on Smoky. In total, there were 786 individual flights associated with the programs conducted during PLUMBBOB. (Harris 1981, 149)

LIVING CONDITIONS DURING OPERATION PLUMBBOB

During PLUMBOB, the maximum on-site population, excluding the participants in Desert Rock Exercises, reached approximately 3,000. Additional sleeping quarters were built between Operations TEAPOT and PLUMBBOB. There were now 35 dormitories for men that could accommodate 1320. There were fewer hutments, which now held only 972. Two hundred thirty-five trailers had been procured that housed 940 men. Thus, men's living quarters could now handle a total of 3232. Three women's dormitories were built for a total capacity of 180. The trailers for women were no longer used.

The allocation of men's housing space during PLUMBBOB was as follows:

Organization	Dormitories		Trailers		Hutments		Total Spaces
	#	Spaces	#	Spaces	#	Spaces	
Test Manager*	3	124	-	-	-	-	124
Test Director	19	722	150	600	88	384	1706
Support Director	10	362	85	340	77	540	1242
DoD	3	12	-		6	48	160

(*Includes space for official visitors and miscellaneous participating agencies)

REECo was still the contractor for housing, feeding, and related personnel services. The food remained very good. Cafeteria #1 was on straight cafeteria-style service for all meals, with breakfast from 5:00 AM to 8:00 AM, lunch from 11:00 AM to 1:00PM, and dinner from 5:00 PM to 8:00 PM. On shot days breakfasts were available three hours before the scheduled detonation. For personnel working odd shifts, arrangements could be made for special food services. Cafeteria #2 was operated on an a la carte basis from 5:00 AM to 4:00 PM daily, and with one-half of this cafeteria opened for the regular cafeteria dinner from 5:00 PM to 8:00 PM. The other half of Cafeteria #2 was operated as a "Steak House" from 5:00 PM to 9:00 PM. Cafeteria #1 was closed on September 28, 1957. The Steak House closed shortly thereafter. Other arrangements for food in the forward areas were about the same as during TEAPOT.

During PLUMBBOB, the Recreation Hall contained a Snack Bar, Soda Fountain, a Lounge Room where weekly dances were held, and a Beer Bar where beer was available for purchase during scheduled hours. The Recreation Hall was popular and often crowded. REECo was instrumental in the promotion of recreational interest and in the scheduling of league and tournament play for a variety of outdoor activities. Movies operated by the DoD continued. Two house trailers were used as

Chaplin offices. Vehicles were made available to rent for private use at the public service station. About two vehicles were rented per week.

Like room accommodations, equipment for telephone and radio communications was assigned to the separate organizations. There were 300 telephone lines in Mercury and 300 in the forward areas. There was a switchboard for incoming and outgoing long distance calls that was manned 24 hours a day 7 days a week by nineteen operators. There were also 400 mobile and 210 fixed radio stations assigned that were used all over the NTS. (Reeves 1980:403-4)

After PLUMBBOB, the Test Manager recommended that 28 additional dormitories, designed for 2 people per room, be constructed to house approximately 1000 persons. He also recommended the construction of a swimming pool, provision for TV reception, improvements in the recreational building, and the organization of recreational programs.

Test Director's Services

The Test Director's office and laboratory spaces were in Mercury (Quonset area and buildings 102 and 111) and at the CP compound. The Test Director had an administrative services group, called S-1, staffed by EG&G. They handled administrative duties for the Test Director's organization. For example, they took care of housing, space allocations, communication and recreation.

The Test Director made several recommendations after the operation to further improve the living and working conditions at the NTS. One was to build a swimming pool in Mercury. Another recommendation was to have REECo rather than the military operate the movie theater in order to avoid the conflict over proper attire in the theater. Presumably the military felt that the civilians were lax in their sartorial standards.

An operations group called S-3 was commissioned in September 1956 to handle the coordination and dissemination of the Test Director's plans, dry run schedules, travel in the forward areas on D-1 and D days, communications frequencies and emergency evacuation procedures. S-3 had an Air Operations Officer to act as liaison between the Air Force and the test staff for aircraft requirements during the test series.

S-6 coordinated the construction of technical facilities in the forward test areas that were required for the experimental programs during the tests.

RADIOLOGICAL SAFETY

The Test Director was responsible for radiological safety at the site. He delegated this responsibility to the AEC Support Director who tasked the REECo Rad-Safe unit with the execution of the on-site rad-safe functions. The Test Director also assigned rad-safe responsibilities to the Test Group Directors who each had a Rad-

Safe Officer who handled the rad-safe functions for their groups. (Johnson 1957: 87)

Rad-Safe personnel used the terminology described next to describe contaminated areas. Contaminated real estate was referred to as a Radex, or radiological exclusion, area. A Full Radex Area was contaminated at a level of 100 mr/hr or higher (or where alpha contamination was above permissible levels). A Limited Radex area was contaminated at a level between 10 and 100 mr/hr. When the level was less than 10 mr/hr the area was considered Non-Radex.

Off-site radiological safety was the responsibility of the U.S. Public Health Service. Interestingly, the career officers were assisted by U.S.P.H.S. volunteer reserve officers who generally welcomed the opportunity to gain valuable training in radiation safety and protection. As a consequence, the public relations aspects surrounding the upcoming test series were aimed largely on trying to assure the public that the AEC would have the fallout situation well in hand.

In fact, PLUMBBOB went very well from this perspective. There was some concern within the AEC, the DoD, and the laboratories over the potential hazards associated with the John shot. This was the 2 kt air-to-air missile test. The concern was centered on plutonium dispersal in the event of a high explosive detonation caused by a malfunctioning missile hitting the ground. There wasn't much danger of a nuclear detonation. The chance of one point high explosive detonations under these circumstances was unknown and would be dependent upon the specific scenario.

Eye Burn Safety Concerns

There was concern about the pre-dawn tests exposing the public to flash blindness. The balloon shots were higher than the tower shots and there was more opportunity for the public to have a line of sight view of the detonating device from off-site roads. Using topographical maps and an assumed height of burst of 2500 feet above the surface, Los Alamos identified those portions of public roads where flash blindness could be of concern. Nevada State Police and local sheriffs were then employed to erect roadblocks prior to shot time to preclude the public from the danger areas. This successfully prevented the public from any possible ill effects of the light flash from the nuclear detonations.

2-12.1

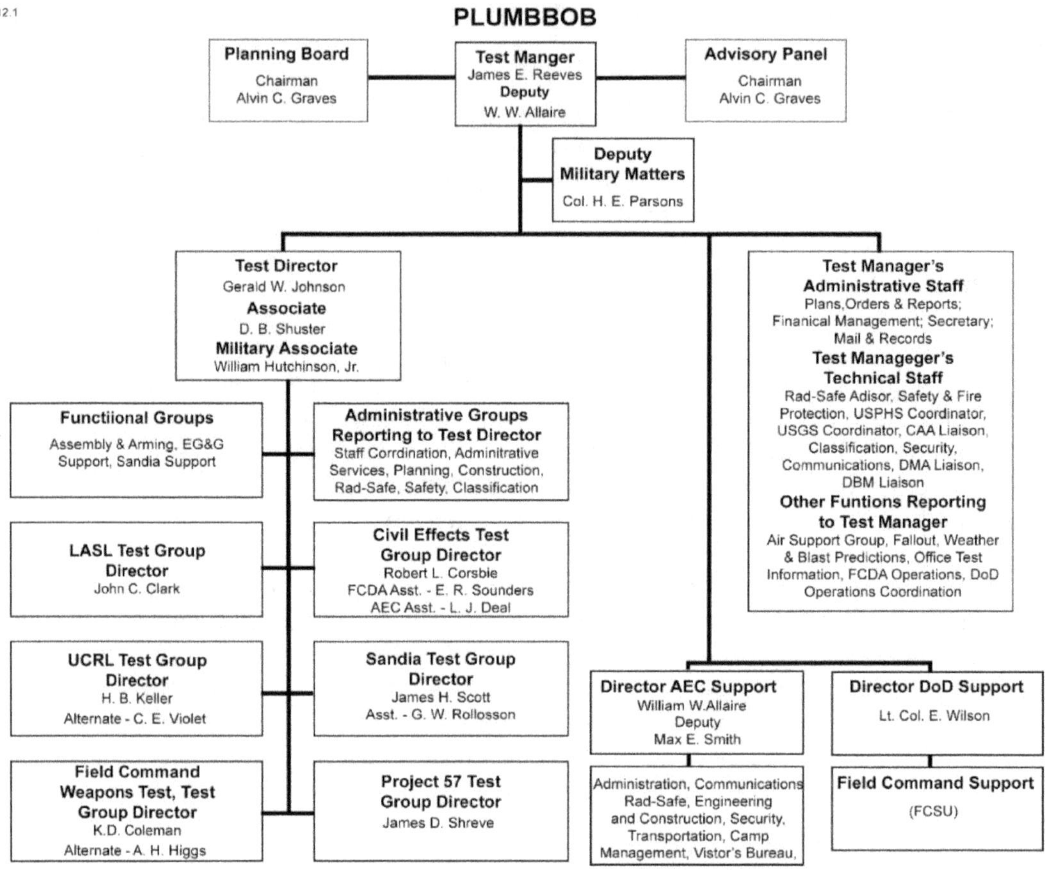

Figure 2-12.1 NTSO for Operation PLUMBBOB.

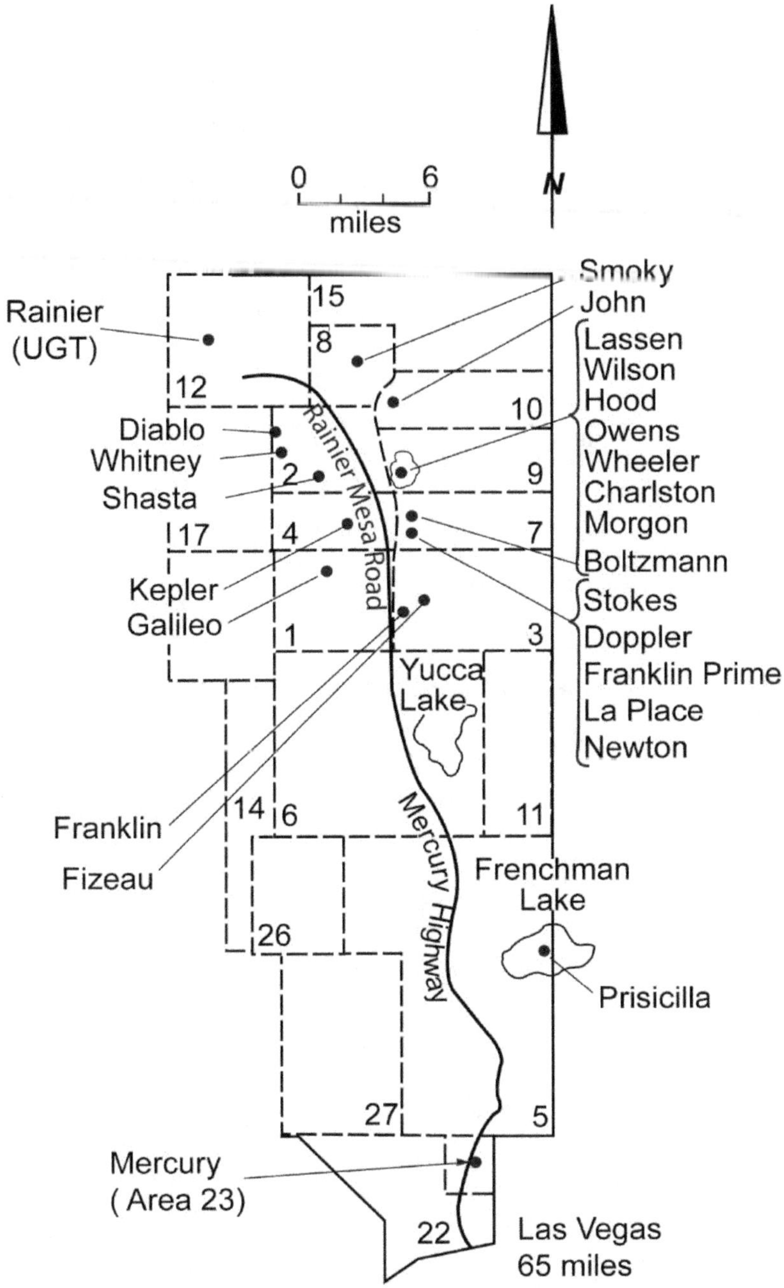

Figure 2-12.2 Location of PLUMBBOB Tests.

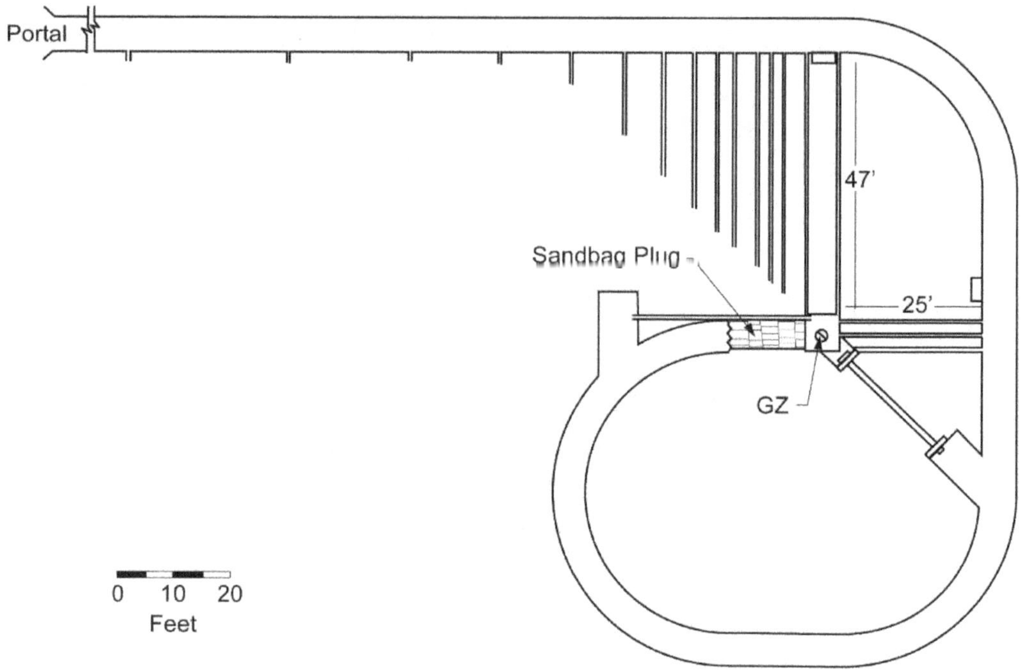

Figure 2-12.3 Rainier tunnel zero room emplacement with turns.

Rear Admiral Edward N. Parker

Blimps are vulnerable to airblast.

Aerial view of Camp Desert Rock, April 1957.

Aerial view of Camp Mercury, April 1957.

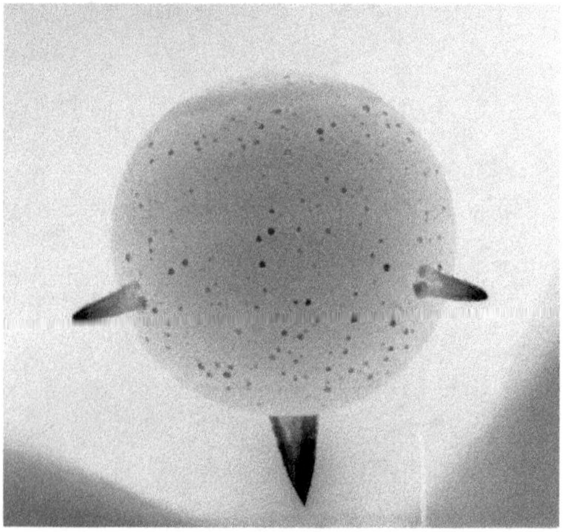

Boltzman

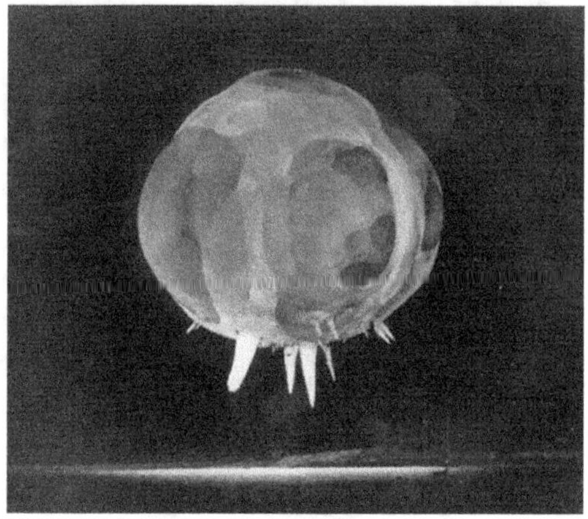

Priscilla

German Structure - One of 9 shelters sponsored by the West German government. The main room had a capacity of 25 people. At the left foreground is the location for the stairwell and main blast door. Size: 9' x 21' x 71/2'. The walls and roof were 2' thick.

The comfort level of conference rooms at NTS encouraged short meetings.

Dome structures on Frenchman Flats - Pre Priscilla.

CHAPTER 13. 1957 - 1958

PROJECTS 58 and 58A

Project 58 consisted of two Los Alamos safety shots. The first was Pascal-C, which was fired on December 6, 1957 in a vertical shaft, with no material above the device to seal the hole. The term for this is that the hole was unstemmed. The shot had a "slight" yield, but no radioactivity was detected offsite.

The second was Coulomb-C, fired the next day on the surface. It had a yield of approximately 500 tons. Unfortunately, the weather conditions were such that the fallout made its way to Los Angeles.

Consequently, Los Alamos proposed to fire some of its one-point tests being planned for MILLRACE* as underground shots. [*Footnote: MILLRACE was the next operation being planned for NTS in 1958.] A few 36-inch-diameter by 500-foot-deep holes were drilled between June 8 and July 19, 1958 so that operations could start by August 1. (Ogle 1985: 104)

Project 58A consisted of two Livermore safety tests. Venus was fired on February 22 and Uranus on March 14, 1958. Both tests were conducted in tunnels at area 12, and both had yields of less than 1 ton with no release of radioactive material.

EARLY 1958

In a January 8, 1958, letter to General Starbird, Norris Bradbury expressed "concern that the Laboratory (LASL) had 'lost control of its own destiny' since it no longer chose what it felt best to work on in the light of its own knowledge, but rather responded to external pressure from the AEC and DoD (brought about partly by the growing strength of Livermore and AFSWP)." He felt the laboratories were now making very little progress per dollar invested, and that perhaps a moratorium would be beneficial.

Bradbury continued,

> If we had to sit down and think, if we had time to sit and think, we might think of something. It is very unlikely that the press of affairs as they are, and with the general attitude of the Commission what it is, and with our own response what it is that we will have the intellectual fortitude to say 'No!' to any proposal, nor will we, with the continual workload (which we will partly bring upon ourselves) find the elusive 'new' idea if it exists at all. ... A moratorium followed by the possibility of further testing would at least force us to take stock of our situation ... It is my own impression that LASL has let itself get slightly too bogged down in mass production of weapons designs, and that we should try to take that aspect of our life a little easier and work a little harder in general research - which is thought to be good for the country too! It is for reasons like this that the thought of a moratorium, cast in the proper context, is not too painful. (Ogle 1985: 97)

In March 1958, Herb York left Livermore to serve as chief scientist of the DoD's Advanced Research Projects Agency (ARPA) and in December he was appointed DoD Director of Research and Engineering. He also became a member of the President's Science Advisory Committee, 1958-1961. York was succeeded at Livermore by Edward Teller who retained the Director's position until July 1960 when Harold Brown became the 3rd Director

Teller, was more positive about Livermore's position during a moratorium than Bradbury was about Los Alamos'. He sent Starbird an extensive list of projects and programs worth pursuing by the UCRL, along with a request for increased funding. During this period Teller and his colleagues at Livermore argued against a moratorium and made the case for underground testing as the way to address the public health issues surrounding atmospheric testing and the attendant fallout.

Meanwhile, on January 22, the AEC approved Operations HARDTACK (in the Pacific) and MILLRACE (at NTS). Presidential approval for Hardtack alone was received on January 31. During January and early February Livermore surfaced a proposal for a so-called clean shot (reduced fallout) named Pine, to be fielded in the Pacific phase of Operation HARDTACK (HARDTACK PHASE I). This was largely a political event, with representatives from the 14 member nations of the UN Committee on Radioactive Fallout invited as observers. Pine would be detonated with a yield of 2 MT on July 26, 1958, at Enewetak with the UN observers present.

Also, during the early months of 1958, it was becoming clear to the AEC and to the Laboratories that underground testing was the wave of the future. It was equally clear that more data and experience were necessary in order to test successfully underground. In this climate, Livermore proposed a 40 kt contained shot for Operation MILLRACE. The AEC was enthusiastic about UCRL's proposal for this 40 kt underground shot, and in early March both General Starbird and AEC Commissioner Willard Libby suggested the December 1958 - January 1959 time frame for the event. Operation TRUMPET, a full-scale test series slated for the spring of 1959, would also give Livermore an opportunity to concentrate on underground-contained shots. Meanwhile, even though Los Alamos was starting to seriously consider underground testing, they were still mainly focused on atmospheric tests.

FRENCH VISIT THE NEVADA TEST SITE

In 1957 the French nuclear weapons program was fast approaching the time when a full-scale nuclear test would be seriously considered. It was only natural for them to focus on the Sahara as a site for their experimentation. France still controlled Algeria, and the population density was low. Furthermore, the area under consideration was extremely dry and remote from centers of population.

In the late 1950s it was in the United States' interest to assist France in safety-related technology but the U.S. could and would not collaborate on other nuclear weapons issues. There is no direct evidence that the French had suggested such a collaborative relationship.

Late in 1957, there were government-to-government discussions in Paris about a visit of a few senior French nuclear weapons experts to the Nevada Test Site. These were followed up with a letter shortly after the beginning of the New Year from the French Government (the Commissariat a l'Energie Atomique (CEA), which is the French equivalent to the U.S. AEC) to the U. S. AEC. The AEC subsequently sent a TWX to the Laboratories suggesting a "visit of French experts to the USA concerning largely the visit to our NTS to observe the meteorological and safety aspects of our test methods and organization." The French also shared an interest with the U. S. in the Civil Defense issues associated with nuclear weapons effects.* [*Footnote: France had shelter designs tested on PLUMBBOB, which were fielded by US personnel, see Part II-Chapter 12.]

The French visit* to the Nevada Test Site did come to pass in February, 1958. [*Footnote: The following visitors were from France: Prof. Yves Rocard, General Charles Ailleret, General Andre Buchalet, Col. Henri Debrabant, Jean Georges Parreins, Poundres Barguillet, Pierre Leon Billaud, Jean Kaufmant, Jean Lenouvel, Andre Gauvenet, Commandant Maral, Lieutenant Colonel Cristian Maurice, Felix Joseph, Darde, Commandant Francis Tyrode, Commandant Leonard, Michel Jogot-Lagoussiere, and Commandant Andre August Aeberhardt. (Los Alamos Archives 230.033 French)]

Clearly this was a sensitive subject in 1958, and in April Starbird sent the labs and EG&G a TWX with the following request: "The February 1958 briefing on our nuclear test operations conducted for the French CEA and Special Weapons Command representatives at NTS is prompting follow-up requests from the French for additional details. I wish to channel such requests, and the answers thereto, through my office..." About 2 years later, the first French nuclear test was Gerboise Bleue, on February 13, 1960 at their site near Reggane, Algeria.

ISSUES REGARDING OPERATIONS AT NTS

Continuous Operation of the NTS Revisited

In March 1958 AEC/DMA Director Starbird revisited the issue of continuous operation of the test site in a message to Los Alamos, Livermore and the AEC Albuquerque Operations Office: "We have considered frequently in the past the desirability of keeping either the NTS or EPG (Enewetak Proving Grounds) opened for a near year-round operation. Such consideration has normally led to the conclusion that: the labs were insufficiently staffed to handle (year-round operation); costs would be higher than under the present arrangement; and important programmatic gain was questionable. I desire to secure certain information from you concerning the programmatic advantage and increased costs that might be involved ..."

Starbird presented 2 scenarios as a basis for discussion. The first was to keep the EPG open continuously after HARDTACK. NTS shots would be on a campaign basis. As far as diagnostics permit, NTS shots should be done underground." The second scenario was to use the EPG for campaign operations (every 2 years) and continuous underground shot operation at NTS. (March 6, 1958 TWX USAEC (Starbird) to LASL (Bradbury), Livermore (Ken Street), ALOO (Reeves). Ref: A-99-019 635 NTS (1/17/1958-9/7/1962) 279-5)

Livermore responded to DMA on March 10, 1958 and essentially wanted both options: "UCRL is heartily in favor of continuous testing. Having major experiments arbitrarily quantized at two year intervals, as with present system, is a very unnatural way of doing research." Livermore, preferred EPG to be kept open continuously after HARDTACK. In addition Livermore liked the idea of one or two short operations each year conducted in Nevada and also suggested that it would be desirable to have NTS available on a continuous basis for underground shots.

Norris Bradbury responded the next day and chose the EPG. After citing reasons for continuous testing, he states: "For these arguments and others involving the somewhat unknown difficulties of extensive underground testing it would seem preferable to consider more frequent or continuous testing at Enewetak with periodic operations at Nevada rather than the reverse arrangement."

Centralization of Operations

Later in the year, on August 10, 1958, Ogle, who was on Johnston Island, sent a TWX to Graves on a topic related to continuous operations - the centralization of operations at NTS. Ogle suggested two choices:

> Put all testing under one central authority who decides what is to be tested, what measurements constitute testing, who should make the measurements, how they should be made, what support is necessary, and then carries out the measurements and the operation. Since there are so many varied interests in the country, and at the moment so many organizations concerned, I consider the transfer of the present effort to this type of organization completely out of the question practically, and not desirable anyway.
>
> The second possibility is, which is considered more desirable, is to maintain the same lines of authority and responsibility that apply between operations. Thus I maintain that the laboratories should have the responsibility and authority for their own portion of an operation necessary to achieve the stated aims of that portion of the operation. Since this does not normally include the functions of support and safety, these are functions that should be the responsibility of the Test Manager. ... Each prime user should deal directly with the test mangers organization in order to obtain a sympathetic coordination of the two. The laboratory or DoD people should not however be placed in the position of being under the orders of or working for the overriding organization. Each user should form their test group.

> The test manager's organization should not have technical or operational responsibility except as concerns support or safety (not including actual detonation of devices). Thus I believe it appropriate for him to have a scientific deputy or a scientific advisor, but not both, and neither of these people should be considered as a Test Director.

Ogle's "second possibility" is essentially the structure that was ultimately developed for NTS.

OPERATION HARDTACK PHASE I

HARDTACK PHASE I opened on April 28, 1958 in the Pacific with a balloon shot, Yucca, at 86,000 feet. It was a weapons effects event with a yield of 1.7 kt that was detonated over the open ocean northeast of Enewetak and northwest of the Bikini Atoll. In addition, twenty-two of the shots were at Enewetak, ten at Bikini, and two above Johnston Island, for a total of 35 shots. The two Johnston Island shots were high yield rocket borne high altitude shots: Teak, with a yield of 3.8 Mt at an altitude of 77 kilometers and Orange, also 3.8 Mt, at 43 kilometers. The last HARDTACK PHASE I shot was the Livermore Fig event conducted on the surface with a yield of 20 tons on August 18 at Enewetak. Interestingly, twenty-six of the events were fired on barges in the lagoons.

OPERATION ARGUS

Immediately following HARDTACK PHASE I, Operation ARGUS was conducted in the South Atlantic. ARGUS consisted of 3 weapons effects tests, each in the 1-2 kt range. The devices were rocket launched and each detonated at an altitude of about 300 miles. Argus 1, August 27, 1958, was at 38.5° S 11.5° W; Argus 2, August 30, 1958, was at 49.5° S 8.2° W; and Argus 3 was at 48.5° S 9.7° W. The purpose of these shots was to explore the geophysics associated with high altitude detonations and their effects on military radio and radar systems. Operation ARGUS was a DoD and ARPA operation conducted by AFSWP. (DOE 2000: 12-3; DTRA 2002: 139-140)

TEST BAN INITIATIVES

After months of diplomatic wrangling with the Soviets regarding test ban verification, there was finally an agreement to have a conference of experts to be held in Geneva beginning the first of July, 1958. The purpose of the conference was to explore verification technologies that could support a nuclear test ban. See Attachment V: "Arms Control" by Milo Nordyke.

The issues were not solely the purview of the United States and the Soviet Union. Britain was testing nuclear devices in Australia and on Malden and Christmas Islands in the Pacific. France was getting ready to test in Algeria. (Mikhailov 1999: 18-23) With respect to Britain, at least, the United States would have to have a closer nuclear weapons relationship to induce them not to continue to test.

However, the Congress's Joint Committee on Atomic Energy was not enthusiastic about the notion of sharing more nuclear weapons information with our NATO allies. (Hewlett 1989: 537) Ultimately, John Foster Dulles gave a strong endorsement for legislation that would permit limited sharing of nuclear information with the British as a necessary step in achieving the United States' arms control objectives. On June 30, 1958 Congress amended the Atomic Energy Act of 1954 to permit sharing of nuclear information and hardware with nations that had "made substantial progress in the development of atomic weapons." (Hewlett 1989: 538)

James B. Fisk, a vice-president of Bell Laboratories and a member of the President's Science Advisory Council, was selected to lead the Western delegation of experts at Geneva. E. O. Lawrence and Robert Bacher were members from the United States.* [*Footnote: Advisors to the United States delegation included: Hans Bethe, Harold Brown (UCRL), Perry Byerly (UC), Stirling Colgate (UCRL), Norman Haskel (Air Force, Cambridge), Spurgeon Keeny (Presidential Science Advisor James R. Killian's Office), J. Carson Mark (LASL), Doyle Northrup (Air Force Office for Atomic Energy), Herbert Scoville, Jr. (Consultant, PSAC), Anthony Turkevich (University of Chicago), Donald Morris, Ronald Spiers and Thomas Larson (State Department)]

Sir John Cockcroft and Sir William Penny represented the United Kingdom. Yves Rocard represented France and Ormond Solandt represented Canada. The Soviet delegation of six senior scientists was lead by Simyon Tsarapkin, an experienced diplomat. There were also scientists from Czechoslovakia, Poland and Rumania. (Ogle 1985: 27)

The Americans wanted to address detection techniques for nuclear tests in all environments: atmospheric, high altitude, in space, underground and underwater. The technical methods included cloud sampling, acoustic detection, seismic wave measurements, and the detection of electromagnetic signals. The Soviets, on the other hand, wanted an agreement on a test ban prior to exploration of the technical challenges. This was contrary to the American team's instructions that were to confine their work to the technical considerations and to leave the political policy decisions to Washington. (Hewlett 1989: 539)

E. O. LAWRENCE DIES

On August 27, 1958, Earnest Orlando Lawrence died. He was born August 8, 1901 in Canton, South Dakota. Lawrence was awarded the 1939 Nobel Prize in Physics for the invention of the cyclotron at Berkeley. The University of California Radiation Laboratory became the Lawrence Radiation Laboratory (LRL) on November 7, 1958. The Livermore branch was referred to as LRL Livermore.

THE RAMP-UP TO HARDTACK PHASE II

Starbird asked the labs in mid June to be finished with MILLRACE by the first of November, assuming no budget limitations. Bradbury, Teller, and Kenner Hertford

(ALOO) replied that it was possible, but would indeed require extra funding. Finally seeing the handwriting on the wall, Los Alamos suggested to Starbird that two new 3,000-foot tunnels be mined for full-scale LASL tests underground. (Ogle 1985: 104)

The test site contractors were also making plans for coping with a moratorium. In mid-June Bernard J. O'Keefe proposed a scenario for EG&G. It assumed an October 1 start date for a moratorium with a readiness program that would permit a resumption of testing with six months warning. O'Keefe's readiness program included restricting hiring and the development of alternate funding sources. (Ogle 1985: 111)

On July 11, 1958, Bradbury and Teller got guidance from Willard Libby, who was then acting AEC Chairman, regarding the potential test moratorium. There were two parts: 1. (a) Test ban only:

> Then your job - on atomic weapons - would be to digest and collate the results from PLUMBBOB and HARDTACK, which are rich sources of basic weapons science that when fully understood and analyzed will enable us without additional tests to materially improve our weapons designs. A period of eighteen months or two years probably could be most profitably employed in this way. Experimental work at sub nuclear yields probably would be involved. In addition, we hope that whatever the nature of a test ban, there would be special exception made of the nonmilitary applications of nuclear explosions ...2.The second part of our guidance would be to advise you to make plans on a strictly confidential basis which you would hold in readiness to reorganize your work and reslot people should a cessation actually occur. (i.e. if testing resumes).

Libby encouraged the lab directors to get those involved in the applied scientific fields to become actively engaged in basic scientific research in order to sharpen their skills. He viewed this as a two way street and recommended that those engaged in basic research become involved in weapons work. (Ogle 1985: 98)

Meanwhile, by mid-July, the AEC and the labs had concluded that it was most likely that only underground testing would be permitted in the future. As a result, Hertford proposed that MILLRACE be publicly announced as an underground test series. (Ogle 1985: 105)

In late August, General Starbird changed the name of the next Nevada test series from MILLRACE to HARDTACK PHASE II. This series would include up to nine nuclear shots, plus up to seven one-point safety tests. In addition to the Nevada shots Starbird mentioned the probability of returning to Enewetak for one additional test for Los Alamos. This possibility was first raised at the end of HARDTACK PHASE I. General Starbird also said "We should eliminate projects directed toward conducting operations with greater economy, capacity, or content at either location." Interestingly, Starbird said that it was not necessary to test underground yet. Atmospheric shots were just fine. On August 28, 1958 President Eisenhower approved an accelerated HARDTACK PHASE II, but disapproved any further shots in the Pacific. (Ogle 1985: 106,107)

The Nevada Planning Board met at Mercury on September 9 to firm up plans for HARDTACK PHASE II. Duane Sewell of Livermore chaired the meeting. They settled on a program that consisted of 6 tunnel shots, 1 tower shot, up to 4 balloon shots, and several one-point safety shots. The AEC appointed James Reeves as Test Manager; Gerry Johnson as Deputy Test Manager; Duane Sewell as Scientific Advisor; and Col. W. S. Hutchinson as Deputy for Military Matters. (Ogle 1985: 107)

EISENHOWER ANNOUNCES U.S. MORATORIUM

On August 22, 1958, President Eisenhower announced a one-year Moratorium to begin October 31, 1958. In a letter to Edward Teller the President wrote:

> I am today announcing that the United States will suspend nuclear weapon tests for a period of twelve months and, under certain conditions of progress toward real disarmament, continue that suspension on a year-to-year basis.
>
> It will, of course, require an extended period to negotiate and install a genuine and assured disarmament arrangement. Even though we will not be doing any weapons testing, it will be necessary that we maintain our weapons development progress during the period and with no less urgency than in the past. It is necessary, in the interest of our country's defense, that the staff of your laboratory, and that of the other weapons development laboratories, continue their research and development in this field with their current vigor and devotion.
>
> I am instructing the Atomic Energy Commission to develop plans to see that these essentials are met and that the vitality of our laboratories is maintained.

Similar letters were sent to James McRae (Sandia President September 1953 - November 1958) and to Norris Bradbury at Los Alamos. (Ogle 1985: 99)

THE SOVIET UNION AGREES TO START NEGOTIATIONS

In late August, Krushchev agreed to start negotiations on October 31 as a step toward a possible nuclear test ban, but he did not initially agree to a moratorium. The Soviets started a test series on September 30 at the Novaya Zemlya test site and continued with 18 more shots at Novaya Zemlya through October 25. They moved to the Missile Test Range* for their last two shots on November first and third when they too began an indefinite moratorium. (Ogle 1985: 6, 30) [*Footnote: The Missile Test Range used by the Soviets Union for these tests is located in the Astrakhan region of the Russian Federation, in the vicinity of Kapustin Yar. Kapustin Yar is about 225 miles north of the Caspian Sea, just west of Kazakstan's western border, and just east of the Volga River. This site was only used for 10 rocket launched nuclear tests between 1957 and 1962. The last two Soviet tests in November were rocket launched with 10 kt yields. (Mikhailov 1999:11, 23)]

READINESS ISSUES
The readiness to resume testing, should the government make such a decision, was a key issue during the spring and summer of 1958. On August 22, AEC Chairman John A. McCone emphasized to the labs that they must be prepared to resume testing (Operation TRUMPET at the NTS and WILLOW in the Pacific in 1960). This was a touchy subject. Ogle noted that on August 22: “Starbird started down the path that was to so infuriate the labs over the next three years. Stating that we should be prepared to revert to testing on short notice if the situation warranted, he (Starbird) went on, ‘We should be prepared to reinstitute TRUMPET ... limiting major expenditures to those essential to readiness, and approved individually by DMA...’” McCone wrote to Teller on August 29: “Your efforts should be so oriented that, in the event the test suspension is not extended or is cancelled, we can revert to testing and ensure consequent advancement of our developments with a minimum of delay.” (Ogle 1985: 111)

As so often happens, the signals were mixed. In August James Reeves “pointed out that the Bureau of the Budget was already tying up most of the construction funds for the two test sites, and putting on pressure to reduce the maintenance and operations costs”. (Ogle 1985: 112)

The AEC was ambivalent about the funding for the nuclear weapons program in light of the pending test moratorium. On September 17 AEC Commissioner Willard Libby suggested that the laboratories be limited to 3,000 persons. This didn’t have much effect, however. Ogle attributed this to the fact that the President had already stated “the laboratories should be kept at peak efficiency, and that every effort should be exerted to maintain the morale of the laboratories.” (Ogle 1985: 99)

In October AEC Chairman John A. McCone asked the laboratories for outlines of their proposed programs during the moratorium. Teller, speaking for Livermore, said that they planned to work on Pluto (nuclear ram-jet), pure research, controlled thermonuclear reactors (CTR), verification techniques for test ban agreements, the weaponization of tested designs, and an expanded peaceful nuclear explosion (Plowshare) program. See Attachment VI: “The Plowshare Program” by Milo Nordyke. It should be noted that on August 22 McCone had told the laboratories that the moratorium did not include Plowshare. (Ogle 1985: 99)

The Los Alamos response, from Bradbury, outlined a program consisting in large measure of projects aimed at catching up with weapons related technologies that had been short changed during the testing period because of the press of higher priority goals. These included some weapons development, test diagnostic development, and basic weapons science without nuclear testing. But he “emphasized that if the moratorium were to continue more than a couple of years the role of LASL in the national picture was not obvious and should receive very careful consideration at that time". He also pointed out the possible diversion of laboratory effort to Rover (nuclear rocket engine R&D), Sherwood (controlled fusion R&D), and Plowshare." (Ogle 1985: 99-100)

Meanwhile, Teller, Gerry Johnson, and others at Livermore were hard at work developing the technology to test underground. "By increasing the yield of the devices tested by a factor of 20 or so each time, it is hoped to reach the megaton range in underground testing by 1959." Building upon the very successful experience with the Rainier shot in September 1957 Teller tried to convince the AEC and the President that fallout associated with atmospheric shots could be largely eliminated by testing underground. See Attachment VII: "Underground Testing and Early containment Concepts" by Cliff Olsen. However, Bradbury was not convinced that the labs could make the necessary experiments to determine device performance. He felt that yield measurements and thermonuclear weapon development could not be accomplished with underground tests. As it turned out, of course, Bradbury was wrong. (Ogle 1985: 100)

Livermore was pursuing plans for a civil engineering demonstration of a nonmilitary application of nuclear explosives. This program was called Plowshare and the shot was named Gnome. The proposal called for a 10 kt device buried 1200 feet down in a salt bed twenty-five miles southeast of Carlsbad, New Mexico. The Plowshare people were interested in the possibility of excavating a large cavity in a salt bed with a nuclear explosive.

Nuclear test verification was also vigorously pursued during the moratorium by persons who had experience at the test site. The military as well as Sandia, Livermore, and Los Alamos were involved. The Vela Program was developed with three parts: Vela Uniform, which focused on the detection of underground nuclear explosions; Vela Sierra, which explored the detection of high altitude nuclear tests using ground stations; and Vela Hotel, which focused on the detection of tests at high altitudes and in space using satellite-borne detectors. See Attachment VIII: "The Vela Program" by Milo Nordyke.

CHAPTER 14. OPERATION HARDTACK PHASE II: SEPTEMBER 12 - OCTOBER 30, 1958

Operation HARDTACK consisted of two phases: PHASE I (in the Pacific) and PHASE II (at NTS). The terms HARDTACK PHASE I and simply HARDTACK I are used interchangeably; similarly the terms HARDTACK PHASE II and simply HARDTACK II are used interchangeably.

PLANNING AND RUNUP

On May 5, 1955, the AEC DMA announced that Operation JULEP, which had been tentatively scheduled for the fall of 1956 at the NTS, would be combined with Operation HARDTACK. HARDTACK had been tentatively scheduled for the PPG for about February 1958. In the latter part of 1955, AFSWP was notified by the AEC that Operation HARDTACK was scheduled for the spring of 1958. (DASA 1959: Part II Chapter 7-5 pg. 8, 11-12)

In October 1955, the Naval Research Laboratory (NRL) presented AFSWP with a proposal for a program that would be code named CLASSICIST. CLASSICIST is an example of the breadth of testing ideas during the mid 1950s. It involved the launch of several balloons with instrumented draglines for test diagnostics. The balloons would be launched at a site where weather conditions were such that there would be a good probability that at least one balloon would pass over an unspecified Nike B site at an altitude of approximately 100,000 feet. A Nike B with an atomic warhead would be fired at a radar beacon at the end of the balloon's dragline. (DASA 1959: Part II, Chapter 7-5 pg 12)

The JCS approved on June 5, 1956 a very high altitude (VHA) weapons effects test for HARDTACK. The inclusion of a VHA shot on Operation HARDTACK marked the end of CLASSICIST as originally suggested. (ibid., 23)

Preliminary planning for two nuclear test operations was under way during 1957: TRUMPET at the NTS during the spring 1959 and WILLOW at the PPG during the spring of 1960. In October, 1957, the AFSWP Test Division sent a letter to the Services requesting project proposals. For planning purposes, it was assumed that in the near future, there would be no international agreement to preclude nuclear testing in the atmosphere.

After PLUMBBOB, planning was initiated by the AEC and the User Agencies for a series of tests to be held at NTS during 1958, 1959, and 1960. By the end of February, 1958, a new code-name, MILLRACE, was added. The schedule and numbers of shots for these NTS operations had been defined as follows:

OPERATION, DATE	TOTAL	TOWER	TUNNEL	BALLOON	SAFETY
MILLRACE, FALL 1958	11		4		7
TRUMPET, SPRING 1959	27	3	6	12	6
MILLRACE "A", FALL 1959	9		6		3
MILLRACE "B", FALL 1960	4		1		3
TOTAL	51	3	17	12	19

In July 1958, the developing probability of a nuclear test moratorium caused the advancement of some events from TRUMPET to MILLRACE. Also, Area 14 was designated as a potential, additional, tunnel testing are.

On August 28, 1958, six days after he announced the test moratorium, President Eisenhower approved an accelerated series of nuclear tests for Operation MILLRACE. These tests were to be completed at the NTS before the start of the moratorium on November 1, 1958. By that time, MILLRACE consisted of 6 nuclear weapons tests and 8 safety tests. (Ponton 1982d: 29)

An AEC directive, dated August 29, announced a revised "crash" test program for NTS with the name of the operation changed from MILLRACE to HARDTACK PHASE II. The thought was that PHASE I and PHASE II would constitute one rather than two separate operations. On September 10, the published Schedule of Events showed an escalation in the number of tests to a total of 20 shots: 6 in vertical holes, 4 suspended by balloons, 9 in tunnels, and 1 in a tower. The number of shots eventually grew to 32 and finally to 38. The number of shots actually fired during HARDTACK II totaled 37 with: 6 in vertical holes, 11 balloon, 7 tunnel, 10 tower, and 3 surface. The planned 38th shot was not conducted,

The main objectives of the tests in Operation HARDTACK PHASE II were to:
- Test nuclear devices for possible inclusion in the stockpile
- Test safety characteristics of nuclear devices, and
- Improve containment techniques for underground detonations.

(Ponton 1982d: 1)

NEVADA TEST SITE ORGANIZATION (NTSO) FOR OPERATION HARDTACK PHASE II

Numerous structural changes were made to the Nevada Test Site Organization after Operation PLUMBBOB. These changes were accomplished so that more efficient channels would be available to the using agencies and to more clearly indicate functional responsibilities of the different organizational segments. The most important changes were:

- Abolishment of a single Test Director's office and organization. The result was the establishment of Test Directors for each organization.
- Consolidation of all coordinating functions under the Test Manager.

- Designation of Test Groups to execute field programs developed by parent organizations.
- Designation of channels for technical support echelons to provide support on a mission basis by arrangement with the contract administrator or by inter-agency agreement. (Reeves 1958:3)

Figure 2-14.1 shows the Nevada Test Site Organization for HARDTACK II. The Test Manager was again James E. Reeves with Gerald W. Johnson of UCRL as his deputy. Col. W. S. Hutchinson, the Deputy Test Manager for Military Matters, served as liaison between the military and the AEC.

The User Organizations comprised six Test Groups. These groups and their directors were:
- Los Alamos Scientific Laboratory Test Group for weapons tests - Director William E. Ogle with Robert H. Campbell as alternate.
- University of California Radiation Laboratory, Livermore, Test Group for weapons tests - Director H.B. Keller.
- DoD Effects Test Group - Director Lt. Col. John W. Kodis.
- AEC-DBM and OCDM Civil Effects Test Group (CETG) and Office of Civil and Defense Mobilization Test Group (OCDMTG)* [Footnote: Prior to July 1, 1958, the OCDM was known as the Federal Civil Defense Administration (see this chapter's section titled CIVIL DEFENSE PARTICIPATION).]
These two groups were directed by Robert L. Corsbie with L. Joe Deal as Deputy.
- Sandia Effects Test Group - Directors George B Strobe and C. B. Campbell. (Reeves 1958:4)

ADVISORY PANEL AND THE PREDICTION GROUP

The Advisory Panel, Figure 2-14.1, was chaired by Duane Sewell with William E. Ogle as alternate. The other members are cited in Appendix F. The Prediction Group provided weather forecasts and blast and fallout advice for the Advisory Panel and Scientific Advisors.

Since the yields of the HARDTACK II tests were very low, the weather, blast, and fallout were not the driving factors that dominated the schedule in previous series. As a consequence only 25 of the 37 detonations were delayed by unfavorable atmospheric conditions. The Advisory Panel met a total of 81 times between September 12 and October 30 to consider the advisability of proceeding with the shots. (Reeves 1958: 69-70)

PLANNING BOARD DURING HARDTACK PHASE II

The purpose of the NTS Planning Board was the same for HARDTACK II as for PLUMBBOB - to recommend and advise the Test Manager in matters relating to the overall planning at the NTS. The Board was chaired again by Alvin C. Graves with Duane Sewell as vice-chairman. The membership of the Board was increased by two with the addition of M.L. Merit, of Sandia, as an Alternate Member and R. R.

Corsbie, representing CETO. The new DoD Field Command representative was Col. W.S. Hutchinson.

The Board met only once for HARDTACK II, at Mercury on September 9, 1958. They considered: schedule, containment of the underground shots, hydrogeology for Area 12, weather expected during September and October, fallout concerns, potential blast effects, balloon capabilities, and possible interference on the Annual Air Force Gunnery Meet scheduled for October 11-19 at Indian Springs AFB.

Some of their conclusions were:

> The overall test schedule as presented by the various groups looks extremely tight. However, no items could be found that look to be impossible to accomplish. Therefore, the Board recommends that the planning for the Operation HARDTACK - Phase II proceed in accordance with this schedule. If the fallout pattern from shots in Areas 3, 7, or 9 is not predicted to go near Area 12, the Board recommends that, if it is operationally advisable, personnel be allowed to continue working in Area 12 during shots in any of these before-mentioned areas. (Reeves 1958: 67)

HARDTACK PHASE II TESTS

The 19 full-scale weapons related and the 18 safety tests conducted during Operation HARDTACK PHASE II are given in Table 2-14.1*. Their locations are shown in Figures 2-14.2 and 2-14.3.

Table 2-14.1 HARDTACK PHASE II Tests.

TEST	SPONSOR	DATE 1958	TYPE	AREA Or HOLE	PURPOSE	YIELD Kt
Otero	LASL	9/12	Shaft, -480 ft	U3q	Safety	.038
Bernalillo	LASL	9/17	Shaft, -456 ft	U3n	Safety	.015
Eddy	LASL	9/19	Balloon, 500 ft	7	Weapons	.083
Luna	LASL	9/21	Shaft, -484 ft	U3m	Safety	.0015
Mercury	UCRL	9/23	Tunnel, -183 ft	U12f.01	Safety	Slight
Valencia	LASL	9/26	Shaft, -484 ft	U3r	Safety	.002
Mars	UCRL	9/28	Tunnel,-140 ft	U12f.02	Safety	.013
Mora	LASL	9/29	Balloon, 1500 ft	7	Weapons	2.0
Colfax	LASL	10/05	Shaft,-350 ft	U3k	Safety	.0055
Hidalgo	LASL	10/05	Balloon, 377 ft	7	Safety	.077
Tamalpais	UCRL	10/08	Tunnel, -407 ft	U12b.02	Weapons	.072
Quay	LASL	10/10	Tower, 100 ft	7	Weapons	.079
Lea	LASL	10/13	Balloon, 1500 ft	7	Weapons	1.4
Neptune	UCRL	10/14	Tunnel, -110 ft	U12c.03	Safety	.115
Hamilton	UCRL/DoD	10/15	Tower, 50 ft	5	Weapons	.0012
Logan	UCRL	10/16	Tunnel, -932 ft	U12e.02	Weapons	5.0
Dona Ana	LASL	10/16	Balloon, 450 ft	7	Weapons	.037
Vesta	UCRL	10/17	Surface	S9e	Safety	.024
Rio Arriba	LASL	10/18	Tower, 72.5 ft	3	Weapons	.090
San Juan	LASL	10/20	Shaft, -234 ft	U3p	Safety	Zero

Socorro	LASL	10/22	Balloon, 1450 ft	7	Weapons	6.0
Wrangell	UCRL	10/22	Balloon, 1500 ft	5	Weapons	.115
Rushmore	UCRL	10/22	Balloon, 500 ft	9	Weapons	.188
Oberon	LASL	10/22	Tower, 25 ft	8	Safety	Zero
Catron	LASL	10/24	Tower, 72.5 ft	3	Safety	.021
Juno	UCRL	10/24	Surface	S9f	Safety	.0017
Ceres	UCRL	10/26	Tower, 25 ft	8	Safety	.0007
Sanford	UCRL	10/26	Balloon, 1500 ft	5	Weapons	4.9
De Baca	LASL	10/26	Balloon, 1500 ft	7	Weapons	2.2
Chaves	LASL	10/27	Tower, 52.5 ft	3	Safety	.0006
Evans	UCRL	10/29	Tunnel, -850 ft	U12b.04	Weapons	.055
Mazama	UCRL	10/29	Tower, 50 ft	9	Weapons	Zero
Humboldt	UCRL/DoD	10/29	Tower, 25 ft	3	Weapons	.0078
Santa Fe	LASL	10/30	Balloon, 1500 ft	7	Weapons	1.3
Blanca	UCRL	10/30	Tunnel, -987 ft	U12e.05	Weapons	22
Ganymede	UCRL	10/30	Surface	S9g	Safety	Zero
Titania	UCRL	10/30	Tower, 25 ft	8	Safety	.0002

(DOE 2000, 12-17; Lewis 1997, 35)

[*Footnote. The locations of tests conducted in tunnels or shafts, were identified by U, for underground, the area, and an identifying letter and sometimes another number. For example, for shafts, U3q means: Underground in Area 3, and shaft (or hole) q. For tunnels, U12f.01 means: Underground in Area 12, tunnel f, drift 01. Surface tests had locations identified by S, for surface. For instance, S9e means: Surface in Area 9, location e. Within an area, the locations a, b, c, etc. were not laid out in any type of regular grid fashion. The letters were chosen somewhat, but not strictly, chronologically in the order in which they were developed (drilled or mined). The drift numbers referred to the order in which they were mined from the tunnel.]

HARDTACK II consisted of 19 Weapons shots and 18 Safety tests. There were no Effects tests. The 18 safety shots were, of course, very low yield. However, of the other 19 weapons tests, only six had yields over 1 kt. The pragmatic DoD therefore focused its programs to fill in information not previously obtained regarding low yields.

The last planned weapons test, a 1,500 foot balloon shot named Adams, was scheduled for detonation on October 31. The device, however, was not fired. It was lowered and disarmed after midnight, on November 1st; because it had been determined that overpressure damage to Las Vegas and Indian Springs could have resulted from the shot. (Ponton 1982d: 29)

Duane Sewell was the Scientific Advisor on Adams. His perspective from a 1981 interview at Livermore is as follows:

> We left one device unfired, and I remember that night very well. I had about fifteen hundred people who really were upset with me because I didn't tell the AEC to go ahead and fire that device. I told them not to fire it. I remember the one story that Dodd Starbird likes to tell. He was the Director of Military Applications at the time that operation was going on. The scheduled deadline was midnight on October 31st, Halloween night. I

remember a lot of masks around the place. It was obvious we were going to have trouble, but not from fallout. The wind pattern was in a direction that was not going to give us trouble, and that last shot was a balloon shot, so there was not going to be a great deal of dirt picked up and local fallout from it. But the wind pattern was such that there was a potential for a pressure impulse into Las Vegas that was strong enough to possibly break plate glass windows. We obviously didn't want to hurt anybody, and didn't want to break windows either.

We were testing with shots of a half-ton of high explosive mounted on one of the hills a short distance from the CP. We'd fire a number of those during the evening, and it was a double bounce. The shock wave bounced down around Indian Springs, then the next bounce was into Las Vegas, and it was rather sharply focused. We had trouble getting enough high explosive; I was blowing up all the high explosive on the site to make those measurements every half hour to forty-five minutes. I got on the phone with Dodd Starbird and I said, 'That's midnight Washington time not Greenwich time when we start the moratorium.' We agreed on that. That gave us an extra five or six hours. When it got to that point I said, 'No, it's really midnight here,' and I got him to agree to that. Then I tried to get him to agree to midnight within the United States, which would mean Hawaii, but he wouldn't buy that. He wouldn't go that far, so Pacific Standard Time was what we finally had to go on.

We fired the last H.E. shot about eleven-thirty that night. I was in the microbarograph room, and we had people out in the field with mobile measuring systems. The statement came back from the people there - 'My God, what did you fire that time?' It really shook them. Apparently we had them just at the focus and I thought, 'Boy, if a half a ton can be heard that far, I'm not going to fire.' The last thing we wanted was to have some sort of damage, or the potential of harming people in Las Vegas. That's why I made the decision I did. I advised Jim Reeves not to fire and he went along with it. That's why we left that thing hanging on the balloon that night." (Lawrence Livermore National Laboratory archives)

HARDTACK PHASE II set a record for the number of non-simultaneous shots conducted on one day. On both October 22 and October 30, four shots were conducted in 4 different areas.

Livermore conducted the three safety tests that were in tunnels, one of which was Neptune at 115 tons. It had the highest yield of their safety shots and was at the shallowest depth below the surface. Neptune resulted in a slight vent, but radioactivity was not detected off site. It was the first underground test to form a subsidence crater.

The four weapons development tests in tunnels were all conducted by Livermore. There was no radioactive release detected on Logan. Tamalpais and Blanca both vented a slight amount. Blanca, at 22 kt, was the highest underground yield prior to the 1958-1961 moratorium.

Ground motion was measured in the medium surrounding the Tamalpais zero room, where the device was emplaced. There were two accelerometers and two strain gauges at each of six stations located in the immediate area. Also, differential motion was measured at various locations in the tunnel. (Ponton 1982d: 98-99)

Ground motion and permanent displacement was also measured on Evans and Blanca. Unfortunately, little useful data were acquired on Evans because of the lower than expected yield. (Ponton: 161-3; 180)

Tunnel Tests

There was a steep learning curve associated with the underground tests in the early days. For example, during an interview by James Carothers many years later, mining supervisor William Flangas comments on a Livermore tunnel shot:

> Tamalpias was where we had the infamous hydrogen explosion. When we shot Tamalpias, because of the short lived products, some of the early readings in the tunnel were up there in the 10,000R range. And so the consensus was, 'Okay, this tunnel is gone.' And we still had not fired Evans.
>
> We had been working seven days a week, twenty-four hours a day, and I never left that tunnel day or night. Most of the time I was sleeping on my desk. By the time we shot Tamalpias some of us were flat worn out. So, once they start reading those kinds of numbers it looked like the ball game was over as far as that tunnel went, and I went home. I got home about nine or ten o'clock that night, and I was still asleep at two o'clock the next afternoon when a call came through that said to hurry on back. The readings were down to 300 or 400 mR, and they were anxious to get started again. By the time I got back up there it was like four o'clock. The Livermore honchos were there, and some of my troops had been assembled and they were there.
>
> I asked the question, 'What have we got?' They said, 'It looks like the highest exposure right now is like 400mR.' We could stand that for reentry. And then, of course, my next question was about explosive mixtures. I was assured that there was no explosive mixture. What had really happened is that due to the inexperience of both the lab people and others, the meters they had in those days got saturated, and so they were reading zero, when in fact the place was loaded with hydrogen."
>
> I went into the tunnel and I went back several hundred feet. The hair was standing up on my head, because I knew there was something wrong, but I couldn't put a finger on it. So, I came back out, and I repeated the question. 'How are we in terms of an explosive mixture, or are there any other gases, or any exotic gases I don't know anything about?' And again I was assured. 'Quit worrying about it. You do not have an explosive mixture.'

I went back in the tunnel. We were doing some preliminary work to get started, because it was important to get ventilation established so we could clear the tunnel out so we could proceed. I came back out again, was reassured again. As I ruled out every possibility, it occurred to me to wonder if my antennae weren't geared to an oxygen deficiency. One of the things copper miners fear the worst is oxygen deficiency, and in those days, in a copper mine, under Nevada state law, you had to provide every miner with a candle. The way you checked for oxygen deficiency was with a candle, because a candle goes out at 16% oxygen, or thereabouts. ...

I lit the candle, and I went all the way back in the tunnel. I was holding it just about chest level, and it was burning, so that ruled out oxygen deficiency. The rad-safe superintendent had climbed up on a sandbag plug, which was at about the 700 station; - 700 feet from the portal. And he says, 'Hey Flangas, hand me that candle.' So, I handed him the candle. Well, being a light gas, and without that environment having been disturbed, the hydrogen had accumulated along the top of the tunnel. He was up in that atmosphere, and Lordy, I was standing in the middle of the drift, at the 700 station, and he was up at the top of that sandbag plug. He said, when we talked to him a couple of days later, that he saw a flame that just went down to the 1200 station, where the other door was, and he was fascinated by the sight. I was standing right on the track there, and the next thing I knew I was head over heels, and when I picked myself up, I was at the 350 foot station.

I have no idea ... it was ... just everything was in motion. We laid plywood along that entire tunnel to protect the cables. That plywood was shredded to sawdust, to small fragments. There was a six inch steel door at the 350 foot station, and fortunately one of my shifters laid the track across there. We had to pull the track out to close the door, so when we opened the door, we put the track back in. That six inch door folded over that track into a U.

The interviewer, asked Flangas about the rad-safe man who was standing on the sandbag plug.

Fortunately, what happened to him is that when it went off the concussion knocked him down to the base of the plug, and when the explosion took place, it blew over him. Now, in that melee I turned around to look for him. My miner's lamp was shattered, and the place was just a bedlam. So, I looked for him for about a millisecond, and then I decided, 'What the hell, it's every man for himself, and I'm getting out of here.'

There were another four or five people in a side drift, and they escaped the blast. It went right past them. After all of this settled down we kind of found one another in the dark there. We finally retrieved this fellow by the name of Wilcox, and he was out colder than a wedge, at the base of the plug. When the blast door folded over it left a hole just barely big enough for a person to squeeze through. We accounted for everybody and got them out. The

> people on the outside were pretty excited. They thought everybody in that tunnel was dead, and that was a pretty good presumption at that time. So they called the ambulance and doctors, and there was a lot of commotion. It was a very unique experience. (Caging the Dragon, James Carothers, U. S. Government DOE/DP and DNA, 1995, 421-423)

After treating nine men at the first aid station in Mercury, two were sent to a Las Vegas Hospital. Fortunately, no permanent injuries were sustained. The explosive gas was believed to be hydrogen, which could have been from the breakdown of the two tons of paraffin used in Tamalpais experiments. (Reeves 1958: 144)

A few hours after the Tamalpias explosion someone suggested that the lights in the tunnel be turned on in order to assess the situation. This was done and promptly resulted in a second explosion. This time no injuries were sustained, but a gate that had been installed to limit tunnel access was blown approximately a hundred yards.

The nuclear test community learned a number of lessons from this experience. The most significant was the need for a formal tunnel reentry procedure. In the new regime "The re-entry inspection group was equipped with anti-contamination clothing, self-contained breathing apparatus, and instruments for the detection of radiation, explosive mixtures, carbon monoxide and oxygen deficiency. .... At the portal of the tunnel, there was a standby rescue crew, a first aid man, and an ambulance. All information from the re-entry group was relayed by telephone line to a recorder at the tunnel portal." (Reeves 1958: 144)

Also, as a result of this incident, Bill Flangas instituted formal mine rescue training. Fortunately, with the new policies in place, there were no additional tunnel explosions over the following thirty-five years of testing in Nevada.

By the end of HARDTACK the testing community in general and Livermore in particular had developed a much better understanding of the challenges associated with underground detonations. Successful containment was one goal, but it was necessary to be able to diagnose the performance of the nuclear device as well. A fair start was made in addressing these issues before the 1958 to 1961 moratorium, but it took the whole decade of the 1960s to perfect the containment and diagnostic technologies required for a successful underground nuclear test program.

There were other interesting incidents. For example, the balloon shot, Socorro, scheduled for detonation at 1500 feet in Area 7, was delayed when it was discovered that the firing cable was broken. Soon after the delay was announced, security guards at station 385 at the north end of Yucca Flat reported that they were questioning three men in a jeep. The three had driven that morning north from Frenchman Flat, on the Nellis Bombing and Gunnery Range, beyond the low hills to the east of Yucca Flat. They had traveled as far as Papoose Lake, due east of areas 7 and 9, when they realized that they were lost. They then drove west until they reached the NTS boundary, marked by a sign and a ditch across the road. They drove around the sign and the ditch and continued on until they reached the

Mercury highway. They then turned north and were eventually stopped at Station 385. Had Socorro been detonated on the original schedule, the three men would have been in the neighborhood of the intersection of the Papoose Lake Road and the Mercury Highway (in Area 2, north of the B-J Y). Socorro had a yield of 6 kt. While the men probably would not have been seriously injured, provided that they had not been looking directly at the device, it would not have been a pleasant experience. No record has surfaced that sheds light on what these men were doing or what action was taken to see that there was no recurrence of inadvertent wandering into the forward area near shot time.

DoD PARTICIPATION ON OPERATION HARDTACK PHASE II

DoD participation on Operation HARDTACK PHASE II was limited due to the operation's accelerated schedule and to its emphasis on weapons development and safety tests. The relatively small number of participants, about 1,375, was composed of approximately: 80 Army, 15 Navy, 320 Air Force, and 960 scientific personnel, contractors and observers. (Ponton 1982d: 11) There were no Desert Rock activities conducted on HARDTACK PHASE II. The main involvement by the DoD was by the Air Force Special Weapons Center (AFSWC) and by AFSWP's Weapons Effects Test Group.

Indian Springs was again used as the principal staging area for AFSWC's operations. As on past operations at the NTS, they provided aircraft and personnel for cloud-tracking and sampling missions, sample courier service, weather-reconnaissance, aerial surveys and security sweeps, and other air support as requested by the NTSO.

DoD NWE PROGRAMS AND PROJECTS

The number and diversity of DoD effects projects on HARDTACK PHASE II were not nearly as extensive as those on previous NTS operations other than RANGER. The 11 projects consisted mainly of measurements that: did not require long-term planning or major instrumentation development; were fairly fast and easy to field; or had been conducted on previous tests in a similar manner.

OPERATION	PROGRAM	# PROJECTS
HARDTACK II	1) Blast and Shock	1
	2) Nuclear Radiation and Effects	5
37 Tests	4) Biomedical Effects	2
	6) Tests of Service Equipment and Materials	2
	8) Thermal Radiation and Effects	1
	Total	**11**

The tests Hamilton and Humboldt, which were both 50' (wooden) tower tests with yields of 1.2 and 7.8 tons respectively, were the ones that had the most DoD effects measurements. But compared to previous operations these efforts were modest. (Reed 1962:18)

There was one blast measurements project, 1.7, fielded on nine shots that was conducted by only 4 people from BRL. One of the tests, Evans, used telemetry while the other tests used self recording gages that were recovered by the crew. (Ponton 1982d: 88-9, 93-4, 102, 108, 114, 138, 147-8, 160-1, 168)

The five projects in program 2 (2.12a, 2.12b, 2.12c. 2.12d, and 2.13) addressed the environment and effects of nuclear radiations from very-low-yield bursts. These projects measured the: neutrons; soil activation by neutrons; gamma-ray dose, and thermal radiation. (Ponton 1982d: 191, 195, 198, 201)

Several dozen mice and about 500 pigs were placed in foxholes, tanks, aluminum containers, and armored personnel carriers for the project 4.2 study of immediate lethality. (Moncerief 1961: 5)

A flash blindness exercise was conducted as project 4.3 with participants who faced in different directions away from GZ during detonation with their eyes open and unprotected by filters or goggles. After the detonation, they were asked to: read a vision chart; distinguish men moving in and out of view; and identify the display of different colored panels. (Verhuel 1960: 5, 24-6)

The operational capabilities of flash-ranging instrumentation for determining location and HOB were evaluated in Project 6.14. (Scarborough 1959: 9-10; Ponton 1982d: 39, 94) Project 6.15 was designed to locate "friendly nuclear detonations" * by their EM signals. [*Footnote: A "friendly nuclear detonation" appears to be an oxymoron but is defined as one where the approximated time and location of its occurrence is known ahead of time.] This project was conducted on 10 weapons development tests with yields of less than a kt and 4 safety tests with yields between 13 and 115 tons: Mars, Hidalgo, Neptune, and Vesta. This was the only NWE project conducted on safety tests during HARDTACK II. An instrument trailer located near Boulder City, NV, was to detect, record, and analyze the electromagnetic pulses from the tests. (Cantor 1960: 5, 9; Ponton 1982d: 191, 195, 198, 201)

For Project 8.8, measurements of total thermal radiation (energy/area) were made with calorimeters, and irradiance (energy/area-time) was measured with a bolometer and spectroscopic detector. Both of these quantities were measured as a function of time and wavelength of the radiation. (Reed 1962: 5, 18-21, 33-6)

CIVIL DEFENSE PARTICIPATION

On July 1, 1958, the major civil defense and emergency preparedness programs at the federal level were once again reorganized. The FCDA and the Office of Defense Mobilization (ODM) were consolidated into a single agency: The office of Defense Civil and Defense Mobilization (OCDM), which was housed in the Executive Office of the President (EOP). On Operation HARDTACK II, the Civil Effects Test Operations, under director Robert L. Corsbie, was comprised of two groups: the Civil Effects Test Group (CETG) and the Office of Civil and Defense Mobilization Test Group (OCDMTG).

Like the DoD, participation of the CD groups was limited by the considerable acceleration of the operation's plans that were undertaken to quickly execute HARDTACK II. In comparison to the previous three operations, the HARDTACK II CD programs were small in terms of the number and extent of the projects and in their complexity.

OPERATION	PROGRAM	# PROJECTS
HARDTACK II	34) Nuclear Effects, AEC Test Structures	3
	37) Further Evaluation of Tower and Balloon Shot Fallout Patterns	1
37 Tests	39) Radiation for Human Exposures	3
	70) Office of Civil and Defense Mobilization Test Group Projects	6
	Total	**13**

The projects were similar to projects undertaken on previous operations. Although some relatively small shelters and some houses were built, no large-scale construction projects were fielded. The CD community did not undertake the conduct of either an open shot or training exercises on HARDTACK II. The four programs that were conducted consisted of only 13 projects that were similar to those conducted on previous operations, and they were limited in scope. Also, last minute changes in shot locations and actual yields being different than pre-shot predictions restricted the amount of data collected.

Project 34.1 recorded and evaluated air blast and thermal damage to pre-existing AEC structures with emphasis on structures that had been exposed to high levels of overpressures and thermal radiation on previous shots. (Cameron 1961: 5, 15, 39) Project 34.2 studied the effects of air-bursts on underground structures used in PLUMBBOB. Accelerations and radiation measurements were made in these structures. (Cameron 1962: 5, 11, 17) Project 34.3 attempted to measure acceleration of underground shelters and the adjacent free-field acceleration. But low yields inhibited data acquisition. (Reeves 1958: 100).

Program 37 was to use aerial surveys to construct iso-intensity maps and to re-sample previously established biological collection areas. Scheduling and yields prohibited data acquisition on most shots. (Reeves 1958: 100)

Project 39.1 obtained neutron and gamma-ray dose distribution in simulated Japanese-type structures for correlation with data obtained by the Atomic Bomb Casualty Commission. (Reeves 1958: 100) Project 39.2 measured pressures and displacements as functions of time in seven Japanese-style houses. Project 39.9a was a continuing project that appraised nuclear effects on vegetation.

The 6 projects of the OCDM's program 70 did not fare well. One project provided instrumentation for pressure valves and shelters for two tests, but the tests went low yield and changed location respectively. Project 70.2 had some success in measuring gamma and neutron activity and fallout as a function of depth below the surface on 4 Area 7 shots.(Reeves 1958: 96) An aerial survey of radiation did not compare with the ground survey, and problems arose in removing self-recording gages from a tunnel.

Measurements of accelerations in the free-field and in shelters were attempted, but the yields of the tests went too low to be of much use. Three family fallout shelters were constructed in Area 9 by Project 70.4, but the shot was subsequently moved to another area.

LIVING CONDITIONS AT THE SITE DURING HARDTACK PHASE II

Camp facilities and management were the purview of the Coordinator for Base Support Operations, Robert E. Georges, who acted in a staff capacity to the Test Manager. Georges was an AEC employee while his deputy was from REECo. This organizational structure, which involved individuals familiar with Test Site operation procedures and facilities, enhanced the efficiency and effectiveness of the service.

No new facilities were built in Mercury between PLUMBBOB and HARDTACK Phase II. In fact, the dreaded plywood hutments were removed. (Reeves 1958:5) However, prior to HARDTACK, a construction camp was set up in Area 12, the Rainier Mesa area, about 45 miles N of Mercury. Seventy-seven trailers from Watertown and Mercury, having a design capacity of about 260 beds, were on hand to meet the demands of the camp. Housing at Mercury consisted of the following facilities: (Reeves 1958:137)

<u>Men's Quarters</u>	
35 Dormitories - Design capacity	1282
203 Trailers - Design capacity	<u>406</u>
Total Design capacity =	1688
<u>Women's Quarters</u>	
5 Dormitories - Design capacity	180

During HARDTACK II, the peak Mercury population of 2400, including 143 women, was on October 9, 1958. The total NTS population, Area 12 plus Mercury, peaked at 2655 on October 2.

With few exceptions, the 510 seats of Cafeterias 1 and 2 fulfilled the demand in Mercury. The popular Steak House was opened on September 8 and remained open throughout the test series. Supplemental dining facilities were maintained at the Area 12 camp, the snack bar at the CP, as well as at Area 400 in Jackass Flats for the Rover program. (Reeves 1958:139)

Church services were conducted in Mercury on Sundays. Protestant services were held in the west wing of the Recreation hall and Catholic services in building 125.

The AEC maintained a contract with the Nye county Sheriff's Department to provide law enforcement. The problems were mostly traffic related.

The offices of a messenger service between Mercury and the CP, a reproduction center, and a labor pool were located in Warehouse #3. The labor pool provided people to test and to support groups for jobs such as clerks, stenographers, secretaries, electronic technicians, machinists, warehousemen, fork-lift operators, and laborers. Among the organizations that drew from this pool were the Test Manager, UCRL, LASL, Sandia and the Visitor's Bureau.

The main medical facility was the Mercury Dispensary, which operated on a 24 hour basis, seven days per week throughout the year. It had a full-time physician and registered nurse and a consulting staff of medical specialists in Las Vegas. The specialists included surgery, orthopedics, internal medicine, radiology, ophthalmology, dermatology, otolaryngology, and neurosurgery. Arrangements were made to use the facilities of the Southern Nevada Memorial Hospital, Las Vegas and the Rose de Lima Hospital in Henderson when necessary. The dispensary was shared with medical personnel of the DoD. Field aid stations manned by first aid attendants were provided at various locations depending upon forward area work activities. During HARDTACK II, these locations included Area 12, BJY, CP Building #2 and Area 400 (the Rover program) CP.

A reception area, doctor's office, treatment room, drug room, and an x-ray machine were available at the Mercury Dispensary. Ambulances and field aid stations were stocked with medical supplies and accessories consistent with the extent of first aid treatment to be rendered at such locations. Isolation quarters were available for those individuals with contagious infections. Radio and telephone communications were available at all aid stations.

The medical personnel and facilities were available to all test participants. The treatment of non-occupational injuries or diseases were provided at no cost to the individual. Medical service statistics for the period August 25 - November 15, 1958 were:

Visits occupational injuries	377
Revisits for occupational injuries	129
Non-occupational visits	712
Revisits for non-occupational	145
Pre-employment culinary examinations	413
Pre-employment examinations non-culinary	26
Total visits =	1802

Pre-employment physicals were provided on request from sponsoring agencies, and for all REECo personnel to be employed for tunnel or culinary work.

The basic criteria for General Safety and Fire Protection are contained in the Standard Operating Procedures for the Nevada Test Organization. They were implemented by Safety personnel under the Support Director. Safety personnel from Los Alamos and Livermore supported their respective organizations.

During HARDTACK II 1.66 million man-hours were worked. In spite of an exhausting schedule, there were no lost time injuries and no motor vehicle accidents involving operational personnel reported. There were 13 lost time injuries sustained by support personnel. Two of these were the result of a single motor vehicle accident and seven were individual accidents in tunnels. (Reeves 1958:144-5)

A total of 4.9 million miles were driven on-site and off-site during the operation with only six motor vehicle accidents.

A recommendation was made In the HARDTACK Phase II Test Manager's report that: "There should be at least two up-to-date, fully equipped ambulances which will allow the proper transportation of more than one patient at a time with space for a first aid man in attendance." (Reeves 1958:142)

Among the recommendations for future facilities at Mercury were:
- Building sidewalks or walkways throughout the base camp and marked cross walks established at intersections with heavy traffic.
- Additional parking facilities for privately owned vehicles.
- Mercury aircraft landing strip should have a taxi lane and parking facilities opposite Building 111 on the outside of the security barricade.
- CP-1 Building should have more sleeping accommodations.
- Construct a steak house that is removed from the main feeding facilities. Included in this facility should be a private room for cocktail parties and general get-togethers.

(Reeves1958:139)

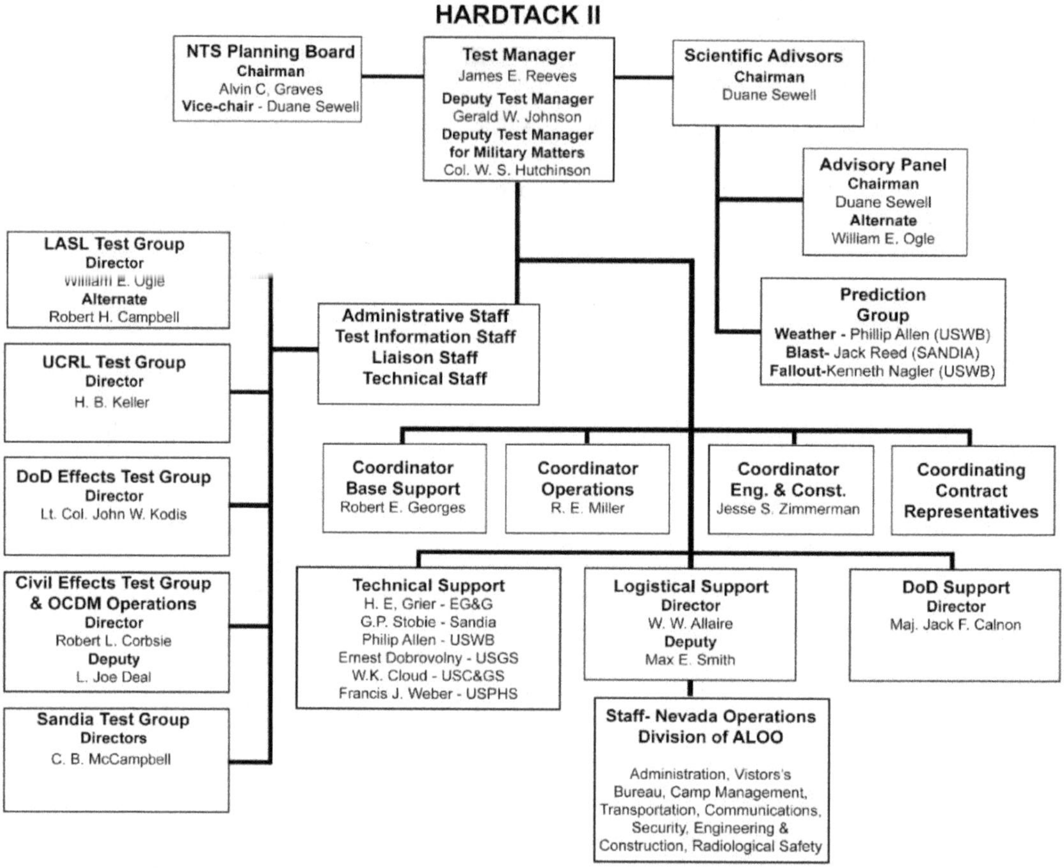

Figure 2-14.1 NTSO for Operation HARDTACK PHASE II.

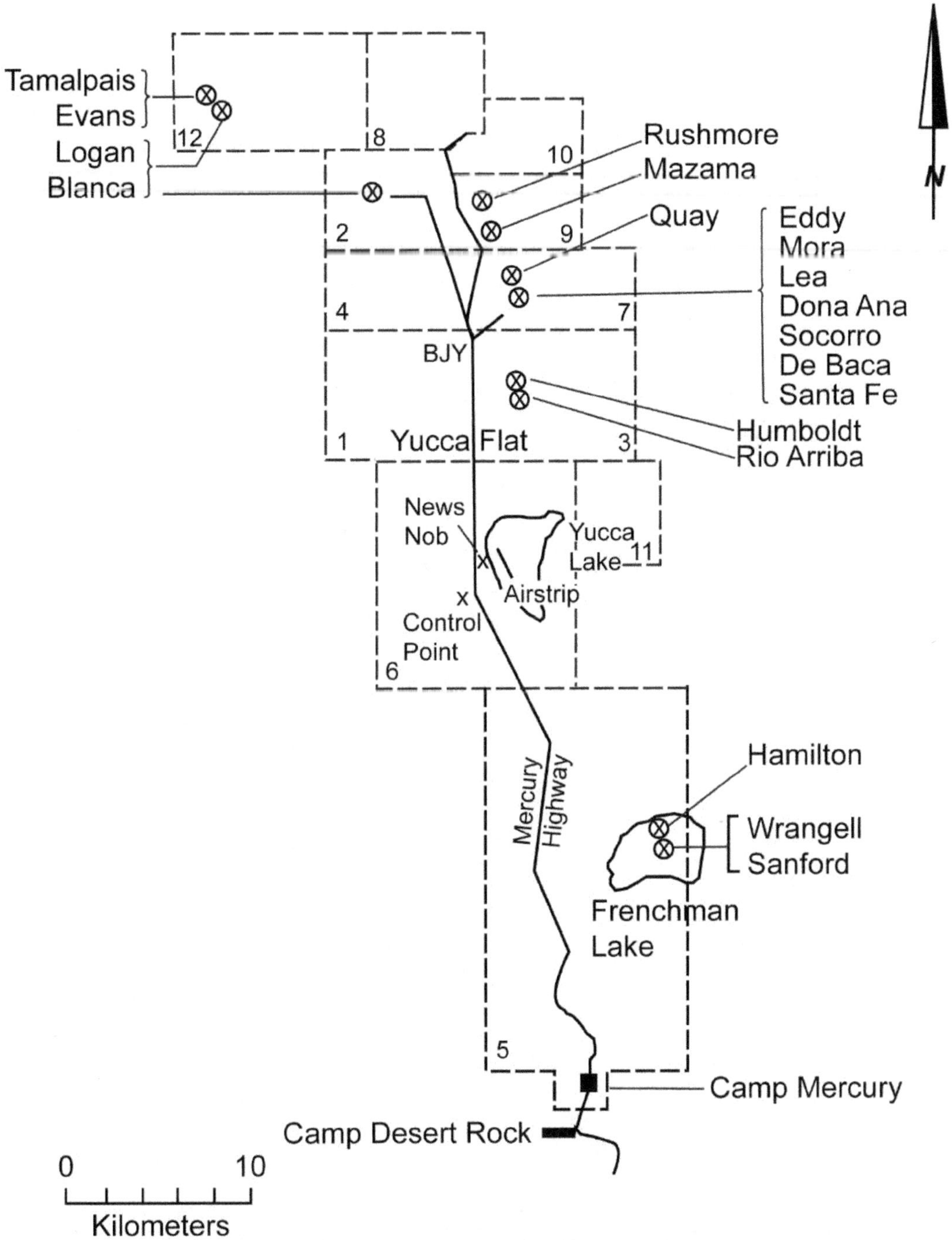

Figure 2-14.2 Location of HARDTACK PHASE II Weapon Development Tests.

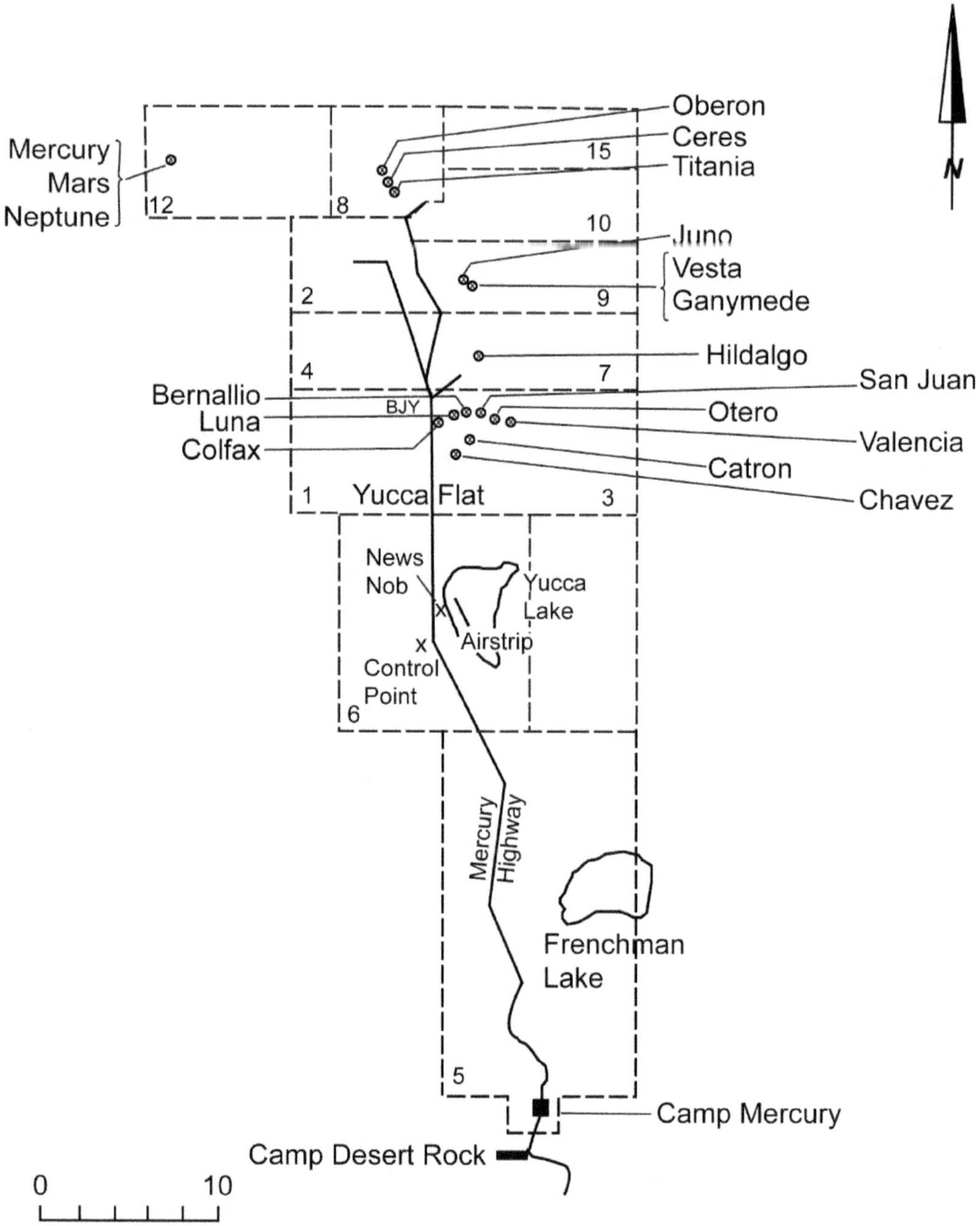

Figure 2-14.3 Location of HARDTACK PHASE II Safety Tests.

CHAPTER 15. NUCLEAR TEST MORATORIUM: 1958 - 1961

READINESS

Major General Alvin R. Luedecke, Commander Joint Task Force Seven, for HARDTACK PHASE I in the Pacific, in an October 1958 conversation with Bill Ogle, was quoted as saying that "Our experience indicates that the U.S.S.R. will resume testing at such a time as the Kremlin considers that it is in their best interests to do so, progress of negotiations or agreements notwithstanding. However, it appears possible, or even likely, that their interests would best be served by cooperating in negotiations to the extent necessary to cause the United States to refrain from testing for an extended period of time."

Ogle continues: "He (Luedecke) proposed that the AEC and DoD could maintain a capability to conduct a limited number (3-4) of proof tests at both test sites within 3 months, and 10-12 developmental tests within 9 months, if (a) continuous plans were maintained; (b) continuous capability to activate a test organization were maintained; (c) necessary plant equipment, and funds were adequate; (d) provisions were made for 'normal service support' by appropriate AEC and DoD agencies; and (e) studies were conducted of alternate means of conducting test operations to effect simplification and economy." (Ogle 1985: 113)

Joe Sanders, of the Albuquerque AEC office, forwarded to the laboratories and major NTS contractors a TWX on October 26, 1958, with a message that Brig. Gen. Starbird had sent to Hertford:

> DMA has submitted to the Commission a staff paper setting forth following policy regarding readiness to resume testing:
>
> a. On 90 days notice, conduct a few 'quick' /minimum diagnostics/test overseas/EPG or other oceanic locations including possibly a ship at sea.
> b. On 90 days notice, initiate conduct at NTS of low-yield tests primarily underground.
>
> ...
>
> This will require as a minimum the following:
>
> a. Maintain the small test organization of ALOO and maintain the EPG and NTS on a standby status utilizing the present contractors.
> b. Assure that necessary physical plant, equipment and resources /including funds/ are adequate and available for testing purposes. In this connection there would be no advance procurement and inventory of readily available 'off-the-shelf' commercial items, or those available in DoD. However, there would be maintained a minimum inventory of long lead-time essential items if not available elsewhere in AEC or DoD.
> c. In cooperation with the DoD and its agencies, continuously maintain up-to-date plans for test programs that are designed to fulfill priority needs of the weapons programs.
> d. Initiate action, for construction of a tunnel, at NTS for LASL testing and clean up the UCRL tunnels. Necessary tunneling and core drilling for

samples and scientific studies of past underground shots should continue concurrently.

e. Study and compare various means of conducting test operations to effect simplification and economy of operation and lessening of required lead-time periods.
f. Provide stand-by costs of approx. $20 million in FY 1960. ... This assumes, of course, that no resumption of tests would occur in FY 1960 or early in FY 1961.

This message contains the readiness theme that figured so prominently during and after HARDTACK. The AEC felt that the testing community should be ready to resume testing reasonably promptly if the federal government chose to do so if and when the moratorium were to go by the boards. They also felt that there should be little or no overt signs of test preparations for fear that such activities would be viewed as inconsistent with official policy. Also, the AEC wanted to preserve the nation's nuclear weapons technology, while simultaneously curtailing expenditures. Of course these various objectives were mutually exclusive and were thus a great source of frustration and annoyance for all of the players*. [*Footnote: The Los Alamos archives (A-99-019, Box 99 folder 19) contain many memos reflecting the frustration associated with maintaining a readiness capability while simultaneously reducing expenditures. See for example memo from Darol Froman to Norris Bradbury dated August 27, 1958. This would be a way of life with more or less frustration indefinitely.]

By mid-spring of 1959 the testing community had arrived at a proposed test series, assuming that testing resumed, consisting of fourteen full yield underground shots and several safety tests. The series was planned to begin just after the end of the one-year moratorium; November 1, 1959, and run through September 1, 1960. (Ogle 1985: 130) This looked a lot like continuous operation at the NTS.

Possible ground water contamination was the primary environmental concern with regard to underground testing. Los Alamos preferred the concept of firing in vertical holes in Yucca Flat, but was ready to join Livermore at Rainier Mesa if the ground water problems proved intractable. In the spring of 1959 Los Alamos gave ALOO Test Manager James Reeves specifications for both vertical holes in Yucca Flat and tunnels in Rainier Mesa. Los Alamos wanted four shot points each for the vertical holes and for the tunnels. The objective was to stay above the water table for all firing points. The vertical holes were to be cased 36-inch diameter by 1000-foot depth. At this stage in the planning Los Alamos only considered alpha measurements as weapons diagnostics. (Letter, R. W. Newman to James E. Reeves, April 15, 1959 Los Alamos Archives A-99-019 Box 100, File 14) (Memorandum James E. Reeves to A. C. Graves March 30, 1959 ref TN: CAG-4085, LANL Archives 353.4)

By June, 1959, Los Alamos was reasonably assured by the geologic studies of William Twenhofel and his colleagues at the USGS and elsewhere that ground water contamination in Yucca Flat would not be a problem. LASL promptly cancelled their request for shot locations on Rainier Mesa and reiterated their request to the ALOO test organization to drill four holes in Area 3. (TWX K. F. Hertford to DMA, May 13, 1959. Los Alamos Archives A-99-019 Box 100 file 14. Also: Ogle 1985: 136)

Questions naturally arose about the future of nuclear weapons research and development and whether underground testing was really feasible. AEC Chairman McCone, Secretary of Defense Quarles, and James Killian, the President's Science Advisor, assembled a group of senior leaders of the nuclear weapons programs to address the issues. Participants included: MLC Chair Herbert Loper; Norris Bradbury, Jane Hall, William Ogle and Carson Mark from Los Alamos; Edward Teller, Gerald Johnson, John Foster, and Roland Herbst from Livermore; Glenn Fowler, Don Shuster, R. A. Bice and M. L. Merritt from Sandia; K. Hertford and James Reeves from ALOO; and Brig. Gen. A. D. Starbird from the AEC/DMA. The general conclusions of the May 1959 meeting in Washington were that there could be substantial improvements in size and weight and in yield to weight ratios to be gleaned from a test program, but that there would probably not be as dramatic developments as were seen earlier in the 1950s. The group also forecast potential improvements in enhancing specific outputs, such as neutrons or x-rays, for certain military applications.

Ultimately the senior weapons leaders concluded that an underground test program was feasible. It was possible to fire shots up to some, as yet, undetermined yield, but at least to 30 to 40 kt. They also suggested looking for alternate sites for yields up to a megaton. (Draft Report of Ad Hoc Committee on Test Requirements. May 14, 1959, communication from Brig. Gen. Alfred D. Starbird to participants.) As it turned out, the largest shot in Nevada was Boxcar at 1.3 Mt, which was fired in April, 1968. The largest U.S. shot underground was Cannikin, fired on Amchitka Island in the Aleutians in November 1971. It was slightly less than 5 Mt.

It is interesting to note that the enthusiasm of the laboratory's weapons leadership for underground testing varied from quite high for Harold Brown at Livermore to moderately low for Norris Bradbury at Los Alamos. (TWX. Bradbury to AEC/DMA May 5, 1959)

Verification of a test ban agreement was a key issue of the Geneva Conference which was a tri-lateral negotiating forum consisting of the US, SU, and UK, see Attachment V. The verification issue stimulated research in seismic signals from underground explosions, both conventional and nuclear. The three labs had initiatives to address the technical questions both theoretically and experimentally, but of course, the moratorium restricted the experimental work to non-nuclear explosions. (TWX. Starbird to AEC/SAN, ALOO, LRL, Info LASL, Sandia, June 12, 1959) Livermore had a program, code-named Cowboy, to use buried high explosive shots to explore the geophysical phenomena that bear on the feasibility of decoupling nuclear explosions. (Los Alamos File 635 Cowboy) Additional related Livermore studies were code named Lollipop and Peacock. Interestingly, the British had a similar decoupling program code-named Orpheus, and the Soviet Union certainly did work in decoupling.

WASHINGTON POLITICS

George B. Kistiakowsky replaced James Killian as the President's Science Advisor in July 1959. This was a setback for the pro-test factions. Kistiakowsky was bound and determined to rein in McCone's (and Sec. Def. Thomas Gates) efforts to win approval for test resumption. He appointed a panel under the chairmanship of James W. McRae, Western Electric VP, to examine the need to test. They concluded that there was no need to test immediately with the possible exception of one point safety tests. (Hewlett and Holl, Atoms for Peace and War 1953-1961, p.557)

By August, 1959, it looked more and more as though the moratorium would be extended, and on the 26th the President announced an additional year's extension in order to continue the test ban negotiations. The British, who in 1958 had been firing shots at Christmas Island, followed suit the next day and agreed to an extension of the moratorium provided the test ban negotiations appeared to be productive. (Ogle1985: 126,138; Mikhailov 1999: 20-2)

The day after the British announced their intentions to observe the moratorium the Soviets announced their intention not to resume nuclear tests if the west refrained from testing. This is a very important point. The first French test was on February 13, 1960 at Reggane, Algeria. They had two additional tests in 1960 and a test in April 1961. The Soviet Union claimed that these shots were tests by a western power and thus served as a basis for Soviet resumption in September 1961.

1960

Nuclear test procedures were exercised during the moratorium while pursing both test readiness and the seismic detection program. The latter was code named Vela Uniform. Livermore also had conventional high explosive experiments in support of the Plowshare Program. For example Scooter, a half-kiloton non-nuclear high explosive shot, was fired at the NTS in October 1960 at a depth of burial of 125 feet. (Ogle 1985: 182)

Bradbury sent Hertford a memorandum in February giving the Los Alamos position with regard to readiness to resume testing: "The LASL is considering a variety of possible nuclear tests for proposal under such circumstances. In general we would probably attempt to resume such testing 3 - 4 months after a presidential announcement to this effect (P-Day)." He went on to suggest that Los Alamos would have several test candidates with yields up to 30 kt, which he assumed would be a reasonable maximum for NTS underground shots at least until more experience had been accumulated. Bradbury also asked for more and deeper holes in anticipation of a resumption of testing. (February 11, 1960 Memo Bradbury to K. F. Hertford, manager, ALO Subject: NTS Preparations for Possible Resumption of Nuclear Testing. (Dir 1547))

The AEC and the labs were planning a test series to begin after the first of July 1960. The DMA suggested that the first shot be a Los Alamos device of about 25 kt. This would be followed by about ten more through November. (February 19, 1960 TWX Starbird to Bradbury and Teller)

In the mean time, the President's Science Advisor, George Kistiakowsky, notified the Chairman of the AEC that President Eisenhower put a high priority on a technical program focused on the detection of underground and high-altitude nuclear explosions. (Ogle 1985:184)

President Eisenhower noted, "On March 19, (1960) the Soviet Union expressed a willingness to: Conclude a treaty halting all nuclear weapons tests in the atmosphere, the oceans, and cosmic space, and all underground tests of seismic magnitude 4.75 or more." In his memoirs, at least, Eisenhower erroneously equated this to a nuclear yield of 4.75 kt. (Eisenhower 1965: 480)

Bradbury felt that the hopes for an early resumption of testing were rapidly fading. "It is my opinion that the probability of nuclear testing in 1960 is so low that the course of the Laboratory should be along the lines 'We aren't going to test in 1960; therefore we will start as if we weren't and planning for the long range as if we weren't, and keep testing and devices for test very much on the back burner.' Then if this assumption turns out at any moment to be wrong, and we are told we can resume testing - well, we fall out and fall in again pointing (and going) in the new direction as fast as possible. Admittedly, on this basis we would not test as soon as we might otherwise after a Presidential directive to resume testing, but we would have a lot more sensible program in the meantime, a lot of people would be driven less rapidly to schizophrenia, and the actual difference in time might not be more than 30 days!" (Ogle 1985: 160)

It was clear that a short lead time test capability would require stockpiling many segments of the test cycle. These included holes or tunnels, emplacement and diagnostic hardware, nuclear test devices, recording equipment, cables, and detectors. The equipment requirements were augmented with the personnel requirements. In order to achieve a very short lead time the test community had to proceed almost to the point of the installation of the device in the test location.

By May 1960 President Eisenhower concluded that the Soviet Union was not serious about test ban verification negotiations and was ready to authorize a resumption of U.S. testing, but felt that he should not tie the hands of the incoming President. "Accordingly, we did no more testing during the remaining few months of my administration, but I emphasized to President-elect Kennedy my conviction that our nation should resume needed tests without delay." (Eisenhower, op. cit. p. 481)

At Los Alamos the last (and 174th) meeting of the Fission Weapons Committee was held on March 21, 1960. H. M. Agnew, G. H. Best, M. L. Brooks, R. Canada, D. K. Froman, J. H. Hall, D. P. MacDougall, J. C. Mark, W. E. Ogle, R. K. Osborne, E. L. Peterson, M. F. Roy, A. R. Sayer, R. G. Shreffler, J. J. Wechsler, D. R. Westervelt, and D. R. Woods attended. The Tech Board, whose membership comprised those laboratory leaders at the division leader level and above, replaced a variety of specialized groups, including the Fission Weapons Committee. It was chaired by Norris Bradbury.

Meanwhile, in Nevada, Test Manager Reeves outlined the appointments to the NTS Organization in March:
Test Manager: James E. Reeves
Deputy Test Manager: W. W. Allaire
Scientific Advisor, LASL: Dr. Alvin C. Graves
Scientific Advisor, LRL: to be announced
Military Deputy, FC/DASA: Col. Leo A. Kiley
NTS Planning Board:

> Chair: Alvin Graves LASL
> Vice Chair: Duane Sewell LRL
> Member: W. E. Ogle LASL
> Member: G. W. Johnson LRL
> Member: Leo Kiley FC/DASA
> Member: R. L. Corsbie CETG, AEC Wash. DC
> Member: R. A. Bice Sandia Corp.
> Member: M. L. Merritt Sandia Corp.

(March 28, 1960 James Reeves, Test Manager, USAEC NTS Organization, to distribution)

In July 1960, Harold Brown became the third Director of LLNL. Edward Teller of course remained active at the laboratory in scientifically important but non-administrative roles. Brown received his PhD in physics from Columbia University at the age of 21 in 1949 and joined LLNL in 1952. He left LLNL in May 1961 to work under Robert McNamara as Director of Defense Research and Engineering (DDR&E) until 1965 when he became Secretary of the Air Force. He was the first scientist to become Secretary of Defense between 1977 and 1981. John S. Foster became LLNL Director in 1961.

1961

Starbird responded in January 1961 to Reeves' November request for guidance regarding test readiness.

> I would like to emphasize, however, that no impression should be conveyed or implied that a resumption of testing is imminent. ... He directed Reeves to ask LRL to redefine Succotash (their readiness plan) program and to also approach LASL, DASA, and others with the thought of producing an overall plan which could then be examined by DMA with the idea of authorizing such portions of the new plan as might be feasible, considering budget

> restrictions and political implications. The Labs responded leisurely, with Graves commenting that it was worthwhile to have some general plan of what weapons tests should have first priority, but that he couldn't see the likelihood of major construction effort, and, furthermore, he would in general rather see planning done on a less formal basis rather than trying to develop a single plan or a set of alternate formal plans. He commented in mid-February that LASL could organize any test effort within the time that would be required for the Lab to prepare the test devices for use.

(Ogle 1985: 205)

During the late winter, spring and early summer of 1961 there was a growing sense that full scale nuclear testing would actually resume. The labs were tasked with planning and preparation for resumption, provided there were no outward signs of preparation. Bradbury noted, "In brief, we are doing everything we can to get ready for testing short of getting ready for testing in Nevada!" (N. E. Bradbury to Brig. Gen. A. W. Betts, DIR-1683 (SRD) (May 8, 1961) 7pp., B9 353.4 Test Moratorium 5/1/59-5/31/63)

The pace increased as the summer wore on. On August 11, 1961 Brig. Gen. Betts sent a letter to the lab directors and AEC area managers with an attachment forecasting imminent presidential authorization of test resumption. The labs were expected to be in a position to exercise the test readiness activities that had been pursued with increasing intensity over the past months. (A. W. Betts to Dr. Norris E. Bradbury with enclosures (SRD) (August 11, 1961), 18 pp., B9 353.4 Test Resumption - 10/31/65) Less than a week later Betts notified the lab directors that he was authorizing the expenditure of up to three million dollars for underground test preparations. (Letter Betts to Bradbury (August 17, 1961) 3pp., A86-003 23-2)

Test Resumption

On August 31, 1961 the Soviet Union announced that they would resume testing. Betts the same day notified the labs that they should proceed with a program to test within two weeks and forecast that there would be no early end to testing.

The Soviet Union fired their first post-moratorium shot two days later, on September 1, 1961 a 16 kt airdrop at Semipalatinsk in what is now Kazakhstan.

In a press release dated September 5, 1961, President Kennedy said "In view of the continued testing by the Soviet Government, I have today ordered the resumption of nuclear tests, in the laboratory and underground, with no fallout. In our efforts to achieve an end to nuclear testing, we have taken every step that reasonable men could justify. In view of the acts of the Soviet Government, we must now take those steps which prudent men find essential." He went on to say that the U.S. offer to negotiate a ban on atmospheric tests would remain on the table until September 9. The Soviet response was more tests.

The Soviets clearly were ready for a full scale resumption of testing and had teams firing shots at Semipalatinsk, Novaya Zemlya and the missile test range near

Kapustin Yar. The Soviets fired fifty-nine shots between September 1 and November 4, when they finally stood down for about three months to regroup.

The U.S. testing community, while not caught totally by surprise, was certainly not ready to resume. Los Alamos and Livermore had a very frustrating three years between 1958 and 1961. They were expected, by the AEC, to maintain a readiness to resume testing with inadequate funding and the constraint that they were not to appear as though they were preparing to test.

However, despite the constraints of the previous 3 years, United States responded within 2 weeks, on 15 September 1961, with Operation NOUGAT, 43 tests at NTS and one near Carlsbad, NM. NOUGAT began with the Antler event and ended with the Livermore Sacramento shot on June 30, 1962. In addition, Operation DOMINIC was underway in the Pacific between April 25, 1962 and November 4, 1962 with 35 tests.

EPILOGUE

Enormous strides were made in weapons development and in understanding weapons effects during the 1950s. At the beginning of the decade the arsenal was relatively small and consisted of fairly simple fission devices. By the end of the decade the arsenal consisted of a large and rapidly growing inventory of sophisticated weapons encompassing a very broad spectrum of yields. The Nevada Test Site played a key role in this. Progress would have been considerably less had Nevada not been available for nuclear tests. While it was important to have the Pacific Proving Grounds for the very high yield shots, before the moratorium, about twice as many tests were done in Nevada as at Enewetak and Bikini (120 vs. 64). After 1963, when the Limited Test Ban Treaty eliminated atmospheric testing altogether by the Soviet Union, the United Kingdom, and the United States, underground testing was perfected to the degree that tests up to 1.3 megatons were safely and fully contained underground in Nevada.

Summary

One hundred twenty nuclear tests (including the safety tests) were fired at the Nevada Test Site between the beginning of January 1951 and the end of October 1958. Most were fired during the seven major operations. Eighty-four were atmospheric shots on towers, suspended by balloons or airdrops. The two crater shots were shallowly buried, low yield, devices. Toward the end of the 1950s the airdrops gave way to balloons as the preferred method of raising the height of burst. Two were what could be called systems tests; a canon (Grable) and a rocket (John).

	Airdrop	**Tower**	**Surface**	**Crater**	**Canon**	**Balloon**	**Rocket**	**Shaft**	**Tunnel**
RANGER	5								
BUSTER-JANGLE	4	1	1	1					
TUMBLER-SNAPPER	4	4							
UPSHOT-KNOTHOLE	3	7			1				
TEAPOT	3	10		1					
PROJECT 56			4						
PROJECT 57			1*						
PLUMBBOB		9	2			13	1	2	2
PROJECT 58			1						
PROJECT 58A									2
HARDTACK PHASE ii		10	3			11		6	7
TOTAL=120	**19**	**41**	**11+1***	**2**	**1**	**24**	**1**	**8**	**11**

[*Footnote: The Project 57 surface test was actually conducted across the NTS eastern border in the Groom Lake area of the Nellis Air Force Bombing and Gunnery Range (NAFR), see Part II-Chapter 12.]

A moratorium on nuclear testing began on November 1, 1958 and lasted until September 1, 1961, when the Soviet Union fired a 16 kt shot at their Semipalatinsk test site. Two weeks later, the United States responded with Operation NOUGAT, the first shot being Antler at NTS on September 15, 1961 with a 2.6 kt yield.

Assessment

The first decade of testing in Nevada was expensive in terms of both the resources expended and the efforts of tens of thousands of participants. The benefits are measured in the ephemeral currency of aggression that was deterred and wars that were not fought. Clearly the military capability of the United States was greatly enhanced by work done at the Nevada Test Site during the 1950s and equally clearly an international confrontation did not ignite a world war. South Korea was attacked by the North in 1950, but that occurred prior to the acquisition of the test site. In any case, the deterrence of regional conflicts is only marginally related to the nuclear capability of the superpowers.

Norris Bradbury argued that the role of nuclear weapons was to buy time for the achievement of diplomatic solutions to international confrontations. The argument can certainly be put forth that nuclear deterrence worked over the past sixty years. The unknown, of course, is what would have happened had nuclear weapons not been part of the equation. The task will remain for future historians to assess the validity, and weights, of the components of the argument. To the extent that nuclear testing in Nevada can be credited with the development of an effective deterrent to a major war, the case can be made that the testing program was a great investment in peace. Probably not surprisingly, the authors feel that a net assessment would unequivocally conclude that the continental nuclear testing described here was during those Cold War years in the best interests of the Nation.

APPENDIX A

BASIC CONCEPTS OF NUCLEAR WEAPON DESIGN

INTRODUCTION

This appendix provides a brief review of the basic concepts relevant to the design and test of nuclear weapons. It is intended to be a reference for the reader, and hopefully it provides a broader understanding of the discussions presented in the main text regarding why and how nuclear tests were conducted at the Nevada Test Site.

Some similarities and differences between conventional high explosives and nuclear explosives are discussed to highlight features of nuclear weapons that require testing information.

An explosion is caused by the rapid release of a large amount of energy. In the case of a conventional explosion (such as TNT) this energy comes from chemical reactions, which involve a rearrangement among the atoms, e.g., of hydrogen, oxygen, and nitrogen in the chemical high explosive material. In a nuclear explosion, the energy is produced as a result of the formation of different atomic nuclei by the redistribution of the protons and neutrons within the interacting nuclei. (Glasstone 1962:1)

The rapid release of energy produces high temperatures and pressures in the materials of the explosive and those near it. These materials vaporize; and the resulting hot and high-pressure gasses expand rapidly, causing a pressure wave that is transmitted into the surroundings. When the pressure wave is transmitted through air, it is often referred to as a blast wave. When the pressure is transmitted through water or ground, it is often referred to as a shock wave. Blast and shock waves from conventional or nuclear explosions are a primary cause of damage in the surroundings. (Glasstone 1962: 1) However, nuclear explosions:

1. Can be many thousands of times more powerful than the largest conventional detonations;
2. Emit a fairly large proportion of their energy in the form of light and thermal radiation;
3. Emit X-rays, gamma rays, and neutrons, and
4. Have debris that are radioactive and emit radiations over an extended period of time. (Glasstone1962:2-4)

All substances are made up of one or more elements. The smallest part of any element that can exist, while still retaining the characteristics of the element, is called an atom. (Glasstone 1962:4) Every atom consists of a relatively heavy central region or nucleus surrounded by a number of very light particles known as electrons. The nucleus of an atom of an element is itself made up of a definite number of particles called protons and neutrons. Protons and neutrons have almost the same mass, but the proton carries a unit charge of positive electricity. The neutron is uncharged electrically, i.e., it is neutral. The electrons that surround the nucleus each carry a unit charge of negative electricity. (Glasstone 1962: 3)

When an atom has no electrical charge, the number of electrons surrounding the nucleus equals the number of protons. When an atom has an electrical charge, due to the loss OR gain of an electron, it is called an *ion*. If the atom loses one or more electrons, the atom is positively charged and is called a *cation*. If the atom *gains* one or more electrons, it is negatively charged and called an *anion*. Chemical processes occur among ions, electrically charged cations combining with anions. A chemical process may require the addition of heat (energy) before it can occur, or the process may release heat (energy) while it occurs. When a great deal of heat is released in a short period of time from a chemical process, an explosion occurs.

The characteristic that identifies atoms of different elements is the number of protons in the nucleus. The number of protons in the nucleus of an element is called the *atomic number* of the element. All atoms that have the same number of protons are the same element. However, atoms of the same element can contain a different numbers of neutrons. The *isotopes* of an element are those atoms whose nuclei contain the same number of protons (equal to the atomic number of the element) but different numbers of neutrons. (Pauling 1954: 74)

The masses of the proton and neutron are each very close to one atomic mass unit (amu), which is equal to 1.67×10^{-24} gm (Pauling 1954:74), and is defined as 1/16 the mass of O16. The sum of the number of protons plus the number of neutrons is called the mass number of the atom. The mass of an electron is only about 1/1800 of the mass of a proton or a neutron, so ignoring the electrons and just summing the number of protons plus neutrons provides an approximation for the atomic weight. (Kaplan 1955: 595) In summary, the *atomic number* is the number of protons in the nucleus, and is what identifies an element. The *mass number* is the number of protons plus the number of neutrons in the nucleus.

Avogadro's number, 6.023×10^{23}, is the factor for converting atomic weights to the number of atoms. The mass in grams of an element that is equal to its atomic weight and contains Avogadro's number of atoms. (Pauling 1954:78) For instance, there are 6.023×10^{23} atoms in 235 grams of U235 and there are 6.023×10^{23} atoms in 1 gram of hydrogen.

Nuclear processes can occur that result in the redistribution of the protons and neutrons within nuclei. However, only those nuclear processes occur in which mass is converted to energy. Einstein found that there is a relationship between mass and energy expressed in the now famous formula $E=mc^2$, Energy = mass x (velocity of light)^2, where the velocity of light is, $c \sim 3 \times 10^{10}$ cm/sec. When a decrease of mass occurs in a nuclear reaction, there is an accompanying energy release, which is related to the decrease in mass by Einstein's formula. (Glasstone 1962:5)

In addition to the necessity for there to be a net decrease in mass, the release of nuclear energy in amounts sufficient to be classified as an explosion requires that the processes occur in a mass with a configuration that allows the reactions to be self-sustaining for some time. Two kinds of self-sustaining nuclear interactions satisfy the conditions for the production of large amounts of energy in a short time. They are known as fission and fusion. The fission process takes place with some of the heaviest (high atomic number and mass) nuclei, whereas fusion involves some of the lightest (low atomic number and mass) nuclei. (Glasstone 1962:5)

The materials used to produce nuclear explosions by fission are certain isotopes of the elements uranium and plutonium. Uranium (U), whose atomic number is 92, is found in the natural state, in two isotopes: U238 and U235. (The number reflects the mass number.) In the natural state, U235 is by far the less abundant (~0.7%); and it is the readily fissionable isotope. U233, which is not found in nature, is also a material that could be used in an atomic weapon. The element plutonium (Pu), which does not occur naturally, is made by bombarding U238 with neutrons. Nuclear accelerators, such as cyclotrons, can make small quantities of Pu, while large quantities are produced in nuclear reactors.

When a neutron enters a fissile nucleus it can cause the nucleus to split into two smaller nuclei plus 2 or 3 neutrons. This is the fission process, and it is accompanied by the release of a large amount of energy. The two smaller resulting nuclei are called the fission products.

The neutrons released in fission can cause additional fissions in other nuclei, be captured without causing fissions, or escape from the surface of the material. If, on the average, one of the fission neutrons causes additional fissions then the reaction is self-sustaining. The greater the volume the more likely this is to happen. On the other hand, if the surface is too large for the volume then more neutrons will be lost, and the reaction process will not be self-sustaining. This is why there is a critical mass for the fission process. It depends on the configuration, or shape and density, of the U or Pu. For a spherical configuration the volume increases as the radius cubed while the surface area increases as the radius squared. At some radius, the number of neutrons that result in fission will outweigh the number lost from the surface; and a self sustaining reaction will occur. On the other hand, a quantity of U235 or Pu 239 that would be a critical mass if in the shape of a sphere would be less than a critical mass if in the shape of a disc. The shape issue is discussed below in this appendix.

The materials used to produce nuclear explosions by fusion are the isotopes of hydrogen (H), deuterium (D or H2), and tritium (T or H3). In nuclear fusion, a pair of nuclei combines to form a nucleus of a heavier atom. Nuclear fusion reactions can be brought about in an environment of very high temperatures and pressures, and are referred to as thermonuclear processes. Energy is released because the resulting nucleus has a smaller mass than the initial nuclei.

The fission and fusion processes are described in the following parts of this section. However, first a comparison of the energies that can be released by chemical explosives and nuclear fission and fusion explosives is considered.

It is customary to express the energy released by a nuclear explosive in terms of the equivalent mass of the chemical high explosive TNT (trinitrotoluene) that would be required to obtain the same energy. The mass of TNT is expressed in terms of metric units (kilograms), not British units (pounds). Some conversion factors for mass are:

1 kilogram (kg) = 10^3 grams (1,000 gm) = 2. 2046 pounds
1 metric ton = 10^3 kg (1,000 kg)
1 kiloton (kt) = 10^3 metric tons = 10^6 kg = 2,204,600 pounds

The energy released by the complete explosion of a 1 kt mass of TNT is simply called 1 kt of energy. In terms of other energy units, 1 kt is equivalent to 10^{12} calories or 4.2×10^{19} ergs*.[*Footnote: the fundamental units of the cgs system are: centimeter, gram, and second. An erg has the cgs units of gm-cm^2/sec^2.]

As will be described later in this section, when one fission reaction occurs, about 3.2×10^{-4} ergs of energy is released. To release 4.2×10^{19} ergs (1 kt) of energy would require 1.3×10^{23} fissions. There are 6.023×10^{23} atoms (fissionable nuclei) in 235 grams of U235. Therefore, to obtain 1.3×10^{23} fissions would require about 51 grams of completely fissioned U235 compared to 1000 metric tons of TNT.

In a nuclear fission weapon, not all of the fissionable nuclei present will fission. But, the difference in required weights between TNT and U235 is obviously so large that fission nuclear weapons offer many possibilities not achievable with conventional explosives.

THE FISSION PROCESS

Distribution of Fission Energy

Fission processes in different materials (i.e. U235, Pu239, or U233) release about the same amount of energy per fission, 200 Mev. The distribution of energy among the fission fragments and the various radiations associated with fission is given below in Table 1. (Glasstone 1962:13)

Table A1 Distribution of Fission energy

Form of Energy	**Mev (milliom electron volts)**
1. Kinetic energy of fission products	165 ± 5
2. Prompt gamma-ray energy	7 ± 1
3. Kinetic energy of fission neutrons	5 ± 0.5
4. Beta particles from decay of fission products	7 ± 1
5. Gamma rays from decay of fission products	6 ± 1
6. Neutrinos from decay of fission products	10
Total energy per fission	**200 ± 6**

where: 1 Mev = 1.6×10^{-6} ergs and 200 Mev = 3.2×10^{-4} ergs

The forms of energy listed in Table A1 are described in the following.

Kinetic Energy of Fission Products
The Table A1 energy distribution is applicable to U233, U235, or Pu239. These are the only three known substances that are reasonably stable so that they can be stored without appreciable decay and that are capable of undergoing fission by neutrons of all energies. Hence, they are the only materials that can be used to sustain a fission chain reaction. U238, the most abundant isotope (~99.3%) in natural uranium, and Th232 can be fissioned by neutrons of high energy (above about 1 Mev) only, but not by those of lower energy. For this reason these substances cannot sustain a chain reaction. However, when fission does occur in these elements, the energy distribution is similar to that shown in Table A1. (Glasstone 1962:13)

There are more than 40 different ways that a nucleus of uranium or plutonium can split when fission occurs, so that more than 80 different nuclei (fission fragments) can be produced. The types of fission fragment nuclei produced depend on the atom that undergoes fission and the energy of the neutron that causes the fission. (Glasstone 1962:19)

Figure A1 shows the fission products produced from U235, U238, and Pu239. It shows the percent of the fission yield plotted versus mass number. The range of mass number for the fission products of U235 ranges from 72 (probably an isotope of zinc with atomic number 30) to158 (probably an isotope of europium with atomic number 63). There are 87 possible mass numbers between 72 and 158, which represent the total number of different nuclides that can be formed as direct fission fragments. (Kaplan 1955:498)

Figure A1 also shows that about 97% of the U235 nuclei that undergo fission, yield fission products that fall into two groups, a light group with mass numbers from 85 to 104, and a heavy group with mass numbers from 130 to 149. The most probable type of fission, which occurs in about 6% of the total, gives products with mass numbers 95 and 139. (Kaplan1955: 496-498) Actually, the mass distribution of fission products is a function of the incident neutron energy, with the double-humped curve of Figure A1 becoming single-humped as the neutron energy increases.

The first energy entry in Table A1 is the kinetic energy of the fission products, 160 Mev. Kinetic Energy is simply ½ x the mass (m) of the fission fragment x the square of the velocity (v) of the fission fragment or

$$K\,E = mv^2/2.$$

From Table A1, with K. E. = 160 Mev of the fission products, one can obtain the velocity of the fission products. To do so, use 1 Mev = 1.6×10^{-6} ergs and the atomic mass of the fission products which is about 235 times that of a proton or a neutron which is 1.67×10^{-24}gm. This gives a velocity for the fission products of a few percent of the velocity of light (which is ~ 3×10^{10} cm/sec).

The velocity of the fission products causes them to exchange momentum with the surrounding material, and that material starts to expand immediately. After a very short time, the expansion makes it less likely that a neutron will interact with a fissionable nucleus; and the fission process stops.

Prompt Gamma-Ray Energy

The second form of energy cited in Table A1 is the prompt gamma-ray energy. A gamma ray is an electromagnetic wave from a nuclear transition. Electromagnetic waves (also referred to as electromagnetic radiation), are waves of electric and magnetic fields propagating at the speed of light denoted by, c (where c ~ $3.0\text{x } 10^{10}$ cm/sec or 186,000 miles/sec). The length of an electromagnetic wave, commonly denoted by λ, is measured in centimeters (cm) or in Angstroms (A) with 1 A = 10^{-8} cm. Figure A2 shows the electromagnetic spectrum from the very short gamma rays through the (extremely low frequency) ELF waves. (Kaplan 1955:71)

Figure A2 also shows the frequency of these waves, commonly denoted by μ. The frequency is measured in Hertz (Hz) = cycles per second or sec^{-1}. The relationship between frequency (μ) and wavelength (λ) is:

$$c = \lambda\mu$$

where c is the speed of light. The longer the wave length, the lower the frequency.

The energy carried by a wave packet, called a photon, is expressed in electron volts (ev) where 1 ev is the energy that can be given to 1 electron by 1 volt, and 1 ev = 1.6×10^{-12}ergs. A more convenient unit used with gamma rays is a million electron volts (Mev), 1 Mev = 1.6×10^{-6}ergs. The energy, e, of an electromagnetic wave of frequency μ is given by:

$$e = h\mu$$

Where h = 6.63×10^{-27} erg-sec is Planck's constant. (EB)

Kinetic Energy of Fission Neutrons

The third energy form in Table A1 is the kinetic energy of fission neutrons. These are the neutrons that are released when the nucleus fissions. The number of neutrons that are released in a given fission is, of course, an integer number. Since the fissile nucleus can split in about 40 different ways, the average value of the number of neutrons released is not an integer. The average numbers of neutrons per fission, induced by low energy neutrons, for U235 and Pu239 are approximately 2.5 and 3 respectively. The number of neutrons per fission increases for increasing incident neutron energy. (For more information on the numbers of neutrons per fission see "Prompt Neutrons from Fission" Nuclear Physics 48:433 (1963) B. C. Diven and John C. Hopkins) The released neutrons have initial velocities of about 19% of the speed of light.

The rate at which the energy is released in a fission explosion can be obtained by treating the fission chain as a series of *generations*. Suppose that a certain number of neutrons are present initially and that these are captured by fissionable nuclei; then, in the fission process other neutrons (2 or 3) are released. These neutrons are, in turn, captured by fissionable nuclei and produce more neutrons, and so on. Each stage of the fission chain is regarded as a generation, and the generation time is the average time interval between successive generations. The time required for the fission of a nucleus after capture of the neutron is extremely short. Also, most of the neutrons are emitted promptly. Therefore, the *generation time* is essentially equal to the average time elapsing between the release of a neutron and its subsequent capture by a fissionable nucleus. This time depends on the energy (or velocity) of the neutron and the separation of nuclei, which determines the mean free path of the neutron. If most of the neutrons are of a fairly high energy (about 1 Mev), usually referred to as fast neutrons, the generation time is about 10^{-8} sec. (Glasstone 1962:14)

The unit of time 10^{-8} seconds is often used in weapons work and is referred to as a *shake*. It is 1/100th as long as a microsecond (μsec):

$$1 \text{ shake} = 10^{-8} \text{ sec} = 1 \text{ μsec}/100.$$

Folklore in the weapons community says that the shake is how fast a lamb can shake its tail.

Some of the neutrons produced in fission are lost by escape from the fissionable material or by capture within the fissionable material by non-fission processes. If the conditions are such that the neutrons are lost at a faster rate than they are released by fission, the chain reaction will not be self-sustaining.

A key issue for a nuclear weapon is how the neutron population changes with time. To determine this, what has become known as "Alpha" measurements are made. (They are often the "#1" measurement for validating calculations of weapon performance.) The rest of this section derives and describes Alpha.

Let R represent the average number of mono-energetic (i.e. same energy) neutrons that are released when nuclei fission and L represent the average number of neutrons lost per fission (by means such as escape or non-fission capture) during the time of one generation. Thus, there will be R-L neutrons at the end of the generation.

If there are N neutrons present at a time representing the beginning of a generation, then N(R-L) will be present at the end of that generation. The <u>increase</u> in neutrons in one generation is the number at the end of 1 generation N(R-L) minus the number at the beginning of the generation, N,

$$N(R-L) - N = N(R-L-1).$$

The following derives Equation 1 from which we obtain Alpha. The reader will miss nothing by going ahead to Equation 1.

If the time required for 1 generation is denoted by a constant g, in a time period Δt, the increase in the number of neutrons, ΔN, is given by:

$$\Delta N = N(R\text{-}L\text{-}1)\ \Delta t/g$$

or

$$dN/N = (R\text{-}L\text{-}1)dt/g$$

Using the mathematical technique of integration which is found in elementary calculus text books, one obtains:

$$\ln (N\text{-}N_0) = (R\text{-}L\text{-}1)(t\text{-}t_0)/g$$

Suppose that the limits on the integrals are: at a time of t_o, $N = N_o$; and at a time of t, $N = N(t)$

$$N/N_0 = e^{(R\text{-}L\text{-}1)(t\text{-}t_0)/g}$$

or

$$N(t) = N_o\, e^{(R\text{-}L\text{-}1)(t\text{-}t_0)/g} \qquad \text{Equation 1}$$

N(t) represents the number of neutrons as a function of time. R, L, and N were defined above. The constant g represents the time for 1 generation and is approximately 1 shake = 10^{-8} sec. The e in Equation 1 is also a constant, with the approximate value of 2.718. It is the base of the system of natural logarithms denoted by ln, and ln(e) = 1 (like 10 is the base of common logarithms denoted by log and log(10) = 1).

Suppose that at a time of $t_0 = 0$, $N_0 = N_0$, Equation 1 can be simplified to:

$$N(t) = N_0\, e^{(R\text{-}L\text{-}1)t/g} = N_0 e^{\alpha\tau} \qquad \text{Equation 1}$$

Where α, called Alpha, is:

$$\text{Alpha} = \alpha = (R\text{–}L\text{-}1)/g$$

At a given time, t, in a fission reaction, αt describes the increase in neutrons during a generation time, g. The value of g is about 1 shake = 10^{-8}sec.

Various properties of fission reactions can be calculated using Equation 1. For example, one might ask, how many generations are required to produce 100 tons, 100 kt, and 100MT of energy? Again, the reader will miss nothing by going ahead to the last two paragraphs of this section.

There are 4.2×10^{19} ergs/kt of energy (as described in the Introduction to this Appendix) There are 3.2×10^{-4} ergs/fission (energy released in one fission from Table A1)

Using these two numbers and calculating for 100 t = 10^{-1}kt = 4.2 x 10^{18} ergs
The number of fissions required to obtain 100tons of energy is

$$n = 4.2 \times 10^{18} \text{ ergs}/3.2 \times 10^{-4} \text{ergs/fission} = 1.3 \times 10^{22} \text{ fissions}$$

Similarly, the number of fissions for 100 kt and 100Mt can be calculated:

For 100 tons, the number of fissions N(t) required = 1.3 x 10^{22} fissions
100 kt, = 1.3 x 10^{25}
100 Mt, = 1.3 x 10^{28}.

Using Equation1, $N(t) = N_0\, e^{(R-L-1)t/g}$ the natural log of each side can be taken

$$\ln(N(t)) = \ln(N_0) + (R-L-1)t/g \times \ln(e)$$

Since ln(e) = 1, $\ln(N(t)) = \ln(N_0) + (R-L-1)t/g$

If we reasonably assume that R = 2.5 neutrons per fission and L = 0.5, R-L-1 = 1

$$\ln(N(t)) = \ln(N_0) + t/g$$

Using this equation, and the above numbers for N(t), the following is obtained for the different yields:

For 100 tons, $\ln(1.3 \times 10^{22}) = \ln(N_0) + t/g$
100 kt, $\ln(1.3 \times 10^{25}) = \ln(N_0) + t/g$
100 Mt, $\ln(1.3 \times 10^{28}) = \ln(N_0) + t/g$

An easy way to find the natural log of a number is to use EXCEL or a similar spread sheet program. For 1.3 x 10^{22}, 1.3 x 10^{25}, and 1.3 x 10^{28}, the natural logs are 51, 58, and 65 respectively. Thus, the above becomes:

For 100 tons, $51 = \ln(N_0) + t/g$ or $t/g = 51 - \ln(N_0)$
100 kt $58 = \ln(N_0) + t/g$ $t/g = 58 - \ln(N_0)$
100 Mt $65 = \ln(N_0) + t/g$ $t/g = 65 - \ln(N_0)$

It is common practice to assume that $N_0 = 1$, and ln(1) = 0. However, it can be argued that to initiate the chain reaction, neutrons are inserted, so N_0 is not = 1. But, it can be assumed that the number of neutrons inserted to initiate the chain reaction is small compared to the number of neutrons prduced, i.e. $\ln(N_o)$ is small compared to 51.

The term t/g is simply the number of generations that have elapsed in the time t and is generally denoted by n. (It is unfortunate that the letter n has so many uses is nuclear terminology.)

$$n = t/g$$

Thus, the number of generations that have elapsed to produce the yield desired can be written:

For 100 tons, n = 51
100 kt n = 58
100 Mt n = 65]

The important point is that it takes the same number of generations, namely 7, to go from 100 tons to 100 kt (58-51) as it does to go from 100 kt to 100Mt (65-58). That is 99.9% of the energy is generated in the last 7 generations. Since a generation time, t, is about 1 shake, 99.9% of the energy is generated in about the last 7 shakes or 0.07 microseconds. (Glasstone 1062:15)

Measurement of Alpha

A measurement of alpha provides information about the multiplication of neutrons. This information is related to how efficiently the weapon performs. In practice, it is usually easier to measure gamma rays than it is to measure neutrons. It has been found that in fission reactions the number of gamma rays that are produced is (approximately) proportional to the number of neutrons produced. Therefore, in diagnostic measurements of a weapon's performance, the gamma rate as a function of time, is measured to determine α. As a practical matter, all that can be measured with prompt diagnostics is the leakage of neutrons or gamma rays from the exploding assembly. This measured quantity must then be related back to what is actually occurring inside of the nuclear device through detailed theoretical modeling. This process called for close collaboration between the diagnostic scientists and the theoretical designers.

Critical Mass

If the term R-L-1 (released neutrons - lost neutrons - 1) becomes too low, an explosion will not occur. Some energy could be liberated, but the amount may not be sufficiently fast to cause an effective explosion. It is desirable to minimize the conditions under which neutrons are lost, i.e., L should be minimized. Neutrons can be lost by non-fissionable capture or by escape from the surface of the fissionable material. Not much can be done about neutron capture except to minimize the presence of any impurities in the fission material. The surface loss can be addressed by minimizing the surface/volume ratio.

For a sphere, the surface/volume ratio is inversely proportional to the radius, r:

$$\text{Surface/volume} = 4\pi r^2/4\pi r^3/3 = 3/r.$$

If the fissionable material is in the shape of a sphere and if the volume (and radius) of the spherically shaped fissionable material is made larger, fewer neutrons per fission will escape (L will be smaller); and the release of a given amount of energy can occur in fewer generations (n will be slightly smaller).

If the volume of fissionable material is small, the surface/volume ratio is large, and L may be so great that the proportion of neutrons lost will be such that propagation of a nuclear fission chain will not be possible. The mass of fissionable material that is necessary to just sustain a fission chain is referred to as a critical mass.

L can be reduced if a suitable neutron reflector surrounds the fissionable material. Also, reflectors of high density provide inertia, thereby delaying expansion of the exploding fissionable material, and provide the same function as tampers in conventional blasting operations. They are also referred to as tampers in nuclear weapons.

The fissionable material in a nuclear weapon is subcritical before employment. To produce a nuclear explosion, the fissionable material must be made super-critical, i.e. (R-L-1) much greater than 1; and this must occur rapidly to preclude relatively slow sub-explosive changes that would increase L. Also, the material must be held together as long as possible to maximize the number of fissions and hence the explosive yield.

Two general methods have been used for quickly converting a sub-critical system into a super-critical one. In the first, two or more pieces of fissionable material, each less than a critical mass, are brought together very rapidly in order to form one piece that exceeds the critical mass. This may be achieved in some kind of gun-barrel device, in which an explosive is used to shoot one sub-critical piece of fissionable material from the breech end of a gun into another sub-critical piece firmly held in the muzzle end.

The gun-barrel design was used in the Little Boy bomb that was dropped on Hiroshima. It was not tested prior to use because of its simplicity. A replica of the Little Boy weapon can be seen at the National Atomic Museum in Albuquerque, NM. It has a length of 120 inches, a width of 28 inches and weighs about 8,900 lbs. This weapon was designated the Mk-1.

The second method for rapidly converting a sub-critical mass into a critical mass is by compression. When a mass is compressed, the density of the material is increased, and the surface area decreased. Free neutrons have less far to travel to find a fissionable nucleus, and the rate of production of neutrons by fission is increased relative to the rate of loss by escape. Compression can be achieved by implosion. An <u>im</u>plosion is the opposite of an <u>ex</u>plosion. When a sphere of mass undergoes an explosion, particles of mass move outward (approximately radially), the volume increases, and the density decreases. When a sphere of mass undergoes implosion, particles of mass move inward, the volume decreases, and the density increases.

The compression of a spherical mass of fissionable material can be achieved by means of a spherical arrangement of specially fabricated shapes of high explosive placed outside the tamper surrounding the fissionable material. One example of such a structure employs high explosive in the shapes of hexagons and pentagons to produce a spherical configuration - like a soccer ball -. (Serber 1992: xvii; Alvarez 1987:131). By simultaneously detonating the 32 points (which represent the centers of the triangles and pentagons), an approximately spherical wave can be generated that compresses the fissionable material and causes it to become supercritical. The nuclear chain reaction is then initiated by the introduction of neutrons from a suitable source.

The compression design is considerably more complicated than the gun-barrel assembly. The challenge was to avoid instabilities in the compression wave with a resultant loss of sphericity. For this reason, a test, code named Trinity, was considered necessary prior to military employment over Japan. The weapon was called "Fat Man." It had a length of 128 inches and width of 60.25 inches with a total weight of 10,300 lbs. The mass of Plutonium was somewhat greater than 6 kilograms. A replica of the Fat Man weapon, designated as Mk-3, can also be seen at the National Atomic Museum in Albuquerque, NM.

The Fission Products and Their Decay

The first three forms of energy in Table A1 are immediately available in an explosion. The last three are not, and arise from the decay of fission products, which occur over a time that is long compared to that of an explosion.

As mentioned previously, there are about 40 different ways in which nuclei can split up when fission occurs, so that about 80 different fragments are produced, see Figure A1. Almost all of these 80 fission fragments are radioactive isotopes of well-known elements. An isotope is referred to as being *radioactive* if it has the property of spontaneously emitting particles of radiation from its atomic nucleus at a well-defined and characteristic rate. The main types of particles of radiation that are emitted from the nuclei of the radioactive isotopes that are formed as a result of fission are: β (beta)-particles, γ (gamma)-rays, and neutrinos. In addition, α (alpha)-particles are emitted as one possible decay mode for very heavy elements, including uranium and plutonium. These decay particles and the decay processes are discussed next.

<u>Alpha (α)-Particles</u>

An *α-particle* is the nucleus of a helium atom. It consists of two protons and two neutrons; and therefore, it has two positive charges and a mass of four units. When an alpha particle is emitted from a nucleus, the resulting daughter nucleus has two less protons and two less neutrons. The daughter therefore has an atomic number of two less and an atomic weight of 4 less. For instance, when polonium-210, with a mass number of 210 and atomic number of 84, undergoes decay by emitting an alpha particle, the daughter is lead-206 with an atomic mass of 206 and an atomic number of 82. (EB) The principal alpha emitters are found among the elements heavier than bismuth (atomic number 83) and also among the rare-earth elements from neodymium (atomic number 60) to lutetium (atomic number 71). (EB)

Alpha particles can travel no more than about 2.5 to 7.5 cm in air before being stopped. (Glasstone 1962:21) An alpha particle of 3 Mev has a range in air of 2.8

cm. (Kaplan 1955:290) It is doubtful whether these particles could get through unbroken skin, and they certainly could not penetrate clothing. (Glasstone 1962:21)

Beta (β)-Particles and Neutrinos

In elements lighter than lead, decay by the emission of a beta particle is the main decay mechanism. Beta decay can occur by any of three processes of radioactive disintegration by which some unstable atomic nuclei spontaneously release excess energy and undergo a change of one unit of electrical charge without a change in mass number. The three processes are called electron emission, positron (positive electron) emission, and electron capture. The orbital electrons that surround the nucleus are not involved in the two processes of electron and positron emission. They are, however, involved in the electron capture process.

In the electron capture process, an electron orbiting the nucleus enters the nucleus and combines with a nuclear proton to produce a neutron. This neutron remains in the nucleus, and a neutrino is ejected. Most commonly, the electron is captured from the innermost, or K, shell of electrons around the atom; for this reason, the process is often called K-capture. The nuclear positive charge and hence the atomic number of the atom decreases by one unit, and the mass number remains the same.

In *electron emission*, also called negative beta decay (symbolized by β-), an unstable nucleus ejects an energetic electron (of relatively negligible mass) and an antineutrino (with no rest mass). A neutron in the nucleus becomes a proton that remains in the nucleus. Thus, negative beta decay results in a daughter nucleus with a proton number (atomic number) that is one more than its parent but with the same mass number. (EB)

In *positron emission*, also called positive beta decay (β+), a proton in the parent nucleus decays into a neutron that remains in the daughter nucleus. A positron (which is positively charged, is like an ordinary electron in mass but of opposite charge) along with a neutrino, are ejected. Positive beta decay produces a daughter nucleus, with an atomic number that is one less than its parent and the mass number that is the same. (EB)

Beta particles have considerably greater range than alpha particles. A 3 Mev beta particle can travel about 1000 cm in air (compared to 2.8 cm for an alpha particle). (Kaplan 1955:290) The emission of β-particles differs from that of α-particles. Alpha particles are emitted from a nucleus of a specific type of isotope at a single energy. The most characteristic feature of the spontaneous beta disintegration of a nucleus is the continuous distribution of energies of the β-particles emitted from a specific isotope. Beta particles can have energies that go as high as nearly 11 Mev where they have a velocity of .999 the speed of light. Low energies of the order of .0025 Mev result in velocities that are about 10% of the speed of light. (Kaplan 1955: 287)

Neutrinos have no charge and do not cause ionization. If they have any mass, it is exceedingly small compared to that of the electron. Neutrinos travel with the speed of light and are the most penetrating of subatomic particles. Only 1 neutrino in 10 billion traveling through matter a distance equal to the Earth's diameter reacts with a proton or neutron.

Gamma (γ)-Rays

Gamma rays are emitted promptly in the fission process and are also emitted over longer periods of time by the fission products. Frequently, but not always, the emission of beta particles is accompanied by the emission of gamma rays. If the nucleus of the daughter of the beta particle decay is still in an excited state, a gamma ray may be emitted to reduce the energy of the excited state. In some cases, only gamma rays are emitted in the decay of a fission product; and they carry off the excess energy, without beta particles being emitted.

Rate of Radioactive Decay of Fission Products

The rate of emission of gamma rays, beta particles, and alpha particles is usually expressed by means of the half-life of the particular isotope involved. The half-life is defined as the time required for the radioactivity of a given quantity of a particular radioisotope to decrease (or decay) to half of its original value. Each individual radioactive species has a definite half-life that is independent of its state or the amount of material. The half-lives of the fission products have been found to range from a small fraction of a second to millions of years.

Although every radioisotope present among the fission products is known to have a definite half-life, the sum of all the isotopes remaining after a nuclear explosion is so complex that it is not possible to represent the decay as a single half-life. However, a rough rule-of-thumb to describe the decrease in the total radiation intensity from the fission products following a nuclear explosion is as follows:

> *For every seven-fold increase in time after the explosion, the dose rate decreases by a factor of 10.*

For example, if the radiation dose rate at 1 hour after the explosion is taken as a reference point, say 1.0, then at 7 hours after the explosion, the dose rate will have decreased to 1/10. At 7x7 = 49 hours after the explosion, the dose rate will have decreased to 1/100. At 7x7x7=343 hours after the explosion, the dose rate will have decreased to 1/1000. Another example is that at the end of 7 days, the radiation dose will be 1/10 of the value at day 1. This rule is accurate to within about 25% up to 2 weeks or so and is applicable to within a factor of 2 up to roughly 6 months after the nuclear explosion. After about 6 months, the dose rate decreases at a more rapid rate than predicted by this simple rule.

THE FUSION PROCESS
In the fusion process, a pair of light nuclei unites (or fuses) together to form a nucleus of a heavier atom, and energy is released. In some fusion processes, an energetic neutron(s) or a hydrogen atom may also be released.

The element of hydrogen is known to exist in three isotopic forms, in which the nuclei have masses of 1, 2, and 3. These are generally referred to respectively as: hydrogen (H1), deuterium (H2 or D2), and tritium (H3 or T3). Each of these isotopes has a nucleus that contains 1 proton. The nucleus of the isotope H1 contains no neutrons; the nucleus of the isotope H2 contains 1 neutron; and the nucleus of the isotope H3 contains 2 neutrons.

Several different fusion reactions have been observed among the nuclei of the three hydrogen isotopes. These reactions involve either two similar or two different hydrogen isotopes. Five thermonuclear fusion reactions are of interest for the production of energy because they occur sufficiently rapidly at realizable temperatures. These are:

Table A2 Fusion Reactions

	Initial Nuclei -->	**Product Nuclei +**	**Energy (Mev)**	**Atomic Mass Units (amu) Of Fusion Material**	**Energy Produced per amu (Mev/amu)**
#1	D2 + D2 -->	He3 + n	3.3	4	0.8
#2	D2 + D2 -->	T3 + H1	4.0	4	1.0
#3	T3 + D2 -->	He4 + n	17.6	5	3.5
#4	T3 + T3 -->	He4 + 2n	11.3	6	1.9
#5	Li6 + n -->	He4 + T3	4.8	7	0.7

He is the symbol for helium, Li is lithium, and n represents a neutron. The energy liberated in each of these reactions is given in millions of electron volts (Mev). (Glasstone 1962:22) Natural helium, He4, has 2 protons and 2 neutrons in it nucleus. He3 has 2 protons and 1 neutron. Li6 has 3 protons and 3 neutrons.

The atomic weights or atomic mass units (amu) of the materials used in each fusion reaction are given in Table A2 as is the energy per amu. The amu of U235 is 235; and when an atom of U235 fissions, it releases about 200 Mev, or about 0.85 Mev/amu. By weight, the amount of energy released by fusion in reaction #1 is comparable to that of fission. Energy released by fusion reaction #3 is about 3.5 times the energy of fission.

Fusion has two other attractive advantages. First, there are no radioactive residual products produced *directly* as a result of most fusion reactions. That is, fusion does not result in radioactive decay products such as the mid-range atomic mass number elements that are produced in the fission process. However, some fusion reactions do produce neutrons, which can activate various materials in the environment in the same manner that fission neutrons cause induced radioactivity. But, the residual radiation is much less for fusion. Second, the raw material costs of most fusion materials are much less than the costs of fissionable uranium and plutonium. (Bridgman: 172)

As seen in Table A2, fusion reactions #1 and #2 both involve the fusion of 2 deuterium atoms. While the fusion reaction rates depend on temperature and pressure, it is assumed for illustrative purposes that these two reactions are about equally likely to happen. Under this assumption, reaction #3 is about 100 times more probable than #1 and #2. The tritium produced by reaction #2 would almost always fuse immediately with another deuterium atom. Therefore, reactions #1, #2, and #3 can be considered as a group that consumes 5 of the initial deuterium atoms. (Bridgman: 172,173)

Another reaction of thermonuclear weapons interest, with tritium as a product, is #5 in Table A2. In #5, Li6 represents the lithium-6 isotope, which makes up about 7.4 percent of natural lithium. Other reactions can occur with lithium-6 or the more abundant isotope lithium-7 (3 protons and 4 neutrons) and various particles produced in the weapon. Reaction #5 is of interest for two reasons: (1) it has a high probability of occurrence; and (2) if the lithium is placed in the weapon in the form of the compound lithium deuteride (LiD), the tritium formed in the reaction has a high probability of interacting with the deuterium by reaction #3. Large amounts of energy are thus released by reaction #3, and additional neutrons are produced to react with lithium-6 in reaction #5. (Glasstone 1977:21)

In a fusion process, there are no gamma-rays or products that radioactively decay (2, 4, 5, and 6 in Table A1). The energy produced by a fusion process resides in just the kinetic energies of the new fused atom(s) that are produced and the neutrons that are emitted. For instance, in Table A2, the energy of 3.3 Mev that is released in reaction #1 resides in just the kinetic energy of the produced He3 atom and the neutron. The neutrons or lighter products of a fusion process have more energy than the heavier products. For instance, in the case of process #1, the He3 has about 0.8 Mev of kinetic energy, and the neutron has about 2.5 Mev of kinetic energy. (Bridgman: 180)

Using these energies, the rest masses (m) for neutrons and protons (see physical constants at end of this section), and the equation for

$$\text{Kinetic Energy} = \tfrac{1}{2} mv^2,$$

one finds that the helium atom and neutron have relativistic velocities (v). The helium atom is traveling at about 5% the velocity of light, and the neutron at about 22% the velocity of light.

Generally, in a fusion process, the neutron (or the lighter atom) produced travels at a higher velocity than the heavier product. This means that most of the energy obtained in a fusion process can be carried away by the neutrons and is not available for local heating of the remaining fuel.

The probability of a fusion reaction occurring when the fusion particles have energies less than about 10 to 20 kev is negligible. In order to provide the fusion particles with such energies, the fusion material is heated. However, the heat energy provided causes the atoms to lose their electrons and to become positively charged nuclei. Thus, a fusion reaction involves bringing together two positively charged nuclei. Either or both of the particles in a fusion reaction must possess enough kinetic energy to overcome the repulsive force generated when particles of the same charge interact. The kinetic energies required for fusion are in the range of tens of kev. (Bridgman: 172-173).

By using accelerator technology, a small number of ions can be accelerated and impact a stationary target (for instance D2). However, in order to obtain the amount of energy that is of interest for weapons, a large number of fusion particles must be energized; and accelerator technology is impractical. The only practical way today to achieve such energies in a sizable number of fusion particles is by means of a fission explosion. (Bridgman: 174)

By heating and compressing a quantity of deuterium or tritium or a mixture of deuterium and tritium with a fission device, it is possible to initiate one or more of the thermonuclear fusion reactions. If these fusion reactions, which are accompanied by a release of energy, can be propagated rapidly through a volume of the hydrogen isotopes, a thermonuclear explosion may be realized. (Glasstone 1962: 22-23)

Neutrons are produced in 3 of the fusion reactions shown in Table A2. Because of their relatively small mass, the neutrons are very energetic and carry off most of the energy from the fusion process. The form of uranium that is found in nature is mostly (~99.3%) U238, and it requires a high-energy neutron (greater than 1 Mev) in order to fission. It is possible to make use of the high-energy thermonuclear neutrons by surrounding the fusion weapon with a blanket of ordinary uranium. The high-energy neutrons can then be captured by U238, which can fission. This contributes to the overall energy yield of the explosion as well as the residual nuclear radiations that arise from the fission products. (Glasstone 1962: 23)

As mentioned above: If fusion reactions, which are accompanied by a release of energy, can be propagated rapidly through deuterium or deuterium plus tritium, a thermonuclear explosion may be realized. (Glasstone 1962: 22-23) Two factors: (1) the rate of burn and (2) the ratio of: energy generated/heating energy, will determine whether or not a sustained fusion burning takes place. That is, the rate of energy generation and the ratio of energy generated to energy that heats more deuterium (or deuterium plus tritium) determine whether or not the fusion fire is propagated such that an appreciable fraction of the fuel is consumed. (Bridgman 2001: 179)

Boosting

Boosting is a technique to increase (i.e. “boost”) the yield of a fission weapon by adding deuterium and tritium to a cavity within an implosion assembly. It generally works as follows: Upon firing the detonators, and activating a neutron source, the high explosive compresses the U or Pu resulting in a run-away chain reaction. This process rapidly increases the energy of the deuterium and tritium gas in the device. Fusion reactions then occur in the gas which in turn produce flood of high energy neutrons that also “boost” the fission yield.

Thermonuclear Device Diagnostics

The objective of the nuclear weapons diagnostic scientists is to match the observable features of the test to the same features as calculated using the theoretical models for the behavior of the device. Thermonuclear weapons, from a test perspective, differ from primaries or fission triggers, in several major ways. Thermonuclear tests are generally of higher yield than primaries, or single stage devices, and it is necessary to separate the observations of the individual stages. The primary stage is separated physically from the secondary stage, but both are confined within a single housing. The challenge to the diagnostic scientist at the test site in Nevada or the Pacific was to measure the outputs (gamma rays, X-rays and neutrons) as functions of time and energy of the emitted radiation for each stage separately. There eventually evolved a desire on the part of the theoretical designers for data on the spatial location of the source of the emitted radiations. In effect the designers wanted a “picture” of the exploding device in terms of neutrons or gamma rays rather than visible light. Of course, it was only a short step from a picture to a motion picture of the exploding device, but these developments occurred in the 1960s and later.

Much of the previous description applies mainly to the so-called prompt measurements using high-speed electronics, but in fact it can also apply to the radiochemistry measurements as well. The radiochemists were very clever in developing techniques to measure not only yield but also details about the behavior of specific parts of a device. This could be accomplished by using specific tracers located at particular locations in the device. The activation of the tracers and subsequent radiochemical analysis of the debris gave a great deal of information on the performance of the device.

PHYSICAL CONSTANTS

Speed of light	$c = 2.998923 \times 10^{10}$ cm/sec
Planck's Constant	$h = 6.6252 \times 10^{-27}$ erg-sec
Rest Masses:	
	neutron = 1.67470×10^{-24} gm
	proton = 1.67239×10^{-24} gm
	electron = 9.1085×10^{-28} gm

APPENDIX A REFERENCES

Diven, B. C. and John C. Hopkins, "Prompt Neutrons from Fission", Nuclear Physics, 48, 433, 1963.

Encyclopedia Britannica CD 1998 Knowledge for the Information Age, Encyclopedia Britannica, Inc. Chicago IL, 1997.

Glasstone, Samuel editor, *The Effects of Nuclear Weapons*, United States Department of Defense and United States Atomic Energy Commission, US Government Printing Office, Washington D.C., April 1962.

Glasstone, Samuel and Philip J. Dolan, *The Effects of Nuclear Weapons*, United States Department of Defense and United States Department of Energy, 1977

Kaplan, Irving, *Nuclear Physics*, Addison-Wesley Pub. Co., Cambridge, Mass., 1955.

Pauling, Linus, *College Chemistry*, W.H. Freeman and Co., San Francisco, CA, 1954.

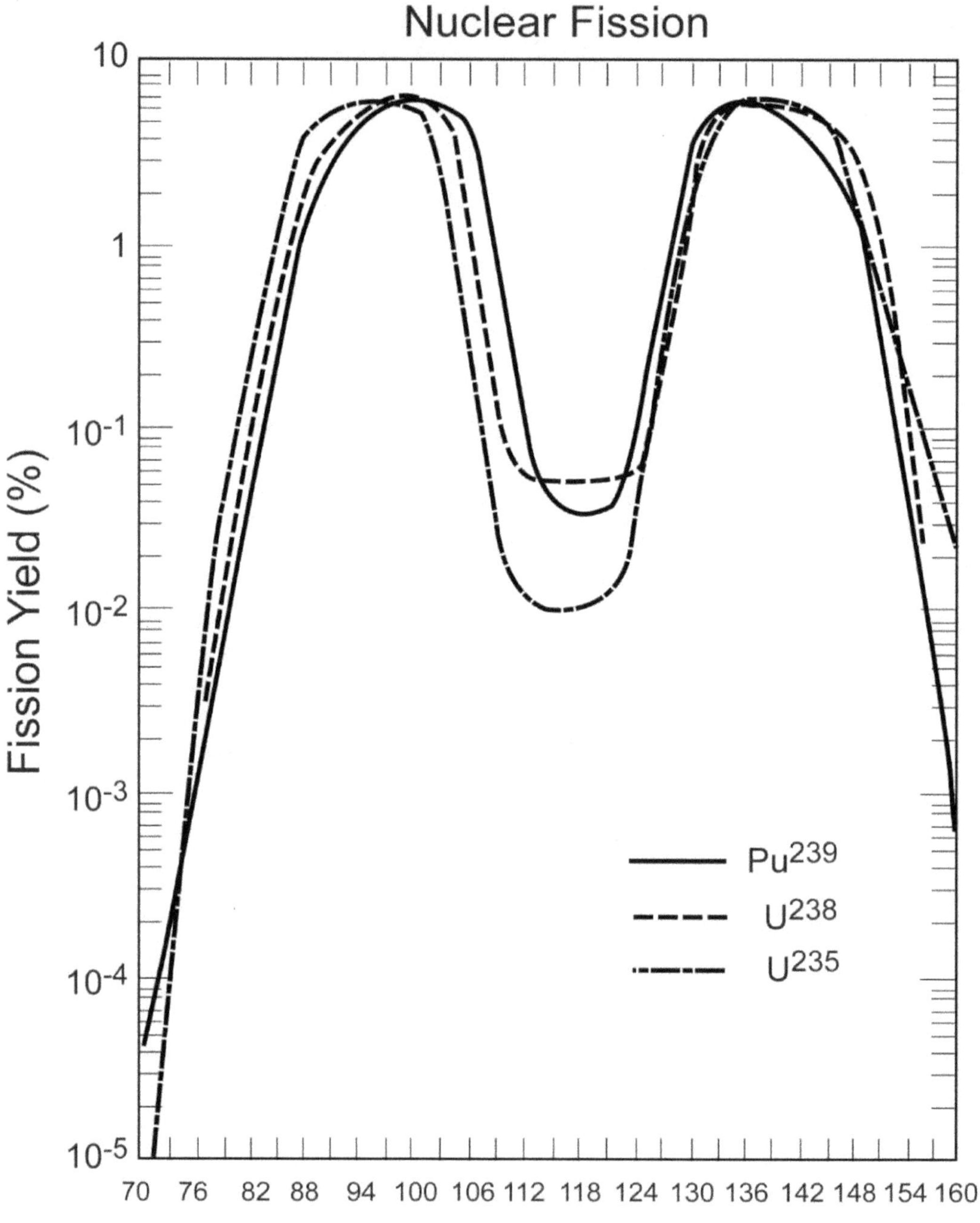

Figure A1 Mass number of fission products produced by U233, U235, and Pu239 versus percent of fission yield. (Kaplan 1955: 498)

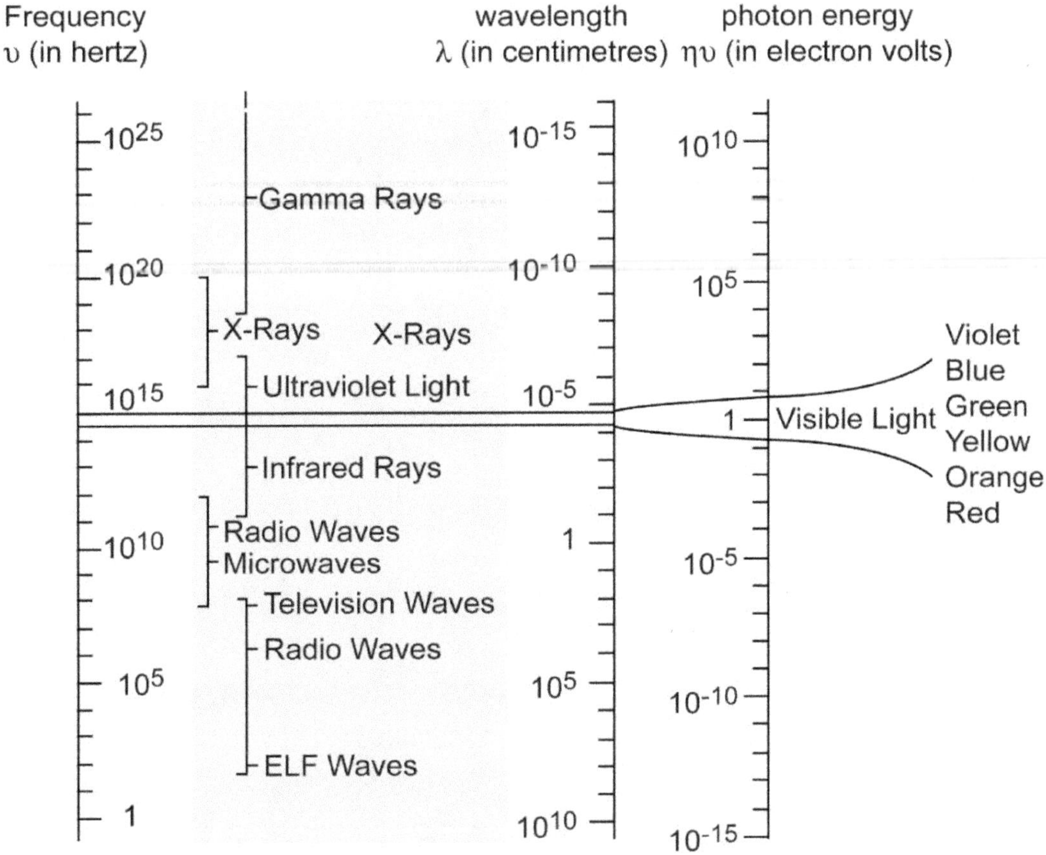

Figure A2 The electromagnetic spectrum in frequency, wavelength, and photon energy

APPENDIX B

LOS ALAMOS' ORGANIZATION DURING WWII AND TRINITY

Introduction

This appendix presents some sidebars to the main text regarding Los Alamos' war-time organization and the management approach, which greatly influenced future weapons work in general and field testing in particular. It also contains some discussion about Trinity and the organization chart that was included in Kenneth Bainbridge's report on Trinity. The organizations used for subsequent tests conducted at NTS had many similarities with that used on Trinity.

Initial Organization and August 1944 Reorganization

When work started at Los Alamos in April 1943, people were organized into Divisions, generally along scientific discipline lines. In addition to an Administration Division (A), there were four main scientific divisions: "Bethe would head the theoretical, or "T" Division. Caltech's Robert Bacher was in charge of Experimental Physics, P Division. Chemistry, C Division, shortly became Chemistry and Metallurgy, CM Division. It was in the hands of two men: Cyril Smith* [*Footnote: Smith, a metallurgist, might be called a Renaissance Man with his careers in science, technology, history, and the arts. He was a member of the AEC's first GAC and later established at MIT the Laboratory for Research on Archaeological Materials] and Joseph Kennedy** [**Footnote: Kennedy was young Berkeley chemist who had predicted that no nation would have enough uranium to build a bomb during the war]. The Ordnance and Engineering, E Division, which would actually assemble the bomb, was headed by William "Deak" Parsons." John Manley became Oppenheimer's right-hand man for planning and staffing. (Herken 2002:75)

Although organized along scientific discipline lines, interdivisional interactions were common in the work being done. The majority of effort in 1943 and early 1944 was devoted to the gun assembly of the fissionable material. Two types of gun weapons were considered: uranium and plutonium. As described in Part I-Chapter 1, there was also some effort in exploring implosion technology under Seth Neddermeyer in group E-5.

Oppenheimer assigned Robert Serber* the job of briefing the dozens of scientists on the current state of knowledge about the atomic bomb. [*Footnote: Serber was a theoretical physicist and protégé of Oppenheimer who was present at the 1942 conference at UC Berkeley (see Chapter 1). His "five stunning lectures - summed up everything the world knew about how to build an atomic bomb", (Serber 1992:cover)] For his April 1943 indoctrination course, Serber had been cautioned about using the word bomb while uncleared construction workers were still crawling about the ceiling above his head. Serber began calling it the "gadget" instead. This name for an atomic bomb was used by the community for many years. Serber named the cylindrical plutonium gun gadget being studied, "Thin Man", after the character in the popular film starring William Powell and Myrna Loy; and the uranium gun was dubbed "Little Boy". Serber also called the implosion gadget "Fat

Man", in honor of the figure played by Sidney Greenstreet in "The Maltese Falcon". (Herken 2002: 84,126)

Segre's group characterized the first reactor-produced plutonium at Los Alamos in early summer 1944. They found that because of the plutonium-240 content (which has a high spontaneous fission rate) in the reactor bred material, a plutonium gun would probably pre-initiate and end up with a low yield. Oppenheimer ordered work on Thin Man abandoned. Also, calculations conducted by John von Neumann, a Princeton mathematician brought to the lab as a consultant, and Edward Teller* had shown that the compression attainable in the implosion design made it potentially far more efficient than the gun. (Herken 2002: 116-117) In August 1944, Los Alamos placed a high priority on the implosion weapon; and in order to expedite work, reorganized the laboratory, see Part I- Chapter 1. [*Footnote: Edward Teller was among the handful of most prominent European physicists at Los Alamos. He had driven Leo Szilard to Albert Einstein's home in Long Island the day Einstein signed the now famous letter to Roosevelt urging him to organize and speed the government's work in chain reactions which could lead to the construction of "extremely powerful" bombs.(Teller 2001:145-8) He was at the 1942 UC Berkeley conference convened by Oppenheimer:

> "After a day or so of discussing the fission bomb in a superficial way, everyone decided that the problem looked straightforward. We proceeded with some excitement to find out whether we could make a thermonuclear reaction proceed. During the next few weeks, we convinced ourselves that it could be done." (ibid., 159)

These few sentences from Teller's autobiography indicate how he would spend his future years - thermonuclear weapons development would be his main objective. Teller was among the first to arrive at Los Alamos, and he helped with recruitment. Although he went to observe Trinity, he was not a prime mover in making it happen due to his overriding interests in thermonuclear weapons.]

There had been a sequence of minor reorganizations since the lab started, but the August 1944 reorganization was drastic. The initial organization around discipline was abandoned. After August 1944, the organization was focused on functional needs of the implosion or gun assembly programs. The reorganization affected the group and division leaders more than the workers, who in many cases just kept doing the same thing, even though the name of their division and possibly leaders had changed. This reorganization, which was to endure throughout the war, is described in Reorganization to Expedite Implosion Work, in Part I-Chapter 1.

Work on implosion now involved hundreds of workers and became a model "big science" effort. It was a new American (not at all European) approach to science. Many research and development organizations would try to emulate this structure after the war with varying degrees of success. Interdisciplinary team collaboration and a belief that almost anything is possible were important attributes. (Hoddeson 2004:3, 247) Also an important feature was that the work was viewed as very important at the highest echelons of the US government, which provided timely and unfaltering support without much undue interference.

Hoddeson describes:
"One prototype for Oppenheimer's approach was Ernest Lawrence's laboratory at the Berkeley Radiation Laboratory in the 1930s, where Lawrence gathered together theoretical and experimental physicists, chemists, and engineers having a wide spectrum of skills and experience." He encouraged the use of empirical solutions to problems, such as: scale models, the Edisonian approach, the shotgun approach, overlapping approaches, iteration, and numerical analysis*. (Hoddeson 2004: 9-10,405)
[*Footnote: Edisonian approach - trying, in the absence of good theoretical guidance, one after another system or material.
shotgun approach - all experimental techniques available and everything known about a particular issue were fired at the problem to be solved, in hopes that one or more techniques would hit on a piece of the problem and reveal some important facet.
overlapping approaches - multiple approaches were taken simultaneously to a specific problem in recognition that any single approach could be incomplete and uncertain by itself, but that together they might be used to build up a consistent picture.
iteration - the systematic generalization of cut-and-dry "tinkering", long characteristic of American science, in which empirical models were progressively improved after testing.
numerical analysis - although messy and unaesthetic, it was more far-reaching than analytic models alone, which were too incomplete and idealized to handle concrete problems. When combined with analytic methods, numerical ones formed a tool of striking power which increased as the speed of computers increased.]

Oppenheimer, like Lawrence, was able to instill in his workers a dogged determination and self-confidence. "Robert Wilson, a Berkley graduate student in the late 1930s, explains that Lawrence taught them to 'think how you want' to design a piece of equipment. And then work 'as hard as you can on any point of weakness until you solve all your problems'. The most important lesson was, 'You don't say no, ever'". (Hoddeson 2004: 405)

The "can do", "let's try different approaches", and "You don't say no, ever" approach found at Los Alamos carried over to Sandia and the AFSWP as well as the Livermore laboratory, which was founded with many people from the Berkeley Rad Lab. This approach was a work ethic that was in the blood of those who conducted field testing of nuclear weapons. (The Berkeley people who arrived in Livermore's early years were often referred to as "the nuclear mafia".)

The results of the Los Alamos approach might be viewed from the perspective of time. From the start-up of Los Alamos to Nagasaki was a mere 30 months (April 1943 to August 1945). In less than a year, about 11 months, after the August 1944 reorganization, a successful test of Fat Man was conducted at Trinity.

Los Alamos' Organization for TRINITY July 16, 1945
Robert Oppenheimer named the first nuclear test Trinity. The test and the work associated with it were often referred to as Project TR. The location of the shot was known as the Trinity Site.

Oppenheimer said in a 1962 letter to Groves that at the time he chose the name Trinity, he was thinking of a poem by John Donne. (Norris 2002: 397) Herken states: "For reasons that Oppenheimer decided to keep obscure, he had named the test *Trinity* - perhaps a secret tribute to Jean Tatlock…."*
(Herken 2002:29-30, 128-9) [*Footnote: From a sonnet by John Donne:

> Batter my heart, three-person'd God, for you
> As yet but knock, breath, shine, and seek to mend;
> That I may rise and stand, o'erthrow me, and bend
> Your force to break, blow, burn, and make me new.

Oppenheimer had met Tatlock in the spring of 1936. She was seen by Oppenheimer's students as having a humanizing influence on him, and she had introduced him to the romantic poetry of John Donne. Tatlock had a circle of radical friends, among them Haakron Chevalier, to whom she introduced Oppenheimer.]

The Hindu trinity consists of Brahma (the Creator), Vishnu (the Preserver), and Shiva (the Destroyer). Oppenheimer had an avid interest in Sanskrit literature, and following the test is reported to have recited the following passage from the Bhagavad-Gita: (Internet, fas.org/Trinity9/15/01)

> If the radiance of a thousand suns
> Were to burst at once into the sky,
> That would be like the splendor of the Mighty One …
> I am become Death,
> The shatterer of Worlds

Being the first nuclear weapon test, Trinity set many precedents for dos and don'ts during a nuclear test operation, many of which would carry into future testing at the NTS.

In March 1944, Kenneth Bainbridge was assigned Group Leader of E-9, High Explosives Development. In the August 1944 reorganization, it became X-2. The duties of this group were to look into all aspects of the feasibility of a nuclear test. As Group Leader of X-2, Bainbridge was also the Director of Project TR. Bainbridge's role as Director of Project TR was similar to that of the combined roles of the later NTS's Test Manager and Test Director. However, he had veto power on suggested experiments. At NTS, neither the Test Manager nor Test Director had such authority; it rested with the sponsoring agency. (Bainbridge 1947: 2)

Site selection (which is discussed in Part I-Chapter 1), planning for the base camp, selection of specific experiments, and the design and construction (through contract) of JUMBO were early activities of the group. On October 14, 1944, plans for the base camp, a survey of the proposed scientific measurements, and the equipment required were submitted to Groves and approved. By early November, contracts were let for initial construction. In May 1945, the initial construction was expanded to take care of the expansion in activities planned for the final test of the gadget in July.

In late December 1944, the camp was completed; and a small detachment of Military Police took up residence. Lt. H. C. Bush was named Commanding Officer of the Trinity Base Camp. It was a "happy camp". The excellent camp morale and military-civilian cooperation did much to ameliorate the difficulties of operation under primitive conditions. (Bainbridge 1947: 3-8)

The firing of the 100 ton shot on May 7, 1945 provided an excellent dry-run for the subsequent gadget test about 2 months later on July 16. However, the organization and manpower involved in May was considerably less than in July.

In the final 2 weeks of preparations for Trinity, about 250 men from Los Alamos were engaged in technical work at TR. For the period 4-2 weeks prior to the test, only 2/3 to 3/4 of this number were engaged full time, and only about 125 were engaged full time for the prior 4 months.

The organization chart for Trinity, which is given below provides a view of the different responsibilities and numbers of people necessary for the conduct of the various tasks involved in a nuclear weapons test. The responsibilities, and numbers of people used on Trinity are surprisingly similar to those that would be used in the field by Los Alamos (and later Livermore) for the assembly, detonation, and measurements of performance for essentially all of the future tests conducted at the NTS.

Detailed operating procedures for personnel at the site at the time of the shot were developed and documented. Procedures were also developed for the arming party, the medical corps, possible evacuation, safety, security, transportation and installation of the device, final assembly, and dry runs.

Some of the recommendations regarding "Preparations and Administration" that were made after Trinity might have been made on almost any of the future NTS operations because few, if any, of these conditions actually existed on any of the NTS operations:
1. A firm directive should be obtained for a test at least six months in advance of CONUS tests. This assumes that a location for the test has been agreed upon.
2. A firm agreement should be obtained regarding procurement of personnel.
3. It is essential to have a first-class man in charge of "services" and to have all services under one head.
4. It is essential to have the base camp installations complete four months prior to the date of the test.
5. The wiring should be complete at the latest one month prior to the test, which means that 90% of the requirements should be known four months prior.
6. No new experiments should be introduced later than six weeks prior to the test.

7. No new equipment of any kind, electrical or mechanical should be installed or removed after the first test rehearsal except as required to minimize pick-up and interference encountered in the first rehearsal.
Etc. of a more specific nature. (Bainbridge 1947)

ORGANIZATION CHART FOR PROJECT TR

K. T. Bainbridge	Administrative head. Over-all responsibility.
Barbara S. Anderson	Secretary.
J. H. Williams (TR-1)	Alternate to Bainbridge.
F. Oppenheimer	Administrative aide. Planning. Safety. Emergency roving center.
Lt. H. C. Bush	Commanding Officer of Trinity Camp. Responsible for camp matters, barracks, mess, Road Maintenance, Guarding, Camp Hygiene.
Lt. R. R. Taylor	Security for Trinity at Los Alamos.
John Anderson	Security for Trinity at Trinity Site (TR).
Capt. S. P. Davaloe	TR US Engineer Detachment Operations.
Lt. R. D. Wholey	US Army Contracts.
Sgt. W. Stewart	Responsible for HE tests and for all instrument calibration, velocity of sound charges, etc.
Consultants	
R. W. Carlson (X-2)	Design of structures, Installation of tower.
P. E. Church	Meteorological Problems, particularly the dilution of active gases.
E. Fermi (F-Division)	All nuclear physics measurements.
J. O. Hirschfelder (T-7)	All problems affecting damage that arises after the nuclear reactions have stopped.
S. Kershaw	Proper safety regulations and procedures.

L. D. Leet	Earth Shock problems particularly at distant points.
W. G. Penny	Blast, Earth Shock and Permanent Earth Displacement problems.
Ens. G. T. Reynolds	Structures.
V. Weisskopf	Nuclear physics measurements, radiation problems. Chairman of Nuclear Measurements Committee.

TR Assembly
Comdr. N. E. Bradbury (X-1, X-6)
G. B. Kistiakowsky,

Pit Assembly (G-1) 13 persons	Over-all responsibility for pit assembly.
Detonator Unit (X-5) 3 persons	
Simultaneity (X-7) 5 persons	Installation and check of special switches and circuits.
E.W. Titterton	Recording Equipment, Testing of HE-actuated switches.

HE Assembly & Detonators

9 persons	HE and mechanical assembly. Installation of detonators. Test of detonators.
TR-1: J. H. Williams	Services.
TR-1A:Lt. Comdr. T. M. Keiller 9 persons	Head of Construction. Electrical Construction, Telephone Services, Motor Generators.
TR-1B: J. L. McKibben 11 persons	Timing, all timing and remote control signals. Arming Party, Searchlight Group, Electronic Time Signals.

TR-1C: R. J. Van Gemert 4 persons	Procurement at Los Alamos, Stock, Shipping, Unloaders, and Clerks.
TR-1D: D. Greene 3 persons	Transportation at/between TR & Y sites.
TR-1E: F. Stokes 6 persons	Head. Radio Communications.
TR-1F: Capt. B. B, Geery 3 persons	Head. Balloon Operations.
TR-1G H. S. Allen 5 persons	Head. High Iron Work and Special Jobs.
TR-2 J. H. Manley (R-3) 4 persons	Air Blast and Earth Shock - Responsible for all measurements in TR-2 group.
Air Blast	
TR-2A: R. L. Walker 11 persons	Piezo Gauges.
TR-2B: W. C. Bright 7 persons	Condenser gauges.
TR-2C: H. H. Barschall 4 persons	Excess Velocity Measurement.
TR-2D: T Jorgensen 3 persons	Impulse Gauge.
TR-2E: H. Sheard 4 persons	Maximum Pressure Gauge.
TR-2F: J. C. Hoogterp 2 persons	Box Gauge.
Earth Shock	
TR-2G: H. M. Houghton 5 persons	Velocity Geophone.
TR-2H: L. D. Leet 11 persons	Displacement Seismographs at 9000N, Tularosa, Elephant Butte, San Antonio, and Carrizozo, NM.
TR-2I: W. G. Penny	Permanent Earth Displacement.

4 persons

TR-3: R. R. Wilson E. Fermi, Consultant V. Weisskopf, Consultant	Physics.
TR-3A: R. R. Wilson 13 persons	Prompt measurements: alpha and shock wave transmission time.
B. Rossi 11 persons	Prompt alpha and shock wave transmission time.
TR-3B: H. T. Richards 8 persons	Delayed Neutron Measurements.
TR-3C: E. Segre 4 persons	Delayed Gamma Rays, Ionization chambers, shelter design.
TR-3D: P. B. Moon 6 persons	Gamma Ray Sentinels and Delayed Gamma Rays.
TR-3E: H. L. Anderson 31 persons	Sampling, counting, radiochemistry, at TR & Y.
TR-4: J. H. Hubbard	Meteorology.
TR-4A: Lt. C. D. Curtis 6 persons	Radar.
TR-4B: Sgt. J. C. Anderson 3 persons	Pilot Balloons (Arming Party).
TR-4C: Sgt. P. A. Tudor 3 persons	Radiosconde.
TR-4D: Sgt. W. Blades	Base Weather and Reporting.
TR-5: J. E. Mack 15 persons	Spectrographic and Photographic.
TR-6: B. Waldman 10 persons	Air Blast - Airborne Condenser gauges.
TR-7: Dr. L. H. Hempelmann 4 persons	Medical Group.

TR-7A: R. Watts 4 persons	Instruments.
TR-7B: P.Aebersold 13 persons	Monitor Group at N-, W-, & S -10,000; Road Monitors, Emergency Medical Aid.
Special Assignments 18 persons	Searchlight Plotting, Plane Crews, Ground-to-Plane Communication and Shelter Announcer, Consultants, Waterological Consultant, Weather.

A total of 125 people were under Lt. H. C. Bush's command charged with the responsibility of guarding and maintaining the camp. An additional 160 were located north of the test are under the command of Major T. C. Palmer with sufficient vehicles to be able to evacuate ranches and towns if fallout descended in dangerous amounts. At least 20 men associated with the Military Intelligence were in neighboring towns and cities up to 100 miles away.

There was also a group of distinguished visitors present between July 10-16. J. R. Oppenheimer; R. C. Tolman; V. Bush; J. B. Conant; Brig. Gen T. F. Farrell; Maj. Gen. L. R. Groves; C. C. Lauritsen; I. I. Rabi; Sir Geoffrey I. Taylor; Sir James Chadwick

APPENDIX B REFERENCES

Bainbridge, K. T., editor, *Trinity*, Volume 24, Los Alamos Laboratory, LA-1012-Del., May 8, 1947.

Herken, Gregg, *Brotherhood of the Bomb - The Tangled Lives and Loyalties of Robert Oppenheimer, Ernest Lawrence, and Edward Teller*, Henry Holt and Company, New York, 2002.

Hoddeson, Lillian, Paul W. Henriksen, Roger A. Meade, and Catherine Westfall, *Critical Assembly, A Technical History of Los Alamos During the Oppenheimer Years, 1943-1945*, Cambridge University Press, Cambridge, England, 1993.

Norris, Robert S., *Racing for the Bomb*, Steerforth Press, South Royalton, Vermont, 2002.

Serber, Robert, *The Los Alamos Primer, The First Lessons on How To Build an Atomic Bomb*, Univ. of CA Press, Berkeley, CA 1992

Teller, Edward, *Memoirs A Twentieth Century Journey in Science and Politics*, Perseus Publishing, Cambridge, Massachusetts, 2001.

APPENDIX C

THE SOVIET UNION'S SEMIPALATINSK TEST SITE (STS)

The material for this appendix was taken from: "The Semipalatinsk Test Site: Creation, Operation, and Conversion", a monograph edited by Professor Vladimir S. Shkolnik. The monograph was prepared and published under the joint Kazakhstani-American project between the Republic of Kazakhstan Institute of Nonproliferation and the Cooperative Monitoring Center of Sandia National Laboratories. The text was translated by Paul B. Gallagher.

The Soviet Union's most important military proving grounds were located in what is now known as Kazakhstan, then known as the Republic Region of Kazakh, (SSR.) "In geophysical terms, you simply could not find a better place. Forty percent of Kazahh land area was desert, 23% was semi-desert, 20% was steppe, 7% was forest-steppe, and 10% was mountainous." So naturally attention turned to that region for nuclear testing." (The Soviet Union also developed a test site at Novaya Zemlya which they started to use in 1957. It was used primarily for their larger yield tests.)

"The developers' main criteria in choosing a location for construction of the nuclear test site were that the area be practically uninhabited, free of agricultural lands, and large in area. In addition, the area had to be close to transportation arteries, and permit construction of a local runway for cargo planes, since they would have to carry in large quantities of cargo, and establish permanent operational communications. Preliminary calculations indicated that the diameter required for the test site should be at least 200 km."

After a long search, taking the main criteria into account, a suitable area was found in the steppes of Semipalatinsk region, Kazakhstan. The nuclear test site was located in the steppes near the Irtysh River, about 140 km west of Semipalatinsk. This part of the Kazakh SSR was and is now an arid steppe with scattered seasonal wells. The southwestern part of the area is low mountains with massifs dissected by valleys and washes. In the east is the valley of the Sagan River (Russian *Hagan*), a left-bank tributary of the Irtysh. Here there are shallow salt lakes that dry up in summer.

The climate is continental. Its principal features are aridity, with a cold, relatively snow-free winter and a relatively short, hot summer. Precipitation is low. Strong winds are frequent. In winter, the temperature reaches -40°C, and in the summer it exceeds 30°C. Annual precipitation ranges from 200 to 300 mm, most of it falling in the summer. Snow depths of 100-200 mm, produce small amounts of melt water and deep freezing of the soil (down to 1.5 - 2 meters). In winter and spring, prevailing winds are from the southeast, averaging 4-5 m/s; in summer, winds are typically from the north, with dust storms. Wind speeds and directions are quite variable in the region, even during a single day."

“Construction of the nuclear test site was begun in 1947.” (Note, this is about 4 years prior to the development of the NTS.) The construction appears to have been of a substantial nature. The Headquarters are buildings are multi-storied, large and substantial looking, like university buildings, with a statue of Igor Vaasilyevich Kurchatov. Igor Vaasilyevich Kurchatov was born January 12, 1903 and died on February 7 1960. He “directed the construction of the first Soviet cyclotron” -- and “the first atomic reactor in Europe.” His team produced the first Soviet atomic bomb in 1949 -- “detonated at STS at 22 kt. The team also detonated their first thermonuclear explosion of 400 kt in August of 1953 at STS. Kurchatov is known as the father of the Soviet atomic bomb. (Encyclopedia Britannica 2006; Mikhailov 1999: 14, 16) The Headquarters is 14 km NE of the test field and is used for the temporary accommodation of testers, see Figure C1.

The community of Kurchatov on the banks of the Irtysh River, 60 km NE of the Test Field, is the site’s housing and administrative center. On the edge of Kurchatov, along the road to the Test Field (not shown in Figure C1) is the site’s experimental research section with substantial looking laboratory buildings.

Also “along an edge of Kurchatov”, is “a military station housing numerous military units with their depots, motor pools, etc.” Military exercises, possibly similar to the Desert Rock exercises, were also conducted at STS.

For a test, “- the Test Field was a relatively flat area. The nuclear item was to be placed at its center.” “The majority of the various measuring instruments and optical equipment was placed in the instrumentation facilities, the so-called ‘geese’” The “geese” (located approximately where the Sh and N circles are shown in Figure C1) “were aimed at the center of the area, where the metal tower was located. The tower, some 30 meters high, with underground and elevator systems, was where assembly and checkout of the physics package took place.”

“Numerous sensors and indicator were installed outdoors on the ground and in the combat equipment, on fortifications, and on other structures, and also at the locations of experimental animals.”

APPENDIX C REFERENCES

Encyclopedia Britannica 2006, *Encyclopedia Britannica Premium Service*. 7 June 2006 http://www/britannica.coim/eb/article?tocid=9046465.

Mikhailov, V.N. Editor in Chief, *Catalog of Worldwide Nuclear Testing*, Begell-Atom, LLC, Begell House, New York, NY, 1999.

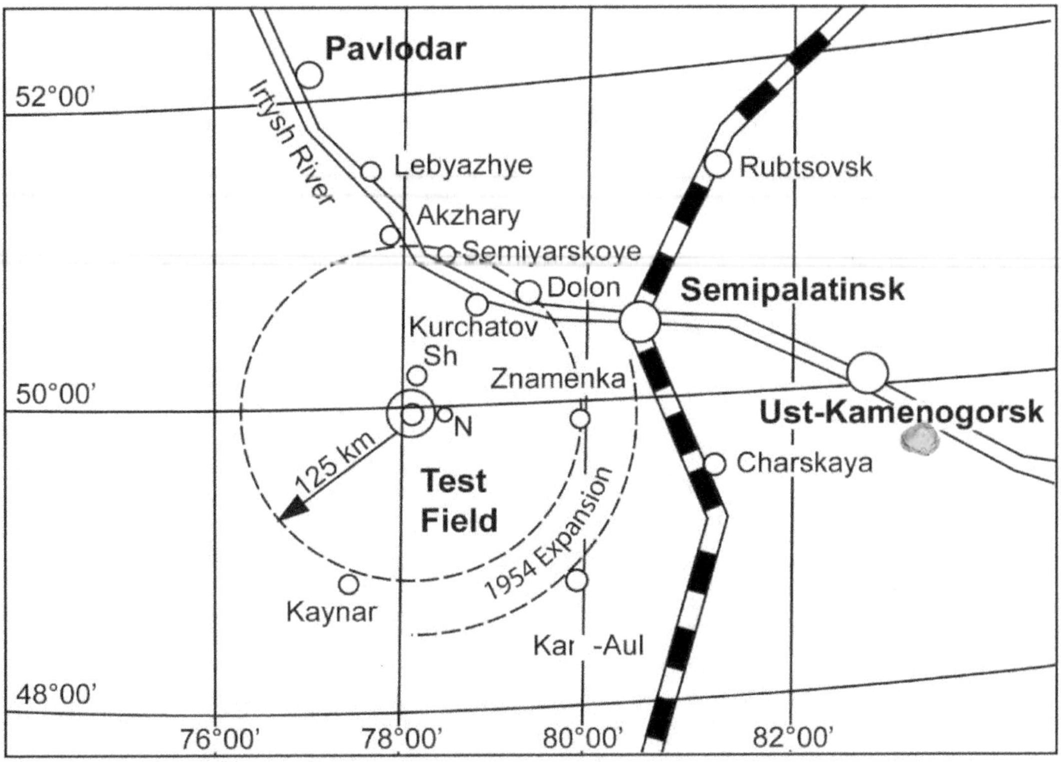

Figure C1 Test Field and Main Areas of the Semipalatinsk Test Site.

APPENDIX D

ACQUISITION OF REAL ESTATE FOR THE NEVADA TEST SITE

From information supplied by Carole Schoengold

The acquisition dates of real estate parcels at the test site are often ambiguous. There is an early date, when a decision is made, and signed-off, by a senior government official and a much later date when the paper work is finally finished. There are even earlier dates when the acquisition was proposed and numerous intermediate milestones. This chronology usually focuses on the earliest date that a firm decision was made. The reason for this is that the AEC effectively started using the land around that time. The land transfer paper work often took years, and the AEC facilities were long established by then.

October 29, 1940 (Executive Order 8578) and January 12, 1942 (Executive Order 9019): These documents describe the withdrawal of "a portion of the public domain for the use of the War Department and subsequently the United States Air Force as a bombing and gunnery range in Nevada." (Gorden Dean letter to Oscar L. Chapman, Secretary of the Interior, October 25, 1951)

December 18, 1950: President Harry S. Truman approves "the development of a portion of the Las Vegas Bombing and Gunnery Range as an atomic weapons test site." The initial area was approximately 12X30 miles or 360 square miles. (NSC Memo dated December 14, 1950 by James S. Lay, Jr. and NSC Memo dated December 19, 1950 by James S. Lay Jr.) This document is considered the "birth certificate" of the Nevada Test Site.

May 15, 1951: The AEC certified the necessity for and the scope of construction of a permanent support camp facility at the Nevada Test Site (AEC 141/17, November 30, 1951 p.27)

June 8, 1951: In a memo from the Air Force to the AEC, the Air Force agreed to release to the AEC the rights to approximately 16 X 40 miles (640 square miles) of the Las Vegas Bombing and Gunnery Range. (AEC 141/12 June 8, 1951)

October 25, 1951: AEC Chairman Gorden Dean sent a letter to Secretary of the Interior Oscar L. Chapman in which he cited coordinates for the 16 X 40 mile tract upon which the AEC and the Air Force had agreed:

NE Corner:	Latitude	37°	15'	07.268"	North
	Longitude	115°	55'	42.268"	West
SE Corner:	Latitude	36°	40'	43.752"	North
	Longitude	115°	55'	37.687"	West

SW Corner:	Latitude	36°	40'	22.391"	North
	Longitude	116°	13'	27.751"	West
NW Corner:	Latitude	37°	15'	12.534"	North
	Longitude	116°	13'	15.615"	West

Note, these corners do not form an exact rectangle. The largest variation is in the southern boundary which corresponds to the southern boundary of the Bombing and Gunnery Range and runs slightly south of due west.
Total area about 432,000 acres or about 675 square miles.

In his October 25, 1951 letter, Dean:
"requested that an appropriate land order be issued withdrawing the land described --- from all forms of appropriation under the public land laws, including the mining and mineral leasing laws, and reserving it for the use of the Atomic Energy Commission. It is also requested that the Atomic Energy Commission be permitted to continue its occupancy and use of the land pending the issuance of such an order.

The Atomic Energy Commission agrees to assume any obligation which the Air Force may have to restore the land described --- to a condition similar to that existing at the date of initial occupancy of the land by the Air Force; provided, however, that this obligation will not be performed by the Atomic Energy Commission until such time as it no longer has a requirement for this land." (AEC 141/15)

November 2, 1951: Another letter from the AEC (by Commissioner Sumner T. Pike) was sent to Secretary of the Interior Oscar Chapman. This letter again requested that a land order be issued withdrawing approximately an additional 2024 acres "for the use of the Atomic Energy Commission for facilities required in connection with the Nevada Test Site. It is also requested that the Atomic Energy Commission be permitted to use the land pending the issuance of such an order."(AEC141/16)

November 30, 1951: The AEC outlined the terms and conditions under which the operation of the Crystal Mine and the Climax Mine, located within the Nevada Test Site, would be permitted. (AEC 141/17, Appendix A) Further rules and regulations are included in a June 9, 1952 memo from George P. Draker, Deputy Manager SFO to Brigadier General K.E. Fields Director AEC/DMA.

1951 and 1952: The AEC negotiated the terms for mine operation and explored the possibility of acquiring the mineral right on the Test Site. There are tungsten deposits on the site. Similar negotiations were pursued for the grazing rights.

February 12, 1952: Public Land order No. 805 withdrew from the public domain approximately 417,459 acres (652 square miles) for the 16X40 mile tract plus the 2000 acre tract.

April 2, 1956: "(T)he USAF has been urged to release excess lands and has announced that 1.8 million acres (about 2800 square miles) of the Las Vegas Gunnery and Bombing Range, which lie to the north and west of the Nevada Test site are available for the use of other government agencies." This area was occasionally subject to fallout from the nuclear tests in the atmosphere. The AEC was interested in controlling the area because occupation by others could end up restricting test operations. (AEC 141/3, p.3)

April 23, 1956: AEC announced that negotiations are underway to transfer approximately 12.2X39.6 miles adjacent to the western boundary of the NTS to the AEC for a special reactor test site (Project 400, Rover, to develop test reactors for nuclear propulsion). Nuclear weapons testing would remain in the earlier tract. Negotiations were also taking place to secure a use permit for an area approximately 24X26 miles in the northwestern portion near Tonopah, Nevada, for Sandia Corporation testing. (USAEC Release LAV-56-18)

No further real estate was acquired by the AEC until after the moratorium. In November 1, 1963: The AEC and the USAF agreed on joint use of Pahute Mesa. The AEC intended to use the land for nuclear weapons tests. This agreement essentially said that the Manager of NVOO controlled the land surface and subsurface while the Commander, Nellis Air Force Base, controlled the air-space. (MOU Between the USAEC and the USAF for use of Pahute Mesa. 11/1/63)

APPENDIX E

ENVIRONMENTAL CONDITIONS AT NTS

From briefing notes by Lawrence S. Germain and Barbara Germain Killian

TOPOGRAPHY

The United States Basin and Range geologic province occupies much of the western and southwestern part of the United States: almost all of Nevada, the western half of Utah, southeastern California, the southern part of Arizona, and into northwestern Mexico. Basin and range topography consists of an extensive network of relatively small tectonic depressions closely akin to rift valleys. The basins are about 5 to 20 miles wide and about 30 to 100 miles long. The basins are separated by ranges of similar dimensions which tend north-south. The basins contain young sediment derived from neighboring ranges and are quite flat. They are generally about 4,000 - 5,000 feet above sea level. The mountain ranges rise 3,000 - 5,000 feet above the level of the basins. (EB 1998)

The mountain ranges found on the site are generally lower in the south and higher in the north. (Fenher and Gosling 2000:5) The sides of the basins can be steep or gentle. Where a major fault separates a basin from a range, the edge of the basin is often steep. Where the edge of the basin is produced by the tilting of the basin down and of the range up, the flank is gentle, with average slopes of from a few to 15°. (EB 1998)

The basins contain an accumulation of alluvial debris washed down from the mountains. This alluvial debris which extends to great depths (over 1000 feet) turned out to be a unique feature which greatly aided the containment of gases generated by explosions which were detonated in the flats during underground testing.

Most of the basins in the basin and range province have no outlets for their drainage. Rainwater accumulates in the form of salt lakes (such as the Great Salt Lake of Utah, and Walker and Pyramid lakes in Nevada) or in playas (such as Yucca, Frenchman and Jackass) which are mud flats occasionally covered by a few inches of water which turns salty and rapidly evaporates. (EB 1998) These playa lakes are fleeting features which last only a few days at most. Persons who worked at NTS and had the rare opportunity of seeing Frenchman, Yucca, or Jackass Flats after a rain or a snow, long remembered it.

Climate

About 10,000 to 12,000 years ago, southern Nevada was cooler than it is today, and there were marsh areas and pluvial lakes in some of the valleys. While there is no evidence to indicate that the basins on the NTS supported lakes during this period, nearby valleys to the east and to the north apparently did. An increasingly arid climate dried up most of the lakes by approximately 8,000 years ago.

The period between 7,500 and 4,500 years ago witnessed a climate that was even hotter and dryer than is currently experienced. Between 4,500 and 1,900 years ago, the climate was cooler and wetter than today. Two notable hot and arid periods occurred between 1,900 and 1,000 years ago and between 700 and 500 years ago. Then, about 500 years ago, heavy winter precipitation began which ended at the end of the Little Ice Age, about 150 years ago. Since about 1850, temperatures have gradually increased and rainfall diminished. (Fehner 2000:6]

Today, the lower elevations of NTS, such as Yucca and Frenchman at about 3,000 ft, have hot, dry summers and mild winters. (The higher elevation Pahute Mesa area of the site was not acquired until after the moratorium.) Temperature extremes at on the site range from about zero to 110° F. Snow occasionally falls in the area, but it generally does not last long at the lower elevations. The mountains surrounding the flats can retain snow for a few days.

Long term climate monitoring has taken place at Indian Springs, about 40 miles from Yucca Flat. The following information is from a weather station at Indian Springs which depicts recent climatic conditions. (WeatherDisc Assoc. 1994)

	INDIAN SPRINGS Elevation 3160 feet, 40 miles from Yucca Flat
Mean Yearly Temperature (°F)	61.0
Yearly Range (°F)	44.4
Daily Range (°F)	38
% Days/year > 90 °F	38.7
% Days/year < 32 °F	32.3
% Days/year < 0 °F	
Precipitation/year (inches)	3.4
% Days/year precipitation > 0.1"	3.3

From the point of view of conducting operations at the test site, the most important feature of the NTS climate is wind magnitude and direction. Wind was of course extensively monitored during operations, and it was essentially the only cause of weather delays. The most delays in an operation were during PLUMBBOB (see Part II-Chapter 12, section PLUMBBOB TESTS, The Grueling Schedule) when about 60% of the 120 delay days were due to "weather" (essentially winds) and about 40% to technical reasons.

The dominate characteristic of the climate at NTS is the lack of water. (Fehner 2000: 5) The high Sierras of California often cause clouds of Pacific origin to drop their moisture before reaching Nevada. Thus, Nevada has a semiarid climate. The level of aridity commonly used to delimit a desert is a mean annual precipitation value equal to 10 inches or less. (Encyclopedia Britannica 1998) As seen in the table above, the 3.4 inches/year of precipitation at Indian Springs certainly qualifies this region as a desert.

There is abundant ground water beneath the NTS, but it is located at considerable depths. Except for the basins such as Yucca, Frenchman, and Jackass Flats, which may occasionally contain salty water for a few days, there is essentially no fresh surface water at the NTS. Springs are the only perennial water sources. They are widely, if not abundantly scattered across the site. However, the flow from these springs is often minimal. (Fehner 2000:10)

Population

As frequently mentioned in the text, the sparse population was a key factor in the choice of the Nevada Proving Grounds as a nuclear weapons test site. This lack of population was of course due to the topographic conditions, severe climatic, and lack of water found there. Few people ever made a living in the area of the site. Some Indian cultures frequented the site but did not make permanent residences there. Later, some marginal ranching and mining occurred.

In the late 1920s at Wahmonie, located on what is now the test site west of Cane Spring and on the eastern edge of Jackass Flats, Nevada's last major mining rush occurred. Mining operations in the area dated back at least to 1905, but the area remained quiet until the discovery of high-grade silver-gold ore in 1927. The Wahmonie mining camp was established in February 1928, and grew to a population of some 500 within a month. Many miners lived in small tents, but Wahmonie soon had boarding houses, tents, streets, and cafes, as well as two saloons: the Silver Dollar and the Northern Club. Its population peaked in early summer at some 1,000 to 1,500; and by the end of the year it was clear that the strike was not as rich as had first been thought. The town went bust, the mines were abandoned, and the equipment moved to other locations. (Fehner and Gosliung 2000: 14-17)

In 1940, the desolation, lack of water, and general un-inhabitability brought the area to the attention of the federal government. With war looming on the horizon, the United States had begun a major rearmament program. Part of this program involved locating bombing and gunnery training ranges for The Army Air Corps. On October 29, 1940, President Franklin D. Roosevelt established the Las Vegas Bombing and Gunnery Range. Encompassing more than 3 ½ million acres north and west of Las Vegas, the range stretched almost to Tonopah and included all of what is now the test site. More than 90 percent of the range was in the public domain, but a number of grazing, homestead, and mining claims made it difficult to take possession. In August 1941, the government began condemnation proceedings against the outstanding parcels of land. (Fehner and Gosling 2000:20) As mentioned in the text, the fact that the Bombing and Gunnery Range was already government property was a positive factor for locating the site there.

An excellent history of the site is the reference by Terrance R. Fehner and F.G. Gosling, *Origins of the Nevada Test Site*, U.S. Department of Energy Report DOE/MA-0518, December 2000.

In 1950, the population of Las Vegas was only about 25,000 (Almanac 1998), and it more or less surrounded the downtown area. In 2007, the population of the metropolitan area of Las Vegas was estimated by Clark County to be 1,836,333. Today, there is continuous, or nearly continuous housing for tens of miles from the old down town area. The Las Vegas area has been one of the fastest growing areas in the United States for many years.

APPENDIX E REFERENCES

EB - *Encyclopedia Britannica Compact Disc* version, 1998.

Erken, E.B., *Geologic Setting of Nevada Test Site and Nellis Air Force Range*, pages 11- 19 in Eckle 1968.

Eckle, Edwin B. editor, *Nevada Test Site*, The Geologic Society of America, Inc., Memoir 110, 1968.

Fehner, Terrence R. and F.G. Gosling, *Origins of the Nevada Test Site*, U.S. Department of Energy Report DOE/MA-0518, December 2000.

WeatherDisc Associates, Inc. 1994, *World Weather Disc Climate Data for the Planet Earth*, Compact Disk Data Storage, Seattle, Washington, 1994.

Basin and Range Topography

18Jan55 - Yucca Flat with snow.

APPENDIX F

AEC and DoD Senior Leadership, Advisory Panels, and Planning Boards for NTS Organization During Atmospheric Testing

ASSIGNMENT	RANGER 1/27/1951 - 2/6/1951	BUSTER-JANGLE 10/22/1951 - 11/9/1951
President	Harry S. Truman	Harry S. Truman
AEC Commissioners (*Chairman)	*Gordon Dean Sumner T. Pike Henry De Wolf Smyth Thomas E. Murray Thomas Keith Glennan	*Gordon Dean Sumner T. Pike Henry De Wolf Smyth Thomas E. Murray Thomas Keith Glennan
AEC General Manager	Marion Boyer	Marion Boyer
Director, Division of Military Applications	Brig. Gen. James McCormack, Jr.	Brig. Gen. Kenneth E. Fields
Manager, Santa Fe Operations Office	Carroll L. Tyler	Carroll L. Tyler
RESPONSIBLE MILITARY AGENT	Commanding General SWC Brig. Gen. John S. Mills	Commanding General SWC Brig. Gen. John S. Mills
MAIN MILITARY LIAISON BETWEEN AEC & DoD	Commanding General SWC Brig. Gen. John S. Mills	Lt. Col E. W. Kesling - SWC
TEST MANAGER'S ADVISORY COMMITTEE (*Chairman)	*Alvin C. Graves - LASL Brig. Gen. James P. Cooney - AEC Col. George F. Taylor - USAF Col Benjamin G. Holzman - USAF Capt. Howard L. Andrews - USPHS Shields Warren - AEC Darol K. Froman - LASL Thomas Shipman - LASL	*Alvin C. Graves - LASL Walker Bleakney - Princeton Curtis Lampson - APG Capt. Howard L Andrews - USPHS Col. George F. Taylor - USAF Comdr. Russell H. Maynard - USN Shields Warren - AEC/DBM John Bugher - AEC/DBM

ASSIGNMENT	TUMBLER-SNAPPER 4/1/1952 - 6/5/1952	UPSHOT-KNOTHOLE 3/17/1953 - 6/4/1953
President	Harry S. Truman	Dwight David Eisenhower
AEC Commissioners (*Chairman)	*Gordon Dean Henry De Wolf Smyth Thomas E. Murray Thomas Keith Glennan Eugene M. Zuckert	*Gordon Dean Henry De Wolf Smyth Thomas E. Murray Eugene M. Zuckert
AEC General Manager	Marion Boyer	Marion Boyer
Director, Division of Military Applications	Brig. Gen. Kenneth E. Fields	Brig. Gen. Kenneth E. Fields
Manager, Santa Fe Operations Office	Carroll L. Tyler	Carroll L. Tyler
RESPONSIBLE MILITARY AGENT	Chief AFSWP Maj. Gen. Herbert B. Loper	Chief AFSWP Maj. Gen. Alvin R. Luedecke
MAIN MILITARY LIAISON BETWEEN AEC & DoD	Commander Test Command - AFSWP	Commander Test Command - AFSWP
TEST MANAGER'S ADVISORY COMMITTEE (*Chairman)	*Alvin C. Graves - LASL Duncan Curry, Jr. - LASL T. N. White - LASL Lt. Col. J.B. Hartgering - LASL Maj. N.M. Lulejinan - RAD-SAFE C.A. Spohn Thomas Shipman - LASL	*Alvin C. Graves - LASL Thomas Shipman - LASL Brig. Gen. James P. Cooney - USA Howard L. Andrews - USPHS Everett F. Cox - Sandia Col. Benjamin Holzman - USAF Capt. Harry Haight - USN

ASSIGNMENT	TEAPOT 2/18/1955 - 5/15/1955	PROJECT 56 11/1/55 - 1/18/56
President	Dwight David Eisenhower	Dwight David Eisenhower
AEC Commissioners (*Chairman)	*Lewis L. Strauss Thomas E. Murray Willard F. Libby John von Neumann	*Lewis L. Strauss Thomas E. Murray Willard F. Libby John von Neumann Harold S. Vance
AEC General Manager	Kenneth D. Nichols (to 3/30/55) Kenneth F. Fields (from 5/1/55)	Kenneth F. Fields
Director, Division of Military Applications	Brig. Gen. Kenneth E. Fields (to 4/30/55) Maj. Gen. Alfred Starbird (from 7/1/55)	Maj. Gen. Alfred Starbird
Manager, Santa Fe Operations Office (TEAPOT) Albuquerque Operations Office (PROJECT 56)	James E. Reeves	James E. Reeves
RESPONSIBLE MILITARY AGENT	Chief AFSWP Maj. Gen. Alvin R. Luedecke	Chief AFSWP Maj. Gen. Alvin R. Luedecke
MAIN MILITARY LIAISON BETWEEN AEC & DoD	Commander Test Command - AFSWP	Commander Test Command - AFSWP
TEST MANAGER'S ADVISORY COMMITTEE (*Chairman)	*Alvin C. Graves - LASL J. C. Bugher Everett F. Cox - Sandia L. Machta Duane Sewell - UCRL C. A. Spohn T. N. White	*Alvin C. Graves - LASL John C. Clark - LASL William E. Ogle - LASL Maj. R. A. McKown - USAF Maj. O.W. Stopinski Oliver Placak - USPHS

ASSIGNMENT	PROJECT 57 and PLUMBBOB 4/24/1957 - 10/7/1957	PROJECT 58 and 58A 12/6/1957 - 3/14/1958
President	Dwight David Eisenhower	Dwight David Eisenhower
AEC Commissioners (*Chairman)	*Lewis L. Strauss Thomas E. Murray (to 6/30/57) John S. Graham Willard F. Libby Harold S. Vance John Forrest Floberg (from 10/1/57)	*Lewis L. Strauss John S. Graham Willard F. Libby Harold S. Vance John Forrest Floberg (from 10/1/57)
AEC General Manager	Kenneth F. Fields	Kenneth F. Fields
Director, Division of Military Applications	Maj. Gen. Alfred Starbird	
Manager, Albuquerque Operations Office	James E. Reeves	
RESPONSIBLE MILITARY AGENT	Chief AFSWP Rear Admiral Edward N. Parker	
MAIN MILITARY LIAISON BETWEEN AEC & DoD	Commander Test Command - AFSWP	
TEST MANAGER'S ADVISORY COMMITTEE (*Chairman)	*Alvin C. Graves - LASL William E. Ogle - LASL Duane Sewell - UCRL Howard Andrews - USPHS Philip W. Allen - USWB Roger Batzel - UCRL Thomas B. Cook - Sandia Charles L. Dunham - AEC/DBM Gordon M. Dunning - AEC/DBM Harvey I. Israel - LASL William R. Kennedy - LASL Col. Clinton Maupin - USA Melvin Merritt - Sandia Lt. Col. DeWitt Morgan - USAF Kenneth Nagler - USWB Jack Reed - Sandia J. D. Shreve - Sandia A. Vay Shelton - UCRL	*Alvin C. Graves - LASL William E. Ogle - LASL William R. Kennedy - LASL

	Kenneth Street - UCRL O.W. Stopinski - LASL Col. Ralph M. Lachausse - USAF Maj. R. E. McKown - USAF C. D. Broyles - Sandia	
NTS PLANNING BOARD (*Chairman)	*Alvin C. Graves - LASL Vice-*Duane Sewell - UCRL William E. Ogle - LASL Gerald Johnson - UCRL Richard A. Bice - Sandia Robert L. Corsbie - DETG Col. H.E. Parsons, FC/AFSWP	

ASSIGNMENT	HARDTACK II 9/12/1958 - 10/30/1958
President	Dwight David Eisenhower
AEC Commissioners (*Chairman)	*Lewis L. Strauss Thomas E. Murray (to 6/30/57) John S. Graham Willard F. Libby Harold S. Vance John Forrest Floberg (from 10/1/57)
AEC General Manager	Kenneth F. Fields
Director, Division of Military Applications	Maj. Gen. Alfred Starbird
Manager, Santa Fe Operations Office	James E. Reeves
RESPONSIBLE MILITARY AGENT	Chief AFSWP Rear Admiral Parker
MAIN MILITARY LIAISON BETWEEN AEC & DoD	Commander Test Command - AFSWP
TEST MANAGER'S ADVISORY COMMITTEE (*Chairman)	*Duane Sewell - UCRL Alternate* William E. Ogle - LASL Howard Andrews - USPHS George Anton - AEC/DBM Roger Batzel - UCRL Carter D. Broyles -- Sandia Thomas B. Cook - Sandia Gordon M. Dunning - AEC/DBM Harvey I. Israel - LASL William S. Johnson - REECo William R. Kennedy - LASL Alfred Klement - AEC/DMB Col. Ralph M LaChausse -- USAF Col. Clinton Maupin - USA Melvin L. Merritt - Sandia Gene T. Pelser - UCRL A. Vay Shelton - UCRL O.W. Stopinski - LASL Benjamini Sussholz - Space Tech Lab
NTS PLANNING BOARD (*Chairman)	*Alvin C. Graves - LASL Vice-* Duane Sewell - UCRL William E. Ogle - LASL Gerald W. Johnson - UCRL Richard A. Bice - Sandia Robert L. Corsbie - CETG Col. W. S. Hutchinson - FC/AFSWP

APPENDIX G

DIAGNOSTICS EXPERIMENTS CONDUCTED BY AEC LABORATORIES 1950 - 1958

Adapted from work by William E. Ogle (1985: 52-83)

INTRODUCTION

This section describes the types of measurements that were made on nuclear weapons tests prior to the 1958 moratorium.

These experiments required many innovative people. Although only a few individuals may be mentioned by name, in fact, teams of people from a number of disciplines and backgrounds were involved.

Some key factors in the period from 1950 to1958 that influenced the diagnostic measurements include the facts that:

- Nuclear weapons technology represented frontier science.
- Rapid developments occurred in the technical disciplines that contributed to nuclear weapons, such as computers, numerical methods for modeling weapons, and electronics.
- Work on nuclear weapons was considered very important by the nation, and new ideas were encouraged.

DEVICE DIAGNOSTICS - STANDARD MEASUREMENTS

Most of these measurements have been discussed to some extent within the main text. However, they are presented here in order that they may be together. The general descriptions given in the following are from the work by William E. Ogle and may differ from a particular measurement described in the main text.

Radiochemistry

Radiochemistry was the primary method of measuring yield. A sample of the condensed particulate debris from the detonation was collected and taken to the laboratory for quantitative analysis. Chemical techniques were used to separate the fissionable material (uranium or plutonium, referred to here as the fuel) and representative radioactive fission fragments. A measurement of the amounts of both the fuel and fission fragments in a single sample then allowed a determination of the fraction of the fuel that had fissioned. From this fraction and the total amount of fissionable material in the device the amount of fuel that had fissioned was obtained. Knowing the energy released per fission, it was thus possible to calculate the total energy released in the explosion.

Uncertainties in this method arise from a variety of sources. For example, the samples that were obtained may not be representative of the debris immediately after the detonation. Therefore, multiple samples were collected and analyzed separately in the laboratory to evaluate the fractionation, which reflected the differences due to the samples not being representative of the bomb debris immediately following the shot.

Samples were obtained from air samplers placed on the ground, as was done for Trinity, or from air samplers installed in aircraft that flew through the detonation cloud a few hours after the shot. An Air Force organization first designated AFOAT and later AFTAC (Air Force Technical Applications Center) worked with the weapons laboratories to develop sampling techniques, and they conducted the aircraft sampling.

Rod Spence and George Cowan at Los Alamos played major roles in the development of the radiochemical analysis of bomb debris during the 1950s.

Fireball Yield

At the time of Trinity, the general equations had been developed for the growth of the fireball as a function of time and yield. However, in 1945, the properties of air at very high pressures were not well known; and the effects of nuclear and thermal radiation on the air prior to shock arrival were not understood. In addition, the material surrounding the device was not just air. It included high explosives, the bomb casing, and the hardware in the cab and tower.

Also, high-speed photography was necessary in order to measure the rate of fireball growth. In fact, a primitive capability was developed and deployed for the Trinity shot in 1945 and further refined for Operation SANDSTONE in 1948 by Berlyn Brixner, Lou Fussel and others.

A concerted effort was made in the late 1940s to understand, and to be able to accurately measure, the rate of fireball growth as a function of device yield.

Fred Reines set up a section that included Fran Porzel and Joe Mullaney to model the fireball theoretically. This work was augmented by contributions from Livermore. The models that they developed accounted for the mass of the bomb and the nearby materials consumed in the fireball.

EG&G played a major role in the development of the diagnostic hardware, which included excellent high-speed photography and accurate timing signals. They developed a bank of fast cameras, which they called the Rapatronic cameras that took pictures at accurately known times during the fireball growth. The first Teller light was used to trigger the camera sequence.

Fireball growth was the main measurement of the yield of a thermonuclear device by 1958, surpassing even radiochemistry.

Reaction History, Alpha
Reaction history, or alpha, is described in some detail in Appendix A. The term is used to characterize the neutron population growth in an exploding device. The basic techniques developed for the Trinity test (see Part I-Chapter 1) were refined and employed during the whole duration of nuclear testing. Major advances were made over the years in detectors and in high speed oscilloscopes for recording the very fast signals. Significant progress was also made in understanding cable attenuation corrections and in techniques for unfolding signals at the very high frequencies observed.

Bhangmeter
A double-humped curve of light intensity as a function of time was observed on the Trinity shot. The minimum, after the first maximum, occurs at "breakaway," when the shock front separates from the fireball. Prior to breakaway, when the shock front is still within the fireball, the light decreases due to three phenomena: (1) the expanding fireball is cooling; (2) nitrogen, oxygen, and hydrogen compounds in the shock front attenuate the visible light, and (3) the heated air within the fireball further attenuates the light. The light increases again as the shock front expands and cools, and becomes transparent to the visible radiation from the inner region of the fireball. The time from first light to the first minimum is dependent upon the yield of the device.

A measurement of light intensity as a function of time was successfully made on SANDSTONE, which stimulated Fred Reines to suggest that they capitalize on this to produce a simple instrument to quickly measure yield. Los Alamos asked EG&G to make such an instrument, which they did, using photocells and oscilloscopes to measure the time to first minimum. Of course, the selection of just the right name for the instrument was essential, and occupied the better part of an afternoon by Reines and his colleagues from J-7. They settled on Bhangmeter, named after bhang, which is a narcotic made from hemp. They were undoubtedly aware that for ever more bhang would be misinterpreted as bang, the noise from a shot. Ogle suggested that bhang should be interpreted to mean that anyone would have to be smoking something to believe a yield taken with such an instrument. (Ogle: 67)

Electromagnetic Pulse (EMP)
Fermi suggested, before Trinity, that nuclear explosions might be expected to produce an electromagnetic signal that was approximately proportional to alpha. The electromagnetic pulse (EMP) signal was indeed observed on SANDSTONE shots and was, eventually, shown to be proportional to alpha, at least at early times. However the uncertainties that persisted regarding EMP throughout nuclear testing history inhibited its use as a measuring technique.

Time Interval
When discussing single stage devices the time interval referred to a time derived from a measurement of the time between the detonators firing and the detection of the first fission gammas. For two stage devices the time interval is derived from a measurement of the time between the detonations of the two stages.

DEVICE DIAGNOSTICS - NON-STANDARD MEASUREMENTS
Other diagnostics were developed during the 1950s. All were aimed at either increasing the accuracy of existing techniques or shedding light on the details of weapons behavior. Those techniques used only in the high yield Pacific shots are not covered.

Thermonuclear Temperature
TENEX (TEmperature of Neutron EXperiment) was a time-of-flight measurement of the neutron spectrum that yielded information on the thermonuclear burn temperature. This technique was used both in Nevada and the Pacific.

Imaging of the Burn Region
PINEX (PINhole EXperiment) used a pinhole collimator to image neutrons from the device onto a detector.

OUTPUT MEASUREMENTS
Measurements of the radiations from a device were important to both the weapons designers and those focusing on the effects of nuclear weapons.

Neutron Flux and Spectrum
Measurements of neutrons and their spectrum were made using activation foils at various distances from the explosion. Spectra were measured using foils with cross section thresholds at known energies. Louis Rosen, of Los Alamos, used photographic plates behind large collimators to record recoil protons and unfold the incident neutron spectrum.

Gamma-Ray Flux and Spectrum
There are three principal sources of gamma rays from a nuclear detonation: (1) gamma rays emitted from direct fission and fusion of the device; (2) fission fragments or activated nuclear in the fireball; and (3) the decay of air nuclei that had undergone neutron capture. Gamma rays and their spectra were assessed using film badges with and without various shielding materials by Ellery Storm of Los Alamos, and H. O. Wycott and L. S. Taylor of the NBS. Not too surprisingly, the gamma dose is approximately proportional to the yield.

John Malik, at Los Alamos, developed gamma detectors that were used throughout the 1950s. They were tested as possible techniques for measuring yield, but other methods were more than adequate for that purpose and the interest in measuring yield from the gamma flux was eventually dropped.

Thermal Radiation
Julian Mack and his colleagues measured the thermal characteristics of the Trinity shot in 1945. Similar measurements were made in the Pacific by Los Alamos and the NRL.

Many different instruments and techniques were used to measure flux and spectra. One of the simpler pieces of hardware was known as the Black Ball. It was a black-painted copper sphere containing air and a pressure transducer. The thermal radiation from a bomb detonation could be deduced from the pressure rise in the black ball.

Cord Experiments and Atmospheric Chemistry

Cord experiments were performed by Herman Hoerlln and his colleagues at Los Alamos and the NRL from 1953, (Upshot-Knothole) on. A bright light was observed by spectrometers using a line-of-sight, through the air, to one side of a fireball. The purpose was to gather data relevant to the atmospheric chemistry associated with a nuclear burst. Some years were devoted to the analysis and interpretation of the data, which ultimately resulted in a good understanding of the molecular processes in the air surrounding a nuclear explosion.

APPENDIX G REFERENCE

Ogle, William E., *An Account of the Return to Nuclear Weapons Testing By The United States After The Test Moratorium 1958-1961*, United States Department of Energy, Nevada Operations Office, NVO-291, (SRD) October 1985.

APPENDIX H

SYNOPSES OF NWE FIELD ACTIVITIES and STRUCTURES ON JANGLE

SYNOPSES OF FIELD ACTIVITIES CONDUCTED FOR SOME TYPES OF NUCLEAR WEAPONS EFFECTS PROJECTS

The following synopses address the field activities at NTS that would typically be conducted by the DoD's nuclear weapons effects projects in several technical areas. The field activities described herein were only a small part of the total activities of a project; small, but essential, for validating concepts and theories regarding nuclear environments.

Exposure Projects

The most numerous of the effects projects conducted during atmospheric testing are termed herein *exposure projects*. These projects consisted of exposing objects to different levels of a nuclear environment and analyzing how the exposures affected the object. Animal, vegetable, and mineral objects were used in the DoD exposure projects; but military equipment was the largest category. It can be argued that ALL projects at NTS, not just those identified as such herein were really exposure projects.

The characteristics of the nuclear environment(s) to which an object was to be exposed was estimated before the shot. Such estimates were the basis of much of AFSWP's work. It should be kept in mind that at the time of BUSTER, such estimates had more uncertainty than would be the case for later operations.

Objects to be exposed would arrive at the NTS at varying times, months to hours before detonation, depending on the objects. Some of the objects might be assembled or partially assembled at the site, others might be sent directly from the home base or laboratory, ready to expose.

Objects would usually be placed for exposure at more than one distance from the detonation. Often three distances were used which represented: over exposure; expected or desired exposure; and under exposure. If objects were placed at only one distance and the actual yield differed appreciably from the anticipated yield, or if the estimates were poor, the objectives of the project probably would not be met. Using at least three exposures was "hedging your bet" and provided the possibility of getting data over a range of environments.

Post-shot, the exposed objects were either examined where they had been placed in the field and/or or retrieved from the field for examination in the laboratory. In some cases, for example when radioactivity was the effect of interest, early time retrieval of the objects was essential, before decay caused the measurements to be less accurate or meaningless. Simple procedures were devised for early time retrieval, such as attaching the objects to cables (or placing the objects on sleds attached to cables) so that they could be pulled away from the hazardous exposure area to an area with a lower level of radiation acceptable for humans. More sophisticated means were also used, such as an army recovery tank with access to objects through its floor.

For technical reasons like data interpretation, objects exposed to radiations often needed to be protected from the subsequent airblast and/or ground motion. A wide variety of innovative devices were developed and used to isolate objects exposed to radiation.

In other cases, the assessment of damage due to airblast was not an issue; and postshot examinations could occur when hazardous radiation conditions had subsided in the field.

In yet other cases (like structures, vehicles, aircraft exposures), exposure and the resulting damage due to airblast and ground motion was the objective of the measurements.

Photography was extensively used in the field and the laboratory to document pre- and post-test conditions. For comparison purposes, a set of "control samples", just like those exposed, might have remained at the laboratory.

Biomedical Exposures, Animals

The use of animals was the most frequent type of biomedical exposure. These projects, with their wide range of objectives, were generally aimed at obtaining information that could be applied to humans. Mice were perhaps used in the greatest numbers. Chester swine were often used in large numbers because their skin is fairly similar to that of humans. Dogs, goats, and sheep were commonly used, and monkeys were used on occasion. In one PLUMBBOB FCDA project, obstinate burros were used to examine effects on large animals.

Animals were brought to the site and cared for by attendants. Often, these attendant jobs were not easy, especially when the unexpected happened - like when the pigs outgrew their uniforms due to schedule delays. Care was considered "highly specialized" at the *Pork Sheraton*. Usually at a few hours prior to the shot, the animals were taken to their containers and/or harnesses. In some cases, the animals were anesthetized prior to the shot.

The animals were usually placed inside of a harness or container of some sort of ingenious design. The containers might be open (constructed of a wire mesh) or closed cages with or without exposure holes. They might be placed in the open field or inside a structure. The containers themselves might be designed to eliminate certain kinds of radiation or other nuclear effect, like air blast or ground shock, but allowing the effect of interest to pass through to the animal. (For instance, if the effect of neutrons was of interest, the gammas might need to be separated out; and the container would include materials that absorb gammas.)

Sometimes the containers and/or buildings would be instrumented with radiation gages, pressure gauges, photographic systems, etc.

Harnesses, again often of ingenious design, also might be used to hold an animal in a particular position during exposure. For instance, it might be important that one side or part of the animal be exposed at the time of detonation. The animals might be "dressed". For instance pigs were sometimes dressed in uniforms made of military textiles, some animals were dressed in outfits that might also serve as a harness.

Post shot, the animals were retrieved and probably photographed, perhaps even given a brief examination. They would then be taken back to Mercury where they might undergo a more detailed examination(s) and/or be transported back to a laboratory. Some projects involved killing the animals, perhaps at specified time intervals, to examine certain inner organs such as lungs, liver, heart as a function of time after exposure. Some projects conducted observations of the exposed animals over a long term, perhaps years, in order to study changes. For the purpose of comparison, a *control group* of animals of the same background as the exposed animals might be maintained in either laboratory or natural environments, but they were not exposed.

<u>Airblast, Ground Motion, and Thermal Measurements</u>
AFSWP sought the capability to be able to predict the environmental conditions resulting from nuclear explosion of different yields and detonation locations. The environmental conditions of pressure, velocity and displacements, and temperature resulting from radiations, airblast and/or ground motion at different distances from the detonation were all of interest. They sought to determine the magnitude of each of the NWEs as a function of distance from GZ and as a function of the yield and height (or depth) of the detonation. To do this, numerous measurements were conducted for each of the NWEs.

The resulting data was analyzed and organized in handbooks for rapid use by field commanders and/or researchers. These handbooks were refined over the years as new data became available and as theoretical models and numerical simulations became available. The combination of predictions refined by measurements was just like that of weapons development. The resulting and ever improving capabilities for being able to predict nuclear environments represented a key product of AFSWP and its decedents.

Starting with JANGLE Sugar and Uncle, the shots on which the DoD conducted a significant number of NWE measurements used *blast lines**. These were straight radial lines from GZ along which instrumentation would be placed for the measurement of NWEs. There might be more than one blast line on a shot; and if so, they would often be run at about 90° or 180° directions from each other. [*Footnote: The blast line along which airblast and/or ground motion measurements were made was usually referred to as the *main blast line.* Although referred to as blast lines, measurements other than airblast and ground motion (e.g., thermal and nuclear radiations) might also be made along them.]

Because only so many folks and so much "stuff" can be in one place at one time, some of the fielding issues for blast lines were: the size of the equipment required for the measurements themselves as well as for their installation (for instance drill rigs, grouting equipment, etc.); how many people were involved when at a station; proximity to other projects or installations; time schedules for what or who was where when, etc. Such issues would be factors in determining whether one or more parallel lines or lines elsewhere would be used.

A number (generally 3 or more, often 6-10, sometimes more) of measurements for a NWE that used the same or similar instrumentation, would be made by one organization at different ranges from GZ along a blast line. Other organizations, probably using different instrumentation, might make measurements of the same or other NWEs along the same or a different blast line.

Line projects were usually fairly extensive. It was not uncommon for scores of measurements of a given NWE to be made by one organization along a blast line for one shot. These projects were usually conducted by one of a few organizations that were the technical leaders in such measurements. More than one organization might conduct the same or similar measurements on a given test. The advantages of what might appear to be duplication are described in the text.

Blast line projects would be responsible for providing their data, in usable form, to the other projects on a test. This saved the time, money, and effort that each project would expend if they had to obtain their own measurements. However, in some instances (such as if an exposure took place inside of a building), a project would require its own measurements away from the blast line. Such measurements would usually be conducted as a separate project(s) by an organization with such expertise.

Locations for measurements would be above ground or underground along the blast line. Above ground locations were usually facilitated by using poles, but "goal posts" and towers were also used. Underground locations might be dug or drilled if significant depth was required; and some protective canister would probably be used to protect the instrument when the hole was backfilled. Generally, the group doing the measurement project had also developed the instrument and its canister.

After detonation, information regarding how the instrument interacted with the nuclear effect it measured, needed to be "recorded". Instruments generally produce information as an electrical signal. Recording of this signal might be done within the canister itself in the case of self recording instrumentation. Or, the information signal might be sent somewhere else for recording. Information signals might be sent by radio waves or over cable, traveling at the speed of light. At early times after detonation, cables were generally but not always more reliable than radio waves. Laying the miles of cable, usually buried in trenches, was a dirty and tedious job, even in the mildest of weather. Cable would connect the instrument to a recording station which often was a trailer that was parked farther from GZ than the measurements and usually in a revetment for added protection. For close-in DoD measurements, and for some of the weapons development experiments, elaborate underground recording stations near the instrument(s) might be required.

The amount of information being sent; the distance from the instrument to the recording station; the recording equipment itself; the equipment required to make the recorded signal intelligible to a viewer; and the methods of signal storage, retrieval, and analysis were all important factors of the project. Thus, a measurement consisted of a system of components and their inter-workings.

It is difficult to separate measurements from instrumentation development. It is probably fair to say that each time a measurement was made, the scientists gleaned ways of improving their measurement system and did so. Each of the components of the measurement systems used during atmospheric testing were undergoing tremendous technical advancement during the 1950s, and changes in the measurement systems were continuously made that took advantage of them.

If a project specifically states instrumentation development or evaluation in its main objectives, it is considered here as being an *Instrumentation Development* Project. (Only 2 NWE projects were so designated here, and they were both on UPSHOT-KNOTHOLE.) Otherwise, it is considered as a *Measurement* project. Instrumentation and system (transmission, recording, etc.) development was generally just done without it being a specified part of the project objectives. Experimenters often referred to their development work as having been "bootlegged".

Nuclear Radiations

Measurements of thermal and nuclear radiations often shared many similarities with the measurements of airblast and ground shock like the positioning of measurement stations along a line and instrumentation development and evaluation. Thermal radiation measurements were often made in close cooperation with the airblast and ground shock programs, and they represented important contributions to solving the non-ideal airblast issues. A key feature of many of the measurements involving nuclear radiations, neutrons and gammas in particular, was the time of recovery of the exposed samples and/or the devices used to make the measurement.

A number of innovative methods were used for rapid recovery of samples and/or measuring devices. Recovery by personnel was perhaps the most common. People would enter the field on vehicles as soon as radiation levels had subsided and stay no longer than prescribed, simply picking up by hand the devices and/or samples. Helicopters would sometimes be used. Trailers containing equipment for analyzing the samples/devices might be moved to the forward area(s), saving travel time. Means of easily removing samples/devices from their exposure positions were devised. If the objects were exposed on a cable, the cable might simply be pulled 1,000 yards or more out of the contaminated area. One project had samples automatically eject from their underground exposure position; and they were then towed on cable to a less contaminated area where radiation levels were acceptable for humans. Objects might be exposed on sleds which were attached to cables that were pulled out of the active areas. Some recoveries used a tank with a lot of shielding and an open able floor area. The success of a nuclear radiation project could depend on its rapid and effective recovery, and many innovative methods were used.

Structures

The structure to be tested might be a complete building, bridge, or shelter, perhaps even containing ventilation and other systems. Or it might be an element of a structure that was commonly used such as columns, walls, footings, or beams. In some cases, scaled models were designed for testing. Also, relatively inexpensive idealized shapes such as cylinders or rectangular parallelepipeds were constructed and tested in groups as a parametric study. A key part of the design process was the determination of the loads a structure would be able to withstand.

For a specific nuclear test, the locations where the design loads would be experienced was estimated, probably with AFSWP aiding the personnel conducting the project. A plan would then be developed to place the structure at one or more locations in the field. Again, it was often decided to use 3 locations and to construct 3 structures, representing loads that were: over design, at design, and under design. For large and expensive structures, only one location generally would be used which was usually near to or somewhat over design loading.

Construction of the structure would be conducted by an AEC contractor(s); but there would usually be continuous communication with the design group. During BUSTER-JANGLE, Reynolds Electric and Engineering Co. or Silas-Mason would probably have been the construction contractor. If possible, the structure might be constructed the same distance from GZ as a measurement instrument on the blast line and/or near the blast line. Instruments such as pressure-time gages or displacement-time might be installed for measuring structure-specific loads.

Post-shot more photographic coverage would take place as well as “hands-on” inspection. Some samples of the structure or its debris might be collected for special analyses back at the laboratory. Comparison of the structure’s behavior in the field would be compared to that expected from the design.

AFSWP STRUCTURES ON JANGLE
The construction in support of the Structures Program on JANGLE was staggering! It consisted of Projects: by the: Navy (Hazzard 1953: ii, 6, 118); by the Army (Hansen 1952b:4-30); and by the Air Force (Armour 1953: 7). There were 65 major structures or structural elements of 26 different types. A summary of these 26 types from George 1979:128-132 is given in the figures below. These figures illustrate the extent and complexity of the effort better than words. To indicate the size scale, a human figure is shown in each figure (sometimes hard to find because of its smallness).

.

(George 1979: 128-133)

Structural Test Projects

Project	Agency	Description	
3.1a,b	Bureau of Yards and Docks	Light steel frame buildings with corrugated metal siding, 40' x 100'. At 4000 and 10,000 feet from zero.	
3.2a,b,c	Bureau of Yards and Docks	Tripod type antenna towers, 100' high. At 2900,3000 and 3100 feet from zero.	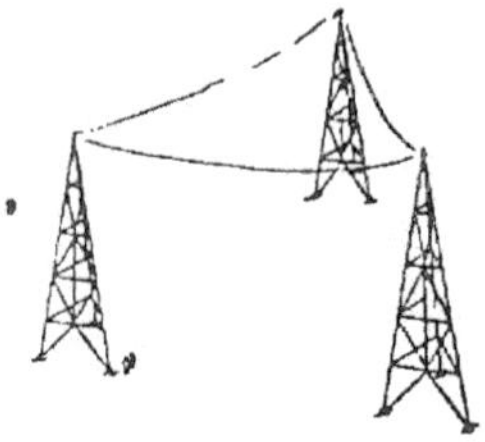
3.2d,e	Bureau of Yards and Docks	Cantilever type antenna towers, 100' high. At 2900 and 3100 feet from zero.	

Project	Agency	Description	
3.3	Office, Chief of Engineers	1/5 size model of heavy steel column target structure 9' x 12' x 6' high. At 504 feet from zero (see 3.7 for prototype).	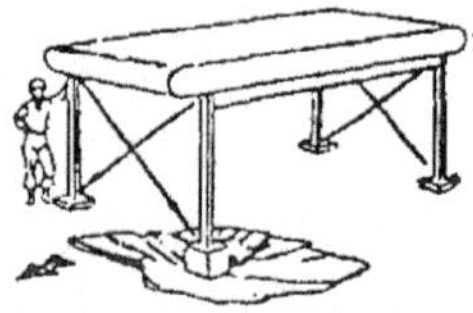
3.4	Office, Chief of Engineers	Reinforced concrete shear wall target structure 33' x 52' x 16' high. At 441 feet from zero.	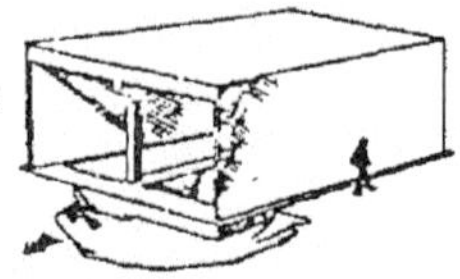
3.5	Office, Chief of Engineers	Reinforced concrete column target structure 11' x 14' x 16' high. At 504 feet from zero.	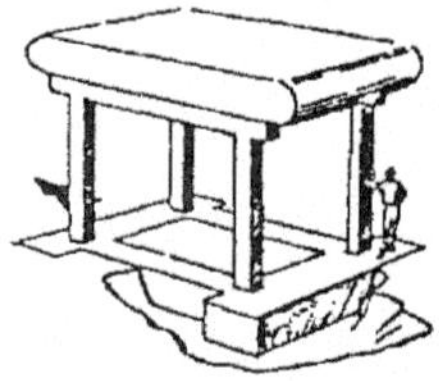

3.6	Office, Chief of Engineers	Three-bay heavy steel column target structure 11' x 47' x 16' high. At 441 feet from zero.	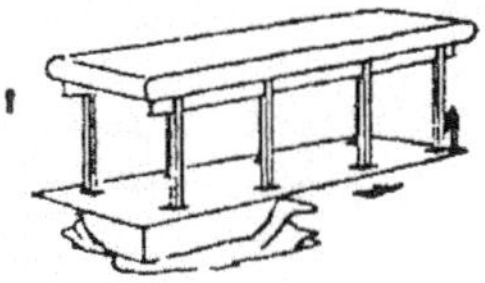
3.7	Office, Chief of Engineers	Single-bay heavy steel column target structure, 24' x 36' x 21' high. At 504 feet from zero.	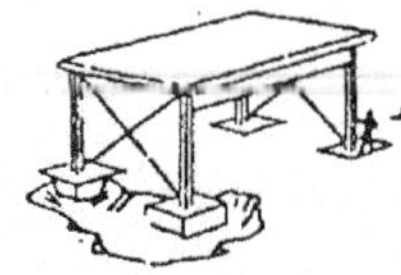

Project	Agency	Description	
3.8a,b	Office, Chief of Engineers	Single-bay light steel column target structures, 24' x 36' x 21' high. At 630 and 882 feet from zero.	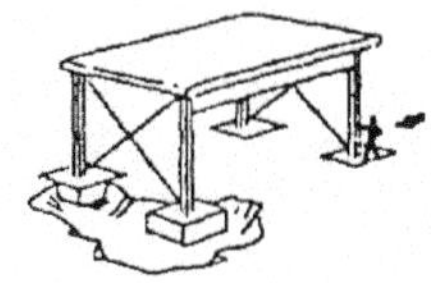
3.9a,b	Air Materiel Command	Reinforced concrete retaining walls, one with loaded face toward zero and one with loaded face away, 12' high. At 900 feet from zero.	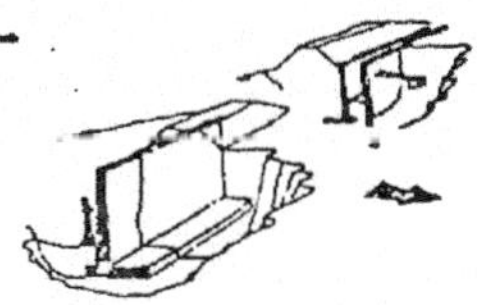
3.10a, b,c	Air Materiel Command	Single-bay, 2-story light reinforced concrete building frames with brick curtain walls, 25' x 27' x 30' high. At 900 feet from zero.	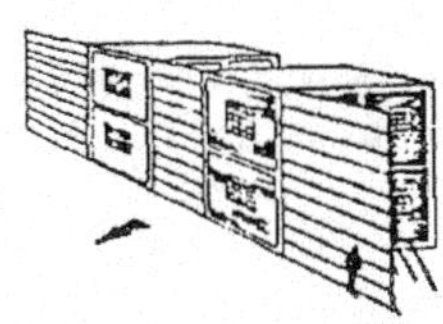
3.11	Air Materiel Command	14" and 18" reinforced concrete pavement slabs, 10' x 150'. At 500 feet from zero.	
3.12a, b,c	Bureau of Yards and Docks	Precast, prestressed concrete slabs, 20' x 60' x 6" thick. Reinforced concrete slabs 6" x 12" thick, 10' x 60'. At 1200 feet from zero.	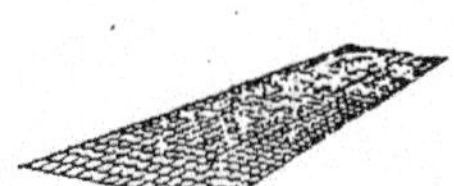

Project	Agency	Description	
3.13	Office, Chief of Engineers	Buried reinforced concrete box, 25' x 25' x 17' high, open top and bottom, front and rear walls 5' thick. At 238 feet from zero.	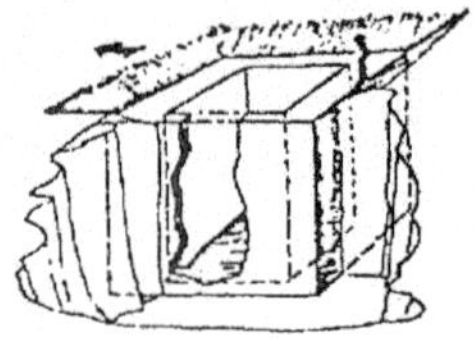
3.15a,b	Bureau of Yards and Docks	Thin-walled rigid frame buried buildings. Precast concrete construction, 20' x 40' x 14' high. At 750 and 1000 feet from zero.	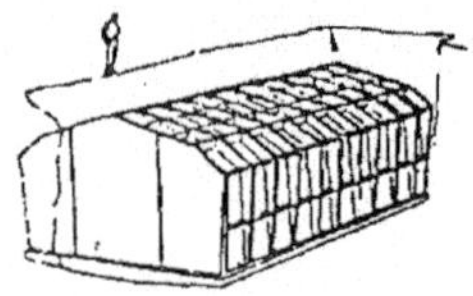
3.16a,b	Bureau of Yards and Docks	Thin-walled circular arch rib buried buildings. Precast concrete construction 20' x 40' x 10' high. At 750 and 1000 feet from zero.	
3.17a,b	Bureau of Yards and Docks	Thin-walled dome-shaped buried buildings. Precast concrete construction, 22' diameter at floor x 10' high. At 750 and 1000 feet from zero.	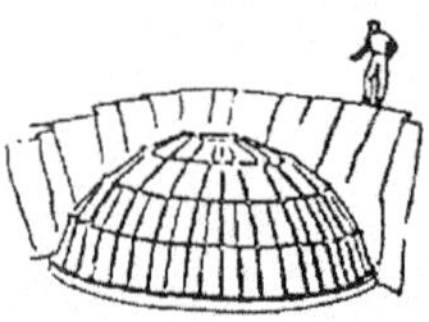
3.18a,b	Bureau of Yards and Docks	Prestressed concrete buried fuel tanks. Precast stave construction, 10' diameter x 8' high. At 750 and 1000 feet from zero.	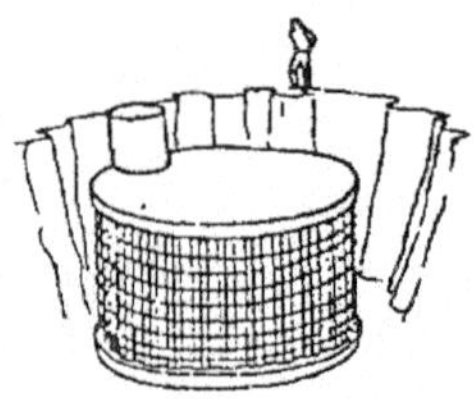

Project	Agency	Description	
3.19	Air Materiel Command	Unloaded concrete footings and 6' concrete cubes, buried. Located adjacent to several structures	
3.20a_1, a_2	Air Materiel Command	Free standing brick walls, 16' long x 11' high. At 900 and 1050 feet from zero.	
3.20b	Air Materiel Command	Brick wall with superimposed load, 16' long x 11' high. At 1050 feet from zero.	
3.20c_1, c_2	Air Materiel Command	50' high brick chimneys. At 750 and 1050 feet from zero.	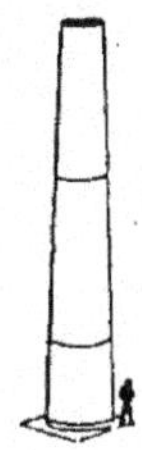
3.21a_1-f_1 3.21a_2-f_2	Air Materiel Command	Two arrays of buried square boxes, open at top and bottom. Reinforced concrete and steel construction, between 8' and 20' square with 4" to 8" walls of reinforced concrete and one of 12 gage steel. One array at 725 and one at 850 feet from zero.	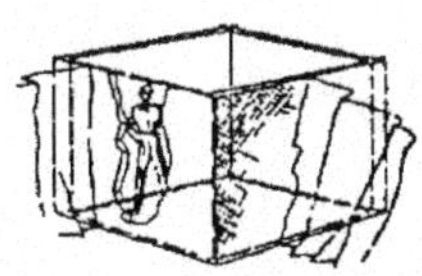

Project	Agency	Description	
3.23a_1, a_2	Air Materiel Command	Circular section reinforced concrete tunnels, 6' in diameter with 5" wall.	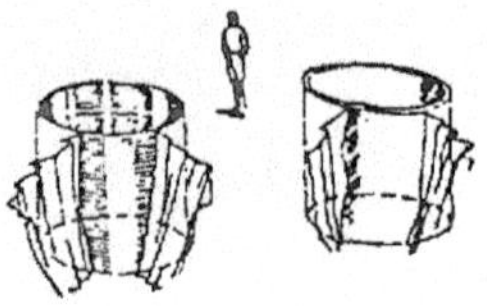
3.24a_1, a_2	Air Materiel Command	Friction pile clusters. 3 timber piles 20' long in each cluster. At 475 feet from zero.	
3.26a-e	Bureau of Yards and Docks	Water, sewer, air, steam and electric lines, all underground. Extended be- tween 600 and 1800 feet from zero.	
3.27	Air Materiel Command	Two-foot concrete cubes, buried. Located adjacent to several structures	

APPENDIX I

A BRIEF HISTORY OF U.S. CIVIL DEFENSE ORGANIZATIONS

The term *civil defense* refers to all nonmilitary actions taken to reduce loss of life and property resulting from enemy action. (1) The history of civil defense in the US is confusing and has been significantly and continuously influenced by such issues as: Federal versus state, local, and personal responsibility and control; constantly changing enemy threats; the level of public fear and outcry; integration or lack of it with federal efforts for natural disasters; and (perhaps most importantly) financing. It was amidst such an ever-changing background of policies that effects measurements at the NTS were undertaken for civil defense purposes. This background is quite different from the stable environment and supportive posture toward the labs and AFSWP of the AEC and DoD during atmospheric testing.

During WWII, the US had a civil defense program that was established by Franklin D. Roosevelt in May 1941, by Executive Order, the Office of Civil Defense (OCD). (2) To head the new agency, FDR selected Farrell H. La Guardia for its director while he was still the mayor of New York City (1934-1946). (3) The OCD coordinated federal, state, and local defense programs. On June 4, 1945, before the end of WWII but when the public's fear of attack from Germany and Japan had greatly lessened, the OCD was terminated.

During the period between 1945 and 1949, not much was done regarding civil defense. However, the National Security Resources Board (NSRB) was created on July 26, 1947 by the National Security Act (NSA) of 1947. This act placed the NSRB under the Secretary who was responsible for defense. The act also created the National Military Establishment with three executive departments (Army, Navy, and Air Force) also under the Secretary of Defense. In 1949, the National Military Establishment was renamed the Department of Defense (DoD). (10)

Within the National Military Establishment, the NSRB had been responsible for Civil Defense planning duties. However, neither the Federal government nor the National Military Establishment considered civil defense as important as military defense. (4)

Truman believed that civil defense was a state and local responsibility, even though there was interest from these levels of government in Federal leadership. Limited defense budgets meant that the armed services had little interest in taking on additional responsibilities, and there was a widespread perception that the Soviet Union would not pose a nuclear threat until 1953 at the earliest. (5) On March 3, 1949, the NSRB was moved out of the military to the Executive Office of the President (EOP). (6)

Public perception changed after the detonation of the Soviet Union's first atomic device on August 29, 1949. Shortly afterwards, the start of the Korean War, and the intervention of forces of the Chinese People's Liberation Army in November 1950 intensified the public's concerns about civil defense. (5) In response, on December 1, 1950, President Harry Truman created the Federal Civil Defense Administration (FCDA) (EO10186) within what was called the Office of Emergency Management (OEM), attached to the Executive Office of the President (EOP).

Prior to December 1, 1950, The OEM had provided the President with a mechanism to monitor emergencies and disasters that affected the US, but it offered no direct assistance to state or local governments. Congress quickly passed the Federal Civil Defense Act of 1950 (64 Stat. 1245). As a result, on January 12, 1951, the FCDA was made an independent agency of the federal government, and absorbed the functions of what had been called the National Security Resources Board (NSRB). (7)

In the early 1950s, Civil Defense programs sought to develop sheltering capabilities to house people in attacked cities. Some of these programs were conducted at the NTS. However, civil defense planners were also developing mass evacuation plans for the supposed target cities. During this period, there were no intercontinental missiles; and it was thought by many that there would be time for the evacuation of city populations. The evacuation versus shelter approaches added a layer of controversy for the planners as well as for those trying to execute meaningful programs. (6)

Just prior to Operation UPSHOT-KNOTHOLE, Val Peterson, former Governor of Nebraska (1947-1953), was sworn in as Administrator of the Federal Civil Defense Administration on March 2, 1953. Prior to his nomination as FCDA Administrator, Gov. Peterson had served as administrative assistant to the President. Peterson was a key leader in the government's civil defense efforts during the years of participation at the NTS. (FCDA 1953a)

On June 12, 1953, the functions of the former NSRB were removed from FCDA, and along with programs for other disaster and emergency relief of the EOP were consolidated into a new Office of Defense Mobilization (ODM) housed within the EOP. (10) The FCDA concentrated solely on preparing the civilian population for a nuclear attack and the new ODM assumed all responsibilities related to domestic emergency preparedness and development of the nation's civilian capability to ramp up and go to war. (7)

The Soviet Union detonated its first thermonuclear explosion on August 12, 1953. This again fueled fears of the potential for a Soviet attack on the US. The development of intercontinental ballistic missile capability and the subsequent launch of the Sputnik satellite on October 1957 further intensified the public's fear and clearly demonstrated that there just would not be time to execute plans for the evacuation of cities. (6)

On July 1, 1958, the major civil defense and emergency preparedness programs at the federal level were once again reorganized. The FCDA and the ODM were consolidated into a single agency: The Office of Civilian and Defense Mobilization

(OCDM), which was to be housed in the EOP. The federal government provided 50/50 matching funds to personnel and administration costs for agencies engaged in civil defense preparedness. The concept of a joint federal-state-local responsibility for civil defense and attack preparedness was also articulated in guidance distributed by the new OCDC. (7)

In 1961, however, President John F. Kennedy, sensing that the overwhelming majority of state and local governments were doing little if anything to develop a sheltering capability, decided to make civil defense preparedness once again a central issue. Kennedy separated "civil defense" functions and other emergency preparedness functions into two agencies. Executive Order 10952 moved the civil defense functions into the Office of Civil Defense (OCD) within the Department of Defense; back where it had been in 1949. A full-fledged nationwide shelter program, funded by the federal government was developed, resulting in engineering studies of existing structures, and the acquisition and deployment of stockpiles of necessities for the shelters.(7)

One wonders how the people involved with civil defense measurements at the NTS ever managed to accomplish as much as they did!

APPENDIX I REFERENCES

(1) Encyclopedia Britannica CD1988, *civil defense.*
(2) Office of Civil Defense, http://www/gwu.edu/~erpapers/abouteleanor/q-and-a/glossary/office-civil-defense.htm, 5/28/03.
(3) Encyclopedia Britannica CD 1988, *La Guardia, Fiorello H.*
(4) Hollings, Fritz, *Online Office of United States Senator Fritz Hollings, South Carolina*, National Security Council
as Created in 1947, 5/28/03.
(5) Electronic Encyclopedia of CD and EM, http://www.richmond.edu/~wgreen/Ecdtruman.htm, 5/25/03.
(6) *The Documentary History of the Truman Presidency*, http://www.lexisnexis.com/academic/2upa/Aph/truman_docs/sampdocs/ sampdocs_26.htm, 5/25/03.
(7) *The History of Civil Defense and Emergency Management in Tennessee*, http://www.tnema.org/Archives/EMHistory/TNCDHistory3.htm, 5/25/03.

APPENDIX J

LOCATION OF TROOPS DURING ATMOSPHERIC TESTS

The following table lists the tests on which observers and troops were located in forward areas to observe a detonation. For each test, the table provides the:

- Yield of the test in kt
- Height Of Burst (HOB) in ft and km, and
- Distance from GZ (as measured on the ground) in km at which the closest trenche(s) for troops were located.

The Slant Distance From Detonation Location is the distance from the closest trench to the actual detonation location, which is at a distance of HOB above GZ. The HOB was usually small compared to the Distance from GZ. Therefore, the Distance From GZ and the Slant Distance From GZ are almost equal in the following table.

The first test on which a detonation was observed by the military was on BUSTER-JANGLE Dog, which had a yield of 21 kt, a HOB of 1417 feet or 0.43 km. Its Distance from GZ was 11 km. The Slant Distance From the Detonation Location was calculated (by using the Pythagorean Theorem - The square of the hypotenuse of a right triangle is equal to the some of the squares of its other two sides.) as 11.01 km.

"Cube Root Scaling" is frequently used for nuclear weapons effects. Cube Root Scaling is only an approximation in the real world. However, it is a useful approximation to use when trying to approximate the distances where nuclear effects are about the same from two tests of different yields. Cube Root Scaling can be expressed by the formula:

$$d1/d2 = (Y1/Y2)^{1/3}.$$

Suppose that you are a troop on TUMBLER Charlie and are sitting at a distance from GZ of 7 km, and that Charlie is at a HOB of 3447 ft = 1.051 km. Your Slant Distance From Detonation Location is 7.08 km. Charlie produces a yield of 31 kt. You want to know how far from BUSTER Dog at 21 kt you would have had to be in order to received the same effects as you just did from Charlie.

Let: d1 = ?, d2 = 7.08, Y1 = 21, Y2 = 31 to obtain:

$$?/7.08 = (21/31)^{1/3} \text{ or,}$$

$$? = 7.08 \times (21/31)^{1/3} = 6.22 \text{ km.}$$

You would have had to be at an Equivalent Slant Distance of 6.22 km on BUSTER Dog to receive the same effects as you did receive on TUMBLER Charlie at 7.08 km.

In the table, Cube Root Scaling is used to calculate the Equivalent Slant Distance from BUSTER Dog for each of the tests on which there were troops or observers in the forward areas.

	Yield	HOB	HOB	Dist from GZ	Slant Dist from Det Loc	Equiv Slant Dist On BUSTER Dog	Equiv Slant Dist On BUSTER Dog
	(kt)	(ft)	(km)	(km)	(km)	(km)	(mi)
BUSTER-JANGLE							
Dog	21	1417	0.432	11	11.01	11.01	6.84
Sugar	1.2	3.5	0.001	9	9.00	23.37	14.52
Uncle	1.2	-17	- 0.005	9.5	9.50	24.66	15.33
TUMBLER SNAPPER							
Charlie	31	3447	1.051	7	7.08	6.22	3.86
Dog	19	1040	0.317	7	7.01	7.24	4.50
Fox	11	300	0.091	7	7.00	8.68	5.40
George	15	300	0.091	7	7.00	7.83	4.87
UPSHOT-KNOTHOLE							
Nancy (Volunteer Officers)	24	300	0.091	2.29	2.29	2.19	1.36
Badger (Volunteer Officers)	23	300	0.091	1.83	1.83	1.78	1.10
Simon (Volunteer Officers)	43	300	0.091	1.83	1.83	1.44	0.90
Annie (Troops & Observers)	16	300	0.091	3.2	3.20	3.51	2.18
Nancy (Troops & Observers)	24	300	0.091	3.66	3.66	3.50	2.18
Badger (Troops & Observers)	23	300	0.091	3.66	3.66	3.55	2.21
Simon (Troops & Observers)	43	300	0.091	3.66	3.66	2.88	1.79
Encore (Troops & Observers)	27	2423	0.739	3.66	3.73	3.43	2.13
Encore with actual GZ	27	2423	0.739	3.48	3.56	3.27	2.03

Harry (Troops & Observers)	32	300	0.091	8.96	8.96	7.79	4.84
Grable (Troops & Observers)	15	524	0.160	4.57	4.57	5.12	3.18
	Yield	**HOB**	**HOB**	**Dist from GZ**	**Slant Dist from Det Loc**	**Equiv Slant Dist On BUSTER Dog**	**Equiv Slant Dist On BUSTER Dog**
TEAPOT							
Wasp	1	762	0.232	14	14.00	38.63	24.00
Tesla	7	300	0.091	2.22	2.22	3.20	1.99
Turk	43	500	0.152	5	5.00	3.94	2.45
Bee	8	500	0.152	3.2	3.20	4.42	2.75
ESS	1	-67	-0.020	8.23	8.23	22.71	14.11
Apple 1	14	737	0.225	3.2	3.21	3.67	2.28
MET	22	400	0.122	10	10.00	9.85	6.12
Apple 2	29	500	0.152	3	3.00	2.70	1.68
PLUMBBOB							
Priscilla	37	700	0.213	3.89	3.90	3.23	2.00
Hood	74	1500	0.457	5.03	5.05	3.32	2.06
Diablo	17	500	0.152	3.89	3.89	4.18	2.60
Kepler	10	500	0.152	3.02	3.02	3.87	2.41
Shasta	14	500	0.152	3.02	3.02	3.46	2.15
Doppler	11	1500	0.457	2.61	2.65	3.29	2.04
Smoky	44	700	0.213	4.02	4.03	3.15	1.95
Galileo	11	500	0.152	3.52	3.52	4.37	2.72
Newton	11	1500	0.457	4.48	4.50	5.59	3.47
Whitney	19	500	0.152	3.38	3.38	3.50	2.17

APPENDIX J REFERENCES

Troop locations from descriptions of Desert Rock exercises in Part II. Yields and heights of bursts (HOB) are from tables of tests in Part II.

APPENDIX K

1954 SHOT CRITERIA FOR NTS

On August 18, 1954, the AEC approved the 1955 nuclear test series, code named TEAPOT, proposed for Nevada. (AEC 1954: Meeting Number 1020, AEC707/5) The operation would be conducted under the criteria approved by the Commission on June 30, 1954 and set forth in AEC141/25*. [*Footnote: G. L. Felt and A. C. Graves prepared a paper, J-19996, "Criteria for Future Continental Tests," that formed the basis for AEC141/25, February 1954.]

The criteria of AEC 141/25 are as follows:

a. The number of nuclear shots at the Nevada Proving Ground in one year should be determined by laboratory requirements as reviewed by the Division of Military Applications in the light of other pertinent considerations and approved by the Commission.
b. Each nuclear shot programmed whether AEC, military or civil defense should be justified individually and the number involved should be held to the minimum consistent with technical requirements.
c. Each potentially hazardous shot should be separately identified and justification for such a shot should include plans for controlling or reducing fall-out from it.
d. Shots should be scheduled with more elasticity, so that non-critical shots may be fired when conditions are not right for more critical or marginal shots. Such elasticity will benefit from addition of new firing areas.
e. Marginal shots should be fired only under satisfactory weather conditions that have a high degree of predictable stability. The possibility of continuing postponements and of resulting extensions of series duration should be accepted. Participating organizations and units should be advised that they must accept the possibility of postponements on such shots.
f. Any airdrop of more than 1 kt projected yield should be scheduled only after thorough evaluation of the reliability of its fuzing system.
g. Shots should be limited as follows with regard to yield and burst altitude, with maximum yield to incorporate a reasonable allowance for error:
    a. Surface and subsurface, 1 kt
    b. 300 foot tower, 25 kt
    c. 500 foot tower, 50 kt
    d. Airdrop, 80 kt (Fireball not to touch ground.)

Prior to detonating a 50 kt weapon from a 500 foot tower the safety factor calculated for such a shot should be confirmed by detonating a shot of lesser magnitude from a 500 foot tower.

(TEAPOT-JDO-229, JOL-65-896, December 7, 1954, Los Alamos DX Archives; Reeves 1955:8-9)

Criterion c, (cited above) regarding the potentially hazardous shots, was addressed analytically using a formalism outlined in the Felt-Graves paper entitled “Criteria for Future Continental Tests.”

D, the peak integrated dose in roentgens at the center of the fallout area on the ground, is proportional to the hazard, H.

$$D \text{ is proportional to } 4H \times (\Delta V/\Delta h) = K(V,h)H$$

Where ΔV is the difference in wind speeds between the top of the cloud and 10,000 feet, and Δh is the difference between the height of the top of the cloud and 15,000 feet. K was referred to as the “multiplying factor.” (James E. Reeves and Alvin F. Graves, Operation TEAPOT, JDO-229, JOL-65-896, December 7, 1954, Los Alamos DX Archives)

Reeves and Graves reproduced plots of the Hazard versus height of burst for yields from 100 tons to 100 kt from the paper by Felt and Graves (ibid.). For example a 24 kt device, Nancy, on a 300 ft tower would have an H of 1.2. A 32 kt device, Harry, at 300 ft. would have an H of 1.7. Both Nancy and Harry were fired during Operation UPSHOT-KNOTHOLE.

The multiplying factors from the actual meteorological and cloud height data are 3.85 for Nancy and 7.85 for Harry.

The measured hot spot doses were 5 roentgens for Nancy and 12.5 roentgens for Harry. The predicted hot-spot doses, knowing the wind data and cloud heights, gave Ds of 4.6 and 13.3 roentgens, respectively. This excellent agreement is probably largely fortuitous, but it does demonstrate the significance of the most important meteorological factors on the hot spot dose.

Reeves and Graves assumed a very conservative multiplying factor of 10 for the TEAPOT shots. The maximum hazard, H, calculated for the TEAPOT shots was 1. The five highest predicted doses for the shots proposed for TEAPOT were 10, 10, 10, 9, and 2 roentgens. The first four were considered critical from the off-site fall-out point of view.

ESS was a 1 kt weapon effects crater shot at a relatively shallow burial depth. The analysis just presented is not applicable for buried shots and it was necessary to examine the JANGLE tests for relevant guidance. Col. H. E. Parsons, of AFSWP, analyzed the previous experience and concluded that if the radioactive cloud height were less than 4500 feet, the significant fallout would occur within two and a half hours, and that if the wind speeds were less than 20 knots up to 10,000 feet no significant fallout would occur beyond fifty miles. The test organization agreed with these conclusions and set as criteria for ESS wind speeds of less than 20 knots and wind directions away from population concentrations.

All of the proposed TEAPOT devices were analyzed and divided into two groups depending upon the stringency of the weather requirements. For the so-called critical shots, from a fallout perspective, the predicted debris cloud trajectory was required to be within 30 degrees of north. This avoided Hiko to the northeast and Tonopah to the northwest. It was also highly desirable to have steady, low-velocity winds to maximize the fraction of the fallout that would end up on the test site, rather than off-site.

The weather criteria for shots in the non-critical category were less stringent and closely resembled the criteria prior to TEAPOT.

In all cases steady low velocity winds were desired in order to restrict the fallout as much as possible to the site.

It was clearly recognized at the time that these criteria could cause lengthy delays and inconvenience for the test participants, but that was accepted as a cost of doing business.

There were a total of 122 formal meetings to evaluate the meteorological conditions and test readiness for the 12 shots actually fired. Fifty-four postponements resulted from the predictions of weather conditions that would be unsuitable for the nuclear detonations. Four postponements were due to aircraft operational problems and three postponements were due to difficulties with key scientific diagnostic equipment. In retrospect no good firing days were missed due to faulty weather predictions. (Reeves 1955: 21)

Criterion d required “elasticity” in the schedule, which meant that non-critical shots could be fired when the weather conditions were not acceptable for the critical shots but were acceptable for the non-critical events. This, of course, complicated the planning and participation in the various shots. Never the less this *modus operandi* was followed quite successfully. In fact the planning was such that two shots, Apple 1 and Wasp Prime, were fired on the same day (March 29, 1955). As a result of having non-critical backup shots for the critical events the overall test series was shorter than it would have been had the previous practice been in place.

Criterion e required weather of “predictable stability” to be acceptable for critical shots. Significant steps were taken to improve the weather forecasting closer to shot time. Previously the forecasts, using data that could be twenty-four hours old, were eight to twelve hours old at shot time. Under the new protocol the forecasts would be only one hour old at shot time with meteorological data taken up to that time.

Criterion f required a “thorough evaluation” of the reliability of the fuzing system of any airdrop of over 1 kt. This was addressed by Sandia Director of Field Test R. A. Bice on December 6, 1954 in a letter to Santa Fe Operations Manager D. H. Leehey. (Sandia 5200(304))

Steps were taken to develop substantially improved off-site emergency response plans in order to: (1) gain and retain public confidence; (2) improve off-site warning and emergency response capability to address unanticipated fallout, and (3) collect improved fallout and contamination data. The AEC and the USPHS collaborated in greatly expanding the off-site monitoring capability and the liaison with public officials out to approximately 250 miles from the test area.

By the time of the August 25, 1954 Test Director's meeting the preliminary schedule reflected a reasonably accurate picture of the tests actually fired in TEAPOT. (Clark 1955:16)

Clark informed Reeves, in October 1954, that:

> The Air Operations Officer had proposed to install an IFF (Identification Friend or Foe) Control Facility in the Air Control Room at the Control Point to permit sampler aircraft operations in poorer weather conditions than on previous continental weapons tests. This facility will function much as did the Air Operations Center (AOC) on Enewetak during Castle. In addition, the IFF facility will provide visual indications of aircraft in the H-hour array.

(Memo from John C. Clark to James Reeves, Test Division SFOO, October 14, 1954)

APPENDIX K REFERENCES

Are cited within this text and found under the REFERENCES heading.

APPENDIX L

PUBLIC RELATIONS AND INFORMATION

Public Relations (PR) and the release of information regarding nuclear tests in Nevada were important activities from the beginning, and continued as such throughout nuclear testing. The public relations mission during atmospheric testing was:

a) To support national policy by accurately informing the public in the US and by helping to create a favorable climate of opinion in the world at large.
b) To increase public knowledge and understanding of the purpose and need for continental tests; help protect life and property by obtaining public cooperation to reduce hazards; allay unfounded fear of damage or injury that may arise from public misunderstanding; and to protect classified data while answering questions from the public; and
c) To meet the public information requirements of the participating agencies. (The FCDA had the greatest requirements in this area and played a very prominent role in PR activities during UPSHOT-KNOTHOLE, TEAPOT, and PLUMBBOB.)

To address this mission, activities were conducted in three areas:

1) Pre-operation public education programs.
2) Coordination of activities in public affairs, public information, official visits, and official briefings.
3) Conduct Open Shots.

During non-operational periods, the AEC Albuquerque Operations Office, ALOO, had responsibility for activities at the NTS and for the release of test-related information to the public. During periods of test operations in Nevada, operating responsibility was assigned to the Nevada Test Site Organization, NTSO.

TEAPOT

Although PR activities were conducted during UPSHOT-KNOTHOLE, during TEAPOT and PLUMBBOB these activities would be even more extensive and better coordinated among the user groups. A Joint Office of Test Information (JOTI) was activated on February 1, 1955, prior to the first TEAPOT test, and was deactivated May 18, a few days after the last test. Approximately 25 individuals served in this office during TEAPOT. It was headed by a Director appointed by the AEC, a Deputy appointed by the military, and personnel from the participating groups. All formal issuances were prepared and distributed through the Las Vegas office, JOTI, and all press contact was there. Records were maintained at the Las Vegas Office, and the photo file was maintained at the Mercury Office.

Prior to the operation, public education was conducted in Nevada and nationally. It was based almost entirely on activities that were recommended by the "Committee to Study Nevada Proving Ground." Twenty thousand booklets, the "A-B-Cs of

Radiation," were distributed to schools during the autumn of 1954. In addition, films such as the USAF's "Target Nevada" and "Atomic Test in Nevada" became publicly available and were used in civic club and TV presentations in Nevada and the surrounding states. Also, about 50,000 booklets, "Atomic Tests Effects in the NTS Region" were distributed to the public. The US Weather Bureau produced a comprehensive article on Nevada tests and weather that was widely circulated. Doctors, veterinarians, public officials, and civic leaders in all surrounding regional communities were contacted. Finally, test officials headed by the Test Manager and the Scientific Advisor visited NTS communities to meet with civic leaders and speak at community meetings.

AEC public relations people met with the press for a briefing on each test as shot day approached. Other meetings were held when there was a media or a Test Organization requirement. Plans for weather briefings and the results were announced immediately. The Mercury and Las Vegas offices were staffed at all times during the night readiness meetings. Dissemination of relevant information to nearby communities was through the radiation monitoring organization.

Members of the Test Organization were interviewed; visits for the media were arranged to Mercury, Desert Rock, or Indian Springs; and public addresses including radio and TV appearances were made. JOTI issued 103 press releases during TEAPOT, exclusive of the Open Shot. Official photographs of personnel, equipment or installations were issued upon media request. Photographs of each shot and the resulting cloud were issued along with photographs of official visitors.

A Visitor's Bureau with offices in Mercury and Las Vegas was established to provide a program for observers and visitors. This included: reception, billeting, arrangement for security clearance and badging, providing briefings and tours, and the many "additional services" visitors need to make their stay "as agreeable as possible." Visitors included AEC and AFSWP personnel, invitees of the Test Manager, special military groups, FCDA observers, Congressional observers, contract employees, and the press. On days when shots were postponed, recreational trips were often conducted. The following table provides the numbers of observers during TEAPOT whose activities were coordinated by the Visitor's Bureau.

	***Official**	***Employee**	**Special Military Groups**	**Total**
Badges Processed For Observers	669	**939	1530	3138
Visitors Attending Briefing and Tour	457	214	***660	1331
Visitors Witnessing Shot	286	207	***1025	1518

* Visitor's Bureau direct responsibility.

** Includes miscellaneous visitors not in organized groups.

*** Briefing and field tours by military.

(Reeves 1955:85-91)

NTS PUBLIC INFORMATION POLICY
The AEC and the Department of Defense got together in September 1956 to draft a "Standard Operating Agreement for Field Conduct of Nevada Nuclear Test Public Information." The purpose of the document was:

> To state the policies, procedures, and organizational responsibilities which govern Nevada nuclear test public information in order to achieve a fully-integrated, joint-agency organization and program; with preplanned, positive, mission-supporting action; with uniform information control and release procedures; and with officially-released material reflecting the interests and contributions of all participants.
>
> ...
>
> Continental nuclear testing is a public-financed government operation involving some degree of public hazard, resulting in a very real public impact, and requiring public acceptance for its continuance. Public understanding and acceptance of the need for continental tests and their accompanying off-site effects is advanced by keeping the public and its representatives promptly and continuingly informed of the necessity, value, nature, and conduct of the test operation. The obligation to inform the public promptly and as fully as classification and security permit is recognized as a basic concept.

(Department of Defense/AEC "Standard Operating Agreement for Field Conduct of Nevada Nuclear Test Public Information." September 14, 1956)

PLUMBBOB
The JOTI was activated briefly for the period April 8 - 24, 1957, by only AEC-ALOO personnel, for the initial safety experiment, Project 57. The joint office, JOTI, was activated May 1 for the full-scale phase of PLUMBBOB and deactivated October 8.

Since the Bravo test of Operation CASTLE in 1954, the biological, political, and moral effects of fallout from high yield tests had been widely discussed. Most of it was critical of the AEC in particular and the United States in general. There was growing sentiment that nuclear testing should be stopped. There were only limited official efforts to explain the necessity for tests and the nature of fallout. During the year proceeding PLUMBBOB, the facts regarding fallout and public health and the question of future testing were hot topics.

Two events were particularly noteworthy in the spring of 1957. One was an article in the Reporter Magazine that criticized testing. The second was the hearings of the Joint Committee on Atomic Energy (JCAE) in the US Congress that were underway. These hearing were airing seemingly authoritative reports critical of testing.

PLUMBBOB began at the end of May and continued for some weeks in an atmosphere of almost panic and fear of fallout and demands that all US testing be stopped. Large segments of the public, their opinion leaders, and officials apparently shared these feelings. Some of this was presumably politically inspired, but much of it was an understandable response of people generally to the charges they had heard and read for more than three years without hearing or seeing equally effective rebuttals.

When the Test Organization became fully activated, an even more vigorous public education effort than that undertaken for TEAPOT was conducted. Booklets were distributed, briefings and reporters' tours were conducted, local as well as national reporting was conducted, and USPHS personnel arrived in nearby communities. Also, PLUMBBOB began with a series of first shot postponements, dramatizing the Test Organization's caution. The actual start of tests also helped to turn news and public interest to phases other than fallout.

By early June, there were more favorable statements made in the JCAE hearings. By late June, the possibilities of a "clean bomb" were being becoming public. Although a "clean bomb" test was not a part of PLUMBBOB, the possibilities seemed to help alleviate fear. Also, there were many statements by prominent scientists and individuals that helped to put the fallout and testing questions into better perspective.

As PLUMBBOB progressed, continued safe operation and proof of the new balloon technique were major factors contributing to the relaxation of public fear. However, it may be noted that there were several instances in PLUMBBOB of fallout approaching close to the 3.9 roentgen criterion, levels which in prior series had attracted wide attention or had resulted in public expressions of concern.

The previous NTS operations (UPSHOT-KNOTHOLE, and TEAPOT) had conducted only one Open Shot. For PLUMBBOB, initially, nine shots were designated and scheduled as Open Shots for on-site observation by newsmen and various other uncleared observers. At the insistence of newsmen and on the recommendation of USAF, the air-to-air rocket shot, John, was added as an Open Shot shortly before the event. Newsmen also wanted the underground test, Rainier, to be an Open Shot, but authorization was not given.

When news media were in the area for an Open Shot which was set aside in favor of an alternate shot, NTO was authorized to admit them to the alternate shot being conducted in addition to the designated shot which would be detonated at a later date. This resulted in a "bonus" Open Shot. In late August, the authorization was changed; and the bonus Open Shot replaced the initially designated Open Shot.

The final count was 11 Open Shots plus the Diablo misfire during PLUMBBOB. The Open Shots, (number of news media registrations, number of news media attendance) were: Boltzman (62,40), Diablo Misfire (166,65), Hood (166,82), Diablo (166,13), John (52,22), Kepler (11,7), Stokes (14,10), Shasta (12,6), Smoky (49,42), Galileo (9,8), Fizeau (12,12), and Newton (28,27). The difference between registrations and attendance was certainly influenced by the weather delays encountered. Also, the generally decreasing numbers as the operation progressed were attributed to news media budget limitations and to steadily decreasing interest.

All six safety experiments were closed, and newsmen were not admitted on-site to 13 full scale shots. The full scale shots were announced in advance and newsmen viewed them from Las Vegas or from Angel's Peak in the Spring Mountains, with direct line-of-sight to NTS firing areas, 35 miles to Frenchman Flat and 55 miles to Yucca Flat. (Reeves 1980: Part IX, Cpt 8)

HARDTACK II

The OTI for HARDTACK II was activated by the AEC on September 12, 1958 the day of the first test of the operation, which was a safety test, Otero. The OTI offices at Las Vegas and Mercury were staffed at minimum strength, largely by AEC-ALOO personnel, ranging from a minimum of 4 to a maximum of 8. It was not a joint effort as was the JOTI that was formed during TEAPOT and PLUMBBOB with significant participation by DoD and FCDA. The OTI was just AEC.

The Las Vegas and Mercury offices provided 24 hour-a-day (probably an answering service was used), seven-day-a-week, service. The offices were staffed throughout normal business hours, including Sundays and holidays. As for the previous operations, the staff were on duty during the hours spanning each evaluation meeting and at shot times.

Newsmen observed on-site the 19 full-scale detonations and one safety experiment. During the series, newsmen were admitted to News Nob in Yucca Flat, to an observer area in Frenchman Flat, and to two observer sites in the tunnel area. Except for the Frenchman Flat site, all observer areas were equipped with nearby telephones, a nearby direct telephone to OTI-Control Point, toilets, benches, and tables. Government goggles were made available for the observation, and a snack bar sold beverages and food.

Twenty news media representatives were on-site observers for the first full-scale shot, Eddy. Newsmen's interest in on-site observation diminished gradually, as expected, during most of the following 17 full-scale shots, then increased to 16 for the last shot of the series, Blanca, on October 30.

Basic educational materials, booklets, films and pamphlets were distributed. Public safety education activities were carried out in the NTS region by AEC, USPHS and contractor personnel, and pre-shot flash warnings were issued for high-altitude balloon detonations.

At the invitation of the AEC, residents of the nearby regions visited the Test Site. They toured Jackass Flats and Frenchman Flat, visited Control Point 1, received a detailed briefings, and were housed in Mercury overnight.

In view of the proposed Gnome Project at Carlsbad New Mexico, about 24 New Mexico residents (civic, news, industrial, and mine worker people) accepted an invitation and visited NTS for briefings and tours. This provided the attendees with a better understanding of the NTS test program and the planning and procedures which would be used for Project Gnome. (Reeves 1958: 119-121)

APPENDIX L REFERENCES

Reeves, James E., "Report of the Test Manager Operation HARDTACK Phase II, August to October 1958", Deleted Version, Office of Test Operations, Albuquerque Operations Office, Opennet: NV0091951, 1958.

Reeves, James E., "OPERATION PLUMBBOB Report of the Test Manager Nevada Test Site May-October 1957", AEC, Extracted version, Opennet # NV0014461, 1 February 1980,

APPENDIX M

DESERT ROCK VI

This discussion uses only the reference: Ponton, Jean, Carl Magg, Martha Wilkinson, Robert Shepanek, "Operation TEAPOT 1955", Defense Nuclear Agency, Washington, D.C., DNA 6009F, 23 November 1981.

Desert Rock VI consisted of three programs: 1) Observers, 2) Troop Maneuvers, and 3) Technical Service. The Observers and Troop Maneuvers were fairly similar to those on past operations. However, new projects for Desert Rock Exercises that were of a more technical nature were conducted in the Technical Service programs.

<u>1) Observers</u>
This program was similar to those on past Desert Rock exercises and had about 4,600 participants. It was by far the largest program in Desert Rock VI. Participants attended lectures, films, and tours of equipment display areas to prepare for the observation of a nuclear detonation. After the shot, the participants viewed the equipment to witness the damage caused by the detonation. Desert Rock observers participated on eight shots: Wasp, Moth, Tesla, Turk, ESS, Apple 1, Met, and Apple 2. There were also about 500 non-Desert Rock military observers from the various branches of the services on Bee.

<u>2) Troop Maneuvers</u>
This program was designed to demonstrate and test military tactics, techniques, and doctrine developed for use with nuclear warfare. While many aspects of the Troop Maneuvers on TEAPOT were similar to those on past operations, the TEAPOT maneuvers used more equipment to assess nuclear effects and communicate in the hostile environment. Five projects were conducted during the Troop Maneuvers:

- The Army Demolition Munitions (ADM) project emplaced the device which was an ADM for ESS.

- On all of the tests except Wasp, HA, and Zucchini, about 50 troops took part in a project aimed at locating the atomic burst and determining its yield. The equipment tested included cameras, Bhangmeters, radar sets, and microphones. The participants proceeded to predetermined locations before the shot and set up 8 to 10 instrument stations about 10 to 13 kilometers south of GZ. The stations were placed in the typical pattern of an artillery observation battery in the field and were manned during the shot.

- In the area of Apple 2, a system of communication stations was installed and tested the day before the shot. It consisted of one regimental and three battalion communications command post networks. The posts were unmanned during the detonation. After the shot, the conditions at the posts and the capability to re-establish communications was tested.

- On Apple 2, an Armored Task Force Exercise code named RAZOR was conducted. The task force marched from Camp Irwin, CA, to the NTS where they bivouacked. On shot day, a full tactical exercise was conducted using tanks, armored personnel carriers, and a helicopter airlift of armored infantry troops in support of an assault. After the exercise the task force marched back to Camp Irwin.

- The Third Marine Corps with troops from various Marine Corps commands sponsored a large troop test at shot Bee with nearly 2,300 troops. This was the largest single project of the TEAPOT series. It provided realistic training in planning and conducting a military assault operation and in air-to-ground task force missions following a nuclear detonation.

3) Technical Service
The numbers of participants was usually not cited for Technical Service. The Technical Service activities were new to the Desert Rock exercises, and projects were of two types: 1) exposures (where objects were placed in the field at different locations and examined after the test for damage) and 2) the use and assessment of instrumentation that would provide information about the nuclear environment. The exposure activities were not too dissimilar from those that had been performed in AFSWP's Nuclear Weapons Effects Projects. Nine projects were undertaken on TEAPOT. These projects are described below.

Navy Passive Defense Training - consisted of mostly civilians drawn from various shipyards across the U.S. and Pacific. These shore-personnel were trained in monitoring operations, tested Radiac equipment and developed organizational units for passive defense. They observed Bee from a trench located 3.2 km SW of GZ and ESS about 8 km from GZ. Postshot, they participated in Emergency Recovery operations at intervals over 3 days.

Chemical, Biological, and Radiological Defense Shelters - evaluated chemical, biological, and radiological protection methods being developed for use in field bunkers (located 420 m from each GZ) and foxholes located between 450 and 1,800 m from each GZ. This appears to be a postshot operation "--individuals were probably responsible for - preparation and retrieval of film badges and dosimeters and for conducting postshot inspections of the bunkers and foxholes."

Engineer Field Fortifications and Equipment - New designs for field works designed for conventional warfare were exposed to nuclear battlefield conditions. Twenty structures (9 gun emplacements, 7 shelters, 2 bunkers, and 2 domes) were built at distances of 300, 345, and 420 m from MET GZ.

Engineer Heavy Equipment - Trenches were also constructed below surface level at distances of 480, 630, and 8,100 m into which tractors, grader, truck-mounted air compressors, cranes, and generators were placed. Inspections and evaluations were made postshot.

Effects on Steel Transporters or Containers - Cargo packaged in different types of containers were exposed at 6 distances on MET and 3 distances on Apple 2.

Damage Effects Evaluation - Camp Desert Rock Support troops assisted in the post shot evaluation of cargo in the different transporters and containers.

Sixth Army Chemical, Biological, and Radiological Defense Team Training - exercised troop teams in performing radiological surveys after nuclear detonations. A team included: an officer, a radio operator, a recorder/plotter, and 4 survey parties of 1 monitor, one driver/radio operator, and one soldier for special assignments. Aerial surveys were also conducted on 4 shots.

Clothing Test - exposed American, Soviet, and Communist Chinese protective clothing to withstand thermal radiation. Three mannequins were used at each of 3 stations fitted with: chemical warfare gas capes, reflective barriers, and standard ponchos.

Ordnance Vehicular Equipment Test - exposed vehicles at 11 locations between 240 and 1,110 m to: examine the effectiveness of roll-over safety bars, gather data for future ordnance equipment, and investigate the radiation shielding effect of armor. M48 tanks, M59 armored personnel carriers, T97 self-propelled guns, ¼ ton jeeps, 2 1/1 ton M211 cargo trucks, and 5 ton cargo trucks. "Project personnel were not required to be in the test area at the time of detonations."

DoD Operational Training Projects

The DoD also conducted operational training projects that were similar to those of Exercise Desert Rock in two respects: 1) Their primary objectives were to test service tactics and equipment, and 2) They were planned so they would not interfere with the AEC diagnostic and DoD military effects tests. However, unlike Exercise Desert Rock, these projects consisted of far fewer participants, and they were under the direction of the JTO and AFSWP. The Director, Weapons Effects Tests had the overall responsibility for implementing and coordinating these projects with the participating armed services: Navy, Strategic Air Command, Tactical Air Command, Air Force Cambridge Research Center, Air Defense Command, Air Force Office of Assistant to Atomic Energy, and Marine Corps Fleet Marine Force Pacific.

APPENDIX N

DESERT ROCK VII AND VIII

This appendix describes in some detail the Desert Rock activities that were conducted during Operation PLUMBBOB. From BUSTER-JANGLE to PLUMBBOB, the Desert Rock exercises had evolved and expanded in terms of scope and complexity. However, there were also many similarities in Desert Rock I through VIII. To describe each Desert Rock exercise in the detail provided here for PLUMBBOB would result in unnecessary repetition. This section describes Desert Rock VII and VIII in more detail than was done for the operations after BUSTER-JANGLE and before PLUMBBOB. A comparison of the PLUMBBOB exercises with those of BUSTER-JANGLE indicates the general trend of the troop exercises toward more complexity and the evolution of other activities toward a more technical nature.

The descriptions for Desert Rock VII and VIII provided here were obtained from Unclassified references. The original military records were not researched by the authors for Desert Rock, but the references used and cited at the end of this appendix did use the military records.

The references used here were written between 1981 and 1983, some 24 years after PLUMBBOB. In the early 1980s, considerable research was being done on the history of Desert Rock exercises, in particular on how many and which troops were where and when. The references cited in the footnote do show some differences in the estimated numbers of people involved in some of the projects. The numbers or information shown here were taken from the reference which generally covers that shot in the most detail.

The Desert Rock projects conducted during PLUMBBOB were of five general categories: 1) Troop Maneuvers; 2) Troop-Observer Indoctrination; 3) Training Projects (TP); 4) Technical Service Projects (TSP); and 5) Operational Training Projects (OTP). These projects, cited in Table N1, were organized into Programs 50 - 53 that represented each U.S. service. The Canadians named their projects Operation BOBCAT. The titles of these projects are descriptive of their activities.

Table N1 Military Programs Conducted During PLUMBBOB

PROGRAM	PROJECT #	TITLE
U.S. Army Program 50	50.1	Troop Maneuvers - Army - Task Force WARRIOR & Task Force BIG BANG
	50.2	Troop-Observer Army Indoctrination Program
	50.3	TSP - Evaluate detonation detection and cloud tracking systems.
	50.4	TSP - Water Decontamination - Evaluate solubility in water of bomb debris and procedures for decontamination
	50.5	TSP - Evaluate Shielding (from fallout and neutron induced activities) for Heavy Equipment and evaluate effective uses of this equipment for decontaminating land areas.
	50.6	TSP - Protection Afforded by Field Fortifications. 27 unmanned fortifications were constructed (machine gun emplacements, various fox-holes, and "hasty" shelters), some were instrumented.
	50.7	TSP - Test of Ordnance Material - Armored vehicles, rocket parts and shell fuzes exposed to blast, thermal, and radiation.
	50.8	TSP - Detection of Burst and Fallout - fallout prediction and equipment testing.
	CBRTP	TP - Chemical, Biological, and Radiological (CBR) Defense Training
U. S. Navy Program 51	51.1	OTP Rad-Safe Monitoring Training
	51.2	Navy Aircrew Troop-Observer Indoctrination Program
	51.3	OTP - Heavy Attack Indoctrination - combat crews observe detonation and fly a simulated bomb run on a target offset from GZ at zero time.
U.S. Marine Corps Program 52	52.1	Troop Maneuvers - Marine Brigade Exercise
	52.2	Troop-Observer Marine Indoctrination
	52.3	OTP - Marine Fly-By Indoctrination - combat crews observe detonation
U.S. Air Force Program 53	53.1	OTP - Air Sampling Mission - Training Air National Guard Units to gather samples.
	53.2	OTP - Ground Motion Studies - using aerial photography.
	53.3	Air Force Troop-Observer Indoctrination Program
	53.4	TP - Radiological Defense Training for radiological monitoring under conditions similar to a nuclear battlefield.
	53.5	OTP - Early Cloud Penetration - witness detonation and penetrate cloud.
	53.6	OTP - SAC Aircrew Training
	53.7	OTP - Indirect Bomb Damage Assessment (IBDA) - of the IBDA B-58 equipment.
	53.8	OTP - Evaluate other IBDA equipment.
	53.9	OTP - Photographic Reconnaissance Training
	53.10	OTP - Passive Defense - detection of large scale disturbances in upper atmosphere by passive means - from detonation of nuclear weapons, super or hyper sonic vehicles, or missiles.
Royal Canadian Army and Air Force	No Numbers	Troop Maneuvers Operation BOBCAT I-IV Troop-Observer Indoctrination Program Radiological Teams

(Harris 1981b: 81-83, 182-186)

The rest of this appendix provides brief descriptions of the activities conducted for these projects. It is emphasized that the numbers given regarding the number of personnel are just approximate and obtained from just the references used for this appendix. They are given here in order to provide the reader with a feeling for the magnitude of the activities.

1) TROOP MANEUVERS

The Army and Marines were the only U.S. forces involved with the troop maneuvers. In addition, the Canadians participated with the U.S. troops on the Smoky maneuvers and the pre-Smoky observations of Stokes and Franklin Prime. Troop maneuvers were conducted to provide battlefield training in actual nuclear environments.

When feasible, participants of both the Troop Maneuver and the Troop Observer Indoctrination Projects would observe a shot from trenches. There were six sets of trenches dug for the PLUMBBOB shots, and three sets left over from the TEAPOT Desert Rock VI exercises that were also used on PLUMBBOB. These are cited in Table N2 along with their distance from GZ. (Harris 1981b, 103) On Operation BUSTER-JANGLE the soldiers dug the trenches. However, it was such a difficult task that the powers that be decided that for subsequent operations the digging would be handled by earth moving equipment. (Ristvet, verbal communication April 2, 2003)

Table N2 Trenches used by Troop Units During PLUMBBOB Shots.

TRENCH	USED FOR SHOT:	TOTAL TRENCH YARDAGE	DISTANCE FROM GZ YARDS
T 1	Priscilla	1200	4250
T 2	Hood & Shasta	2349	5500 & 3300
T 3	Diablo	2500	4250
T 4	Doppler	760	2850
T 5	Smoky	950	4400
T 6	Newton	Not Available	4900
T 7	Galileo	2060 [Teapot trench]	3850
T 8	Kepler	1100 [Teapot trench]	3300
T 9	Whitney	540 [Teapot trench]	3700

The three maneuvers conducted on PLUMBBOB were: 1) The Marines on Hood; 2) Task Force WARRIOR with Army troops and AFSWC airlift support on Smoky; and 3) Army Task Force BIG BANG conducting Project HumRRO on Galileo. The Galileo HumRRO activities were not classified as an actual maneuver, but it had many of the same features. These three maneuvers are described next.

Hood Troop Maneuvers

On Hood, the Marines conducted the largest troop maneuver exercise ever performed at the NTS. It engaged 2,025 Marines and about 300 members of the Army's Camp Desert Rock Support personnel who drove the troops to and from the forward area. The maneuvers were originally planned to take place during Diablo scheduled for June 27, 1957. It was planned that some of the troops would take part in a Command Post Exercise (CPE) and others in a Troop Maneuver. The CPE was to start at noon on the day before the shot and end no later than one hour before the detonation. Participants in the CPE would then join the rest of the Marines in observation trenches. Following the detonation, CPE personnel were to be transferred to an equipment display area to view the effects of the nuclear detonation on a wide variety of Marine equipment and uniforms.

The CPE began at 1330 hours on June 26 and completed at 0130 on June 27. The June 27 firing of Diablo was delayed until June 28 because of weather. On June 28, the CPE personnel were transported to the trenches to observe Diablo with the rest of the Marines who were to take part in the Troop Maneuver. At shot time, 0430, Diablo failed to detonate because of electrical problems, and the Marines returned to Camp Desert Rock.

Diablo was rescheduled for July 12, 1957. Because this would have prolonged the Marines' stay, it was decided to perform the troop maneuver on Hood, scheduled for July 3. The equipment display was transferred to the Hood GZ area, and the trenches that were dug for use on Shasta were deepened (because of Hood's higher yield) to about 6 feet.

The CPE exercise was not repeated on Hood. The troop exercise began at 2230 on the second of July. Half of the CPE personnel were trucked with the maneuver troops to the trenches and half were sent to the Command Post area to observe Hood. The maneuver troops left Camp Desert Rock at 2233 hours on July 2 in a five-unit march-column. The last unit arrived at the trench area at 0032 on July 3. The trenches were located about 5.5 km from Hood's GZ, see Figure N1. The firing of Hood was then delayed until 0440 hours on July fifth. The script just described was repeated starting at 2230 hours on the fourth.

On the fifth, after arrival again at the trenches, the personnel rehearsed trench procedures. They would take position in the assigned trenches 23 minutes before the detonation. They would don their gas masks and crouch 2 minutes before the shot. There were also additional observers who were not in trenches at the Vehicle Assembly Area and at News Nob, which is about 20 km south of Hood.

Hood was detonated at 0440 on July 5 at 74 kt (the NTS's largest yield in the atmosphere). "The heat of the detonation ignited many brush fires, and the shock caused some of the trenches to collapse; however, there were no serious cave-ins or personnel casualties. In the mining communities north of the NTS, windows shattered and the buildings shook. The light from the detonation was seen in San Francisco, and the blast was felt in Los Angeles." (Maag 1983: 31)

Fifteen minutes after the detonation, the maneuver troops left their trenches. One company, F, marched west to Helicopter Loading Zone Two in order to wait for the helicopter airlift. Company H waited in the trench area for the truck convoy that would transport it to the Vehicle Assembly Area, behind the Command Post in Figure N1. Company E formed two columns and marched NE toward GZ. They were preceded and accompanied by radiological safety monitors. They stopped at a distance that has been reported as being between 370 and 1,000 meters form GZ. They spent 5 to 10 minutes in the area then marched back to Helicopter Loading Zone Two, near the observation trenches.

Due to dust that obscured visibility in Landing Zone Two, the helicopters delayed their departure (to pick up company F) from Yucca Pass one hour. At 0615, the helicopters landed at the loading zone and began the airlift of Company F and the battalion command elements (who had also been in the trenches). After Company E returned from their march toward GZ, they also joined the airlift. Elements of Company G boarded helicopters at Loading Zone One which is near News Nob.

Each helicopter had a crew of 3, and each carried 5 to 7 Marines. They carried Company G to Landing Zone Blue. Companies E and F were flown to Landing Zone Pink, see Figure N1. The airlift from Loading Zone Two to the two landing zones was completed by 0821. By 0900, the convoy with Company H had linked up with the helicopter force in the area of the Objective.

During the helicopter airlift, air support was provided by 24 F9F Marine aircraft, which were each flown by one pilot. The first of these aircraft arrived at the NTS at 0510. The fighters, flying in groups of four, shuttled between Mojave, CA and the NTS for approximately seven hours.

Upon arrival at Landing Zone Pink, Company F joined Company H in the attack against the Objective, while Company E remained at Landing Zone Pink to back up the attacking companies. Upon landing at Landing Zone Blue, the remaining unit, Company G, also attacked the objective.

According to one participant, Company G had about 30 "casualties" due to heat exhaustion. These men were taken to a special assembly area by helicopter and allowed to rest for one hour. They then marched to the helicopter landing zone to wait for the end of the assault.

While the tactical exercise was taking place, about 300 Marines toured the equipment display area. Upon completion of the helicopter lift and before returning to Camp Desert Rock, 90 pilots and crew members were taken through the display area.

The tactical exercise concluded when the Objective was seized at 1030,.

Smoky Troop Maneuvers

The troop maneuvers on Smoky were conducted by Task Force WARRIOR as Army Project 50.1. WARRIOR was intended to test, and to a lesser extent, to provide information, for the development of infantry air-landed tactics and techniques for the atomic battlefield. In the late 1950s, the Army restructured its fighting forces primarily for nuclear-weapons-supported warfare on the plains of Europe. Smoky, which was originally planned for August 19, provided the Army with a unique opportunity to exercise a restructured Army fighting force. (ibid., 26)

The WARRIOR soldier and helicopter units arrived at Desert Rock during late July and trained in air-landed operations. The soldiers prepared about 115 defensive positions on August 12 and 13, which ranged from 820 to about 1,850 meters W and N of GZ, see Figure N2. Communications equipment, vehicles, and weapons were installed in and around the defensive positions. These items were examined after the shot to determine how effectively the prepared positions protected the equipment. The pre-shot activities were completed by Friday August 16. But, due to contamination from Shasta fired on August 18, Smoky was delayed.

During the delay, as well as during the interval between their arrival and the detonation of Smoky, some Task Force WARRIOR troops observed nuclear detonations.

Because WARRIOR troops had observed detonations before Smoky, their participation in a project known as Project HumRRO* was compromised. [*Footnote: Project HumRRO was sponsored by the Human Resources Research Office. This project was to compare soldiers' performance of several basic military tasks before and immediately after they observed a nuclear detonation <u>for the first time</u>. The tasks included: dissembling and reassembling a rifle, throwing a hand grenade, and traversing an infiltration course. While traversing the infiltration course, the men's reaction to crawling through an area which they thought was contaminated with fallout was also to be observed.] Therefore, a new task force, BIG BANG, was organized to conduct Project HumRRO on Smoky. Due to the predicted fallout pattern, BIG BANG's participation on Smoky was canceled the evening before detonation. BIG BANG troops, however, did observe Smoky at News Nob. The project was completed during Galileo, though the fact that the troops had already observed a nuclear test may have compromised the results of the project. (Harris 1981a: 25)

Trenches were prepared 4400 yards SE of GZ for the troops to observe Smoky. However, the predicted fallout pattern would have blanketed this area as well as the HumRRO sites. Therefore, the WARRIOR troops and other observers witnessed the shot from an area off of the Pahute Mesa Road, approximately 8 miles from GZ. On shot day, the main body of troops left Camp Desert Rock by 0100 and was in the observer area by 0330. The helicopter battalion, which provided airlift support and Pathfinder functions for WARRIOR, was positioned (at the helicopter assembly area) on the S side of the saddle between Yucca and Frenchman Flats, approximately 20 miles S of GZ and 2 miles S of News Nob. The airlift support was provided by AFSWC. (ibid., 27)

Smoky was fired at 0530 on 31 August 1957. At 0545, the group designated as "Pathfinders" boarded the helicopters accompanied by radiological safety personnel. They landed in the objective area at 0617, after conducting a preliminary aerial radiological reconnaissance. The Pathfinders delineated safe landing sites for the main body of the airlift to follow. Rad-safe monitoring showed the maneuver area to be safe, and the Pathfinder team relayed this information to the Task Force WARRIOR commander.

At 0550, Task Force WARRIOR began moving from the observation point to the loading area, arriving at 0605. Helicopters (from the helicopter assembly area) arrived in the WARRIOR loading area between 0700 and 0711.

The airlifts would take the ground troops to a landing zone in the Objective Area which was generally NW of GZ. The troops would then secure an Objective position.

The 1st airlift consisted of fourteen H-34 and eight H-21 helicopters and carried three rifle platoons and the weapons platoon. It left the loading area beginning at 0704 and reached the landing zones beginning at 0715.

The 2nd airlift, which carried the remainder of the task force, consisted of seventeen H-34 and eleven H-21 helicopters. It included some from the 1st airlift.

With the exception of the mortar platoon squads being landed at the wrong place and one aircraft picking up the weapons that platoon soldiers left in the loading area, this landing completed the troop lift. Several observers, who were with Task Force WARRIOR at shot time, flew to the objective area early in the airlift to observe the landing operations.

The supplies accompanying the task force were provided by its normal ground vehicles. Follow-up supplies were transported by truck to the Aerial Supply Distributing Point where they were held until helicopters arrived. Re-supply and evacuation helicopter flights began after the first landings. As each helicopter landed, a truck moved the appropriate load to the aircraft. The load, which was in boxes, was either placed into the helicopter or on a rigged sling. The Aerial Supply Distributing Point contained 27 tons of supplies for distribution to the task force. 2.5 tons (3 sling loads) were automatic re-supply, delivered between 0757 and 0818. The remaining 24.5 tons were prepared for delivery as on-call supplies. Of these, 6 tons were delivered between 0829 and 0940 at the request of the task force commander. After the 0940 delivery, all additional requests were denied in anticipation of the termination of the exercise. All of the supplies except water were simulated by sand-filled boxes. (ibid., 37-38)

With the completion of the airlift at 0830, the ground maneuvers began to seize objectives. At 0915, the task force commander reported that his 2nd and 3rd platoons had advanced to the points permitted by Rad-safe personnel and had been halted prior to seizure. "The exact location and the radiation level for this halt is not specified in the troop test report (27)". (ibid., 37)

"It has not been determined how and when the troops were moved out of the area following the exercise." "--- responses to Army questionnaires indicate that the troops walked to Landing Zone ECHO and were taken from there by truck."

Project HumRRO on Galileo
As mentioned under the Smoky maneuvers, Task Force BIG BANG had been organized to conduct Project HumRRO on Smoky. When the weather predictions prohibited the WARRIOR and BIG BANG troops from taking their positions in the trenches, they viewed Smoky from News Nob. Although viewing Smoky compromised the results of HumRRO, it was decided to conduct the maneuvers associated with it on Galileo.

Because there was not enough time to construct new trenches or clear new test areas for Galileo, it was decided that the Task Force would witness Galileo in open terrain approximately 4,500 yards from GZ. The rifle disassembly-assembly test would be conducted at the observation point, and the infiltration course maneuver would be performed at the original Smoky test area, if radiation from that shot had decayed to an acceptable level. The minefield clearing exercise was cancelled.

The actual locations of the Galileo observation area, the Smoky trench location, and the HumRRO test area for the Task Force have been the source of some uncertainty.

Task Force BIG BANG troops and the team of monitors from HumRRO left Camp Desert Rock at about 0130 hours on 2 September 1957. The Task Force and the HumRRO team were not at full strength; the HumRRO team had reduced from 10 monitors to 3 because of early departures. Some of the military personnel failed to return from weekend pass. Only 110 of the 167 servicemen scheduled to participate actually took part in the troop test. The troops were carried into the forward testing area by one of the transportation support units stationed at Camp Desert Rock. At about 0245 hours, the Task Force arrived at the Galileo observer area.

Galileo was detonated at 0540 hours. The troops witnessed the detonation in open terrain 4500 meters E of GZ. The blast wave caused momentary winds of about 36 knots at the observation area, and raised considerable dust. Troops performed the rifle disassembly-assembly test immediately after the blast wave passed. Then, radiological safety monitors and one member of the HumRRO team left the Galileo observer area for the infiltration course to determine whether residual radiation levels from Smoky, 2 days earlier, were low enough to allow troop entry. The troops had breakfast of assault rations. By 0710, the radiological safety monitors returned and reported that the infiltration course was considered safe for an one-hour stay. The BIG BANG troops departed from the observer area by truck convoy and arrived at the Smoky trench parking area at 0740. After leaving the trucks, the troops went to the infiltration course, located about 1,400 meters to the NW of the parking area and gathered in groups of 4 and 5. Each group moved to the starting line, where, on a signal from a HumRRO monitor, they began the 63 m (70 yard) course. Each group walked 9 m, crawled under a barbed-wire barrier, and continued to crawl for 14 m and under a second barbed-wire barrier. After the soldiers had cleared the second barbed-wire barrier, they sprinted about 6 m to a fox-hole, where they remained for ten seconds. After the 10 seconds, the soldiers sprinted 8 more m to a third barrier, crawled under it, and sprinted 14 m to a "wall" of smooth wire, where they each threw two practice hand grenades at a 1.2 m square pit, located 11 m away. The infiltration test ended with each soldier giving his name to the monitor waiting at the end or the course. He then went to the truck parking area. When the first group of men had completed the test and the course was clear, the next group started the maneuver.

The last group finished the course at 0855. By 0915, the last group had returned to the truck parking area, 1,400 m SE of the infiltration course. The trucks left the Smoky area for the Decontamination Station near Yucca Pass, about 24 km S of the infiltration course. It is estimated that the Task force arrived at the Decontamination Station at about 1000 hours. At about 1030, the troops and vehicles had been monitored, decontaminated, and clothing changed.

2) TROOP-OBSERVER INDOCTRINATION PROJECTS

These projects included 50.2, 51.1, 52.2, 53.3, and BOBCAT. Observer personnel began to arrive three days prior to the scheduled shot date (which could be much longer than 3 days before the actual date of detonation) and stayed at Desert Rock. These projects consisted of an 8-hour orientation in special weapons and voluntary follow-on classes. The number of participants in these classes were:

Army	2849
Navy	93
Airforce	246
Marine Corps	106
Civilian	56
Canadian	316
Total	**3666**

The number of participants who attended classes may be less than the number of observers counted for the shots, see Table N1 Troop-Observer Indoctrination, because one individual who was present for several shots would have been counted several times. Visits to equipment displays before and after the shots were included, as were visits to areas of earlier shots.(Harris, et.al 1981b: 96-7)

At the time of a shot, observers as well as the troop units would be in trenches located a "minimum safe distance in accordance with established criteria". Originally, observers were scheduled to be in the trenches identified in Table N3 during one of the six primary shots: Priscilla, Hood, Diablo, Kepler, Franklin Prime, and Smoky. Table N2 indicates that these shots had the largest number of Troop-Observers, but other shots also had observers. (Mathewson 1958:15, 91; Harris, et.al 1981b:102)

Some persons who participated in the Troop-Observer Indoctrination Programs also participated in the Technical Service Projects. (Harris, et.al 1981b: 96-7)

3) TRAINING PROJECTS (TP)

Three radiological training projects were established to familiarize the participants with radiological monitoring techniques under the realistic nuclear battlefield conditions available at the test site. The largest of these projects was the establishment of the Camp Desert Rock Rad-safe School which trained Army, Marine, and Canadian personnel as radiological monitors. There were 18 hours of formal instruction followed by several days of practicing monitoring techniques in contaminated areas.

Permanent Party	84
9 Army Chemical, Biological, and Radiological (CBR) Survey Teams	89
Marines	196
Infantry Battle Group	30
Canadian Infantry	3
Airborne Pathfinders	14
AEC	1
Total	**417**

Some of the participants of this school, as well as the two other training projects mentioned below, subsequently served as monitors on PLUMBBOB shots. The shot participation by the Army CBR teams is listed in Table N2 under Training Projects, "Army CBR Training".

The Navy's Project 51.1 trained approximately 120 individuals from all parts of the world. They were to participate on post-shot monitoring of Boltzman. However, the shot was delayed; and they practiced at an old TEAPOT site. All but 3 of the participants left the site without witnessing a shot.

The Air Force also had radiological survey training, Project 53.4. Students for continental air bases attended classes at Nellis AFB, after which they viewed a shot at the NTS. After the shot, the students conducted radiological survey monitoring in the target area of the shot. Areas to be surveyed were first marked with stakes. Monitors proceeded down the staked sectors in vehicles toward GZ reporting dose rates by radio to control stations. The control stations integrated the reported dose rates into isodose rate maps. The trainees of this project participated in shots under Training projects, "53.4". (Harris 1981b: 100-101)

4) TECHNICAL SERVICE PROJECTS (TSP)

During Desert Rock VI on TEAPOT, some technical projects had been placed under the supervision of the Desert Rock Exercise Director. He was responsible for the overall supervision, coordination, general administration, and the logistical support of such tests. This was continued on PLUMBBOB.

The TSPs generally involved: the determination of the environmental conditions in a nuclear battlefield and the evaluation of damage to protective materials and field fortifications. The titles and terse descriptions of the projects are given in Table N1. About 30 persons participated on nearly all of the shots for Project 50.3; about 5 for all of the shots thru Stokes for Project 50.4; $\leq \sim$ 5 for 50.5 thru Stokes. For 50.6 only 1-2 persons participated thru Owens. Project 50.7 had 1-5 thru about Smoky with 10 on Hood. Project 50.8 had the most participants with over 550 on each of the tests thru Smoky and about 105 from Galileo thru Fizeau. The last shot with Project 50.8 participation was Newton with 40.

5) OPERATIONAL TRAINING PROJECTS (OTP)

The primary aims of the operational training projects were: to test service tactics and operational equipment and to train and indoctrinate aviation personnel. Table N1 cites the project number and provides a brief statement of the OTPs conducted during PLUMBBOB. Project 51.3 was conducted on Fitzeau and Project 52.3 on Hood. Project 53.1 was conducted on 11 shots and 53.2 on only John. Project 53.5 was conducted on 8 shots. For nearly all of these projects, the number of participants is not available from the references used. (Harris 1981b:183)

Projects 53.6 thru 53.10 were for Strategic Air Command (SAC) Crew Training and the use of equipment for acquiring information about the shot. Projects 53.7 and 53.9 were conducted on nearly all of the flights thru Smoky and Galileo respectively. Hood, Laplace and Fizeau witnessed project 53.8. Details and numbers for the SAC operations are not provided in the references used. However, if these activities were like those described in Part II-Chapter 8, the skys were again busy.

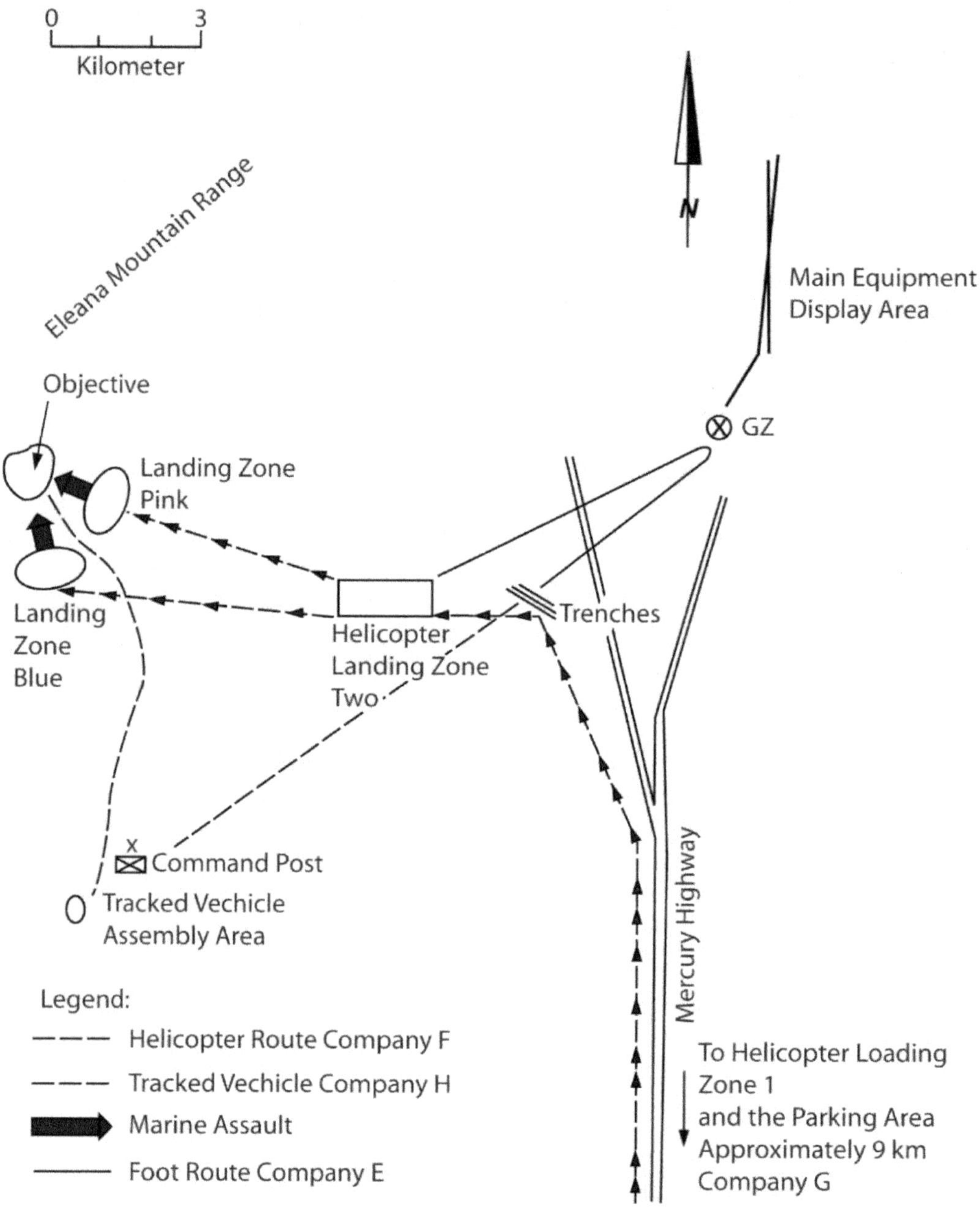

Figure N1 Marine Troop maneuvers during Desert Rock Project 52.1 on Hood.

APPENDIX N REFERENCES

Harris, P. S., C. Lowery, A. G. Nelson, S. Obermiller, W. J. Ozerofff, and E. Weary, *A Shot Smoky, A Test of the Plumbbob Series, 31 August 1957*, DNA 6004F, Defense Nuclear Agency, Alexandria, VA, 31 May 1981(a).

(covers all shots) Harris, P. S., C. Lowery, A. G. Nelson, S. Obermiller, W. J. Ozerofff, and E. Weary, *PLUMBBOB Series, 1957*, DNA 6005F, Defense Nuclear Agency, Washington, D.C., 15 September 1981(b).

Maag, Carl, Martha Wilkinson, James Striegel, and Burt Collins, Shot Hood, A test of the PLUMBBOB Series, DNA 6002F, Defense Nuclear Agency, Washington, D. C., May 13, 1983. (supersedes First Edition of February 12, 1981)

Maag, Carl and Jean Ponton, *Shots Daiblo to Franklin Prime, The Mid-Series Test of the PLUMBBOB Series*, 15 July - 30 August 1957, DNA 6006F, Defense Nuclear Agency, Washington, D. C., September 29, 1981.

Massie, Jeannie and Jean Ponton, *Shots Wheeler to Morgan, the Final Eight Tests of the PLUMBBOB Series, 6 September - 7 October 1957*, DNA 6007F, Defense Nuclear Agency, Washington, D. C., September 30, 1981.

(covers all shots) Mathewson, L. Lt. Gen, *Exercise Desert Rock VII and VIII. Final Report of Operations*, Headquarters Sixth United States Army, Presidio of San Francisco, California, 9 January 1958.

Ponton, Jean, Martha Wilkinson, James Striegel, and Burt Collins, *Shot Galileo, A Test of the PLUMBBOB Series, 2 September 1957*, DNA 6001F, Defense Nuclear Agency, Washington, D. C., February 27, 1981.

Viscuso, Mary Jo, Steven Geller, Martha Wilkinson, James Striegel, and Burt Collins, *Shot Priscilla, A Test of the PLUMBBOB Series, 24 June 1957*, DNA 6003F, Defense Nuclear Agency, Washington, D. C., February 27, 1981.

West, Jaton, Martha Wilkinson, Carol Simpson, and Jeannie Massie, *Shots Boltzmann to Wilson, the First Four Tests of the PLUMBBOB* Series *28 May - 18 June 1957*, DNA 6008F, Defense Nuclear Agency, Washington, D. C., September 28, 1981.

ATTACHMENT I

EARLY WORKING RELATIONSHIPS BETWEEN LABORATORY, MILITARY, AND AEC PERSONNELL

by ROBERT DUFF

After their initial training at Sandia Base on bomb assembly and non nuclear component testing, some of the military personnel were selected for assignment to the AEC where they served in a dual capacity as AEC-AFSWP inspectors and certifiers of weapons for the stockpile. They would also use their AFSWP training in the development of capabilities for the manufacture and assembly of specific weapon components.

As an example of this practice one of the young captains trained in the first Air Force Squadron at Sandia Lab was a Mechanical Section Supervisor (high explosives and detonator assembly) named Bob Duff. He was reassigned from the 502nd Aviation Squadron to the Iowa Ordnance Plant (IOP) to function as an AEC/AFSWP Inspector. In that capacity, he reported to the resident "AEC chief inspector", James K. Hasson. What this really meant at the time was they were responsible for the development of procedures for the inspection and certification of each weapon going through the production line which was being set up by the AEC's contractor, Silas Mason. This was a typical pilot plant operation in preparation for the full scale production of nuclear weapons which was to come later at the AEC operated plant at Pantex in Amarillo, Texas.

In order to facilitate this development, coordination and adjustment process, the Los Alamos Lab set up a resident representative at the IOP and Sandia Lab. Staff assisted from Albuquerque on a daily basis, keeping in touch by telephone and teletype, along with many visits to the IOP at Burlington. In line with this effort at the IOP, similar procedures were being established at the Bendix plant in Kansas City Missouri to produce the electrical and electronic components for the fusing and firing systems of the weapons. These components had to come together with the nuclear and high explosive components to make up the complete weapon entering the stockpile.

This part of the melding of the military and the AEC was a very significant step in the successful transition that took place as production MK III and MK IV weapons began to flow into the stockpile. It was essential that steps be taken to match up weapons with the strategic bombers that would place them on targets according to war plans being developed at that point in time. The Strategic Air Command was developing plans to place airplanes with weapons loaded on both airstrip and airborne alert. This made it necessary for production/stockpile weapons to remain in AEC custody at the storage sites operated by the Air Force according to the provisions of the Atomic Energy Act, and this was made possible by assigning AEC Custodians to each of the sites. These custodians performed inventory, surveillance and weapon modification duties with the help of the military personnel at the sites who functioned as joint AEC/AFSWP inspectors and inventory agents for the AEC and AFSWP. The military teams at the sites were trained by Sandia Lab Tech in the performance of surveillance inspections and weapon modifications. This transition went very smoothly. It was facilitated by the close association and coordination between the AEC, Sandia Lab, Field Command AFSWP, and Air Force personnel involved at the time.

It is interesting to note here that these professional associations developed into some close individual and family friendships that have lasted for the lifetimes of the people involved.

ATTACHMENT II

BYRON AT NEVADA TEST SITE

by BYRON MURPHEY

Participation by Sandia on Operation GREENHOUSE included the instrumentation of the Military Structures on shot Easy. Observation included rounded pressure vs. time signals that signaled the need for more overpressure-time vs. distance measurements. Thus, Sandia fielded such pressure measurements on airdrop shots of Operation BUSTER at the Nevada Test Site in October and November 1951. The lower than anticipated peak pressure observations caused considerable dismay because blast effectiveness was based on theoretical height-of-burst curves. (Refs. 1, 2, 3, 4)

One phone call that I received from someone in DMA of AFSWP stated: "Do you know that you just cut the stockpile in half?" At any rate, we developed plans for measurements on the upcoming TUMBLER-SNAPPER tests scheduled for the Spring of 1952.

As part of the effort to figure out what was happening, the decision was made to develop experimental height-of-burst curves using 250 lb. spheres of high explosive. Although Sandia could do the experiments at Coyote Canyon south of Sandia Base, AFSWP requested that we do some tests at two Nevada Test Site locations: Test Area 7 under the BUSTER air drops, and Frenchman Flat where the TUMBLER air drops were planned to take place. Burst heights of 22, 37, and 47 feet were specified! Consequently personnel from our Field Test organization and I (with help from NTS!) did the area T-7 bursts and pressure measurements on February 13 and 14 and the work at Frenchman Flats on February 16 and 17. After similar bursts at Coyote Canyon on February 28, 29, and March 4, I published the results in a memo dated March 13, 1952. (Ref. 5) It was apparent that mechanical effects alone could not account for the low pressures observed on BUSTER.

Of course the effectiveness of nuclear explosions had already been well-established in 1945. Nonetheless, a great effort went into understanding the early thermal interaction with the ground and the distorted close-in shock waves. A number of us spent considerable time at the test site where we seemed to be able to accomplish whatever sensible experiments we wished. On event day, we had our chance - the event was always spectacular. And usually we did retrieve the data we sought. We all became familiar with the Rad-safe structure near the Control Point, where we donned protective clothing and returned to become relieved of it. Great care was taken to avoid unnecessary radiation exposure.

My interest in nuclear energy dates back to my arrival in Graduate School at the University of Minnesota Physics Department in the Fall of 1939. The then editor of Reviews of Modern Physics, John Tate, gave me a proof copy of the paper by

Meitner, Hahn and Strassman. Also, I began working in the lab of A. O. C. Nier at the time he was doing the first quantitative separation of the uranium isotope 235. But, in the Spring of 1941 along with several other professors and graduate students, I went off to Naval Ordnance Laboratory at the Navy Yard in Washington DC. Later, Prof. John Williams went to Los Alamos and Prof. Al Nier went to New York City to work on isotope separation. Hence, I missed participation in the Manhattan Project.

At Naval Ordnance Laboratory, I became acquainted with Ev Cox, under whom I worked at Sandia from 1949 to 1953. Also, at NOL I worked at times with Bob Campbell, of LASL Nevada Test Site Fame.

Back to TUMBLER-SNAPPER and UPSHOT-KNOTHOLE, I participated as an experimenter on TUMBLER and authored WT-304 on air blast observation (ref. 3). Ev Cox reported on distant air shocks. Some of the other in Ev's Weapons Effects Department at that time were: Mel Merritt, Tom Cook, Jim Shreve, Jack Reed, Carter Broyles, and Frank Shelton. All participated in UPSHOT-KNOTHOLE with Frank working especially on understanding the precursor. Frank went to AFSWP in 1955 and has written about effects testing in some detail (ref. 1). We worked with various LASL people, for example Fran Porzel and Jack Whitener.

From the middle of 1953 until the middle of 1958, I worked at 3M Company in St, Paul Minnesota as head of the Physics Group in the Central Research Laboratory.

Upon my return to Sandia in 1958, I soon found myself becoming reacquainted with the Nevada Test Site in the role of crater experimenter. With H. R. MacDougall, we exploded a number of 256 lb. spheres of TNT at various depths of burst in desert alluvium, and later in tuff, to find the depth to optimize crater dimensions. Important also, we learned what depth was required to provide complete containment.

The results of these and other high explosive cratering experiments at NTS were summarized by Luke Vortman and me at the Geophysical Laboratory- Lawrence Radiation Laboratory Cratering Symposium held at the Geophysical Laboratory in Washington, D. C. on March 28 and 29, 1961.

The cessation of nuclear device testing at the end of October, 1958, took place without having any agreed-upon methods of verification. For underground bursts, seismic detection methodology was thought to be quite certain for yields above a few kilotons. In 1959, A. L. Latter and others at Rand Corporation proposed that small explosions under one or two kilotons could be "hidden" by exploding in an underground cavity. A plan developed to test this hypothesis using TNT explosions in salt. Livermore was assigned overall responsibility and Sandia was asked to do the ground motion measurements. Although the experiments were done in Louisiana, James E. Reeves of the Albuquerque Operation Office of the AEC was Project Manager; and the operation was carried out as though we were at the Nevada Test Site. Project COWBOY was performed in Carey Salt Mine property near Winnfield, LA, with seventeen event days from December 17, 1959 to March 4, 1960. My Livermore contact was Frank Adelman.

Data obtained from COWBOY show that decoupling in halite by a factor of from 40 to 100 can be obtained from high explosives. Up to 2000 lbs. of explosive were used. The results were published by me in the Journal of Geophysical Research in March 1961. The practicality (or not) of decoupling continued to be discussed and evaluated throughout the Comprehensive Test Ban deliberations.

Sometime in 1959, I was promoted to supervisor of the Sandia Underground Burst Physics Division and then in May 1961, to the Applied Research Division which had no NTS connection.

ATTACHMENT II REFERENCES

1. Shelton, Frank H. *Reflections of a Nuclear Weaponeer*, Shelton Enterprise, Inc., Colorado Springs, CO, 1988. Pages: 4-20, 4-22, 5-23, and 5-24.
2. Furman, Necah, *Sandia National Laboratories- The Postwar Decade*, University of NM Press, Albuquerque, NM, 1990. Pages 584, 604, 608-9.
3. Banister, John R., *Historical Sketches of Sandia National Laboratories Nuclear Field Testing*, Sandia National Laboratories, Albuquerque, NM, SAVD93-7054/2 UC-700, October 1994. Page 12, Appendix A for referenced reports.
4. Johnson, Leland, *A History of Exceptional Service in the National Interest*, Sandia National Laboratories, Albuquerque, NM, 1997. Page 44 blast effects, page 85 ground motion.
5. Murphey, B. F., *Pressure-Distance-Height Data for 250-lb. H.E. Spheres, Operation TUMBLER" Project 1.10*, Ref. Sym: 51111 (65), March 13, 1952. Later, a more complete report was written by J. D. Shreve: WT-529, *Pressure-Distance-Height Study of 250 lb. TNT Spheres*, 1952.
6. Murphey, B. F. and L. J. Vortman, *High-Explosive Craters in Desert Alluvium, Tuff, and Basalt*, Journal of Geophysical Research, Vol. 66, #10 October 1961, pages 3389-3404.
7. Murphey, Byron F., *Particle Motions Near Explosions in Halite*, Journal of Geophysical Research, Vol. 66, #3, March 1961, pages 947-958.
   An earlier Sandia Corporation publications was:
   Murphy, B. F., *Particle Motions Near Explosions In Halite*, SC-4440(RR), June 1960.

ATTACHMENT III

THE TURK DEVICE IN TEAPOT
Designed and Produced Under the Linda Program at UCRL Livermore

by RAYMOND A. "RAY" GILBERT

June 1952, I received a Master of Science in Theoretical Physics from The Ohio State University and was assigned to Air Force Special Weapons Center at Kirtland Air Force Base until I received an AEC security clearance. At that time, I was to be assigned to Los Alamos Scientific Laboratories.

The University of California Radiation Laboratory at Livermore was established on 2 September 1952 and was actively seeking military officers with technical backgrounds. With the concurrence of Los Alamos, my assignment was changed to Livermore.

Ralph Pennington and Alfred C. "Carl" Haussmann, two U. S. Army Captains, had arrived from Project Matterhorn at Princeton, University and were there to greet me. U. S. Air Force Lieutenant Jasper A. "Jack" Welsh was assigned there in 1954 and played an important role in the development of the Tesla device under John "Johnny" Foster's leadership.

I became aware of two Navy Lieutenant Commanders - whose names I have forgot - and later, Navy Captain Robert E. "Bob" Odening. Air Force Major Joe Duval was a chemist there. In addition to those assigned to work at the lab, the Field Command of the Defense Atomic Support Agency "DASA" in Albuquerque assigned liaison people. Air Force Lieutenant Colonel Robert "Bob" Colligan was one of the first. The symbiotic relationship was great for the military people assigned there.* [*Footnote: The laboratory people felt it was great for them also.]

When I arrived at UCRL Livermore on 5 January 1953, Ernest O. Lawrence was the Director of UCRL Berkeley and UCRL Livermore had been established under his aegis. He had selected Herbert F. York as the Director and Duane C. Sewell and as the Deputy Director. York was "Mr. Outside" and Sewell was "Mr. Inside." York generally handled technical matters and was the principal interface with AEC management; Sewell handled everything else. Each kept the other well informed.

I was greatly impressed by the informality at the laboratory. Everyone, even the guards and secretaries, called everyone else - except E.O. Lawrence - by his or her first name. Herb York usually came to work in a white shirt with sleeves rolled up and without a tie. It was an informality based on mutual respect that was the great strength of the laboratory.

Initially, there were only a few Departments. For the most part, they consisted of the standard housekeeping functions plus the Theoretical Department that was headed by Edward Teller. There were also a few Offices consisting of the usual functions such as security, personnel, accounting and budget, plant maintenance, etc.; there were several other technical Groups, Shops and organizations such as Physics, Chemistry, Mechanical Equipment, etc.

This was still the state of the formal organization in 1955 because the Laboratory was more interested in doing what needed to be done to achieve success as an alternate weapons laboratory than it was in establishing a formal structure. People were assigned a job or a piece of it because of their previous experience or because they had ideas that seemed worth pursuing; everyone who could help did so eagerly.

York had a very free hand in running the laboratory. E.O. Lawrence, although greatly interested in what was taking place in the laboratory did not interfere. He almost never volunteered his counsel unless York, Teller or someone else from the Livermore laboratory specifically asked him for advice. He did visit the laboratory almost weekly. When York was out of town for an extended period he would spend a significant portion of several days there.

Teller acted as a senior advisor and helped to define the technical problems that needed to be solved in the areas of laboratory interest. These included fission and fusion weapons, controlled fusion research, alternate diagnostic methods, etc.

Harold Brown was the senior technical leader. York and Teller depended on him to see that the appropriate technical studies were performed and thoroughly debated before hardware decisions were made. Many of the brighter young physicists and mathematicians were working under his general guidance.

The acquisition of the Univac under the leadership of Sidney "Sid" Fernbach was a great tool for calculating* how weapons and other devices might work. [*Footnote: The Univac represented the state of the art at that time. However, calculations that took all night to run on it could be completed shortly after loading them onto a typical home computer of today. Also today's computers do not require the extensive climate control facilities that were required for machines like the Univac. (Riley 2003)] Cecil E. "Chuck" Leith was responsible for most of the computer programming. He was able to acquire some programs from AEC contractors; he supervised the development of the others within the lab. A number of Card Punch Computers were also available for less sophisticated programs.

The laboratory also had a role in Project Sherwood, the controlled thermonuclear program. Richard F. "Dick" Post was the leader of the UCRL Livermore effort. A significant effort was also devoted to Health Physics and the effects of low-level radiation on people. Among the research and testing facilities were the 90-inch cyclotron and the Cockcroft-Walton particle accelerator.

Arthur J. "Art" Hudgins was the principal recruiter of scientific and technical personnel. Later, Gerald W. "Jerry" Johnson also helped before he took over L Division - Operations.

UCRL Livermore's first two tests, Ruth and Ray, in Operation UPSHOT-KNOTHOLE in the spring of 1953 had not been successful. In 1954, the laboratory fielded a device in Operation CASTLE in the Pacific. Herb York was so confident of success that he established two small groups to work with Sandia Corporation to produce two weapon versions with differing yields. Jerry Johnson and I were asked to lead the groups. A short time later, Jerry Johnson was asked to help with the recruitment of scientific and technical personnel. On Jerry's recommendation, I was asked to head the combined group.

In the fall of 1954, I proposed a device to test a concept that appeared to have the potential to improve the yield-to-weight ratio of thermonuclear weapons. After senior technical people in the laboratory had made some preliminary calculations and discussed the idea, the Linda Program was established to design and build a fusion device and to arrange for it to be tested at the NTS in the early spring of 1955. I do not know how the name Linda was picked. I was asked to be program director.

I was asked to recommend someone to take over the work of the Linda Group when I left and I recommended Carl Haussmann. I do not know how long the group retained the Linda name or whether any other devices were fielded by the Linda Program per se. Carl remained in a senior position at the lab for many years.

Instead of establishing a line organization, I chose one that was project oriented. Except for a small staff to help keep track of schedules and continually to ask "Have we missed something?" everyone else working on the project remained in his or her own organization. Key people were identified as the principal points of contact in each of the relevant technical organizations. These people were responsible to see that the resources were available and that the work of the Linda Program was completed on schedule. It was indeed a "matrix" type organization patterned after the Douglas Aircraft Company "Douglas Project Method." These organizations were assigned the people and funding for routine work. I had to approve some of the major or extraordinary expenses but much was done informally at that stage of the laboratory development and I received absolutely unbelievable advice, cooperation, help and support from throughout the laboratory. It is eloquent testimony to the enthusiasm and dedication of the laboratory leadership and staff that not one time was it necessary for me to confer with a division leader to resolve a problem.

These key people were also responsible for coordinating with appropriate people within and outside the laboratory. The principal job of the program office was to ensure that all necessary coordination took place. It was largely from frequent checking with responsible people within the laboratory that I formed my opinions about the support of the people at Los Alamos, at Sandia, at NTS, at the on-site contractors and, of course, at UCRL Livermore.

The key people of the Linda Program and their areas of responsibility in early 1955 are listed below.

Linda Program

		Bldg.	Room	Phone
Device	Ray Gilbert	162	264	483
Physics	Harold Brown	161	120	209
Mechanical Engineering	Art Hoffman	151	270	628
Mechanical Equipment	Roy Elke	151	218	631
Chemical Engineering	Barney Rubin	101	108	267
		102		321
Electrical Engineering	Duke Daly	155	260	467
Radio Chemical Diagnostics	Ken Street	101	105	261
Chemistry	Bob Goeckermann	101	121	264
Mechanical Engineering	Art Hoffman	151	270	628
Mechanical Equipment	Roy Elke	151	218	631
Electronic Diagnostics	Fuzzy Wouters	162	228	251
Physics	Boris Ragent	162	223	251
Mechanical Engineering	Herb Weidner	151	170	638
Electrical Engineering	Harvey Owren	155	252	467
Mechanical Equipment	Burt Barrows	151	218	631
Fast Photography Diagnostics	Harry Keller	151	135	246
Physics	Don Born	151	135	246
Mechanical Engineering	Bill Platt	151	170	638
Mechanical Equipment	Ray Morton	151	220	631
Nuclear Emulsions Diagnostics	Steve White	143	172	535
Physics	Chuck Violet	143	176	535
Mechanical Engineering	Carlo Herrala	151	170	638
Mechanical Equipment	Don Walmsley	151	214	631
L-Division - Operations*	Vern Denton	162	163	433
L-1	Tom Brockett	162	177	648
L-3	Val Smith	162	172	206
L-4	Dan Murphy	162	181	649
L-6	Cliff Bacigalupi	162	176	206

* Livermore's L Division was somewhat similar to Los Alamos' J Division

Device

Physics - Harold Brown, as head of the Physics (primarily theoretical physics) group, took responsibility for the detailed physics calculations for the device. Arthur T. "Art" Biehl, Robert "Bob" LeLevier, Ernest A "Ernie" Martinelli, Michael "Mike" May and others who worked in that group made important contributions. Sid Fernbach was responsible for the Univac, Chuck Leith was the lead programmer and Jim Frank was in charge of the principal burn calculation.

Mechanical Engineering - James "Jim" Bell was head of the group, but James A. "Art" Hoffman was the principal mechanical engineer for the device.

Mechanical Equipment - I believe Kenneth W. "Ken" Copenhagen was head of the group but Roy F. Elke was the device principal and Malcolm W. "Bud" Loveland was his assistant. They worked very closely together to assure that the physical items were produced to specifications and arrived on time at the lab and at the test site.

Chemical Engineering - I believe chemical engineering was a part of Chemistry under Kenneth "Ken" Street. Barney Rubin was the device chemical engineer.

Radio Chemical Diagnostics - Ken Street was the head of Chemistry and Roger E. Batzel was his deputy. Robert H. "Bob" Goeckermann was responsible for specifying the diagnostic materials. Since these materials had to be physically incorporated in the device, Art Hoffman handled the mechanical design and Roy Elke made sure they were procured, fabricated and included.

The other diagnostics programs were not physically attached to the device so each area had its own designated principal together with physics and engineering staff assigned from among the regular organizational groups. These staffs were kept informed on the device design and kept me informed on their progress.

Operations - Vern Denton was originally the head of this group that was responsible for coordinating with the many organizations at the NTS to ensure that everything would be ready when the devices furnished by UCRL Livermore were tested.

The primary for the device was provided by Los Alamos. Staff and technicians there were extremely helpful in making suggestions and they willingly provided answers to any questions about installation of the primary. The role of Los Alamos was obviously critical to the success of the Linda Program.

The professional cooperation and help extended also to the areas of radio chemical detectors, fast photography, nuclear emulsions and electronics diagnostics, all of which were used to help determine how well the device performed. Perhaps we could have done the diagnostics without the help of experienced Los Alamos and Sandia people, but their assistance certainly made it easier and added to our confidence.

When it became apparent that UCRL Livermore was probably going to have a device to test in TEAPOT, it was officially designated Turk.

From early in the Linda Program until a few weeks before TEAPOT, Vern Denton and his Operations staff were responsible for the coordination with NTS personnel and with on-site contractors. This included moving the device to the test site, placing it atop the 500-foot tower in Area 2, assembling it and preparing to fire it. It also included all the preparations for the diagnostics such as locations, power, timing signals, etc.; making arrangements for cloud sampling and gathering the results of the bhangmeters; overpressure and dynamic pressure gauge measurements; and other measurements ordinarily provided at the request of the Test Director. The Operations Staff also arranged for face-to-face meetings with appropriate onsite people when Linda Program personnel thought it might be useful. Again, without exception, the people at the test site, whether Los Alamos, Sandia, NTS, Albuquerque Operations or the on-site contractors, were unfailingly cooperative and dedicated to making sure everything was done and done correctly to ensure that the device test and all the diagnostics were successful. Shortly before TEAPOT began, Jerry Johnson became head of L-Division and the cooperation continued.

I visited the test site in January 1955, and it appeared to be ready to receive Turk. I returned in early February and remained there until about 10 March. The billeting and mess facilities were better than I had imagined. They were quite adequate from my point of view.

Turk was to be the first shot of TEAPOT on 15 February and it was ready. Because the calculated yield of the device was approximately 90% of the maximum allowable for a 500-foot tower shot, there was understandable concern for fallout. On several occasions the test countdown was begun and then aborted because of a later winds aloft forecast. Finally, on 7 March, Turk was successfully tested with a yield of 43 kilotons.

I do not wish to imply that there were no problems; there were. What intrigued and impressed me so much was the way professional, dedicated, highly motivated people simply got together without fanfare and solved the problems. My focus on testing at the NTS was understandably narrow; but I can honestly say that because the people there and at Livermore did such a great job that my association with the Turk device and its test was one of the most enjoyable experiences of my life. I am happy to have the opportunity again to express my admiration and appreciation for the wonderful teamwork that made my involvement so memorable.

In addition, to the Linda Program that led to the test of Turk, the laboratory produced two other devices that were also successfully tested in TEAPOT. John S. "Johnny" Foster headed that program, and his group was responsible for the Tesla and Post shots. It is my recollection that he agreed with my general conclusions regarding the cooperation of people at the NTS.

Singing was <u>not</u> a part of the evenings at NTS (except perhaps at parties). The following ditties that were taken from a bulletin board were composed primarily out of a good-natured frustration at the continued delays because of weather. I included them just to show that some people had a nice sense of humor.

<u>To the tune of "Every Day Is Ladies Day"</u>
Oh, every day is MINUS ONE for us
It's neither MINUS THREE, nor is it PLUS
It may be that the 'Rems' are high
Or that the blast is strong
It may be that the troops will fry
Or else the wind is wrong

Oh, every day is MINUS ONE for us
We now accept this fact without a fuss
Although the panic bell rings true
It's quite the same refrain
And when the shot is finally through
It's MINUS ONE again

<u>Ode To A Southwester</u>
Scarlet hues, dimming desert dusk
Upon the mountains fall.
Sage and sand and prairie flower
Aglow with fiery wall -

Amid the finery, barracks cold
Live scientist and engineer.
Their mission known the world around
Peace and plenty ever near.

Despite a power so immense,
Seemingly we have sinned,
For UCRL must wait while
Unfavorable blows the wind.

I do not feel qualified to describe a "typical work night" at the site. After the first few nights "guarding the site," I spent most of my evenings helping J-Division people assemble mimeographed schedules, site instructions, etc. at the CP. I do remember that the electronic and photo diagnostic people spent several evenings rechecking their equipment.

ATTACHMENT IV

Excerpts from Interview with

MAJOR GENERAL EDWARD B. GILLER

These excerpts were taken from tape recordings of an interview the authors had with Major General Edward Giller who has reviewed them with his approval.

Background
Edward Giller was born in Jacksonville, IL in 1918. He graduated from Kemper Military School, MO, and from the University of Illinois with a Bachelor of Science degree in chemical engineering in 1940. After graduation, he worked for the Sinclair Oil Refining Company as a chemical engineer. In September 1941, he entered active military duty and flying school. He earned his pilot wings and commission as second lieutenant in April 1942.

He served in the European theater of Operations as a highly decorated fighter pilot in P-38 and P-51 aircraft with 115 aerial combat missions. After WWII, Giller again attended the University of Illinois under the Air Force Institute of Technology Program where he earned his doctorate in chemical engineering in 1950.

His first assignments, described here, were at Headquarters AFSWP and at the Air Force Special Weapons Center, Kirtland AFB. Griller's subsequent career with the Air Force was also in matters related to nuclear weapons, involving the DoD, AEC, and the CIA. He was assistant general manager for Military Applications with the AEC prior to his retirement from the Air Force in 1972. Since then, he has served: the AEC and ERDA; the Joint Chiefs of Staff, which included nuclear test ban negotiations; and the national weapons laboratories. His career has witnessed active participation throughout the span of US nuclear weapons history.

Excerpts From Interviews
Dr. Scoville* got me into AFSWP. He became my mentor and friend. [*Footnote: Herbert Scoville, Jr. was Head of the Radiation Branch of AFSWP headquarters in 1950. He was Deputy Test Director for Military Effects Tests during TUMBLER-SNAPPER]

In the Radiation Branch, anything that had to do with thermal and radiation measurements and fire measurements that were proposed for all of the tests came through the shop. My part of the technical field was thermal radiation. We were interested in the outputs of the bomb that would be used as inputs to the targeteer. We collected all the ideas the services had about what they wanted or needed. We decided on what measurements made sense to do in upcoming tests, what was the best test for the measurements, etc. We organized the measurements into projects then tried to take a technical look at what was needed to conduct the measurements in the field. The result was a package that was about 90% complete when it went to the field. The field had authority to make changes as necessary.

About a dozen of us from Headquarters would go to the field for various reasons. We worked with the military people who were a part of the Nevada Test Site Organization. We were not in the chain of command in the field, but still we had a "heavy hand" when we needed it.

Some of the effects tests we did looked odd at the time and looked even odder afterwards. The Army wanted to put out one of their transportation locomotives to see what would happen. But they didn't want it hurt. The Yucca National Railroad is only about 100 feet long, and it is still out there. Not much happened to the locomotive. A window was crashed, but that was about all.

LASL thought that AFSWP's experiments were "sophomoric". At first, we got a lot of guff from LASL. Jane Hall would quiz us to great extents before she would allow us to come see people at LASL. She always wanted to know "why". It was not serious. I interacted a lot with Bill Ogle. I liked Ogle - especially after I saw him shave with a straight razor up in Amchitka. Wild Indian! He had a lot of sense, got things done, etc. Also Graves had a lot of common sense. I had no problems with Jack Clark.

Spilhaus was a consultant to AFSWP and was a part of the Joint Task Force (JTF) for Windstorm. Spilhaus was hired as a consultant to interface with Los Alamos because we were getting too damn much guff from Los Alamos about our not knowing what the hell we were doing. Spilhaus was an interesting guy; he stood up to Jane Hall. Jane was tougher than Norris (Bradbury). Spilhaus interacted with the military people running the projects as well as the Los Alamos side of the house.

Three of us from AFSWP went by train out to Las Vegas and the site for an operation. I had top-secret documents so I had a compartment to myself. They had given Scoville a gun, but he didn't like guns. He liked fishing poles. So he put the ammunition in one place and the gun in another. After that three-day trip, we said, never again. It was a <u>long</u> haul.

During BUSTER-JANGLE, we stayed in wooden barracks at Mercury. Movies were shown outdoors in the evenings, and it was real cold. The film would break, it seemed to take forever to fix it. There was usually a lot of good food in the mess hall. At one point, people were getting ill. A medical doctor who also ran some experiments, made an inspection of the messing facility and found that sanitation was unacceptable. He ordered all military personnel to eat at Desert Rock, standard Army stuff. Well, they got their act cleaned up.

Did you see a lot of change in the living conditions at Mercury? Oh, sure, by the time they got done, it got pretty high class. At first living spaces the facilities were built as fast as they could build them, temporary buildings.

Do you remember the silver dollar turn style at the cafeteria? When did it first come? I remember that, I don't remember when it started. I remember, early on, the waiters looked like they were hung over drunks off the strip. They had their finger off in the middle of whatever they were serving. At first, we ate "family style". They'd put this big platter down on the table. The ones closest to the platter would take 2 steaks, 4 eggs, --- these were the workers. If you were at the end of the table, you had to wait until the next batch.

The Civil Defense folks also came to NTS and wanted to find out about things. They had a set of proposals and paid for their own. They built a lot of houses, even with brass hardware, so that they were exactly like a Middle West house with everything. Some even had piles of rubbish in them. We had bets on which ones would catch on fire how fast.

The Forest Service, the research branch, did all of our fire research work. They were under contract by us to do this. They were really a pragmatic bunch, and I worked a lot with them. They are the ones who put up the Yucca National Forest, with trees from Mount Charleston.

We had lots of effects experiments. Some would be repeated on a number of shots during an operation and/or on subsequent operations. There were so many jeeps put out in the field that people said they could determine blast pressure (and thus yield) by the number that were turned over.

The Army had a uniform protection program. How well does the uniform protect you against burn? Pigs have skin closest to that of humans, and they were used in exposure experiments. The Quartermaster had to design uniforms for pigs. We bought some shoats, small pigs. You had to put them out at night before the shot. They had to be anesthetized but you wanted them to wake up in the morning when the shot went off. Of course you put out cameras and everything. Then, we discovered that bobcats, wild cats, would get into the pigs. So we had to have electric fences to keep out the cats. The pigs were so cold at night that we had to put up heat lamps. So we had to turn the lights on and then turn them off a few seconds before the bomb went off. Things took so long that the pigs outgrew the uniforms. We had a lot of pigs on our hands. Somebody said let's have a luau. Someone else said lets fly in palm leaves from Hawaii to wrap the pigs. We gulped and thought that that was a bit far out, so we built a fire, wrapped the pigs in gunnysacks and put them down on the racks. Then everybody went off to the beer bar. We came back and dug up the pigs. But, there is nothing like hot pork with creosote. All of the cooked pigs were thrown in the dump. Thereafter I think that aluminum foil was used for pig luaus.

How did Headquarters and Field Command work with each other? Pretty well. Our job was a lab program, their job would see that it got done. Headquarters made arrangements for Sandia and Stanford to make pressure measurements. They put up these poles around the place with blast gauges and radiation instruments, and probably high speed cameras by EG&G. Measurements were made right at the test unit. A lot of cameras, a lot of cameras!

The biggest problem was that everyone wanted to get home. They'd take their data and get out. They had signed up and agreed to do reports. Thirty, sixty, ninety days later, I was on their backs.

On Friday night we'd go to Las Vegas for the hell of it. Ed Doll would get a case of gin, which he kept under his bed. Every night we'd have a dry martini. He'd take the cap off of the gin bottle and pour vermouth into the rest of the space. That was it. Doll and Spilhaus had essentially the same types of roles.* [*Footnote: A. J. Spilhaus was an AFSWP consultant and was designated the Scientific Director of the WINDSTORM Task Force. WINDSTORM was "postponed" because its objectives could be met on JANGLE, and Spilhaus served as the Deputy to the Test Director for Effects Tests on BUSTER-JANGLE, see Part II-Chapter 2. Ed Doll served as Director of the Military Effects Test Group for UPSHOT-KNOTHOLE and TEAPOT. Spilhause and Doll indeed had the same types of roles, see the operations chapters. Doll was also at Los Alamos during the Manhattan project and became a founding member of its Z Division management in September 1945, see Part I-Chapter 4.]

How we operated. Headquarters people felt free to get into whatever they wanted to do as long as they did not go around the command structure. Everybody wanted to get the job done. We worked long hours. It was exciting. It was new. None of us agonized over radiation in any form particularly.

Were there differences in the interactions between DoD and Livermore versus DoD and Los Alamos? We thought that LASL was pretty conservative. They were a little bit hard to live with, not seriously so. Graves and Ogle were good. We had no role in the second lab. We supported it in sort of an intellectual or emotional way. Livermore came in with a lot of ideas about nuclear weapons and everything. We sent Jack Welch out to Livermore, and Ray Gilbert was there. We sort of liked the gung ho attitude at Livermore. We felt that the competition of Livermore was good for Los Alamos. Mike May (who was later director of Livermore) was a tougher cookie to deal with than Norris ever was.

When you were at headquarters, did you spend much time with Field Command? I'd fly myself out, spend a few days and fly back - get my flying time in. We went back and forth a lot. Most of our interactions were with Field Command, some might have been with the Air Force side of the fence.

We went back and forth to the test site a lot. I flew myself. I flew to Albuquerque from Washington quite often to get my flying time in. It had nothing to do with the business of testing. I did not fly from Washington to the test site. When I was stationed at the Test Site, I sometimes would go over to Nellis and get a plane - fly over the Grand Canyon.

I arrived in Albuquerque in 1955. My job description was the deputy to the flying unit - they flew the planes and dropped the nuclear bombs. I had come from AFSWP Headquarters where Military people in uniform were still doing some science. Several military people were still up at Los Alamos, Hugh Lehman for instance. I wanted to continue doing science, so I talked the Chief of Staff into letting me take the place of the research director of 40 people who was just leaving.

Then we had to decide what the hell we were going to do. I said, let's see if we can do experiments. We picked up a bunch of young first Lieutenants with their degrees and Ph.D.s in mostly physics and engineering sciences, had been postponed during the Korean War. Art Gunther was one of them. They came out to Kirtland, and this group of young kids wanted to do everything. So I said, go do everything. Jack Welch, Lew Allen, and Hugh Lehman had come down from Los Alamos. And some people from the life support facility at Cambridge came. We started little things. Then, when the Van Allen belt was proposed, we designed an experiment, injecting a bomb in the Van Allen belt on Operation ARGUS (see Chapter 13). We went to Washington, but AFSWP blew its stack. Two generals were fighting over it. We ended up with an experiment something like it. When they wanted Orion (a space craft powered by nuclear bombs), we said oh we'd do that one. The military scientists then faded away because the services were not providing them and civilians were taking over. Art Gunther stayed and is probably responsible for the fact that it kept going after I left. Actually that is sort of how the Phillips lab, named for Lt. Gen. Sam Phillips, got started.

When I got to Albuquerque, we were worried about bomb crews flying through radioactively contaminated clouds from the previous drop. Radiation doses could be picked up by the crew from the plane. We also had flash blindness* where we'd see the flash from the previous drop for some time. [*Footnote: AFSWAP conducted some flash blindness projects that are described in Part II-Chapters 2, 8, 12, and 14.] One advance we tried was goggles that turned black instantly. We started the program of how the nuclear weapons effects would affect the flying air force. This became an active program at the NTS during atmospheric testing.

What was the military's feeling about going underground for testing, Rainier and things like that? Change is always to some people terrible, other people fun, others wait and see. Mostly wait and see in this case. I think we thought that we had most measurements from airbursts mostly under our belts.

ATTACHMENT V

ARMS CONTROL: 1945 - 1961

by MILO NORDYKE

Not only is the history of NTS intimately linked with the cold war, it is also intimately linked with international arms control. While none of the important arms control treaties were completed during the period between 1945 and 1961, this era witnessed the birth of arms control concepts that were to predominate during the cold war. A number of important meetings and exchanges occurred during this period in United Nations fora and separately between the nuclear powers: US, Britain, and the Soviet Union. This appendix attempts to summarize these meetings and exchanges which led to the moratorium on nuclear testing that started on November 4, 1958 and ended abruptly on September 1, 1961.

EARLY ATTEMPTS AT ARMS CONTROL THROUGH THE UNITED NATIONS

The scientists involved with the early concepts of atomic energy were international in backgrounds and interests; even before the Manhattan Project, many had early views regarding the international control of atomic energy. On June 26, 1945, just three weeks prior to Trinity (July 16, 1945), the charter of the United Nations was signed in San Francisco. The foundation of the United Nations stimulated further thinking regarding the concepts of international control of atomic weapons. After Trinity, Hiroshima, and Nagasaki, concerns regarding the control of atomic weapons were heightened among the scientists and the general public worldwide.

During November 1945, the United States, Britain, and Canada formulated the Truman-Attlee-King plan for a United Nations Atomic Energy Commission (UN AEC) that would be introduced to the UN General Assembly, which was scheduled to meet in London in January.(Hewlett and Anderson 1962: Chapter 13) After a conference in Moscow, the three ministers: US - James Byrnes; Britain - Ernest Bevin; and the Soviet Union - Vyacheslav Molotov, proclaimed, on December 27, 1945, their intention of recommending a resolution that the UN General Assembly establish a commission to address the problems arising from the discovery of atomic energy. The ministers agreed that the proposed UN AEC would have the same function that Truman, Attlee and King had spelled out in November. The new commission would proceed by separate stages, a key issue in the Truman-Attlee-King plan that was considered very important for the United States. The UN AEC would make proposals for: exchanging basic scientific information; confining atomic energy to peaceful purposes; eliminating atomic and other weapons of mass destruction from national armaments; and effectively safeguarding the states that complied. (Hewlett and Anderson 1962: 476)

On January 24, 1946, Byrnes introduced to the UN General Assembly the Byrnes-Bevin-Molotov resolution for an UN AEC. No nation raised its voice in dissent.

Thus, the UN AEC came into effect almost a year prior to the US AEC. The UN Atomic Energy Commission would consist of all members of the Security Council plus Canada. (Hewlett and Anderson 1962: 532)

On March 16, 1946, President Truman and Secretary of State Byrnes nominated Barnard Baruch, a political and personal friend of Byrnes, as the US representative to the UN AEC. This nomination required senate approval, which was forthcoming April 5, 1946, amid a background of debate and meetings regarding US policy and national and international roles, and control. At Hunter College in New York, on June 14, 1946, Baruch presented the US proposed plan for the charter of the UN AEC to the assembled delegates of the UN AEC and their staffs. This plan had been two years in evolving and was the work of many men; however it became known as the Baruch Plan. (Hewlett and Anderson 1962: Chapter15)

The Baruch Plan proposed the elimination of existing stockpiles of atomic bombs only after a system of international control was established and prohibited veto power in the Security Council on the commission's decisions. On June 19, Andrei Gromyko of the Soviet Union proposed what became known as the Gromyko Plan. A cornerstone of this plan was the destruction of stockpiles before an agreement on an international supervisory scheme. It also retained Security Council veto power over the commission. Ultimately, the conflicting positions of the US and SU prevented agreement.

In 1952, the UN General Assembly abolished the UN AEC and formed the UN Disarmament Commission, which consisted of the members of the Security Council plus Canada. It was directed to prepare proposals that would regulate, limit, and balance reduction of all armed forces and armaments, and would eliminate all weapons of mass destruction. In addition, the commission would propose ways to ensure international control of atomic energy for peaceful purposes. Nuclear weapons issues would be handled by a subcommittee of the new commission. With its broad charter, little progress was made by the commission or the subcommittee. In 1953 President Eisenhower again made a proposal for an international agency under the aegis of the U.N. be established that would be responsible for "... apply(ing) atomic energy to the needs of agriculture, medicine and other peaceful activities", including the provision of electrical power to the world. His proposal ultimately led to the establishment of the International Atomic Energy Agency (IAEA) in 1957 that promoted the peaceful uses of atomic energy. This agency later became responsible for on site inspections under the Non-Proliferation treaty.

On February 28, 1954, during Operation CASTLE at the Bikini Atoll, the Bravo test of an experimental thermonuclear device was conducted. There were a number of uncertainties regarding the design; and the device gave a yield of about 15 Mt, significantly greater than the expected. Fallout was carried over a number of atolls in the Marshall Island Group east of Bikini, requiring evacuation of the residents. The fallout was also carried over an unlucky Japanese fishing trawler called the Lucky Dragon which was just north of the evacuated atolls. The Lucky Dragon sailed directly back to Japan where the crewmen were hospitalized. One subsequently died.

The BRAVO incident, served to focus attention on the world-wide radioactivity contamination problem of nuclear testing in the atmosphere and to stimulate movements for the elimination of such tests. During the 1950s, the Soviet Union was conducting atmospheric tests at Semipalatinsk and Novaya Zemlya, and the United Kingdom had started testing in Western and South Australia. These were less visible to the public than the U.S. tests at the NTS.

In the Fall of 1956, nuclear testing became a presidential campaign issue when Adlai Stevenson called for a ban on all H-bomb testing. Japan and India renewed their efforts to gain a world-wide test ban; public interest and religious groups began to agitate for action. In April 1957, Nobel laureate Albert Schweitzer joined the growing clamor for a ban on nuclear testing with an appeal through the Nobel Peace Prize Committee in Oslo, Norway. His appeal was endorsed by the Pope and a growing list of public officials and governments. On May 15, 1957, the U.K. carried out its first thermonuclear tests at Christmas Island in the Pacific, further exacerbating world opinion. The same day Linus Pauling began circulating petitions among the scientific community calling for a world-wide cessation of nuclear weapons tests, gaining over 2000 signatures within 3 weeks and over 9000 within 7 months from 44 countries and 36 Nobel laureates.

Seizing the opportunity presented by the growth of public clamor over atmospheric contamination, on June 14, 1957, the Soviet Union announced they would change their long-standing position and accept a verification regime for a nuclear testing ban that would include monitoring stations within the nuclear weapons states, including the Soviet Union. They also announced their willingness to accept a temporary suspension of tests for a 2-3 year period.

The four Western Powers, the U.S., U.K., Canada and France, welcomed this change in Soviet position; and the U. S. indicated they might accept a moratorium of up to two years during which time the verification regime would be organized and put into place. However, the Western receptiveness to the idea of a moratorium was conditional on it being part of a larger package of disarmament measures, a position the Soviet Union would not accept. The U.K. also initiated a suggestion, later adopted by all the Western Powers, that a possible first step might well be the establishment of a committee of technical experts: “to consider possible methods of limiting nuclear test explosions and to investigate the requirements of effective supervision over an agreement to limit such explosions”. The Subcommittee of the UN Disarmament Commission adjourned in September without reaching an agreement.

EXCHANGES BETWEEN THE EAST AND THE WEST

In an exchange of letters with President Eisenhower in December 1957 and January 1958, Chairman Nikolay Bulganin, then premier of the Soviet Union, proposed an immediate 2-3 year moratorium, coupled with a number of other arms control measures related to nuclear weapons in Europe. This Soviet initiative culminated in a decree adopted by the Supreme Soviet on March 31, 1958, announcing the Soviets' intention: "to discontinue the testing of all types of atomic and nuclear weapons in the Soviet Union". The decree also contained the caveat that: "If the other Powers possessing atomic and hydrogen weapons continue tests of these weapons, the Government of the USSR will naturally be free to act in the matter, ... having regard to the interests of its security".

The decree was followed up on April 4, 1958 by letters to President Eisenhower and Prime Minister Macmillan from the newly installed Soviet Premier, Nikita Khrushchev, calling for the U.S. and U.K. to follow suit. The proposal was clearly an attempt to capitalize on the world's concern regarding nuclear testing and was too vague to be acceptable to the U.S. and U.K. However, it was received very favorably by many in the public and the press; and it gave new impetus to the movement for the elimination of testing.

Some months earlier, on January 6, 1958, Eisenhower had asked his Presidential Assistant for Science and Technology, James Killian, to set up a special panel under Hans Bethe to look at the question of a nuclear test ban. Their report, of March 27, 1958, concluded that: "a practical detection system" involving some number of permanent monitoring stations and teams of roving inspectors could be devised that would provide adequate verification of such a ban, although there was no such thing as a perfect verification system. They also concluded that while the US could benefit from additional testing, U.S. nuclear weapons technology was sufficiently mature relative to the capabilities of the Soviet Union and that a nuclear test ban could be entered into without prejudice to the U.S. national security. The Bethe Panel also recommended a meeting of technical experts from the two sides: "to study together and advise as to what specific control measures are necessary".

In response to the April 4, 1958 letter from Premier Khrushchev proposing a moratorium on nuclear testing, President Eisenhower responded four days later, on April 8, with his own letter rejecting Khrushchev's call for a suspension of nuclear testing, calling it: "peculiar that the Soviet Union, having just concluded a series of tests of unprecedented intensity, should now, in bold headlines, say that it will not test again, but add, in small type, that it may test again if the United States carries out its already long announced and now imminent series of tests" However, Eisenhower also called for a meeting of technical experts from the two sides as recommended by the Bethe Panel.

After one more exchange of letters, Khrushchev responded on May 9 that the Soviet Union was agreeable "to having both sides designate experts who would immediately begin a study of methods for detecting possible violations of an agreement on the cessation of nuclear tests". After several more letters it was agreed to begin the conference of technical experts in Geneva on July 1, 1958.

THE CONFERENCE OF EXPERTS

The Conference of Experts met during July and August of 1958. Its purpose was to examine the technical questions of how a comprehensive ban on nuclear testing could be verified. The Conference involved two panels of well-known scientists: the Western Panel representing the United States, United Kingdom, France, and Canada and the Eastern Panel representing the Soviet Union, Poland, Czechoslovakia and Rumania. The report of the Conference of Experts served to define the basic technical problems involved in the detection and identification of nuclear explosions as well as the technical equipment of the control systems necessary for the detection and identification of nuclear explosions. The general descriptions of the nature of the detection and identification problems in the Conference of Experts' report has stood the test of time very well, especially when it is considered that in 1958 Rainier had been the only underground nuclear explosion with any significant yield.

THE 1958-1961 MORATORIUM

The day following the release of the report of the Conference of Experts, President Eisenhower announced the United States' willingness to enter into negotiation of "an agreement with other nations which have tested nuclear weapons for the suspension of nuclear weapons tests and the establishment of an international control system ..." He proposed that these negotiations begin on October 31, 1958 and that they be based on the report of the Conference of Experts. Reflecting the growing enthusiasm for the Plowshare Program, Eisenhower also made the point that any agreement on the cessation of nuclear weapons testing "should also deal with the problem of detonations for peaceful purposes, as distinct from weapons tests".

As part of his offer to begin negotiations, Eisenhower also offered to "withhold further testing on its part of atomic and hydrogen weapons for a period of one year from the beginning of the negotiations", subject to the Soviet Union not resuming testing. The Soviet Union accepted Eisenhower's proposal for negotiations to begin on October 31.

However, prior to October 31, both the U.S. and the S.U. actively conducted testing. During September and October 1958, the U.S. conducted Operation HARDTACK II at the NTS with 37 tests.(USDOE/NV 2000: 12-16) Starting on 30 September, the S.U. began a test series at Novaya Zemlya that consisted of 19 tests plus two tests at their Missile Test Range (now in continental Russia) on November 1 and 3 1958 (Mikhailov 1999: 20-23).

The West chose not to make a *cause celebe* out of the two tests after the deadline and the conference began on schedule. During the period November 4, 1958 thru August 31, 1961, the U.S., the U.K., and the S.U. did not conduct nuclear weapons tests. However, the French tested four nuclear weapons at their Reggane Proving Grounds site in French Algeria in February, April, and December 1960 and in April 1961.

THE CONFERENCE ON THE DISCONTINUANCE OF NUCLEAR TESTS - GENEVA CONFERENCE AND THE MORATORIUM

The name chosen for the negotiating forum that began on October 31, 1958 was The Conference on the Discontinuance of Nuclear Weapons Tests, but it has generally come to be referred to as the Geneva Conference. This tri-lateral forum that represented the 3 states that were then nuclear powers, consisted of members from the U.S. and the U.K., who worked closely together, and members from the Soviet Union. They very quickly became enmeshed in the technical details of the scope and character of the verification regime that would be required to police a ban on the testing of nuclear weapons.

By 1959, the United States had analyzed the seismic results of the 2 underground tests conducted during Operation HARDTACK II, Logan and Blanca. In addition significant theoretical advances had been made on the decoupling concept (see Attachment VIII). Data from HARDTACK II indicated that the detection and identification of underground nuclear explosions would be much more difficult than earlier studies. In an effort to provide a forum for technical discussions of these advances with the Soviet scientists, the U.S. proposed the formation of a new experts group - Technical Working Group II. Through this new forum, the U.S. hoped to convince the Soviet Union of the validity of the new data and to define modifications to the system recommended by the Conference of experts. Technical Working Group II met from November 25 to December 18, 1959. Carl Romney presented the HARDTACK II seismic data, and Hans Bethe and Albert Latter presented the decoupling theory.

The Soviets strongly rejected the introduction of the HARDTACK II data and argued that the HARDTACK II data were not representative due to instruments and procedures used to determine seismic magnitudes and that the system recommended by the Conference of Experts based on only the data point from the U.S. Rainier event was adequate. (The Soviets did not conduct their first underground test until October 11, 1961 in a tunnel at Semipalatinsk.(Mikhailov 1999: 24)) They also argued that decoupling was only conceptual and would be impractical to implement because of the enormous cavities required.

When the Technical Working Group II and the Geneva Conference adjourned in mid-December 1959, the atmosphere was acrimonious. On December 29, 1959, Eisenhower announced that, while the U.S. considers itself: "free to resume nuclear weapons testing, we shall not resume nuclear weapons testing without announcing our intention in advance of any resumption". He released the Soviet Union from their commitment to match the U.S. Moratorium but took no steps to prepare the U.S. to resume testing. In January 1960, Khrushchev responded by reiterating an earlier pledge not to resume nuclear weapons testing until the Western powers tested.

During the summer and fallof 1959, while the Geneva discussions were rapidly developing into a deadlock, the Eisenhower administration began the search for a new policy. On February 11, 1960, with the resumption of the Geneva talks,

Ambassador Wadsworth laid the new U.S. "Phased" proposal on the table. The proposal provided for: "the cessation of all nuclear weapons tests in the earth's atmosphere, in the oceans, and in outer space". It suggested limiting underground testing to a threshold level of seismic magnitude 4.75 and below with 20 on-site inspections per year until a comprehensive test ban could be negotiated. The new proposal included reference to the new U.S. seismic research program (see Attachment VIII) and proposed to make this a program of joint research between the U.S., U.K., and the S.U. Among the key issues where the S.U. differed from the U.S. and U.K. regarding this proposal were: (1) whether any tests should be allowed below this threshold (the S.U. wanted a complete moratorium by unilateral decisions), and (2) the number of on site inspections that would be conducted. A meeting of the seismic experts was convened on May 11, 1960; where Frank Press from Cal Tech presented plans for the Vela Uniform series of experiments.

However, on May 1, 1960, a Soviet holiday, the Soviet Union shot down a U-2 reconnaissance plane near Sverdlovsk on the western slope of the Ural mountains. The plane was on an intelligence gathering mission and was piloted by Gary Powers who was captured. At the mid-May summit meeting in Paris, Khrushchev presented a tirade and walked out. The Soviet positions in Geneva rapidly hardened. While the Geneva conference continued to meet through December 1960, little progress was made. With the general worsening of U.S. - Soviet relations and the approach of the presidential elections in the U.S., the Geneva negotiations slowed to a crawl and became moribund by the end of 1960.

In March 1961, after the new Kennedy administration extensively reviewed the U.S. negotiating position, the Geneva negotiations resumed with the new U.S. Ambassador Arthur H. Dean. Dean tabled a number of new positions regarding: monitoring, on site inspections, moratoriums on events below 4.75 seismic magnitude, and the testing and use of Peaceful Nuclear Explosives (PNEs).

France, a major partner in the Western alliance, conducted its first fission nuclear test on February 13, 1960 and two others in 1960 at Reganne, Algeria in the Sahara Desert. On March 21, 1961 the head of the soviet delegation to the Geneva negotiations, Semyen Tsarapkin, said: "if this development of events is not checked ... it will (be) much more difficult to reach agreement on the discontinuance of nuclear tests". A month later on April 25, 1961, immediately after the fourth and last French test at Reganne, Tsarapkin again condemned the French testing and warned that: "the continuance of nuclear weapons tests by France places the Soviet Union in a situation which may compel it to resume atomic and hydrogen bomb tests". At the Kennedy-Khruschev meeting in Vienna in early June 1961, Khrushchev revealed a distinctly harder Soviet line on all the issues being discussed in Geneva. Little further progress was made in the Geneva Conference during the spring and summer of 1961.

On September 1, 1961, the S.U. resumed atmospheric testing carrying out a series of 57 tests within the next 64 days in the atmosphere and high altitude at Semipalatinsk, Novaya Zemlya, and their Missile Test range, including an atmospheric explosion with a yield of 50 Mt. They also carried out their first underground nuclear explosion. (Mikhailov 1999: 23-25) The sudden resumption of testing by the Soviet Union came as a complete surprise to the Western Powers. The magnitude and breadth of the Soviet test program was a particularly severe shock, indicating that the test program had been in planning for many months.

The U.S. was making preparations for the Plowshare nuclear test Gnome and had publicly announced it, and the Vela seismic research tests were being prepared. However, the U.S. had made no significant preparations for the resumption of nuclear weapons testing, even though Eisenhower had announced a year and a half earlier that it felt free to do so. Even so, the U.S. resumed testing with Operation NOUGAT, the first test being Antler on September 15, 1961, conducted underground in a tunnel.(USDOE 2000)

The trilateral Geneva negations continued following the resumption of testing, but little progress was made. They were formally adjourned in January 1962.

ATTACHMENT V REFERENCES

Hewlett, Richard G. and Oscar E. Anderson, Jr., *Volume I A History of the United States Atomic Energy Commission, The New World, 1939/1946*", The Pennsylvania State University Press, University Park, PA, 1962.

Mikhailov, V. N. Editor-in-Chief, *Catalog of Worldwide Nuclear Testing*, Begell House, Inc., New York, NY, 1999.

US Department of Energy, Las Vegas Operations Office (US DOE/NV), *United States Nuclear Tests July 1945 Through September 1992*, DOE/NV 209-Revision 15, Las Vegas, NV, December 2000.

ATTACHMENT VI

THE PLOWSHARE PROGRAM

by MILO NORDYKE

Early Concepts

As the scientists at Los Alamos worked on developing the world's first atomic bomb, they had thoughts of how this tremendous new source of energy could be used for peaceful purposes. During the Manhattan Project, Otto Frisch, one of the pioneers of the nuclear fission process in the 1930s, first suggested using an atomic explosion as a source for large quantities of neutrons which could used in scientific experiments designed to expand the understanding of nuclear physics.

After the war, many grandiose ideas appeared in the popular press on how this new source of energy should be harnessed to serve mankind. However, very few of the articles written in the late 40s and early 50s had concrete ideas on how the explosive force of the bombs themselves could be used for scientific purposes or to transform the landscape and alter the character of geological formations deep under the earth. One of the first was written by Fred Reines which examined the possibilities of using atomic explosives for a few large-scale earth-moving applications such as making canals, mining, breaking up icebergs and melting the polar icecap. He provided a brief description of the characteristics of nuclear explosions, including the types and amounts of radioactivity produced and a general description of the physical effects from such large cratering and underground explosions. In general, his outlook was rather pessimistic, concluding that "such uses appear at best to be extremely limited in scope, owing to the radioactivity hazard associated with atomic explosions."

With the development of thermonuclear devices, new ideas began to ferment in the minds of the bomb-designers. Thermonuclear devices still required a small fission trigger, but it could be as small as 10-20% of the total yield. Since the thermonuclear fuel consisted of relatively cheap deuterium and lithium and produced almost no long-lived radioactive by-products, it offered the possibility of a 5- or 10-fold decrease in both the cost of an explosive and the amount of radioactivity associated with a given total yield.

In 1954 Fred Reines again took up his pen and wrote a brief internal memo examining ideas for containing a nuclear explosion in some kind of containment structure with a view toward recovering a large fraction of the thermal energy released in the explosion. His idea was to somehow surround the explosive with a thick layer of water and then pump the steam produced in the explosion to a power plant. In principal, his containment concept should also make it possible to collect the plutonium bred by the thermonuclear explosion. He very quickly came to the conclusion that the only way to build such a container was to put it underground where the surrounding medium would provide the ultimate containment. The scale of the problem he examined was beyond practical consideration (a 100 kt bomb in a 1000 foot diameter steel lined cavity); but it served to start people thinking about the problems and possibilities of such schemes. Reines continued to be interested in the idea of using nuclear explosions for such purposes for a few years, but soon turned his attention to high energy nuclear physics. In 1995, he became a co-winner of the Nobel Prize in Physics for his detection of the neutrino.

A few months later, Ted Taylor and John von Neumann approached the AEC Division of Research with the idea of using thermonuclear explosions to breed plutonium and tritium. A small research effort, called Project MICE, was set up at Los Alamos to look into the general features of such a scheme. There was little doubt that the neutrons from a thermonuclear device would produce the plutonium in a thorium blanket or tritium in a lithium blanket. The main problem was how to collect the irradiated blanket after the explosion.

One of the more imaginative approaches suggested by Taylor was to detonate 10 megatons a few thousand feet deep in an Alaskan glacier and recover the bomb debris, including the plutonium, from the surface runoff the following spring. If a glacier five or ten thousand feet thick could be found, the bomb could be fired at a large enough depth to be completely contained; and the newly-bred material could be recovered from the resulting cavity. In what is probably a classic understatement, Taylor observed that such an idea "involves such formidable political problems that any such proposal would probably be killed by the political problems".

During the early years of the weapons laboratories, a unique culture existed that welcomed new ideas. What might appear to be bizarre ideas could be put forward without the fear of being branded a *kook.* Sometimes such ideas would work or they might foster other ideas. Almost anything was thought to be either possible or at least worth a try. People were not afraid to have their experiments fail because failure often brought more understanding of a topic than did success. Unfortunately, this culture has since disappeared at the weapons laboratories.

EISENHOWER'S ATOMS FOR PEACE TALK AND THE GENEVA ATOMS FOR PEACE CONFERENCE

The detonation by the Soviet Union of their first thermonuclear explosion, on August 12, 1953, led President Eisenhower to the determination that he needed to take the initiative in dealing with the political aspects of the nuclear arms race. On December 8, 1953 President Eisenhower delivered his now-famous Atoms for Peace speech at the United Nations. He called for the nuclear weapons powers to reduce or eliminate their nuclear arsenals and to turn them "into a great boon for the benefit of all mankind."

As a specific measure, Eisenhower called for the establishment of an International Atomic Energy Agency to be set up under the aegis of the U.N. which would be responsible for "... apply(ing) atomic energy to the needs of agriculture, medicine, and other peaceful activities. A special purpose would be to provide abundant electrical energy in the power-starved areas of the world." The idea for an International Atomic Energy Agency became a reality in 1957 with the establishment of the IAEA in Vienna by the U.N.

Following up President Eisenhower's Atoms for Peace speech, in early 1954, the U.S. proposed that the U.N. sponsor a Conference on the Peaceful Uses of Atomic Energy. The first of four such conferences was ultimately was held in Geneva, Switzerland, in August of 1955. It was a tremendous scientific and political success for the U.S. It was the largest scientific meeting in the world held up to that time with over 2500 participants in attendance; over 1000 technical papers were presented. For many Soviet scientists, it was their first opportunity to attend a scientific meeting outside the Soviet Union and to meet their colleagues from the West

Eisenhower's speech also signaled an important shift of emphasis within the U.S. Government in support of the development of nuclear reactors. A key step in this regard was the enactment in August of 1954 of an amendment to the Atomic Energy Act of 1946, which led to the declassification of large quantities of scientific data needed for development of peaceful applications of nuclear energy.

In the Spring of 1956 a French scientist, Camille Rougeron, wrote a monograph conjuring up images of a wide variety of applications for thermonuclear explosions - building dams, changing the course of rivers, melting glaciers, breaking-up ice jams, changing the climate, constructing underground power plants driven by the heat of thermonuclear explosions, and breaking rock for mining . Rougeron's "dreams" added little in the way of quantitative analysis of such applications, but they did serve to raise the expectation of the general public for some peaceful benefit from the nuclear tests being fired.

FIRST PLOWSHARE SYMPOSIUM

In the summer of 1956, Harold Brown, the 29-year old head of the thermonuclear weapons device design group at UCRL, proposed to the AEC that a symposium be held to discuss a few ideas that were floating around the nuclear weapons community for non-military uses of nuclear explosions. Brown was reacting to the results of informal discussions among a small group of Livermore scientists that included Arthur Biehl, Gerald Johnson, Edward Teller, and Director Herb York and himself regarding possible non-military uses for the nuclear devices they were designing.

Brown's motivation was to generate more and perhaps better ideas, develop a stronger scientific basis for some of the existing ideas, examine what nuclear device designs requirements might grow out of such applications, and to see if there were any ideas really worth testing with an actual field experiment. While he was genuinely interested in the idea of finding non-military uses for nuclear explosions, Brown also hoped that such a program could attract more top-notch scientists to the laboratories as well as improve the acceptability of nuclear weapons and nuclear testing with the general public.

In the Fall of 1956, while an agenda and papers for the symposium were being developed, Egypt suddenly blockaded the Suez Canal, initiating a world crisis over control of this vital mid-East waterway. The small group at Livermore that had been meeting informally on preparations for the symposium quickly turned their attention to using nuclear excavation to develop an alternative to the bottleneck represented by the Suez Canal. Examination of the most logical alternative, a 200 kilometer canal through Israel from Gaza to Aqaba, revealed the difficulties of such a project. "The plan was eventually dropped" but "the fundamental idea of using nuclear explosives to peaceful ends remained."

The Joint AEC Weapons Laboratory Symposium on Non-Military Uses of Nuclear and Thermonuclear Explosions was held on February 6-8, 1957 at the Livermore Radiation Laboratory. Brown invited a number of scientists from the Los Alamos and Sandia weapons laboratories as well as the RAND Corporation and General Atomics whom he knew had interesting ideas on the subject. Because many of the papers were expected to deal with nuclear device designs and nuclear weapons effects, the Symposium was classified as was the first summary of the meeting published in the summer of 1958. An unclassified summary was published about a year and a half later. (Zodtner 1958)

Edward Teller was the keynote speaker. Some 24 papers were presented covering a broad array of ideas. Most were very sketchy concepts with only "back-of-the-envelope" calculations (almost literally in some cases) to support any quantitative conclusions. The advantages promised by thermonuclear explosives of much lower cost and radioactivity than was possible with fission explosives directed most of the attention of the Symposium to three applications: earthmoving using the blast effects of large cratering explosions; production of power utilizing the tremendous quantities of thermal energy released in a nuclear explosion; and breeding of new fissile material such as plutonium with the enormous burst of neutrons released by a thermonuclear explosion.

The discussions were severely hampered by the lack of actual data on the effects of explosions set off under the surface of the earth. Out of a total of 85 nuclear explosions conducted by the US prior to the symposium, only two had been underground earth moving cratering events, not in tunnels: the Uncle event in Operation JANGLE in November of 1951 and the ESS event in Operation TEAPOT in March 1955. Both of these events had a small yield of 1.2 kiloton (kt). JANGLE Uncle was buried only 5.2 meters under the surface in the alluvial soil of Yucca Flat at NTS. TEAPOT ESS was fired nearby at a greater but still relatively shallow depth of only 20.4 meters. The shallow depths of burial resulted in a large fraction of the energy and radioactivity of the explosions being vented to the atmosphere and relatively small craters. However, using these results and data from craters made by much smaller chemical explosions, rough estimates were made of how many nuclear explosions of various yields in different kinds of rocks would be needed to carry out various earthmoving projects using nuclear excavation.

Luke Vortman of Sandia Corporation presented a key paper with an analysis of the costs and problems of using nuclear excavation to dig a new sea-level Isthmian Canal to replace the present Panama Canal. As world commerce recovered in years following World War II, the Panama Canal Company, a creation of the U.S. Government that was responsible for operating the Panama Canal, began to worry about the ability of the canal to handle the volume of traffic projected for the 1960s and 1970s. A study in 1947 indicated that the cost of converting the present canal to sea-level and widening it to provide the capacity required to handle the projected traffic volumes would cost about 2.5 billion (1947) dollars and take 10-15 years. Vortman looked at three routes previously analyzed for a sea-level canal by the Panama Canal Company in 1947: one in Panama 100 miles east of the present canal and two in Colombia. He concluded that between 20 and 30 nuclear explosives with yields ranging between 250 kt and 5 megatons (Mt) could do the job at a savings of about two billion dollars over conventional methods. Vortman also made a crude estimate of the safety problems that nuclear excavation would present: ground shock, air blast, and fallout, and radiation levels on land areas near and inside the produced craters. Vortman concluded that to be practical, excavation explosives much cleaner than the 40 kt of fission per bomb which he used in his study were necessary if nuclear excavation was to have a future.

Building on the concepts briefly sketched by Fred Reines in his 1954 paper, there were several papers describing various ideas for capturing the thermal energy released by nuclear explosions in underground caverns and its subsequent use for electric power generation. The general ideas envisaged that the thermonuclear power plant would operate like a gigantic diesel engine. On a continuous basis, once every few hours or days, a "bomb" would be dropped into a large steam-filled underground cavity, one that had perhaps been lined with concrete or steel, and detonated. The steam would be continuously circulated between the cavity and a power plant on the surface where the thermal energy would be extracted from the steam as in a conventional power plant. With a 1 megaton thermonuclear explosion providing thermal energy at a cost up to one-tenth of other sources, it was argued that enormous capital costs for the power plant and other facilities could easily be absorbed. In order to minimize the build-up of fission product radioactivity in the power station, again *clean* explosives would be required.

One of the most important problems was how to keep the constant drum-beat of the explosions from destroying the power plant. The answer was supplied by Dave Griggs, a geophysicist from UCLA who, in his academic career, had specialized in the behavior of rocks under high temperature and high pressure such as in the crust and mantle of the earth. As a founder of the RAND Corp in the late 1940s and as Chief Scientist of the Air Force in the early 1950s, Griggs had become very interested in the role of nuclear weapons in the nation's arsenal as well as their effects on geologic materials. He also became a close associate of Edward Teller and the two collaborated on many technical as well as political projects.

One of the problems that Griggs attacked was predicting the surface motions from underground explosions. In a paper presented at the Symposium, he laid out his ideas for the seismic motions that could be expected from deeply-buried nuclear explosions. They showed that only 0.1-0.5% of the explosions energy would end up in seismic motion. He calculated that a 1 Mt explosion would be equivalent to a 6.1 magnitude earthquake. He concluded that detonating the explosion in a large cavity, as was planned for power production, would reduce the seismic signal by as much as a factor of 50, depending on the size of the cavity and the yield of the explosion. Griggs' concept of using large cavities for *decoupling* the seismic energy from underground explosion would be more fully developed a year and a half later by a fellow RAND scientist Albert Latter as a means of hiding nuclear tests from detection.

Chuck Violet presented a paper describing plans being developed for the first contained underground nuclear explosion in the Fall of 1957 on Operation PLUMBBOB. The plans were based on a paper published early in 1956 by Dave Griggs and Edward Teller.(Griggs & Teller 1956) As described by Violet, it was planned to carry out a wide variety of experiments on this contained underground test to explore the problems of diagnosing the performance of nuclear devices fired underground as well as the effect of the explosion on the surrounding medium. A thorough post-shot drilling program was planned to map the cavity and fractured region expected from the explosion.

In summarizing the results of the Symposium, Harold Brown emphasized the key role that new, "cleaner" thermonuclear explosives could play in advancing the practicality of non-military uses discussed at the Symposium. He also emphasized the unique scientific role that nuclear explosions could play in providing sources for probing the earth as well as the higher reaches of the earth's atmosphere. Noting that a number of tests were being planned for these environments, he urged that as much information as possible be made publicly available regarding the time and place of these tests and that the study of their effects include non-military purposes as well as their military uses.

Brown summarized the consensus of the Symposium that there was a wide range of opinions on the practicality of any one of the ideas discussed but that a substantial majority believed that an organized effort should be undertaken to study them further. Because of the necessity to involve experts in the particular fields involved in the non-military uses such as mining and civil engineers, chemists, and geologists, he felt it was necessary to find a way to make much of the data on the effects of nuclear weapons more widely available.

ESTABLISHMENT OF THE PLOWSHARE PROGRAM AT LIVERMORE

Responding to this positive appraisal of the prospects for finding peaceful uses for nuclear explosions from the scientists at the laboratories, on June 27, 1957 the AEC formally established a program for non-military uses of nuclear explosions within the AEC's Division of Military Applications. As proposed by Livermore, the emphasis in the first year was to be on studies aimed at further fleshing out the physics and engineering aspects of using underground nuclear explosions for power generation, on beginning nuclear design work on very low fission explosives especially designed for nuclear excavation, and on finding a suitable site for a near-term demonstration of this new technology.

Edward Teller relates that during the summer of 1957 Harold Brown told I. I. Rabi, the famous physicist and quick wit from Columbia University, about his new program. “Rabi responded to Brown’s enthusiasm with a dry remark: “So you want to beat your old atomic bombs into plowshares”. Brown had no reply, but he now had a name - Project Plowshare.

As Herb York pointed some years later, Rabi chose to cite the famous passage in Isiah 2:4 that says:

> "They shall beat their swords into plowshares and their spears into pruning forks; nations shall not lift up sword against sword, neither shall they learn war any more". (This became the maxim of the Plowshare Program)

but apparently Rabi didn't mention the somewhat less familiar contrasting admonition in Joel 3:10:

> "Beat your plowshares into swords, and your pruning hooks into spears: let the weak say, I am strong."

This intimate interrelationship between "plowshares" and weapons, first recognized in the Bible, would prove to be one of the curses to plague the peaceful uses of nuclear explosions and the attempts to control the development and testing of nuclear weapons over the next three decades.

There was a strong feeling that the nuclear excavation application was the closest to practical realization. Livermore, with the strong support of the AEC, directed much of the early effort to finding a field project, which would demonstrate this new technology and catch the imagination of the public. A target date of the fall of 1959 was selected for planning purposes. In the Fall of 1957, the US Air Force was in the midst of constructing a string of distant early-warning (DEW) radar stations on the northern perimeter of the North American continent. Logistic support for these stations was a significant problem; and the idea was seized upon that a nuclear excavated harbor in this area of the world would probably be more acceptable than in more populated areas and serve a useful objective of the U.S. Government at the same time. Studies were focused on having a specific site selected by the summer of 1958.

THE RAINIER EXPLOSION

The names for events in Operation PLUMBBOB were based on mountains on the west coast of North America. A particularly spectacular mountain in Washington state, Rainier, was selected as the name for the first contained underground nuclear explosion. The location selected for the Rainier test was a high mesa at the northwest corner of the Nevada Test Site, which subsequently was given the name Rainier Mesa. The geologic medium surrounding the explosion point was a bedded tuff, a fairly low-density rock with a total porosity of 20-30%. However, about half of the pores in the rock were filled with water, which meant that the explosion would be moderately well-coupled into the rock. An existing device with a yield of 1.7 kt that had been approved for the stockpile would be used.

There was considerable uncertainty among scientists on how deep a one or two kiloton nuclear explosion should be buried to ensure that it would be completely contained. The plan was to mine a tunnel in from the side of the mesa which would place the explosion at least 880 feet under the mesa.

Developing a method for measuring the yield was obviously a key issue. For atmospheric tests, two main methods were being used for determining yield; the so-called "radiochemistry yield" and "fireball yield" (see Appendix G on Diagnostics), both of which gave yield estimates that were regarded as being accurate to within $\pm$ 10% for tests in the atmosphere. Unfortunately, the theory and methodology of both of these methods would have to be changed if they were to be useful for underground tests.

Griggs speculated that the same general approach as used for fireball yield might be used for an underground explosion except some other method for observing the position of the shock wave versus time would be needed. He developed an approximation for how the geologic material surrounding an underground explosion would behave under the influence of a high pressure shock wave. Somewhat fortuitously, on the Operation JANGLE Uncle event in 1951, the first underground nuclear explosion, measurements had been made of the arrival time of the underground shock at a series of buried switches. When Griggs compared his estimated arrival times with the measured Uncle times, they had agreed fairly well. On the basis of this comparison, Griggs and Teller suggested that the yield of their underground explosion should be determined from measurement of the shock wave arrival time close to the explosion. They also were confident that the radiochemists would be able to make an accurate yield determination from samples of the melted rock recovered by drilling back into the near vicinity of the explosion some time after the shot.

To contain the explosion and prevent any radioactive products from escaping, the tunnel used to emplace the explosive and measurement hardware was an elaborate "fishhook" design in which the tunnel doubled back on itself. In this way, the shock wave from the explosion would travel through the rock and crush the tunnel before the explosion byproducts could escape through the tunnel.

Rainier was fired on September 19, 1957. At the control point, only a weak ground motion was felt. The most observable effect of the explosion was the breaking loose of a number of rocks from the top edge of the mesa which then rolled down the face of the mesa, raising a large amount of dust. Radiological surveys of the area failed to detect any radioactivity. Rainier went into the record books as the first completely contained underground nuclear explosion.

Over the following six months, a series of holes were drilled from the mesa above the explosion and from the original tunnel into the region around the Rainier explosion. From the cores removed from the holes and the geophysical logging in the holes, it was possible to construct a picture of what happened inside the mesa during the explosion. More details were provided from a tunnel, mined into the bottom of the explosion cavity itself about 15 months later after the radioactivity lining the cavity had decayed to manageable levels. From these post-shot investigations, scientists were able to develop the *phenomenology* of an underground explosion - the sequence of events and the effect of the explosion on the surrounding environment. The phenomenology of underground nuclear explosions developed through the Rainier event remains essentially unaltered to this day.

Some of the samples from the drill holes contained re-solidified glassy melt that had once lined the cavity during its expansion and contained the remnants of the nuclear device. Using these samples, the radiochemists found that they were able to easily reconstruct the chemistry of the explosion and to use the radionuclides in the samples to determine the yield of the explosion, 1.7 kt $\pm$ 10%. Measurements of the arrival time of the shock wave close to the explosion were made which were regarded as successfully characterizing the behavior of the shock wave at early times. However, there was no analysis published in which these data were used to calculate the yield of the Rainier event

In order to gain as much knowledge as possible about the seismic signal from Rainier, Dave Griggs urged that an announcement of the shot be made in advance so that seismologists throughout the U.S. could be ready to record its signal. He urged that the announcement include the date, time within 0.1 second, location and approximate yield anticipated from the detonation. The AEC approved such a step and Gerry Johnson, the technical director for the test, made such an announcement at a geophysical meeting a few weeks ahead of time. This announcement triggered press reports that the AEC was going to set off an "earthquake maker" leading AEC Chairman Lewis Strauss, ever protective of the AEC image, to immediately call Johnson and demand assurance that Rainier would not produce an earthquake. Some 40 university and government seismic stations in the U.S. and Canada attempted to record the signal from the explosion.

The decision to announce the event and to release its planned time and location was a critical step in ultimately securing AEC approval to treat all the data on the results of the Rainier test as unclassified and acceptable for public disclosure. As a result, all of the data on the effects on the test on the surrounding geologic environment were released as they became available. Ultimately, the decision to declassify all the data from Rainier laid the ground work for the subsequent decisions to declassify all the data gained from the Plowshare Program, protecting only details of the nuclear explosives. Rainier validated many of the concepts that had been only sketchy ideas in scientists' minds and gave new confidence that these ideas would work.

All of the underground weapons test explosions in the 1958 Operation HARDTACK II at the Nevada Proving Grounds were announced beforehand. However, after the end of the moratorium in 1961, the AEC re-imposed secrecy on the scheduling of weapons tests at NTS.

INTEREST BY THE PUBLIC

In February 1958, Harold Brown made his first comprehensive proposal to the AEC for the Plowshare Program over the following two fiscal years. It was an ambitious program, including excavation of the Alaskan harbor in the Fall of 1959 and perhaps a harbor in Chile in 1960 or 1961. It also included a proposal to begin the development of a special explosive for excavation applications. This explosive would have an extremely low fraction of its yield from fission with the remainder from thermonuclear reactions. The initial concept for this "ditchdigger" device envisaged a tunnel the entire length of the channel or canal to be excavated in which would be placed a series of thermonuclear secondaries. Each secondary would be designed to put out enough energy to excavate the desired channel in that region as well as to explode the succeeding secondary, resulting in a continuous chain of discrete explosions, which would excavate the channel. The proposal also included work on several underground applications such as power and special isotope production, and mining.

In reaction to the public announcement of the establishment of the program and the results from the Rainier test, the AEC and the Livermore Laboratory began to receive a flood of inquiries and ideas from the public, Congress, and American industries about the possibilities of using this new tool. Kennecot Mining expressed an interest in the use of nuclear explosions for the mining of copper. Humble Oil Company proposed a joint AEC/Humble Oil evaluation of energy production in their salt domes. Dow Chemical began an intense series of studies looking at a variety of applications including the making of fresh water and retorting the oil from the tar sands of Canada or the oil shales of Colorado. By April 1958, the AEC reported receiving as many as eight concrete proposals for industrial applications from the power, mining and oil industries.

In April, a site for Project CHARIOT, the first nuclear excavation experiment, was tentatively selected by Livermore. The site was the broad Ogotoruk Valley on the northwest coast of Alaska 100 miles north of the Arctic Circle and about thirty miles from the nearest village, Point Hope. The preliminary plan envisaged detonating four 20 kt charges in a row to produce a channel and one 200 kt charge to produce a harbor suitable for ocean-going ships. The AEC quickly approved the selection and authorized studies of the site and authorized a broad array of bio-environmental and ecological studies of the region in addition to the geophysical and geological studies.

In July, the engineering firm of Parsons, Brinckerhoff, Hall and MacDonald had just received a contract from the Panama Canal Company to do a preliminary study of options for building a new sea-level canal across the American Isthmus to replace the existing Panama Canal. Two routes were being considered, one in Panama and the other in Columbia. They contacted AEC Commissioner Libby for information on the possible use of nuclear explosives for constructing the new canal. Libby provided them information that was presented at the First Plowshare Symposium and put them in contact with Gerry Johnson at Livermore.

A study group that included representatives from Livermore and Sandia Corporation was set up in early November to work with Parsons in preparing a preliminary cost estimates for these two routes. For both routes, the estimates were substantially less than the cost of constructing a sea-level canal on these routes by conventional methods. In their preliminary report to the Panama Canal Company on February 25, 1959, Parsons et. al concluded that these results "indicate clearly the great economic potential in the nuclear method. We recommend that further studies be undertaken and - to the extent possible - additional test and exploratory data be obtained." Over the next decade, a continuing relationship between the AEC and the Panama Canal Company evolved that provided a continuing level of support and funding for the nuclear excavation portion of the Plowshare Program.

Livermore and the AEC realized that it was essential to develop some method for providing industry or any user with an estimate of the cost of using nuclear explosives. As a result, in September, the AEC released a "price list" or "service charge" for providing nuclear explosions for peaceful purposes. The charge for fabricating and firing a device with a yield of a few kilotons would be about $500,000. If the yield were a few tens of kilotons, the charge would be about $750,000. These devices, presumably for contained applications such as oil stimulation or breaking ore for mining, were described as being 30" in diameter and as being all fission explosives. For excavation or perhaps power production, a cost of about $1,000,000 was given for thermonuclear devices with a fission contribution of 5% and yields of up to 5 Mt. The diameter for such devices was given as 60". "In the event of multiple firing in the same location, or in using large numbers of devices, the charge for firing would be substantially reduced." These charges were to apply "only to the fabrication of the device, emplacing it in its firing location, making the firing attachments, firing, and studies to assure public safety and to determine the results of the detonation." The AEC was clear that the charges did not include the cost of the hole or tunnel and any other construction required for emplacement of the charge or any industrial activity related to the utilization of the explosion.

The ideas generated in the early days of the Plowshare Program and the data from the Rainier explosion provided the technical foundations for many of the concepts and technical judgments that were to be applied to the issue of banning nuclear tests - a storm which was rapidly gaining strength and would soon break on the scene (See Attachments V and VIII).

THE MORATORIUM YEARS

In late 1958, agreements were reached with the US Geologic Survey and the Bureau of Mines to participate and support AEC studies of potential Plowshare projects. In an effort to avert building criticism within the arms control community that Plowshare was only a scheme to undermine the moratorium and negotiations, in December 1958, the AEC created a new Branch of Peaceful Nuclear Explosions within the Division of Military Applications to oversee Plowshare program development.

In January 1959, AEC Chairman John A McCone announced plans to carry out the first underground PNE test, Gnome. The primary focus of Gnome was to gain data on the feasibility of recovering the thermal energy and special isotopes produced by an underground nuclear explosion. It had been decided that a salt deposit would best meet the technical and safety requirements for a dry, stable, and chemically "simple" environment. With the cooperation of the Dow Chemical and Humble Oil Companies, possible sites had been narrowed to four areas, including the bedded salt deposits near Carlsbad, New Mexico, the site ultimately selected. A wide variety of experiments were being planned to document the results. For planning purposes, a yield of 10 kt at a depth of 1216 feet was assumed with a date in the last half of 1960. On March 16, 1960, the AEC publicly announced plans for carrying out the Gnome explosion sometime early in 1961; and site preparations, including digging the emplacement shaft and tunnel for the Gnome explosion, were begun. Gnome was not executed until December 10, 1961, after the end of the moratorium, on October 12, 1961 with a 3.1 kt yield.

Meanwhile, in January 1959, the Bureau of Mines sponsored a joint symposium in Dallas, Texas focused on the use of nuclear explosions to recover oil from the enormous oil shale deposits in Colorado. Oil shale contains an organic compound kerogen, which when heated to temperatures greater than 800° F, is converted into a very high quality oil. Plowshare envisaged using nuclear explosions to fracture thick sections of the oil shale rock when the nuclear cavity collapsed and then starting a controlled burn at the bottom to retort in-situ and recover the oil from the oil shale.

In May 1959, the AEC and Livermore Lab sponsored a three-day symposium in San Francisco to make public the results of the first two year's work and thought on Plowshare as well as to generate public support for its continuation. The Second Plowshare Symposium was open to the general public. This symposium provided an opportunity for the scientists working in this new field to interact and share results with each other and with workers outside the AEC labs. It was well attended by representatives of U.S. industry, as well as foreign observers from a number of countries, including the United Kingdom, France, and Sweden. The entire first day was dedicated to describing the results of the Rainier and HARDTACK II explosions and the new understanding of the phenomena involved in an underground explosion. The remaining two days were spent discussing various nuclear excavation studies, plans for Projects CHARIOT and GNOME, ideas for the underground recovery of oil, gas, minerals, special isotopes, and thermal energy.

Following selection of a site in the Cape Thompson area for Project CHARIOT, in mid-1958, on-site geological and topographical data collection began immediately. Based on preliminary data from the site, the concept envisaged two 200 kt. explosions to produce the harbor and three 20 kt. explosions for the entrance channel. Summer of 1961 was the planned detonation schedule. Further studies, including initiation of the bio-environmental survey of the Cape Thompson area were planned for the summer of 1959. In 1959, the AEC stated that it could not find a customer for the harbor and stated that Chariot would be conducted to study environmental issues. By 1960 the local inhabitants, the Point Hope Inupiats, began to protest the project based on their concerns regarding increased radioactivity in their food supply of caribou and fish. Their protest gained momentum in the lower 48 as well as internationally; and by 1962, Project Chariot was set aside by the AEC.

Meanwhile, interest of the U.S. Army Corps of Engineers and the Panama Canal Company in the use of nuclear excavation as on option for constructing a new Isthmian canal remained high. In the Fall of 1959 the Corps of Engineers pulled one of their most promising young engineers out of Korea and assigned him to join the Plowshare group at Livermore to assist them with their nuclear cratering studies and to provide the Corps with experience and knowledge of nuclear excavation. Second in his class at West Point, Major Ernie Graves had been one of the first directors of the Army's Nuclear Power Reactor Program at Ft. Belvoir, Va. His participation and that of the group assembled by the Corps provided Livermore with solid experience in the many civil engineering and geological aspects required by Plowshare.

The device designers at Livermore continued to develop concepts and carry out calculations for ways to make nuclear explosives with large total yields but very low fission yields such as the Ditchdigger concept. The initial opposition of the Soviet Union to allowing any test ban exceptions for Plowshare explosions and their later proposals for restrictive schemes of device inspection steadily eroded support in Washington for the program as Plowshare increasingly appeared to be an obstacle to reaching an agreement on nuclear testing.

However, support within the AEC and the Joint Committee of Atomic Energy remained strong; and field activities continued. After the moratorium, the US would conduct an active Plowshare program of 12* Plowshare technology tests, 11 device development tests and 5 heavy element experiments. The Soviet Union would conduct a much more vigorous program with 122 Plowshare tests and 40 device development tests. (Mikhailov 1999: 13) [*Footnote: Technically, Danny Boy was a DoD shot.]

ATTACHMENT VI REFERENCES

Mikhailov, V.N. Editor in Chief, *Catalog of Worldwide Nuclear Testing*, Begell House, New York, NY, 1999.

Reines, Frederick, *A Discussion of Radiological Hazards Associated with a Continental Proving Ground for Atomic Bombs*, Los Alamos Scientific Laboratory, Los Alamos, NM, LAMS-1173, September 1, 1950.

Zodtner, Harlan, *Industrial Uses of Nuclear Explosions*, UCRL-5253, 1958.

ATTACHMENT VII

UNDERGROUND TESTING AND EARLY CONTAINMENT CONCEPTS

by CLIFF OLSEN

In general, containment in the beginning of the pre-Moratorium era was of little concern. In early 1956 Al Graves at LASL and D. T. Griggs and Edward Teller at Livermore began investigating the possibility of doing underground testing. There were several reasons, most of which revolved around the problems caused by radioactive contamination. There were on-site concerns regarding site usage, dependence on weather, and scheduling difficulties. Off-site exposure concerns in the down-wind areas, particularly in Southern Utah were becoming more serious, and there were other problems such as disruption of aircraft travel.

Studies at Livermore estimated that a Scaled-Depth-of-Burial (depth/cube-root of energy) of 1000 ft/kt^1/3 should contain >90 % of radioactive debris. Bob Brownlee at Los Alamos did some rudimentary modeling calculations using Equations-of-State for water, air, aluminum, and uranium on an IBM704.

The first actual underground test was Pascal-A on 26 July 1957 with a yield of 1 ton at a depth of 152 m. The hole was not backfilled. The test was conducted in the pre-dawn hours, and all present remember it as the world's biggest Roman Candle. Spectacular as it was, it did little to enhance the knowledge of containment phenomena.

Before the Moratorium there were twenty underground tests, but only four had yields exceeding 0.1 kt. The four larger yields were all in tunnels on Rainier Mesa. As was determined after testing resumed, the problems of containing sub-kiloton explosions differ from higher yields, hence there was little pre-Moratorium experience that was of real use. There are, however, several marked exceptions; very valuable data were obtained.

The best known of these four events is Rainier, a 1.7 kt test in tunnel U12b at a depth of 274 m on 19 December 1957. Rainier was designed to be contained with the explosive at the end of a "button hook." The energy of the explosion was planned to close the stemmed tunnel before any flow past the collapsed portion of tunnel. Serendipitously, the Moratorium allowed extensive re-entry to the zero room regions, and data were obtained that are still unique. Had testing continued, it is unlikely that the extensive re-entry would have occurred.

The most important of the other three events is Blanca, a 22 kt test in tunnel U12e.05 at a depth of 301 m. It was the highest yield of the pre-Moratorium underground events. Blanca had a prompt, dynamic release of radioactivity. The

SDOB of 107.5 $m/(kt)^{1/3}$ is now known to be just on the ragged edge of safety for that yield.

Another event of significance was Bernalillo, a 15 ton event in drill hole U3n at a depth of 140 m. This was probably the most studied of the vertical events; Brownlee did some computational studies to model event behavior.

The nine Los Alamos events were all in vertical holes in Area 3, with yields ranging from <1 to 38 tons. The eleven Livermore tests were all in Area 12 tunnels, with yields ranging from essentially zero to 22 kt. Such a set of data points provides little information to help devise containment schemes for the type of events conducted after the Moratorium. Nevertheless, some valuable data were obtained on ground shock and cavity conditions on Rainier, a valuable point on determining scaling was obtained on Blanca, and an approach to a systematic study of containment was begun on Bernalillo.

The pre-Moratorium era did provide clues for containment design when testing resumed. When testing resumed in fall of 1961, there were not yet any legal or treaty constraints on radioactive releases, so what would in later years be considered containment failures were useful learning experiences.

ATTACHMENT VIII

THE VELA PROGRAM

by MILO NORDYKE

Background
The February 6-8 1957 Joint AEC-Weapons Laboratory Symposium on Non-Military Thermonuclear Explosions (see Attachment VI) contained the seeds of what was to become the Vela Program for seismic monitoring in support of nuclear testing treaties. These seeds were contained in the symposium papers on: 1) the contained underground Rainier test (conducted on September 19, 1957) and 2) the concept of decoupling of seismic energy by conducting explosions in large cavities.

As discussed in Attachment V, the first step in bringing the scientific community into the arms control process was made by President Eisenhower when he established the Bethe Panel in January, 1958, to look at the technical issues involved in verifying a nuclear test ban. Based primarily on their analysis of seismic data from the Rainier event in the Fall of 1957, they concluded that "a practical detection system" involving some number of permanent monitoring stations and teams of roving inspectors could be devised that would provide adequate verification of such a ban, although there was no such thing as a perfect verification system.

Acting on their recommendation that there be a meeting of technical experts from both sides of the table, in April 1958, Eisenhower proposed a meeting of the Conference of Experts, which met during July and August in 1958. Their generally positive report ultimately laid the ground work for the Conference on the Discontinuance of Nuclear Tests - the Geneva Conference - which began October 31, 1958.

THE BERKNER PANEL
Shortly after the beginning of the Geneva Conference in the fall of 1958, it became apparent within the U.S. government that there was a serious need to improve the basic understanding of explosion seismology to support the U.S. position in the negotiations that any test ban agreement must be verifiable. In response to this concern, on December 28, 1958, James R. Killian, the President's Special Assistant for Science and Technology, appointed a special Panel on Seismic Improvements: to review what was known about the seismology of nuclear explosions, the adequacy of the network recommended by the Conference of Experts, and to recommend a program of research to improve understanding in this critical area. Lloyd V. Berkner, President of Associated Universities was named chairman of the panel which included many eminent seismologists and scientists with experience in the effects of nuclear weapons. Over the years the Panel came to be known as the Berkner Panel.

In the course of their study over the next two and a half months the Berkner Panel reviewed, in particular, the seismic results from Rainier and the Operation HARDTACK II tests. The relationship of seismic magnitude and yield and the analyses required to identify nuclear events from earthquakes were key issues in this review. The Berkner Panel also reviewed the decoupling theory developed by Latter.

The analyses from the Berkner indicated that the detection and identification of underground nuclear explosions would be significantly more difficult than earlier estimates based on Rainier that had been presented to the Soviet Union at the Conference of Experts in 1958. As discussed in Attachment V, when they were presented at the Geneva Conference in early 1959, the Soviets raised a number of objections and rejected the new data from HARDTACK II. This ultimately led to the establishment of Technical Working Group II to bring together scientific experts from both sides to try to resolve differences. However, little progress was made in this venue over the next 6-12 months as the general atmosphere of the overall negotiations deteriorated.

THE CREATION OF PROJECT VELA

The Berkner Panel's deliberations also recommended a vigorous research program in explosion seismology, "a new level of seismological research," funded by the government, "... as a means of realizing the optimum contribution of seismology to the detection problem." They recommended a program of seismic research and system development that included both nuclear and chemical explosions.

The recommendations of the Berkner Panel led to a memorandum of understanding dated April 23, 1959, signed by James Killian, John McCone, Chairman of the AEC and Donald Quarles, Secretary of Defense, assigning the Department of Defense the overall responsibility for a program of research directed at nuclear test detection with support from the AEC and National Aeronautics and Space Agency (NASA). The DoD subsequently assigned the overall programmatic responsibility for the program to a new DoD research agency, the Advanced Research Projects Agency (ARPA) with support to be provided by the AEC and NASA.

The name given to this new program of research on nuclear test detection was Project Vela. The charter was somewhat broader than that envisaged by the Berkner Panel to include detection of nuclear explosions at high altitude by means of ground stations (Vela Sierra), detection of nuclear explosions at high altitudes and in space by satellite-borne detectors (Vela Hotel), as well as detection of underground explosions (Vela Uniform). Initial funding for the Vela Uniform program was made available in September 2, 1959.

The Vela Uniform program included many different elements including: a major program of research directed at understanding the generation and propagation of seismic waves by explosions; and the development of criteria for the detection and identification of explosions and earthquakes. It also provided for the development of prototype seismic stations that might be used to monitor a test ban, looking both at the design recommended by the Conference of Experts and at designs for Unattended Seismic Observatories (USOs), so-called black-boxes.

THE GENEVA CONFERENCE

Meanwhile, at the Geneva Conference, in an effort to encourage technical understanding between U.S. and the Soviet Union, in February, 1960, Ambassador Wadsworth laid out a new U. S. proposal for making this new Vela research program a joint research program among the three parties to the Geneva Conference. The Soviet Union was prepared to accept the proposal for a program of joint seismic research to improve identification capabilities for seismic events with magnitudes less than 4.75, but only under the condition that there be a moratorium on underground nuclear tests below 4.75 and that all the research be done with chemical explosives. Such an approach was very different from the U.S. proposal for a threshold on underground tests with no restrictions on testing below the threshold.

After some debate within the U.S. Government, on March 24, 1960, President Eisenhower decided to accept the Soviet concept for a moratorium on all underground nuclear explosions below 4.75, but only for a limited period of one or two years. In the end the U. S. agreed to start with a one year proposal. However, a Joint Declaration issued by Eisenhower and Macmillan on April 29, 1960 makes no reference to a specific time period. Specifically, the Joint Declaration proposed that

> "as soon as this treaty has been signed and arrangements have been made for a coordinated research program for the purpose of progressively improving control methods for events below a seismic magnitude of 4.75, (the U.S. and U.K.) will be ready to institute a voluntary moratorium of agreed duration on nuclear weapons tests below that threshold, to be accomplished by unilateral declaration of each of the three powers."

The use of the term "coordinated" instead of "joint" to describe the research program was carefully chosen to ensure that the Soviet Union would not have a veto power over direction of the research program or the design of the experiments. The U.S. was presumably prepared to argue that nuclear explosions carried out as part of the research program should be regarded as "peaceful nuclear explosions" which were reserved for special arrangements under the Control Commission then being considered. In addition. Since the moratorium was to be "unilateral", it could also be defined in any way the declaring party wished

On May 3, Tsarapkin provided the Soviet response to the Joint Declaration. It accepted the proposal that the moratorium should be independently and unilaterally declared by each of the three parties. The Soviet response also accepted the use of a "strictly limited number of joint underground nuclear explosions as part of the "joint program of research", abandoning their insistence that all explosions be chemical. Tsarapkin argued that "underground nuclear explosions for research purposes must be carried out jointly ..." because, if not, it would be possible for a country, "on the pretense of carrying out explosions officially designated as 'for the purpose of improving methods', in reality to use and to test new types of nuclear weapons."

In his response, Tsarapkin also conveyed the agreement of the Soviet Union to a convening on May 11 of an experts meeting to be called the Seismic Research Program Advisory Group to exchange information on the seismic research activities of the three countries. The American delegation under Frank Press from Cal Tech presented the plans for the Vela Uniform seismic research program, only then taking shape in Washington involving some $65 million in Fiscal Year 1961 including both nuclear and chemical explosions. The British delegation under H. R. Hulme also described their planned research activities which were on a much smaller scale than the U.S. The planned Soviet research program was much more diffuse and non-specific than the American program. Although there were several references to the use of nuclear explosions in the Soviet program, during later discussion this concept was withdrawn and the statement made that only U.S. nuclear explosions would be used. The Soviet program would only have chemical explosions

In general, these sessions between the scientists went much better than the contentious meeting of the Technical Working Group II six months before. Tthe Soviet scientists showed great interest in upgrading their seismic instrumentation and using the political interest in a test ban to improve seismology. However, they denied any Soviet interest in investigating the decoupling theory.

Unfortunately, on May 1, while these discussions were going on, the Soviet Union shot down Gray Power's U-2 plane over Sverdlovsk in the Ural mountains, an event which ultimately led to a stalemate of the Geneva Conference. The Soviet Union returned to many of its earlier positions based on the Conference of Experts and Tsarapkin stated that they saw no reason to undertake any new joint seismic research programs since they had no doubts about the conclusions reached by the Conference of Experts in 1958. However, they said they would be happy to participate in any Western experiments, particularly those involving nuclear explosions in order to ascertain that they were not weapons development tests

The Geneva Conference continued to meet through December, 1960, BUT little progress was made on the "joint" or "coordinated" research programs. With the general worsening of U.S. - Soviet relations and the approach of the presidential elections in the U.S., the Geneva negotiations slowed to a crawl and became moribund by the end of 1960. The U.S. would conduct the Vela Program with scientists from the U.K., but without Soviet participation.

THE VELA LEGACY

The Vela Program continued as a strong and vigorous scientific program, primarily directed at seismic research. Over the next 10 years it sponsored a number of experimental nuclear tests directed at improving our understanding of the seismic results of large underground explosions in a variety of geologic environments. It also used a nuclear explosion to create a large cavity in a salt deposit in Mississippi which was used for a decoupled explosion to investigate, on a large scale, the decoupling theory.

Perhaps the most long-lasting contribution of the Vela Uniform program was a significant upgrading of basic seismological research in the U.S. universities, supported almost an entire generation of seismologists. Many of the programs started in the late 50s and early 60s under the Vela sponsorship continue to the present day. It was a unique program undertaken by the U.S. government which was dedicated to improving our nation's ability to verify a treaty that had not yet been defined and that had significant opposition within the DoD and AEC. Yet it has survived. Throughout the nearly 50 years since its founding, it has supported an impressive array of research and development projects in support of nuclear test monitoring.

Vela Uniform also had a project to develop on-site inspection techniques that could be used to try to find and identify suspicious seismic events. Of crucial importance to all of these research projects was the plan for a program of nuclear and conventional explosions in a variety of geologic media, including granite and salt. These explosions were to be extensively instrumented to provide a data base for the development of an understanding of the phenomena of explosions and of explosion seismology.

The Vela space program has also led to significant contributions through the development a number of satellites for the monitoring of earth from space to assure compliance with nuclear test ban treaties.

ACRONYMS

A-Bomb	Atomic Bomb
ACC	Army Chemical Corps, also sometimes used for Army Chemical Center
ADM	Atomic Demolition Munitions
ADWD	Assistant Director for Weapons Development (Los Alamos position)
AEC	Atomic Energy Commission
AF	Air Force
A&F	Arming and Firing
AFB	Air Force Base
AFCRL	Air Force Cambridge Research Laboratory also known as AFCRC
AFCRC	Air Force Cambridge Research Center, also known as AFCRL
AFLML	Air Force Lookout Mountain Laboratory
AFOAT	Air Force Office for Atomic Testing
AFSWC	Air Force Special Weapons Command
AFSWP	Armed Forces Special Weapons Project
AFTAC	Air Force Technical Applications Center
AKA	Also Known As
AMC WPAFB	Air Material Command Wright Paterson Air Force Base
ALO	Albuquerque Office of AEC
ALOO	Albuquerque Operations Office of AEC (sometimes called AOO)
AMA	Automobile Manufacturers Association
AOC	Air Operations Center
ARDC	Air Research and Development Command located in Inglewood, CA
ARPA	Advanced Research Projects Agency
ASW	Anti-Submarine Weapon
ATSD/AE	Assistant To the Secretary of Defense for Atomic Energy
AT&T	American Telephone and Telegraph
AWS	Air Weather Service
BCT	Battalion Combat Team
BoB	Back of Book
Brig. Gen.	Brigadier General
BRL	Ballistics Research Laboratories
CAA	Civil Aeronautics Administration
Capt.	Captain
CBR	Chemical, Biological, and Radiological
CD	Civil Defense (Compact Discs had not yet been invented)
CDLB	Civil Defense Liaison Branch
CDR	Camp Desert Rock
CEREL	Civil Engineering Research and Engineering Laboratory - Navy

CETG	Civil Effects Test Group, a test group under the Test Director during operations TEAPOT and PLUMBBOB
Col.	Colonel
Comdr.	Commander
CONUS	CONtinental US
CP	Control Point
CPE	Command Post Exercise, see Desert Rock VII & VIII, see Appendix M
cps	cycles per second
CTR	Controlled Theromonuclear Reactors
DASA	Defense Atomic Support Agency
DBM	Division of Biology and Medicine (in AEC)
DCS, WE&T	Deputy Chief of Staff, Weapons Effects and Tests (AFSWP position)
DDR&E	Director of Defense Research & Engineering
DMA	Division of Military Applications (of the AEC)
DNA	Defense Nuclear Agency
DOB	Depth Of Burial - the depth below the surface of the ground at which an object is placed.
DoD	Department of Defense
DOE	Department of Energy
DSWA	Defense Special Weapons Agency
DTMB	David Taylor Model Basin
DTRA	Defense Threat Reduction Agency
DTRIAC	Defense Threat Reduction Information Analysis Center, (at Kirtland AFB, Albuquerque, NM)
DWET	Directorate, Weapons Effects Tests (AFSWP)
EG&G	Edgerton, Germeshausen, and Grier
EM	Electromagnetic
ENS	External Neutron Source
EOP	Executive Office of the President
EPG	Enewetak Proving Grounds
ERDA	Energy Research and Development Agency
ERDL	Engineer Research and Development Laboratories, Army
ESL	Evans Signal Laboratory
ESS	Effects Sub Surface, a nuclear test on OperationTEAPOT
FC	Field Command
FCDA	Federal Civil Defense Agency
FCWT	Field Command Weapon Test (AFSWP)
FCSU	Field Command Support Unit (AFSWP)
FWC	Fission Weapons Committee (Los Alamos)
GAC	General Advisory Committee (to the AEC)
Gen.	General
GPS	Global Positioning Satellite
GZ	Ground Zero
H&N	Holmes and Narver
HA	High Altitude, a nuclear test on Operation TEAPOT

HE	High Explosive (a chemical explosive)
HF-AM	High Frequency – Amplitude Modulation
HOB	Height of Burst, the distance above the ground surface at which a detonation occurs.
HumRRO	Human Resources Research Office, a private contractor to the Army who interviewed Desert Rock Troops to assess if/how their experience changed their thinking about nuclear weapons
IBDA	Indirect Bomb Damage Assessment
ICBM	InterContinental Ballistic Missile
ID	Inside Diameter
IFC	Intermittent Fallout Collector
IFF	Identification Friend or Foe
IGZ	Intended GZ
INS	International News Service
JCAE	Joint Committee on Atomic Energy (in the US Congress)
JCS	Joint Chiefs of Staff
JOTI	Joint Office of Test Information
JTF	Joint Task Force
kt	kilotons
kw	kilowatt
LAAO	Los Alamos Area Office
LANL	Los Alamos National Laboratory
LASL	Los Alamos Scientific Laboratory
LORAN	LOng Range Aids to Navigation
LPG	Liquefied Petroleum Gas
LRL	Livermore Radiation Laboratory
Lt.	Lieutenant
Lt. Gen.	Lieutenant General
LTBT	Limited Test Ban Treaty
LVBO	Las Vegas Branch Office (of REECo)
LVT	Landing Vehicle, Tracked
MED	Manhattan Engineering District
MEG	Military Effects Group
MET	Military Effects Test, a nuclear test on Operation TEAPOT
MIT	Massachusetts Institute of Technology
MLC	Military Liaison Committee
MSL	Mean Sea Level
Mt	Megaton
MTA	Materials Testing Accelerator
NADA	National Automobile Dealers Association
NASA	National Aeronautics and Space Agency
NATO	North Atlantic Treaty Organization
NBS	National Bureau of Standards
NDRC	National Defense Research Committee
NEL	Naval Electronics Laboratory
nm	nautical miles

NME	National Military Establishment
NML	Naval Material Laboratory
NOL	Naval Ordnance Laboratory
NRDL	Naval Radiological Defense Laboratory, San Francisco, CA
NRL or NARDL	Naval Research Laboratory
NPG	Nevada Proving Ground
NRC	National Research Council
NSA	National Security Act
NSC	National Security Counsel
NSRB	National Security Resources Board
NTS	Nevada Test Site
NTSO	Nevada Test Site Organization
NWE	Nuclear Weapons Effects
NYOO	New York Operations Office
NYT	New York Times
OCAFF	Office, Chief of Army Field Forces
OCD	Office of Civil Defense
OCDMTG	Office of Civil and Defense Mobilization Test Group on HARDTACK II
OCE	Office of the Chief of Engineers (of Army Corps of Engineers)
OD	Outside diameter
ODCM	Office of Defense and Civilian Mobilization
ODM	Office of Defense Mobilization
OEM	Office of Emergency Management
ORNL	Oak Ridge National Laboratory
OTI	Office of Test Information, formed by AEC for HARDTACK II
PDT	Pacific Daylight Time
Ph.D.	Philosophiae Doctor (Doctor of Philosophy degree)
Pmax	Peak or maximum Overpressure
PNE	Peaceful Nuclear Explosives
POL	Petroleum, Oil, and Lubricants
PPG	Pacific Proving Grounds
PR	Public Relations
psi	pounds per square inch, sometimes written as lbs/in^2
PST	Pacific Standard Time
PX	A store on a military base that sells consumer goods, as would be found in a department store, to military personnel, usually at reduced prices.
Radef	Radiological Defense
RADIAC or Radiac	RAdiation Detection Indication And Computation Instrumentation for measurement of the magnitude of a gamma ray flux and/or directional properties of the gamma-ray flux
RAYPAC	RAY Path Analog Computer
RCA	Radio Corporation of America
RDB	Research and Development Board, of the DoD
REECo	Reynolds Electric and Engineering Company

R&D	Research and Development
RF	Radio Frequency
SAC	Strategic Air Command of US Air Force
SAE	Society of Automotive Engineers
SAN	AEC Office in San Francisco
SCEL	Signal Corps Engineering Laboratories, Fort Monmouth, NJ
SFNM	Santa Fe New Mexico
SFO	Santa Fe Office of AEC
SFOO	Santa Fe Operations Office of AEC
Sgt.	Sergeant
SNM	Special Nuclear Materials
SNPC	Service National de la Protection Civiler, of France
SP	Supply and Property
SSR	Soviet Socialist Republic, one of which was the Republic of Kazakh, now Kazakhstan
STS	Semipalatinsk Test Site, in Soviet Union, now Kazakhstan
S.U.	Soviet Union
SWC	Special Weapons Command
TAC	Tactical Air Command of the US Air Force
tba	to be announced
TDP	Tactical Defense Position
TNT	Trinitrotoluene, a chemical high explosive. The complete explosion of a 1 kt mass of TNT is called 1 kt of energy and is equivalent to 10^{12} calories of energy.
TOA	Time of Arrival
TVA	Tennessee Valley Authority
TWX	TeletypeWritter eXchange - A message sent by teletype
UCLA	University of California at Los Angeles
UCRL	University of California Radiation Laboratory, at Berkeley and the Livermore branch
U.K.	United Kingdom
UN	United Nations
UNAEC	United Nations Atomic Energy Commission
U.S. or US	United States
USAEC	Same as AEC but sometimes referred to as USAEC
USA	United States Army
USAAF	United States Army Air Force
USAF	United States Air Force
USGS	United States Geologic Survey
USN	United States Navy
USPHS	United States Public Health Service
USWB	United States Weather bureau
VHA	Very High Altitude
VHF-AM	Very High Frequency – Amplitude Modulation
VHF-FM	Very High Frequency – Frequency Modulation
VLP	Very Low Pressure
WADC	Wright Air Development Center

WET	Weapons Effects Tests
WE&T	Weapons Effects and Tests
WPAFB	Wright-Patterson Air Force Base
wrt	with respect to
WT	Weapons Test - Weapons Test reports that were made for NWE projects conducted during a test operation.
WWII	World War II

REFERENCES

This section contains references from Parts I and II of the main text. The references used in the Appendicies and Attachments are cited in those sections.

Much of the research for this work was conducted in the archives located at the Los Alamos National Laboratory, Los Alamos, NM. Archival information used here from Los Alamos often consisted of memos, minutes of meetings, TWXs, notes of private communications, etc. Such sources are not amenable to standard formats as are books or reports. Therefore, a Los Alamos source of archival information is generally referenced right in the text where it is used, rather than here in the References. The reference information provided in the text is the information an individual would need to acquire the source from the Los Alamos National Laboratory archives.

The Defense Threat Reduction Information Analysis Center (DTRIAC), Kirtland Air Force Base, Albuquerque, NM was also extensively used. Most of the DoD and FCDA sources are in the form of reports which are referenced here. Two key historical sources, AFSWP 1954 and DASA 1959, have not been published but are available for research at DTRIAC. The information cited in the text should enable a researcher to identify the appropriate location of the source cited.

A few references, which retain a classification, were used to extract unclassified information presented in the text. These references are SRD (Secret Restricted Data) and are identified as such in the following. In some instances, an unclassified version of these reports was later developed and may also be cited in the following.

A number of the first authors cited below, had more than one reference used herein published in a given year. In such cases, the first one published in, for instance 1953, is cited as 1953(a). The second one published in 1953 is cited as 1953(b), etc.

AEC, Atomic Energy Commission, Director of Military Application (AEC DMA), Expansion of Permanent Camp, AEC 141/17, November 30, 1951

AEC DMA, *Acquisition By Condemnation of Land and Outstanding Mineral Rights At Nevada Proving Ground*, AEC 141/20, August 12, 1952.

AEC, FCDA, DoD, Joint Test Information Office (AEC-FCDA-DoD) Press Release, Opennet #0322551, March 24, 1953.

AEC Secretary, W.B. McCool, *Atomic Energy Commission Proposed FCDA Program For Spring, 1955 Test Series Note by the Secretary*, AEC 707/9, Opennet #NV74121, November 1, 1954(a).

AEC, *Memorandum of Understanding between FCDA and AEC for FCDA Participation In Full-Scale Nuclear Tests*, AEC 707/10, Opennet # NV136811, November 23, 1954(b).

AFSWP, *Project 'NUTMEG' (Deleted Version),* Opennet, NV0032697, December 31, 1948.

AFSWP, *AFSWP Study report on 'Possible Site for Conducting Full Scale Tests*, Opennet #NV0028597, November 24, 1950.

AFSWP, *First History of Armed Forces Special Weapons Project, 1947-1954*, Part I, 6 volumes, Armed Forces Special Weapons Project, 1954. Located at Department of Defense Threat Reduction Information Analysis Center, Kirtland AFB, Albuquerque, NM.

Agnew, Harold and Raemer E. Schreiber, "Norris E. Bradbury May 30 1909 - August 20, 1997), http://www.nap.edu/readingroom/books/biomems/nbradbury.html, 8/3/2004.

Air Force, Biographies: http://www/af/mil/bios/bio.asp?bioID=7055 2/12/2005.

Allen, Philip W., Lester Machta, Kenneth M. Nagler, Harry L. Hamilton, Jr., Lester F. Hubert, and Robert J. List, *Operation BUSTER, Project 7.1, Transport of Radioactive Debris from Operations BUSTER and JANGLE*, Headquarters, USAF (AFOAT-1) and US Department of Commerce. Weather Bureau, WT-308, March 1952.

Anders, Roger M., Forging the Atomic Shield, Excerpts From the Diary of Gordon Dean, University of North Carolina Press, Chapel Hill, NC, 1987.

Andrews, Thomas J., Edward J. Bryant, Paul H. Lorrain, and Nicholas M. Masich, *Operation JANGLE Project 1.2a-2, Transient Ground Mechanical Effects from HE and Nuclear Explosions*, Ballistic Research Laboratories, Aberdeen Proving Ground, MD, WT-385, June 1952.

Armour Research Foundation of Illinois Institute of Technology, *Air Force Structures Program 3.3 of Operation JANGLE*, Volume I, Illinois Institute of Technology, Chicago, Illinois, WT-405 (Vol. 1), November 1952.

Arnold, Lorna, *A Very Special Relationship, British Atomic Weapons Trials in Australia*, Her Majesty's Stationary Office, London, 1987.

Aronson, C. J., J. F. Moulton, Jr., J. Petes, E. J. Culling, and E. R. Walthall, *Operation TUMBLER, Projects 1.3 and 1.5. Free-Air and Ground Level Pressure Measurements*, U. S. Naval Ordnance Laboratory, White Oak, Silver Spring, MD, WT-513, November 1952.

Bainbridge, K. T., editor, *Trinity*, Volume 24, Los Alamos Laboratory, LA-1012-Del., May 8, 1947.

Banister, J. R. and F. H. Shelton, *Operation TEAPOT Project 1.11, Special Measurements of Dynamic Pressure Versus Time and Distance*, Sandia Corporation, Albuquerque, NM, WT-1110, 1955.

Banister, John R., *Historical Sketches of Sandia National Laboratories Nuclear Field Testing*, Sandia National Laboratories, Albuquerque, NM, SAND93 7054/2 UC-700, October 1994

Baker, W. D., R. W. Newman, F. B. Porzel, and E. J. Zadina, *Jangle Underground Time-of-Arrival Measurements*, in: Ogle, William E., *Handbook of Diagnostics Experiments Operation BUSTER-JANGLE*, Los Alamos Scientific Laboratory, Los Alamos, NM, WT-424, October 1951. (SRD)

Baurmash, L., J. W. Neel, W. K. Vance III, H. M. Mork, and K. H. Larson, *Operation TEAPOT Project 37.2, Distribution and Characterization of Fall-out and Airborne Activity From 10 to 160 miles From Ground Zero, Spring 1955*, UCLA School of Medicine, Los Angeles, CA, WT-1178, September 1958.

Berkhouse, L. H., S.E. Davis, F. R. Gladeck, J. H. Hallowell, C. B. Jones, E. J. Martin, F. W. McMullan, M. J. Osborne, W. E. Rogers, *Operation SANDSTONE: 1948*, Defense Nuclear Agency, Washington D.C., DNA 6033F, 19 December 1983.

Berkhouse, L. H., S. E. Davis, F. R. Gladeck, J. H. Hallowell, C. B. Jones, E. J. Martin, F. W. McMullan, M. J. Osborne, *Operation CROSSROADS 1946*, Defense Nuclear Agency, Washington D.C., DNA 6032F, 1 May 1984.

Binns, John J., Major General, *Report of Test Infantry Troops Test Exercise Desert Rock VII and VIII*, DR7-15, December 1957.

Bohn, J. Lloyd, Ralph J. Cowie, Jr., and Marcus D. O'Day, *Operation UPSHOT-KNOTHOLE Project 8.2, Measurement of Thermal Radiation With A Vacuum Microphone*, Air Force Cambridge Research Center, Cambridge, Mass., WT-767, March 1954.

Bond, V.P., R.E. Carter, J.S. Reed, H.H. Hechter, R. J. Veenstra, E.P. Cronkite, R. A. Conard, H. A. Schlang, and W.T.S. Thorpe, *Operation UPSHOT-KNOTHOLE Project 23.17, Biological Effectiveness of Ionizing Radiation Within Shelters*, U. S. Naval Radiological Defense Laboratory, San Francisco, CA, and Naval Medical Research Institute, Bethesda, MD, WT-793, September 1953.

Bourton, E. H., C. S. Elder, J. S. Kemper, and E. F. Wilsey, *Operation TUMBLER Project 1.9, Pre-Shock Dust*, Chemical and Radiological Laboratories, Army Chemical Center, WT-519, October 1952.

Bowen, Gerald I., Allen F. Strehler, and Mead B. Wetherbe, *Operation TEAPOT Project 33.4, Distribution and Density of Missiles from Nuclear Explosions*, Lovelace Foundation for Medical Education and Research, Albuquerque, NM, WT-1168, March 1956.

Brittain, G. H. and E. H. Scharres, *Operation PLUMBBOB Project 30.51, Dome Structure Response Instrumentation*, Armour Research Foundation of Illinois Institute of Technology, ITR-1525, July 1957.

Broida, T. R., A. Broido, and A. B. Willougyby, *Operation TUMBLER Project 8.2, air Temperatures in the Vicinity of a Nuclear Detonation*, US Naval Radiological Defense Laboratory, San Francisco, CA, WT-542, September 1952.

Broido, A., C. P. Butler, and R. W. Hillendahl, *Operation BUSTER Project 2.4-1, Basic Thermal Radiation Measurements*, US Naval Radiological Defense Laboratory, San Francisco, CA, WT-409, June 1952.

Broido, A., C. P. Butler, R. P. Day, R. W. Hillendaho, S. B. Martin, and A. B. Willoughby, *Operation TUMBLER-SNAPPER Project 8.3, Thermal Radiation from a Nuclear Detonation*, US NRDL, San Francisco, CA, WT-543, March 1953.

Brooks, J. W., W. T. Ham, Jr., R. C. Williams, and E.I. Evans, *Operation BUSTER Project 4.2, Thermal Effects on Animals (Dogs),* Medical College of VA, Richmond, VA, Dept. of the Army, Office of the Surgeon General, Washington, DC,WT-362, June 1952.

Brown, A. A., R. K. Arnold, W. L. Fons, F. M. Sauer, and W. E. Reifsnyder, *Operation TUMBLER-SNAPPER Project 3.3 Blast Damage to Trees Isolated Conifers*, Division of Fire Research, Forest Service, U S Department of Agriculture, WT- 509, Janurary 1953.

Bryant, E. J., N. H. Ethridge, and J. H. Keefer, *Operation TEAPOT Project 1.14b, Measurements of Air-Blast Phenomena With Self-Recording Gages*, Ballistic Research Laboratories, Aberdeen Proving Ground, MD, WT1155, 1955.

Bryant, E. J. and J. D. Day, *Operation PLUMBBOB Project 1.8b, Effects of Rough Terrain on Drag-Sensitive Targets*, Ballistics Research Laboratories, Aberdeen Proving Ground, MD, WT-1408, November 1959.

Bryant, E. J. and J. H. Keefer, *Operation PLUMBBOB Project 1.1, Basic Airblast Phenomena*, Ballistics Research Laboratories , Aberdeen Proving Ground, MD, WT-1401, June 1962.

Bullock, Brig. Gen W. C., *Final Report, Exercise DESERT ROCK V,* Headquarters Camp Desert Rock, Las Vegas, NV, July 1953.

Bultmann, E. H. Jr., G. F. McDonough, and G. K. Sinnamon, *Operation PLUMBBOB Project 1.7, Loading on Simulated Buried Structures at High Incident Overpressures*, University of IL, Urbana IL, and AFSWC Kirtland AFB, Albuquerque, NM, WT-1406, April 1960(a).

Bultmann, E. H. Jr., T. G. Morrison, and M. R. Johnson, *Operation PLUMBBOB Project 3.6, Fill-Scale Field Tests of Dome and Arch Structures*, Mechanics Research Division American Machine and Foundry Co., Niles, IL, and AFSWC Kirtland AFB, Albuquerque, NM, WT-1425, August 1960(b).

Byrnes, Joseph B., UPSHOT-KNOTHOLE Project 21.2, *Effects of an Atomic Explosion on Two Typical Two-Story-And-Basement Wood-Frame Houses*, FCDA, Washington, D.C., WT-792, September 1953(a).

Byrnes, Joseph B., UPSHOT-KNOTHOLE Project 21.1, *Effects of an Atomic Explosion on Underground and Basement Types of Home Shelters*, FCDA, Washington, D.C., WT-801, September 1953(b).

Byrnes, Victor A., *Operation BUSTER Project 4.3, Flash Blindness*, USAF, School of Aviation Medicine, Randolph Field, TX, WT-341, March 1952.

Cahn, Robert, *When an A-Bomb Misfires*, Colliers Magazine Pages 17-19, August 9, 1952.

Cameron, R. A. and P. H. Huff, *Operation HARDTACK II Project 34.1, Physical Damage Survey of AEC Test Structures*, Holmes & Narver, Inc., Los Angeles, CA, WT-1701, May 1961.

Cameron, R. A. and P. H. Huff, *Operation HARDTACK II Project 34.2, Radiation Shielding and Response Studies of AEC Test Structures*, Holmes & Narver, Inc., Los Angeles, CA, WT-1723, June 1962.

Cantor, G. and A. Farnochi, *Operation HARDTACK II Project 6.15, Electromagnetic Pulses from Low-Yield Bursts*, US Army Signal Research and Development Laboratory, Fort Monmouth, NJ, WT-1662, August 1960.

Carothers, James, et. al.,*Caging the Dragon*, Lawrence Livermore National Laboratory, DOE/NV-388, DNA-TR-95-74, 1995.

Cathcart, Brian, *Test of Greatness, Britain's Struggle for the Atom Bomb*, The University Press, Cambridge, England, 1994.

Chapman, Edwin S., *Operation UPSHOT-KNOTHOLE Project 3.27, Effects of Atomic Explosions on Field Medical Installation Equipment* , Medical Field Service School, Fort Sam Houston, San Antonio, Texas, February 1954.

Christman, Al, *Target Hiroshima, Deak Parsons and the Creation of the Atomic Bomb*, Naval Institute Press, Annapolis, MD, 1988.

Clark, John C., *Test Director's Summary Report BUSTER-JANGLE*, Los Alamos Scientific Laboratory, JOL-65-897 and JDO-555, 1951.(SRD)

Clark, John C., *Report of the Deputy Test Director Operation RANGER*, WT-206, Los Alamos Scientific Laboratory, Sept. 1953. (SRD) (The SRD version is used herein, but there is an unclassified version WT-206 (EX).)

Clark, John C., *Operation UPSHOT-KNOTHOLE. Report of the Deputy Test Director,* Los Alamos Scientific Laboratory (SRD), WT-816, June 1954. (SRD) (The SRD version is used herein, but there is an unclassified version WT-816 (SAN).)

Clark, J. C., *Report of the Test Director Operation TEAPOT*, LA-1966, October 1955. (SRD)

Clark, Walton C., *Operation TEAPOT Project 31.4, Comparison of Responses of Structural Slabs to Static and Atomic Blast Loadings*, FCDA, Battle Creek Michigan, Public Building Service, Washington, D.C., May 1955.

Claus, Walter and Joe Deal, *Operation HOT ROD* Pages 249-258 of LASL, *Operation RANGER Vol. 4 Program Reports*, University of California, Los Alamos Scientific Laboratory, Los Alamos, NM, June 1952. (SRD)

Cohen, E. and A. Bottenhofer, *Operation PLUMBBOB Project 30.7, Test of German Underground Personnel Shelters*, Ammann & Whitney, New York, NY, WT-1454, July 1962.

Cohen, E., E. Laing, and A. Bottenhofer, *Operation PLUMBBOB Project 30.2, Response of Dual-Purpose Reinforced-Concrete Mass Shelter*, Ammann & Whitney, New York, NY, WT-1449, April 1961(a).

Cohen, E., E. Laing, and A. Bottenhofer, *Operation PLUMBBOB Project 30.4, Response of Protective Vaults to Blast Loading*, Ammann & Whitney, New York, NY, WT-1451, April 1961(b).

Cohen, Edward and Norval Dobbs, *Operation PLUMBBOB Project 30.6, Test of French Underground Personnel Shelters*, Ammann & Whitney, New York, NY, WT-1453, June 1962.

Colson, E. A. and H. E. Grier, *Operation BUSTER Project 7.2, Long-Range Light Measurements*, EG&G Inc., Boston, Mass., WT-379, January 1952.

Cook, George W. and W. P. Kiley, *Operation JANGLE Project 1.5b, Detection of Time of Arrival of First Earth Motion*, David Taylor Model Basin, Washington, DC, WT 326, April 1952.

Corfield, Guy, *Operation TEAPOT Project 35.4b, Effects of a Nuclear Explosion* on *Typical Natural and Manufactured Gas Under-ground and Above-ground Installations, Including Appliances in Houses*, American Gas Assoc., NY,NY FCDA, Battle Creek MI, WT-1176, 1965.

Corsbie, Robert L., *Operation BUSTER Project 9.1b AEC Communal Shelter Evaluation*, Atomic Energy Commission, Washington, D.C., WT-360, March 1952.

Corsbie, Robert L., *Operation PLUMBBOB, Civil Effects Test Group, Project Summaries 1957*, Mercury, NV, Defense Documentation Center, Defense Logistics Agency, Cameron Station, Alexandria, VA, May 1957.

Cosenza, Charles J., WADC, Wright-Patterson AFB, Ohio, Richard G. Coy and Donald A. Kahle, Univ. of Dayton Research Inst., Dayton, OH, Thomas E. Pascoe and Paul C. Wing, Allied Research Assoc., Boston, Mass., *Operation PLUMBBOB Project 8.3b, Instrumentation for Measuring Effects Phenomena Inside the Fireball*, WT-1443, June 1961.

Costrell, L., *Operation JANGLE Project 2.1a, Gamma Radiation As A Function of Time and Distance*, National Bureau of Standards, Washington, D. C., WT-329, April 1952.

Crocker, J. Allen, *Operation BUSTER Project 7.5 and Operation JANGLE Project 7.2, Seismic Waves From A-Bombs Detonated Over A Land Mass*, Headquarters, USAF AFOAT-1, Washington, D. C., WT-321, March 1952.

Crocker, J. Allen, *Operation UPSHOT-KNOTHOLE Project 7.4, Seismic Measurements,* Headquarters, USAF AFOAT-1, Washington, D. C., WT-764, May 1955.

Dahl, Adrian H., et. al., *Operation UPSHOT-KNOTHOLE Project 29.4, Effective Energy of Residual Gamma Radiation*, Univ. of Rochester Medical School, WT-814, January 1954.

Davies, John, M. and Alfred H. Parthum, Jr., *Operation BUSTER Project 2.4a, Protective Value and Ignition Hazards of Textile Materials Exposed to Thermal Radiation*, Office of the Quartermaster General, Washington, DC, WT-312, June 1952.

Davis, F.J., *ORNL-1081,* Oak Ridge National Laboratory, November 27, 1951.

Deal, L. J., H. H. Rossi, and G. S. Hurst et al., *Operation UPSHOT-KNOTHOLE Project 24.2, Physical Measurements of Gamma and Neutron Radiation in Shelter and Instrumentation Evaluation*, Radiation Instruments Branch, Division of Biology and Medicine, USAEC, Washington, D. C., !T-789, August 1953.

Dean, Gordon, Letter from Gordon Dean, Chairman of AEC to Oscar L. Chapman, Secretary of the Interior, AEC 141/15, NV 72845, October 25, 1951.

Deeds, F. E., Felix W. Fleming, and Robert K. Stump, *Operation PLUMBBOB Project 6.1, Mine-Field Clearance by Nuclear Weapons*, Midwest Research Institute, Kansas City, MO, US Army Engineering Research and Development Laboratories, Fort Belvoir, VA, WT-1435, August 1960.

Defense Atomic Support Agency (DASA), *History of the Armed Forces Special Weapons Project, 1955-1958*, Part II 2 volumes, Defense Atomic Support Agency, 1959. (Located at Department of Defense Threat Reduction Information Analysis Center, Kirtland AFB, Albuquerque, NM.)

DASA, *Operation PLUMBBOB Project 2.2, Neutron-Induced Activities in Soil Elements*, Sandia Base, Albuquerque, NM, WT-1411(EX), July 1959

DASA, Office of Chief of Staff for Weapons Effects Tests, *Operation TEAPOT Technical Summary of Military Effects, Programs 1-9*, Sandia Base, Albuquerque, NM, WT-1153 (EX), February 1960(a).

DASA, *Operation PLUMBBOB Project 2.3, Neutron Flux From Selected Nuclear Devices*, Sandia Base, Albuquerque, NM, WT-1412(EX), April 28,1960(b).

DASA, *Operation PLUMBBOB Project 2.1, Soil Activation by Neutrons,* Sandia Base, Albuquerque, NM, WT-1410(EX), May 16,1960©.

Defense Nuclear Agency, *Operation PLUMBBOB Technical Summary of Military Effects Programs 1-9*, WT-1445(EX), 1 October 1979.

Defense Threat Reduction Agency, *The Defense's Nuclear Agency, 1947-1997*, DTRA, US Department of Defense, Washington, D.C., 2002.

DeMarre, Martha, Private Communication, CD disc from Bechtel Nevada Opennet Archives, with NTS Organizations for Operations RANGER thru Flintlock, November 3, 2003.

Dickey, Richard K, Louis B. Silverman, and Mary Lee Griswold, *Operation TEAPOT Project 37.2, Beta Skin-dose Measurement By Specially Designed Film*-Pack *Dosimeters*, UCLA School of Medicine, Los Angeles, CA, WT1178A, May 1957.

DOE/NV (US Department of Energy Nevada Operations Office), *United States Nuclear Tests July 1945 thru September 1992*, DOE/NV - 209-Rev 15, Nevada Operations Office, Las Vegas, NV, December 2000.

Duval, Belmon, Charles S. Adler, Willard J. Turnbull, *UPSHOT-KNOTHOLE Project 9.6,Production Stabilization*, Waterways Experiment Station, Vicksburg, MS, WT-780, December 1953.

Earl, J. R., R. H. Heiskell, G. L. Smith, Jr., R. L. Stetson, W. E. Strope, L. B. Werner US Naval Radiological Defense Laboratory; J. G. Maloney, G. C. Smith Chemical and Radiological Laboratories; R. H. Reitman, ERDL; E. H. Dhein Office of the Chief of Engineers, *Operation JANGLE Project 6.2, Protection and Decontamination of Land Targets and Vehicles*, WT-400, June 1952.

EG&G, *UPSHOT-KNOTHOLE Project 9.1, Technical Photography*, Boston, Mass., WT-779, 1953.

Eisenhower, Dwight D. *Waging Peace, The White House Years A Personal Account 1956-1961*, Doubleday & Company, Inc., Garden City, NY, 1965.

Engquist, Elmer H., *Operation JANGLE Project 6.3-2, Evaluation of the Potential Respiratory Hazard to the Crews Required to Operate in Contaminated Areas*, Army Chemical Center, MD, WT-402, July 1952.

Engquist, Elmer H. and Charles W. Worsthoff, *Operation UPSHOT-KNOTHOLE Project 8.4-2, Evaluation of a Thermal Absorbing Carbon Smoke Screen*, Army Chemical Center, MD, WT-769, February 1954(a).

Engquist, Elmer H., *Operation UPSHOT-KNOTHOLE Project 8.4-1, Protection Afforded By Operational Smoke Screens Against Thermal Radiation*, Army Chemical Center, MD, WT-768, March 1954(b).

FCDA, *Operation DOORSTEP, AEC Atomic Proving Ground Yucca Flat, Nevada*, Opennet #14434, March 17, 1953.

Fehner, Terrence R. and F.G. Gosling, *Origins of the Nevada Test Site*, United States Department of Energy, Nevada Operations Office, Las Vegas, NV, 2002.

Feldman, David and J. Fred Oesterling, *Operation UPSHOT-KNOTHOLE Project 8.6, Performance Characteristics of Clothing Materials to Thermal Radiation*, Quartermaster Research and Development Laboratories, Philadelphia, PA, WT-771, July 1955.

Felt, G.L., *Report to the Test Director, High Speed Photography, Operation TUMBLER-SNAPPER*, LASL, Univ. of CA, September 1952.

Felt, G. L. and A. C. Graves, *Criteria for Future Continental Tests*, Los Alamos National Laboratory, J-19996, February 1954.

FitzSimons, Neal, *Operation PLUMBBOB Project 30.3, Evaluation of FCDA Family Shelter, Mark I, For Protection Against Nuclear Weapons*, Federal Civil Defense Agency, ITR- 1450, August 1957.

Flynn, Archie P., *Operation BUSTER Project 9.1a, FCDA Family Shelter Evaluation*, Federal Civil Defense Administration, Washington, D.C., WT-359, March 1952.

Forbes, Merwin B., Peter Brown, Amory H. Waite, Jr., Stanley H. Ungar, and Gerald Carp, *Operation JANGLE Project 2.6a, Remotely Controlled Sampling Techniques*, Signal Corps Engineering Laboratory, Fort Monmouth, NJ, WT-334. February 1952(a).

Forbes, Merwin B., Ross G. Larrick, and Edward J. Fuller, *Operation JANGLE Project 2.3-1, Total Dosage*, Signal Corps Engineering Laboratories, Fort Monmouth, NJ, WT-331, March 1952(b).

Froman and Bradbury to Col. Kenneth E. Fields B9 471.6 January 1951.

Furman, Necah Stewart, *Sandia National Laboratory, The Postwar Decade*, University of NM Press, Albuquerque, NM 1990.

Gannon, William F., *Operation JANGLE Project 1.2b, Close In Ground Measurements*, Field Command AFSWP, Albuquerque, NM, WT-364, March 1952.

Garwin, R. L., *Predetonation Effects on Fission Bombs from Preceding Fission Bomb Bursts*, Los Alamos Laboratory report LA-1141, September 7, 1950.

George, M.S. and A.F. Spilhaus, *Operation JANGLE Summary Report: Weapons Effects Tests*, Defense Nuclear Agency, WT-414(EX), October 1979.

Gilroy, John E. with Appendix by Ralph E. Peters and edited by Lloyd J. Breidenbach, *Operation BUSTER Project 3.8, Effects of an Atomic Detonation on Aircraft Structures on the Ground*, Aircraft Laboratory of the Aeronautics Division, Wright-Patterson AFB, Ohio, WT-384, January 1952.

Gilstad, D. A. and Christian G. Weeber, *Operation PLUMBBOB Project 5.2, Structural Response and Gas Dynamics of an Airship Exposed to a Nuclear Detonation*, US Navy Bureau of Aeronautics, Washington, D.C., and Aeronautical Structures Laboratory, US Naval Air Material Center, Philadelphia, PA, WT-1431, April 1960.

Glasstone, Samuel and Philip J. Dolan, *The Effects of Nuclear Weapons*, Third Edition, USDoD and USDOE, 1977.

Goeke, Roscoe H., *Operation TEAPOT Project 38.2, Indoctrination and Training of Radiological Defense Personnel*, FCDA, Battle Creek, MI, WT-1165, November 1955.

Goode, T. B., W. G. Shockley, R. W. Cunny, and W. E. Strohm, Jr, *Operation PLUMBBOB Project 3.8, Soil Survey and Backfill Control in Frenchman Flat,* Corp of Engineers, Vicksburg, MS, WT-1427, October 1959.

Goodwin, Harold L., *Description of FCDA Technical Program*, Joint Information Office, Las Vegas, NV, (NV#323084), 1953

Gordon, M. G., J. F. Stoudt, and A. B. Francis, *Operation TEAPOT Project 1.13, Dust Density Versus Time and Distance in the Shock Wave*, Chemical Corps, US Army Chemical Warfare Laboratories, Army Chemical Center, MD, WT-1113, June 1957.

Graves, Alvin C., *Formation of Committee*, LASL memo to Distribution, January 19, 1951(a).

Graves, A, *Designation of Responsibilities and Procedures Governing Flow of Information, Operation RANGER*, LASL Memo SD-8092 to Distribution, January 20, 1951b.

Greene, Jack C., *UPSHOT-KNOTHOLE Project 22.2, Various Aspects of Nuclear Radiation Measurements for Civil Defense Purposes*, FCDA, Washington, DC, WT-805, November 1953.

Haas, P. H., F. N. Wimenitz, J. C. Hoadley, and J. S. Wicklund, *Operation PLUMBBOB Project 6.2, Measurement of the Magnetic Component of the Electromagnetic Field Near A Nuclear Detonation*, Diamond Ordnance Fuze Laboratories, Washington, D. C., WT-1436, May 1962.

Hacker, Barton C., *Elements of Controversy The Atomic Energy Commission and Radiation Safety in Nuclear Weapons Testing 1947-1974*, University of CA Press, Berkeley, CA, 1994.

Hall, D.B. and J.H. Hall, *Pre-detonation Effects From Preceding Burst, An Experiment Performed at Operation RANGER*, LA 1384, Los Alamos National Laboratory, Los Alamos, NM, March 1952.

Halsey, J. F. and M. V. Barton, *Operation PLUMBBOB Project 1.9, Spectra of Ground Shocks Produced by Nuclear Detonations*, ARDC, Inglewood, CA, WT-1487, August 1959.

Hanlon, P, J. S. Ives, S. E. Cooper, and G. S. Scholl, *Operation PLUMBBOB Project 1.2, Field Test of a System for Measuring Blast Phenomena by Airborne Gages*, American Machine and Foundry Company, Alexandria, VA, and NOL, Silver Spring, MD, ITR 1402, October 1957.

Hanscome, T. D., et. al., *Operation PLUMBBOB Project 2.7, Investigation of Effects of Nuclear Detonations on Electromagnetic Wave Propagation and Nuclear Radiation Detector Design*, Headquarters Field Command, DASA, Sandia AFB, Albuquerque, NM, WT-1416(EX), May 1962.

Hansen, Robert J. and Harry C. Saxe, *Operation JANGLE Project 1.7, Ground Acceleration (Shock Pins),* Massachusetts Institute of Technology, Cambridge Massachusetts, WT-357, June 1952(a).

Hansen, Robert J. and John S. Archer, *Operation JANGLE Project 3.2 Army Structures Test*, Massachusetts Institute of Technology, Cambridge Massachusetts, WT-387, October 1952(b).

Harland, William E, *Operation JANGLE Project 2.1c-1, Aerial Survey of Distant Contaminated Terrain*, Headquarters US AFOAT-1, WT-330, June 1952.

Harris, P. S., C. Lowery, A. G. Nelson, S. Obermiller, W. J. Ozerofff, and E. Weary, *A Shot Smoky, A Test of the Plumbbob Series, 31 August 1957*, DNA 6004F, Defense Nuclear Agency, Alexandria, VA, 31 May 1981(a).

Harris, P. S., C. Lowery, A. G. Nelson, S. Obermiller, W. J. Ozerofff, and E. Weary, *PLUMBBOB Series, 1957*, DNA 6005F, Defense Nuclear Agency, Washington, D.C., 15 September 1981(b).

Haskell, Norman A. and James O. Vann, *Operation JANGLE Project 1.3c, The Measurement of Free Air Atomic Blast Pressures*, Air Force Cambridge Research Center, Cambridge, Mass., WT-325, April 1952.

Haskell, Norman A. and James O. Vann, *Operation TUMBLER-SNAPPER Project 1.1, The Measurement of Free Air Atomic Blast Pressures*, Air Force Cambridge Research Center, Cambridge, Mass., WT-511, February 1953.

Haskell, N. A., James A. Fava, and Richard M. Brubaker, *Operation TEAPOT Project 1.1, Measurement of Free-Air Atomic Blast Pressures*, Air Force Cambridge Research Center, Cambridge, Mass., WT-1101, April 1958.

Hatlem, John C., *History of Participation in Operation BUSTER-JANGLE 1951*, Special Weapons Command, Kirtland AFB Albuquerque, NM, 1951.

Hazzard, C. B. Jr., Lt. Comdr. , CEC, USNR and P.J. McEleney, Lt. (jg), CEC, USN, *Operation JANGLE Project 3.1 Navy Underground and Surface Structures*, Bureau of Yards and Docks, Navy Department, WT-404, May 1953.

Herken, Gregg, *Brotherhood of the Bomb,TheTangled Lives and Loyalties of Robert Oppenheimer, Ernest Lawrence, and Edward Teller*, Henry Holt and Company, New York, 2002.

Hewlett, Richard G. and Oscar E. Anderson, Jr., *Volume I, A History of the United* States *Atomic Energy Commission, The New World, 1939/1946,* The Pennsylvania State University Press, University Park, Pennsylvania, 1962.

Hewlett, Richard G. and Francis Duncan, *Volume II A History of the United States Atomic Energy Commission, Atomic Shield, 1947/1952*, The Pennsylvania State University Press, University Park, Pennsylvania, 1969.

Hewlett, Richard G. and Jack M Holl , *Atoms for Peace and War 1953-1961*, University of California Press, Berkeley and Los Angeles, 1989.

Hillendahl, R. W. and F. I. Laughridge, *Operation TEAPOT Project 8.4b, Basic Thermal-Radiation Measurements*, US NRDL, San Francisco, CA, WT-1146, July 1959.

Hines, Neal O., *Proving Ground, An Account of the Radiobiological Studies in the Pacific, 1946-1961* Univ. of Wash. Press, Seattle, WA, 1962

Hirsch, F. G., Joan Longhurst, D.R. McGiboney, and H.H. Sander, *Operation TEAPOT Project 33.2, The Effects of Noise in Blast-Resistant Shelters*, Sandia Corporation, Albuquerque, NM, WT-1180, May 1956.

Hoddeson, Lillian, Paul W. Henriksen, Roger A. Meade, and Catherine Westfall, *Critical Assembly, A Technical History of Los Alamos During the Oppenheimer Years, 1943-1945*, Cambridge University Press, Cambridge, England, 1993.

Houghten, Richard A. and Richard B. Harvey, *Operation PLUMBBOB Project 6.4, Accuracy and Reliability of a Short-Baseline NAROL System*, Air Force Cambridge Research Center, Bedford, Massachusetts, ITR-1438, March 3, 1958.

Howard, W. J. and R. D. Jones, *Operation JANGLE Project 1.4, Free Air Pressure Measurements*, Sandia Corp., Albuquerque, NM, WT- 306, February 1952.

Hughes, Donald C., Alfred H. Parthum, Jr., Howard James, and John C. Mowhorter, Jr., *Operation JANGLE Project 6.7, Clothes Decontamination and Evaluation of Laundry Methods*, Office of the Quartermaster General, Washington, DC, WT-347, April 1952.

Inn, E.C.Y., *Operation TEAPOT Project 8.4e, Air Temperature Measurements Over Several Surfaces*, US NRDL, San Francisco, CA, WT-1149, September 1957.

J-Division 1952: Office of the Test Director, *Administrative Order No. 1 TUMBLER-SNAPER*, Los Alamos Scientific Laboratory, Los Alamos, NM, J-9889, February 21, 1952. (SRD)

Jackson, Robert E., *Guide to U.S. Atmospheric Nuclear Weapons Effects Data,* Daman Sciences Corporation, Santa Barbara, CA, December 1993.

James, Frank E., *Operation BUSTER Project 6.5 and 6.4, Operational Tests of Techniques for Accomplishing Indirect Bomb Damage Assessment,* Aircraft Radiation Laboratory, Wright-Patterson AFB, Ohio, WT-344, March 1952.

Johnson, Gerald W., *Operation PLUMBBOB Test Director's Report*, University of California Radiation Laboratory, Livermore, UCRL-5166, November 1957.

Johnson, Leland, *Sandia National Laboratories A History of Exceptional Service In The National Interest*, Sandia National Laboratories, Albuquerque, NM, 1997.

Johnson, Richard W., *Operation UPSHOT-KNOTHOLE, Project 28.1, Test of a Radiation Telemetering System*, Division of Biology and Medicine, USAEC, Washington, D. C., WT-796, August 1953.

Johnston, Bruce G., *Operation TEAPOT Project 31.2, Damage to Commercial and Industrial Buildings Exposed to Nuclear Effects*, FCDA, Battle Creek Michigan; Engineering Research Institute, Univ. of Michigan, Ann Arbor, Michigan, WT-1189, February 1956.

Karstens, Eugene H., *Operation BUSTER, Project 8.2, Air Weather Service Participation in Operation JANGLE*, Air Weather Service, WT-342, December 1951.

Keeling, Gerald G., *Operation UPSHOT-KNOTHOLE Project 6.3, Interim IBDA Capabilities of Strategic Air Command*, SAC Offutt AFB, Omaha, NB, WT-752, January 1955.

Kelly, Armand W, *Report to the Test Director STAFF REPORTS Operation Tumbler-Snapper*, WT-553, Los Alamos Scientific Laboratory, Los Alamos, New Mexico, 1955.

Killian. B. C. and A. H. Emmons, *Field Radiological Defense Technical Operations*, Federal Civil Defense Administration, WT-1482, March 1958.

Killian, Barbara Germain, *DoD and Civil Defense Nuclear Weapons Effects Programs and Projects Conducted at the Nevada Test Site, 1951-1958*, DTRA, Alexandria, VA, 2011.

Kirkpatrick, M.D., *Operation GREENHOUSE, Army Structures Test*, Vol. 1, Office of Engineers, Washington, D.C., 1951.

Kornhauser, M. and J. Petes, *Operation TEAPOT Project 1.12 Drag-Force Measurements*, U.S. NOL, White Oak, Silver Spring, MD, WT-1111, June 1958.

Kuhn, U. S. Grant III and Roy E. Kyner, *Operation PLUMBBOB Project 39.6a, Large Animal Neutron-Gamma Irradiation Experiment*, Univ. of Tennessee-AEC Research Laboratory, ITR-1476, May 1958.

Lamoureux, V. B., *Evaluation of Training Program for Radiologicl Defense Personnel,* WT-808, Federal Civil Defense Administration , Washington, D. C., December 1953.

Laug, Edwin P., *Operation UPSHOT-KNOTHOLE Project 22.4, Exposure of Drugs to Nuclear Explosions* Division of Pharmacology, Food and Drug Administration, Dept. of Health Education and Welfare, Washington, D. C. November 1953.

Laug, Edwin P., *Operation TEAPOT Program 32, Exposure of Foods and Foodstuffs to Nuclear Explosions (A Summary of Results),* Food and Drug Administration, Department of Health, Education , and Welfare, Washington, D. C., FCDA, Battle Creek, Michigan, WT- 1222, May 1956.

Laughlin, Kyle P., *Operation TEAPOT Project 31.5, Thermal Ignition and Response of Materials*, Office of Civil Defense Mobilization, Battle Creek, MI, WT-1198, December 1957.

LASL, *Operation RANGER Vol. 4 Program Reports*, Los Alamos Scientific Laboratory, NM, June 1952.

Lenander, Harlan E., *Operation JANGLE Project 3.28, Structure Instrumentation*, Sandia Corp, Albuquerque, NM, WT-406, October 1952.

Lewis, John G., *Operation TEAPOT Project 1.6, Crater Measurements*, Engineering Research and Development Laboratories, Fort Belvoir, VA, WT-1105, July 1958.

Lewis, John G, *Project GREYBEARD Guide The History of Atmospheric Nuclear Testing, Part 1: Shock Physics Experiments, Volume 1: Introduction and Overview,* Defense Special Weapons Agency, Alexandria, VA , DRAFT June 1997.

Lilienthal, David E., *The Journals of David Lilienthal, the Atomic Energy Years, 1945-1950*, Vol. II, Harper and Rowe Publishers, New York, NY, 1964.

Lindberg, Robert G., James T. Scanlan, James C. Watson, William A. Rhoads, and Dermit H. Larson, *Operation UPSHOT-KNOTHOLE Project 27.2, Environmental and Biological Fate of Fallout From Nuclear Detonations in Areas Adjacent to the Nevada Proving Grounds*, University of California School of Medicine, Los Angeles, CA, WT-812, February 1954.

Lindberg, R. G., E. M. Romney, J. H. Olafson, and K. H. Larson, *Operation TEAPOT Project 37.1, Factors Influencing the Biological Fate and Persistence of Radioactive Fall-out*, University of California School of Medicine, Los Angeles, CA, WT-1177, January 1959.

Lindsten, Don C. and Charles E. Hansen, *Operation BUSTER Project 3.9, Effects on Selected Water Supply Equipment* Sanitary Engineering Branch, Engineering Research and Development Laboratories, U S Army, Fort Belvoir, VA, April 1952.

Magg, Carl, Jean Ponton, *Shots DIABLO to FRANKLIN PRIME The Mid-Series Tests of the PLUMBBOB Series, 15 July - 30 August 1957*, Defense Nuclear Agency, Washington, D.C., DNA 6006F, 29 September 1981.

Maag, Carl, Stephen Rohrer, Robert Shepanek, *Operation RANGER Shots Able, Baker, Easy, Baker-2, Fox, 25 January - 6 February, 1951*, Defense Nuclear Agency, Washington, D.C., DNA 6022F, 26 February 1982.

Malik, Vivian, *Minutes of the Second Meeting of the Planning Board for* Nevada *Weapons Tests*, Los Alamos JDO-584, November 23, 1956.

Massie, Jeannie, Carl Magg, Stephen Rohrer, Robert Shepanek, *Shot Simon A Test of the UPSHOT-KNOTHOLE Series, 25 April 1953*, Defense Nuclear Agency, Washington, D.C., DNA 6016F, January 1982.
Maxwell, Charles R., Stewart H. Webster, Ervin J Liljegren, Nicholas H. Cox, and Carl H. Menzer, *Operation JANGLE Project 2.6c-3, Retrievable Missiles for Remote Ground Sampling*, Laboratory of Physical Biology, National Institute of Arthritis and Metabolic Diseases, National Institute of Health, Public Health Service- Federal Security Agency, Bethesda, MD, WT-363, June 1952.

McDonnel, G. M., William H. Crosby, Carl F. Tessmer, William H. Moncrief, Jr, Hinton J. Baker, Joseph D. Goldstein, Kent Woodward, James N. Shively, Harry W. Daniell, Alexander Horava, and Harry A. Claypool, *Operation PLUMBBOB Project 4.1, Effects of Nuclear Detonations on Large Biological Specimen (Swine),* Walter Reed Army Medical Center, Washington, DC, WT-1428, August 1961.

McLoughlin, R. C., *Operation TEAPOT Project 1.5, Preshock Sound Velocities Near The Ground in the Vicinity of an Atomic Explosion*, US Navy Electronics Laboratory, San Diego, CA, WT-1104, March 1955.

Merritt, M. L., *Operation UPSHOT-KNOTHOLE Project 1.1c-2, Air Shock Pressures As Affected by Hills and Dales*, Sandia, Corp., Albuquerque, NM, WT-713, September 1954.

Meszaros, J. J., E. G. Schwartz, H. S. Burden, J. D. Day, and C. H. Hoover, *Operation PLUMBBOB Project 30.5. Instrumentation of Structures for Air-Blast and Ground Shock Effects*, Ballistic Research Laboratories, Aberdeen Proving Ground, MD, WT-1452, March 1960(a).

Meszaros, J. J., C.H. Hoover, J. D. Day. E. G. Schwartz, and H. S. Burden, *Operation PLUMBBOB Project 3.7, Instrumentation of Structures for Air-Blast and Ground Shock Effects*, Ballistic Research Laboratories, Aberdeen Proving Ground, MD, WT-1426, March 1960(b).

Meszaros, J. J., E.G. Schwatz, H. S. Burden, J.D. Day, and C. H. Hoover *Operation PLUMBBOB Project 30.5b, Instrumentation of Structures for Air-Blast and Ground*

Shock Effects, Ballistic Research Laboratories, Aberdeen Proving Ground, MD, WT-1535, January 1961.

Mikhailov, V.N. Editor in Chief, *Catalog of Worldwide Nuclear Testing*, Begell-Atom, LLC, Begell House, New York, NY, 1999.

Miller, Harvey, *Operation BUSTER Project 2.4b, Thermal Radiation Effects On Paints Plastics and Coated Fabrics*, Engineering Research and Development Laboratories, Fort Belvoir, VA, WT-407, July 1952.

Moncerief, W. H., M. P. Dacquisto, J. Fitzpatrick, H. A. Claypool, and W. E. Rothe, *Operation HARDTACK II Project 4.2, Effects of Very-Low-Yield Bursts on Biological Specimens (Swine and Mice),* Walter Reed Army Institute of Research, Walter Reed Army Medical Center, Washington, D. C., WT-1663(EX), September 1961.

Morris, W. E., *Operation JANGLE Project 1.1, Ground Acceleration Measurements*, US Naval Ordnance Lab, White Oak, Silver Spring, MD, WT-388, June 1952.

Moulton, J. F., Jr, E. R. Walthall, and P. Hanlon, *Operation JANGLE Project 1.3b, Peak Pressure vs Distance In Free Air Using Smoke-Rocket Photography*, US Naval Ordnance Lab, White Oak, Silver Spring, MD, WT-389, June 1952.

Moulton, J. F., Jr, and E. R. Walthall *Operation TEAPOT Project 1.2, Shock Wave Photography*, US Naval Ordnance Lab, White Oak, Silver Spring, MD, WT-1102, May 1958.

National Records Management Council, Inc., Research Staff, *Effects of a Nuclear Explosion on Records and Records Storage Equipment*, National Records Management Council, Inc., New York, NY, WT-1191, April 1956

Neidhardt, G. L., R. S. Koike, T. G. Morrison, and W. Tuggle, *Operation PLUMBBOB Project 30.1, Field Test of Reinforced-Concrete Dome Shelters and Prototype Door*, American Machine & Foundry Company, Chicago, IL, ITR-1449, July 1957.

Nichols, K.D., *The Road To Trinity*, William Marrow & Co., Inc. New York, 1987.

Norris, Robert S., *Racing for the Bomb*, Steerforth Press, South Royalton, Vermont, 2002.

Northrop, Paul A. *Instrumentation for Structures Program, 1951*, Part I, Sandia Corp, Albuquerque, NM, WT-1, January 1951a.

Northrop, Paul A. *Instrumentation for Structures Program, 1951*, Part II, Sandia Corp, Albuquerque, NM, WT-10, August 1951b.

Office of Test Information, OTI-57-9, Nevada Test Organization, Opennet #0143864, May 10, 1957.

Ogle, W.E., Editor, *A Handbook for Diagnostic Experiments of Operations BUSTER and JANGLE*, Los Alamos Scientific Laboratory, (SRD), 1951.

Ogle, William E., *An Account of the Return to Nuclear Weapons Testing By The United States After The Test Moratorium 1958-1961*, United States Department of Energy, Nevada Operations Office, NVO-291, October 1985. (SRD) (The DOE released an unclassified version of this work in about 1996.)

Ohio River Division Laboratories, OCE, *Operation JANGLE Project 1.6, Earth Displacement (Shear Shafts),* Ohio River Division Laboratories, Office of Chief of Engineers, Mariemont, Ohio, WT-353, April 1952.

Oleson, M. H., *Operation SNAPPER Project 7.1a, Electromagnetic Effects* From *Atomic Explosions*, Headquarters Air Force Office for Atomic Testing (AFOAT), WT-537, January 1953.

Olmsted, G. B., *Operation BUSTER Project 7.6 and Operation JANGLE Project 7.3, Detection of Airborne Low-Frequency Sound From The Atomic Explosions of Operations BUSTER and JANGLE*, Headquarters USAF, AFOAT-1, WT-322, March 1952.

Olmsted, G. B. and E. H. Nowak, *Operation UPSHOT-KNOTHOLE Project 7.3, Detection of Airborne Low-Frequency Sound From Nuclear Explosions*, Headquarters US AFOAT-1, DCS/0 , WT-763, February 1954.

Olsen, Clifford W., *The Containment of Underground Nuclear Explosions - A Sourcebook*, National Nuclear Security Technologies, LLC. October 21, 2008.

Perkins, Betty L., *AFAUST: First Airdrop United States Test, Operation RANGER*, LA-12454-H, Los Alamos National Laboratory, December 1, 1992. (SRD)

Perret, W. R. and V. L. Gentry, *Operation UPSHOT-KNOTHOLE Project 1.4, Free-Field Measurements of Earth Stress, Strain, and Ground Motion*, Sandia Corporation, Albuquerque, NM, WT-716, February 1955.

Perret, William R., *Operation PLUMBBOB Project 1.5, Ground Motion Studies at High Incident Overpressure*, Sandia Corporation, Albuquerque, NM, WT-1405, June 1960.

Peterson, Val, *OPERATION CUE The Atomic Test Program of the Federal Civil Defense Administration, OBSERVER HANDBOOK* Opennet #0143766, Nevada Test Site of the Atomic Energy Commission, Spring 1955(a).

Peterson, Val, Harold L. Goodwin, and Leonard H. Lieberman, *Operation CUE the Atomic Test Program*, Federal Civil Defense Administration, Nevada Test Site of the Atomic Energy Commission, Opennet #0143768, Spring 1955(b).

Pickering, J. E., R. W. Zellmer, R. E. Benson, and Q. L. Hartwig, *Operation PLUMBBOB Project 39.6, Biological Effects of Nuclear Radiation on the Monkey (Macaca Mulatta),* USAF School of Aviation Medicine, Randolph AFB, TX, , WT-1505 September 1958.

Pickering, J. E., D. B. Williams, G. S. Melville, Jr., A. A. McDowell, T. P. Leffingwell, and R. W. Zellmer, *Operation PLUMBBOB Project 39.6, Biological Effects of Nuclear Radiation on the Monkey (Macaca Mulattta): Two-Year Evaluation,* USAF School of Aviation Medicine, Randolph AFB, TX, WT-1542 June 1959.

Plough, H. H., and C. W. Sheppard editors, *Operation UPSHOT-KNOTHOLE Programs 23.4 to 23.14 and 23.16 Genetic Effects of Fast Neutrons From Nuclear Detonations*, Division of Biology and Medicine, USAEC, Washington, D. C., WT-820, January 1954.

Ponton, Jean, Carl Magg, Martha Wilkinson, Robert Shepanek, *Operation TEAPOT 1955*, Defense Nuclear Agency, Washington, D.C., DNA 6009F, 23 November 1981.

Ponton, Jean, Jeannie Massie, Carl Magg, Robert Shepanek, Stephen Rohrer, *Operation UPSHOT-KNOTHOLE , 1953*, Defense Nuclear Agency, Washington, D.C., DNA 6014F, 11 January 1982(a).

Ponton, Jean, Carl Maag, Mary Francis Barrett, Robert Shepanek, *Operation TUMBLER-SNAPPER 1952*, Defense Nuclear Agency, Washington, D.C., DNA 6019F, 14 June 1982(b).

Ponton, Jean, Stephen Rohrer, Carl Maag, Robert Shepanek, Jean Massie, *Operation BUSTER-JANGLE, 1951*, Defense Nuclear Agency, Washington, D.C., DNA 6023F, 21 June 1982(c).

Ponton, Jean, Stephen Rohrer, Carl Maag, Robert Shepanek, Inara Gravitis, *Operation HARDTACK II, 1958*, Defense Nuclear Agency, Washington, D.C., DNA 6026F, 3 December 1982(d).

Preuss, Paul T. and James S. Russell, *Project NUTMEG, Feasibility of Continental Proof Test Site*, Memorandum Symbol MAW: PTP, NV28600, DMA Files 3865, February 10, 1949.

Preuss, Paul T., *Operation UPSHOT-KNOTHOLE, Operational Summary, Nevada Proving Grounds, 1 March - 9 June 1953*, Headquarters Field Command, AFSWP, Sandia Base, Albuquerque, NM, 1953.

Purkey, Glen F., W. R. Lounsbery, and Staff of Radiation, Inc. , Orlando, FL, *Operation TEAPOT Project 5.1, Destructive Loads on Aircraft in Flight*, Radiation, Inc., Orlando, FL, and WADC, Dayton, Ohio, WT-1132, January 1958.

Rainey, Charles T., James W. Neel, Harold M. Mork, and Kermit H. Larson, *Operation UPSHOT-KNOTHOLE Project 27.1, Distribution and Characteristics of Fall-out at Distances Greater Than 10 Miles From Ground Zero, March and April 1953*, University of California Los Angeles, Los Angeles, CA, WT-811, February 1954.

Randall, Philip A., *Operation TEAPOT Nevada Test Site February - May 1955, Project 31.1, Damage to Conventional and Special Types of Residences Exposed to Nuclear Effects*, Civil Effects Test Group, Office of Civil and Defense Mobilization, Federal Housing Administration, Housing and Home Finance Agency, WT-1194, April 12, 1961.

Rankowitz, S, R. L. Chase, and J. B. H. Kuper, *Operation JANGLE Project 1.3a, Free Air Shock Arrival Times*, Brookhaven National Laboratory, Upton, NY, WT-324, May 1952.

Rawlings, John W., *Operation UPSHOT-KNOTHOLE Project 6.11, Introduction of Tactical Air Command Air Crews in the Delivery and Effects of Atomic Weapons,* TAC Langley AFB, VA, WT-759, December 1953.

Reed, J. W., J. R. Banister, and F. H. Shelton, *Operation TEAPOT Project 1.3, Ground-Level Microbarographic Pressure Measurements From A High-Altitude Shot*, Sandia Corporation, Albuquerque, NM, WT-1103, 1957.

Reed, J. W., J. R. Banister, J. A. Beyeler, and F. H. Shelton*, Operation TEAPOT Project 1.9, Material Velocity Measurements of High-Altitude Shot*, Sandia Corporation, Albuquerque, NM, WT-1108, June 1956.

Reeves, J. E., *Memorandum of Understanding Between FCDA and AEC Participation by Private Industry and by Other Governmental Agencies in Civil Defense Atomic Effects Tests*, Office of Test Operations, SFOO, Opennet #79200, May 12, 1954.

Reeves, James E., *Report of the Test Manager Joint Test Organization, Operation TEAPOT Nevada Test Site Spring 1955*, Santa Fe Operations Office, Opennet #NV0094137, 1955

Reeves, James E., *Report of the Test Manager Operation HARDTACK Phase II, August to October 1958*, Deleted Version, Office of Test Operations, Albuquerque Operations Office, Opennet #NV0091951, 1958.

Reeves, James E., *OPERATION PLUMBBOB Report of the Test Manager Nevada Test Site May-October 1957*, AEC, Extracted version, Opennet # NV0014461, 1 February 1980,

Reines, Frederick, *A Discussion of Radiological Hazards Associated with a Continental Proving Ground for Atomic Bombs*, Los Alamos Scientific Laboratory, Los Alamos, NM, LAMS-1173, September 1, 1950.

Reines, Fredrick, *A Summary of Test Results Operation RANGER*, Los Alamos Scientific Laboratory, LAMS 1240. (SRD)

Reines, Fredrick, *Operation RANGER* vol.2, Los Alamos, WT-202, July 1952. (SRD). (There is an unclassified version, WT-202 (EX).)

Roberts, J. E., C. S. White, and T. L. Chiffelle, *Operation UPSHOT-KNOTHOLE Project 23.15, Effects of Overpressures in Group Shelters on Animals and Dummies*, Lovelace Foundation, Albuquerque, NM, WT-798, September 1953.

Rogin, Leo, Alden C. DuPont, and Christian G. Webster, *Operation UPSHOT-KNOTHOLE Project 5.1, Atomic Weapon Effects on AD Type Aircraft in Flight,* Naval Air Material Center, Philadelphia, PA, WT-748, March 1954.

Rollosson, G. W., *OPERATION TEAPOT Project 39.2, Static and Dynamic Overpressure Measurements*, Sandia Corp., Albuquerque, NM, ITR-1192, 1955.

Rossi, Bruno, *Moments in the Life of a Scientist*, Cambridge University Press, Cambridge, England, 1990.

Ruhl, E. W. , *Operation UPSHOT-KNOTHOLE Project 24.3, AEC Shelter Instrumentation,* Vitro Corp. of America, New York, NY, WT-790, August 1953.

Russell, Demetrius H., *Weather Support for Operation RANGER*, Operation RANGER, LASL WT-204, March 1, 1951.

Rutledge, Jay D. Jr., *Memorandum for the Record, Subject: Joint AEC-AF Agreement on Use of Las Vegas Gunnery Range*, Office for Atomic Energy, Headquarters USAF, 22 December 1950.

Sachs, D. C., L. M. Swift, and F. M. Sauer, *Operation TEAPOT Project 1.10, Airblast Overpressure and Dynamic Pressure over Various Surfaces*, Stanford Research Institute, Menlo Park, CA, WT-1109, September 1957.

Sachs, D. C., and L. M. Swift, *Operation TEAPOT Project 1.7, Underground Explosion Effects*, SRI, Menlo Park, CA, WT-1106, March 1958.

Sakarov, Andrei, *Memoirs*, Alfred A. Knopf, Inc., New York, NY, 1990.

Salmon, V., *Operation TUMBLER Project 1.2, Air Pressure Vs. Time,* Stanford Research Institute, Stanford, CA, WT-512, February 1953.

Sauer, Fred M., W. L. Fons, and Theodore G. Storey, *Operation UPSHOT-KNOTHOLE Project 3.19, Blast Damage to Coniferous Tree Stands by Atomic Explosions*, U. S. Dept of Agriculture, Forest Service, Division of Fire Research, WT-731, 1954

Scarborough, G. D. and J. S. van Scott, *Operation HARDTACK Preliminary Report Project 6.14 Proof Test of AN/TVS-1 (XE-3) Flash-Ranging Equipment.* US Army Artillery Board Fort Sill, OK and US Army Signal Research and Development Laboratories, Fort Monmouth, NJ, WT-1661, 1959.

Schlatter, George F., *Administrative Information Concerning RANGER*, AEC-DMA, January 17, 1951.

Schraut, K. C., Staff Division of Research University of Dayton, and L.J. Breidenbach, *Operation SNAPPER, Project 3.1 Vulnerability of Parked Aircraft To Atomic Bombs*, Aircraft Laboratory Wright Air Development Center, Dayton, Ohio, WT-525, February 1953.

Schreiber, R. E., Los Alamos Scientific Laboratory, *W-Division Lecture Series IV, Lecture XIII*, AW-343, January 1953.

Schuster, Sheldon, *Guide to Nuclear Weapons Technology Information - Cratering and Ejecta* DTRAIC, Kirtland Air Force Base, Albuquerque, NM, August 3, 2007.

Sevin, Eugene, *Tests On the Loading and Response of Railroad Equipment*, Office of Chief of Transportation, Dept. of the Army, Washington, D. C. and Wright Air Development Center, Wright-Patterson Air Force Base, Ohio, WADC TN-55-422, WT 725, September 1955.

Seward, P. D., *History of Field Support and Logistics Division*, Appendix H of Banister, John R. "Historical Sketches of Sandia National Laboratories Nuclear Field Testing, SAND93-7054/2 Sandia Report, Sandia National Laboratories, Albuquerque, NM October 1994.

Shaw, Ebe R (Project 36.1) and Frank P. McNea (Project 36.2), *Operation TEAPOT Project 36.1 and 36.2, Exposure of Mobile Homes and Emergency Vehicles to Nuclear Explosions*, FCDA, WT-1181, July 1957.

Shelton, Frank H., *Reflections of a Nuclear Weaponeer*, Shelton Enterprise, Inc., Colorado Springs, CO, 1988.

Shockley, W. G. and W. J. Turnbull, *Operation UPSHOT-KNOTHOLE Project 9.7, Experimental Soil Stabilization*, Waterways Experimental Station, Vicksburg, MS, WT-781, February 1954.

Shreve, J. D. Jr., *Operation PLUMBBOB Preliminary Report, Summary Report Test Group 57*, Sandia Corp., Albuquerque, NM, 1958.

Siglevich, W. and E. M. Douthett, *Operation BUSTER Project 7.3 and Operation JANGLE Project 7.1, Radiochemical, Chemical, and Physical Analysis of Atomic Bomb Debris*, Headquarters, USAF AFOAT-1, Washington, D. C., WT-320(Ex), May 1952.

Silverman, Myron S. and Victor P. Bond, *Operation UPSHOT-KNOTHOLE Project 23.2 Bacteriological Studies On Animals Exposed to Neutron Radiation*, U. S. Naval Radiological Defense Laboratory, San Francisco, CA, WT-794, July 1953.

Sparks, L. N., *Operation TEAPOT Project 34.4, Nuclear Effects on Machine Tools*, Santa Fe Operations Office AEC, Albuquerque, NM, WT-1184, March 1956.

Stalk, G, R. E. Gee, J. P. Bednar, and Staff of Northrop Aircraft, Inc., *Operation PLUMBBOB Project 5.5, In-Flight Structural Response of an F-89D Aircraft to a* Nuclear, WADC, Wright-Patterson AFB, OH, WT-1434, March 1960.

Stanford, Leland H., *Operation BUSTER Project 6.9, Effects of Atomic Detonations On Radio Propagation*, Office of the Chief Signal Officer, US Army, WT-318 February 1952.

Stanford Research Institute, *Operation JANGLE Project 1(9)a, Ground Acceleration, Ground and Air Pressures for Underground Test*, Stanford Research Institute, Menlo Park, CA, WT-380, 1952.

Steadman, Frank M., *Operation RANGER, VOL 4 Program Reports and Gross Weapons Measurements*, Los Alamos Scientific Laboratory, WT-201, June 1952. (SRD) (WT-201 may have been declassified, comment by Herb Hoppe.)

Strope, W. E., *Operation PLUMBBOB, Project 32.3 Evaluation of Counter Measure System Components and Operational Procedures*, Civil Effects Test Group, WT-1464, September 15, 1959.

Swift, L. M. and D. C. Sachs, *Operation UPSHOT-KNOTHOLE Project 1.1b, Air Pressure and Ground Shock Measurements*, Stanford Research Institute, Stanford, CA, WT-711, January 1955.

Swift, L. M., D. C. Sachs, and F. M. Sauer, *Operation PLUMBBOB Project 1.4, Ground Acceleration, Stress, and Strain of High Incident Overpressures*, Stanford Research Institute, Menlo Park, CA, WT-1404, May 1960 (1960a).

Swift, L. M., D. C. Sachs, and A. R. Kriebel, *Operation PLUMBBOB Project 1.3, Air-Blast Phenomena in the High-Pressure Region*, Stanford Research Laboratory, Menlo Park, CA, WT- 1403, December 1960 (1960b).

Szasz, Ferenc Morton, *The Day the Sun Rose Twice*, Univ of NM Press, Albuquerque, 1984.

Taborelli, R. V., I. G. Bowen, E. R. Fietcher, *Operation PLUMBBOB Project 33.3, Tertiary Effects of Blast Displacement*, Lovelace Foundation, Albuquerque, NM, WT-1469, February 1959.

Taplin, G. V., S. C. Sigoloff, C. H. Douglas, D. E. Paglin, C. J. Heller, and D. F. Wales, *Operation UPSHOT-KNOTHOLE Project 29.1, Comparison and Evaluation of Dosimetry Methods Applicable to Gamma Radiation*, School of Medicine, University of California at Los Angeles, WT-802 (EX), September 1953.

Taplin, George V., Norman S. MacDonald, S. Wayne McFarland, C. William Beckner, Melvin Martin, Peggyann Campbell, Harcey Steinberg, and Kenneth Ferrin, *Operation TEAPOT Project 39.6, Measurement of Initial and Residual Radiations by Chemical Methods*, School of Medicine, UCLA, Los Angeles, CA, ITR-1171, May 1955.

Taylor, Benjamin C., *Operation UPSHOT-KNOTHOLE Project 3.29 Blast Effects of Atomic Weapons Upon Curtain Walls and Partitions of Masonry and Other Materials*, FCDA, Battle Creek, Michigan, WT-741, August 1956.

Terry, John H. and Gene D. Robertson, *Operation BUSTER Project 6.4, Airborne Radiac Evaluation*, Dept of the Navy, Bureau of Aeronautics, Washington, DC, Dept of the Air Force, Air Research and Development Command, Wright-Patterson AFB, WT- 318, May 1952.

Thomson, David B., *A Guide to the Nuclear Arms Control Treaties*, Los Alamos National Laboratory, LA-UR-99-31-73, July 1999.

Thurston, R. D. and Thomas Bardeen, *Operation BUSTER Project 3.5, Minefield Clearance*, ERDL, Fort Belvoir, VA and Gulf Research & Development Company, Pittsburg, PA, WT- 313, March 1952.

Tochilin, E., *Operation UPSHOT-KNOTHOLE Project 23.17, Neutron-Flux Measurements In AEC Group Shelters and Lead Hemispheres*, U. S. Naval Radiological Defense Laboratory, San Francisco, CA, WT-795, September 1953.

Tolan, John H., *Operation TEAPOT Project 38.3, Evaluation of Civil Defense Radiological Defense Instruments*, FCDA, Battle Creek, MI, WT-1190, April 1957.

Tucker, Paul W. and George R. Webster, *Operation TEAPOT Project 35.4a, Effects of a Nuclear Explosion on Typical Liquefied Petroleum Gas (LP gas) Installations and* Facilities, LPG Assoc., Chicago, IL, FCDA, Battle Creek, MI, WT-1175, December 1956.

Tyler, Carroll L., "Administrative Summary Report, Operation RANGER Nevada Test Site, January - February 1951", vol. 6, WT-205, Los Alamos Scientific Laboratory, Los Alamos, NM, July, 1952.

USAEC-DoD Joint Office of Test Information, *The Nevada Test Site and Supporting Installations*, AEC Office in Las Vegas, February 14, 1955. Opennet NV0324116.

Vaile, R. B. Jr., and V. Salmon, *Operation JANGLE: Project 4.2 Cratering Effects of Underground and Surface Detonated Atomic Bombs and Influence of Soil Characteristics on Crater WT-399; and Project 4.5, Characteristics of Missiles from Underground Nuclear Explosions WT-338*, Stanford Research Institute, Stanford, CA, WT-375 (which combines WT-338 and 399), May 1952.

Vaile, R. B., Jr. *Operation PLUMBBOB Project 3.5, Isolation of Structures from Ground Shock*, Stanford Research Laboratory, Menlo Park, CA, WT-1424, April 1960.

Verheul, R. H., Austin Lowrey, and L. E. Browning, *Operation HARDTACK II Project 4.3, Effect of Light From Very-Low-Yield Nuclear Detonations on Vision (Dazzle) of Combat Personnel*, Headquarters US Continental Army Command, Fort Monroe, VA, WT-1664(EX), April 1960.

Vortman, L. J., Harold Birnbaum, Edward Laing, Frank G. Ort, and Ralph V. Schumacher, *Operation TEAPOT Project 34.1 and 34.3, Evaluation of Various Types of Personnel Shelters Exposed to an Atomic Explosio*n, Sandia Corp., Albuquerque, NM, FCDA, Battle Creek,. MI, WT-1218, May 1956(a).

Vortman, L. J. and W. J Francy, *Operation TEAPOT Project 34.2, Effects of a Nonideal Shock Wave on Blast Loading of a Structure*, Sandia Corp., Albuquerque, NM, WT-1162, May 1956(b).

Walls, J.H. and the Staffs of: Douglas Aircraft Co. Inc., US Navy Bureau of Aeronautics Airframe Design Division, and Douglas Aircraft Co. Inc. Advanced Design Section *Operation PLUMBBOB Project 5.4, In-Flight Structural Response of the Model A4D-1 Aircraft to a Nuclear Explosion*, Douglas Aircraft Co. Inc., Airframe Design Division, El Segundo, CA, WT-1433, July 1958.

Walsh, Thomas G., *Operation BUSTER Project 2.6, The Protective Effects of Field Fortifications Against Neutron and Gamma Ray Flux*, ERDL, WT-383, May 1952.

Walls, J. H. and N. C. Heslin, *Operation PLUMBBOB Project 5.1, In-Flight Structural Response of an HSS-1 Helicopter to a Nuclear Detonation*, Bureau of Aeronautics Dept of the Navy, Washington, D.C. and Sikorsky Aircraft Division, United Aircraft Corp., Stratford, Connecticut, WT-1430, July 1960

White, C. S., et. al., *Operation TEAPOT Project 33.1, Biological Effects of Pressure Phenomena Occurring Inside Protective Shelters Following a Nuclear Detonation*, Lovelace Foundation, Albuquerque, NM, WT-1179, October 1956.

Williamson, Raymond H., John E. Young, and B. Curtis Gladney, *Operation TEAPOT Project 35.2, The Effects of a Nuclear Explosion on Commercial Communications Equipmen*t, Radio-Electronics-Television Manufacturers Assoc. Washington, D. C. and FCDA, Battle Creek, Michigan, ITR-1193, May 1955.

Wood, Ralph V. H., Arthur C. Werden, Jr., and Russell L Berg, *Operation TEAPOT Project 35.1, Effects of Atomic Weapons on Electric Utilities*, Edison Electric Institute, New York, NY, FCDA, Battle Creek, MI, WT-1173, June 1965.

York, Herbert F., *The Advisors Oppenheimer, Teller, and the Superbomb,* Stanford University Press, Stanford, CA, 1976 and 1989.

York, Herbert F., *Arms and the Physicist*, The American Institute of Physics Press, Woodbury, NY, 1995.

Young, George A. and Mary L Milligan, *Operation JANGLE Project 1(9)-b, Base Surge Analysis for Nuclear Tests*, US Naval Ordnance Lab, White Oak, MD, WT-390, June 1952.

ABOUT THE AUTHORS

John C. Hopkins
John C. Hopkins is a nuclear physicist with a 1960 Ph.D. from the University of Washington in Seattle. He retired in December 1993, after 34 years at the Los Alamos National Laboratory. The first third of his career was spent in applied nuclear physics research that resulted in over 40 technical publications and election to fellowship in the American Physical Society. The second third was spent in nuclear weapons testing, mostly as the leader of the program. The final third was spent as the leader of nuclear weapons development and finally as leader of the entire nuclear program at Los Alamos. Throughout most of this period Hopkins was involved in national security policy issues and was leader of the Center for National Security Studies at Los Alamos. Just prior to retirement he was a visiting scholar at the University of California's Institute on Global Conflict and Cooperation where he edited a 1994 book on the nuclear policies of France, Britain, and China*. For over 40 years Hopkins has served on governmental and international boards and panels and is a former member of the Chief of Naval Operations' (CNO) Executive Panel for the United States Navy. Hopkins participated in a National Research Council panel examining the future of deterrence. The National Academy Press published the study, with the title *Post-Cold War Conflict Deterrence,* in 1997. He served on senior review committees for the Los Alamos National Laboratory and has been a technical advisor to the Arms Control and Disarmament Agency in Washington and in Geneva. Hopkins has worked closely with the State, Energy, and Defense Departments and has participated in numerous special programs for the U.S. Government.

Dr. Hopkins has been a member of the Cosmos Club of Washington, D.C. since 1979. He resides in Los Alamos, New Mexico, and has two grown daughters.

*"Strategic Views from the Second Tier - The Nuclear Weapons Policies of France, Britain, and China." Edited by John C. Hopkins and Weixing Hu. Transaction Publishers New Brunswick (U.S.A.) and London (U.K.) 1995.

Barbara Germain Killian
Barbara received her B. A. in physics from San Diego State University in 1957. She pursued graduate work in physics, mathematics, and aeronautical engineering at San Diego State, Stanford, and the University of California at Berkeley. She married W. Patrick Crowley in 1956 and with him joined the University of California Radiation Laboratory at Livermore in 1958. Her initial work at Livermore was on calculations of thorium breeder reactors. She then joined the Livermore work on Plowshare where she contributed to advances in the development and use of numerical calculations for air blast and ground shock. Much of this work was utilized for the design and analysis of Marvel, a nuclear test at the NTS which was designed as a nuclear shock tube, and for the containment of nuclear explosions that were conducted underground. During this era, her technical publications were under Barbara K. Crowley.

In 1975, she married Lawrence S. Germain, a Livermore physicist; and in1976, they were hired into Los Alamos by John Hopkins where they both worked for him in the Test Division. At Los Alamos, Barbara continued to work in the areas of airblast, ground shock, and containment. She also served twice as DOE representative for technical assistance on the US delegation to the United Nations Committee on Disarmament. She helped to form the new Geosciences division at Los Alamos; and in 1983, she obtained a Masters Degree in Management from the Robert O. Anderson Graduate School of Management at the University of New Mexico.

In 1985, Barbara and Larry left Los Alamos to work in support of the Department of Defense's Defense Nuclear Agency, mainly with R and D Associates. Her work supported the agency's multitudinous activities in the design, execution, and analyses of underground nuclear tests (some of which were in large constructed underground cavities) that were conducted to understand the coupling of energy from a nuclear explosion into air and earth materials. She also contributed to the determination of nuclear yields by the analysis of ground shock data that was obtained underground, near the explosion. These activities contributed to the Joint Verification Exercises that were conducted with the Russians just prior to the cessation of nuclear testing.

Both Barbara and Larry retired from a formal work schedule in late 1993 to spend more time on new experiences, travel, reading, and playing. Their home is in Albuquerque, New Mexico.

www.ingramcontent.com/pod-product-compliance
Lightning Source LLC
LaVergne TN
LVHW081054220926
848484LV00022B/2439
9780988767904